VLBI AND COMPACT RADIO SOURCES

INTERNATIONAL ASTRONOMICAL UNION
UNION ASTRONOMIQUE INTERNATIONALE

SYMPOSIUM No. 110
HELD IN BOLOGNA, ITALY, JUNE 27 – JULY 1, 1983

VLBI AND COMPACT RADIO SOURCES

EDITED BY

R. FANTI

*University of Bologna
and Istituto di Radioastronomia CNR,
Bologna, Italy*

K. KELLERMANN

*National Radio Astronomy Observatory,
Green Bank, U.S.A.*

and

G. SETTI

*European Southern Observatory,
Garching bei München, F.R.G.*

D. REIDEL PUBLISHING COMPANY

A MEMBER OF THE KLUWER  ACADEMIC PUBLISHERS GROUP

DORDRECHT / BOSTON / LANCASTER

Library of Congress Cataloging in Publication Data

Main entry under title:

VLBI and compact radio sources.

At head of title: International Astronomical Union, Union astronomique inter-
nationale.
Includes indexes.
1. Very long baseline interferometry–Congresses. 2. Radio sources (Astronomy)–
Congresses. 3. Ceccarelli, Marcello, 1927-   . I. Fanti, R. (Roberto), 1941-
II. Kellermann, Kenneth I., 1937-   . III. Setti, G. (Giancarlo), 1935-
IV. International Astronomical Union. V. Title: VLBI and compact radio sources.
QB479.3.V53   1984   522′.682   84–6915
ISBN 90–277–1739–7
ISBN 90–277–1740–0 (pbk.)

Published on behalf of
the International Astronomical Union
by
D. Reidel Publishing Company, P. O. Box 17, 3300 AA Dordrecht, Holland

All Rights Reserved

Sold and distributed in the U.S.A. and Canada
by Kluwer Academic Publishers,
190 Old Derby Street, Hingham, MA 02043, U.S.A.

In all other countries, sold and distributed
by Kluwer Academic Publishers Group,
P.O. Box 322, 3300 AH Dordrecht, Holland

Printed in The Netherlands

DEDICATION

It is with the deepest sorrow that we learned of the death of Marcello
Ceccarelli on 30th January 1984 in Bologna, after many years of
suffering. Marcello will be remembered not only for his scientific
work, first as a particle physicist and later as the radio astronomer
who developed the 'Northern cross' radiotelescope and effectively founded
radio astronomical research in Italy, but also for his fine intelligence
and human character which brought him so many friends all over the
world. To his memory we wish to dedicate these proceedings.

R. Fanti
K. I. Kellerman
G. Setti

TABLE OF CONTENTS

Titles of Invited Papers are given in capital letters

PREFACE

I.A.U. Symposium No. 110 on VLBI and Compact Radio Sources was held in Bologna, Italy from June 27 to July 1, 1983. 166 participants from 19 countries were registered and 106 invited and contributed papers were registered.

The scientific presentations and discussion concentrated on VLBI observation and interpretation of galactic and extragalactic radio sources, including topics as diverse as quasars and galactic nuclei, interstellar masers, pulsars, and astrometry. Geodetic applications and technical development were treated only briefly, as these topics have been the subject of other recent international symposia.

Since the first VLBI observations in 1967, sensitivity, resolution, and image quality have improved dramatically. Radio maps shown at the symposium were of comparable quality to conventional synthesis maps being made at the time of the first VLBI experiments 15 years ago, but with a resolution more than a factor of 1000 better.

We wanted to accommodate the large number of contributed papers in this rapidly developing field, but there was inadequate time for normal oral presentations and discussion. We therefore asked that all contributed papers be put on display for at least 24 hours prior to a brief oral summary. A question and discussion period followed groups of oral presentations on the same or similar topic. In this way the opportunity for interactive discussion, not available in conventional poster displays, was preserved.

The Symposium was sponsored by IAU and the National Research Council of Italy and to these bodies the organisers express their grateful thanks.

On June 29th the group had the opportunity to see the new 32 meter Italian antenna at Medicina – the first in the world dedicated to VLBI research. Following the observatory visit, more than 150 were the guests of the Mayor and village of Medicina for a never-to-be-forgotten "home-cooked dinner" followed by some hours of merriment.

We thank the authors and participants who have contributed to the symposium and to these proceedings. Special thanks go to the local participants who worked so hard to keep the symposium running smoothly. We also wish to express our gratitude to the secretarial staff and, in particular, to Mrs. S. Buckley, to Mr. L. Baldeschi and to N. Primavera, of the Istituto di Radioastronomia, for their invaluable help.

R. Fanti
K.I. Kellermann
G. Setti

SCIENTIFIC ORGANIZING COMMITTEE

        Booth R.
        Bridle A.H.
        Hayakawa S.
        Jauncey D.L.
        Kellermann K.I. (Co-Chairman)
        Pauliny-Toth I.I.K.
        Scheuer P.A.G.
        Schilizzi R.T.
        Setti G. (Co-Chairman)

LOCAL ORGANIZING COMMITTEE

        Fanti C.
        Fanti R. (Chairman)
        Felli M.
        Grueff G.
        Lari C.
        Padrielli L.
        Tofani G.

LIST OF PARTICIPANTS

P. ALLAN, K.P.N.O., TUCSON, U.S.A.

H. ALLER, University of Michigan, ANN ARBOR, U.S.A.

M. ALLER, University of Michigan, ANN ARBOR, U.S.A.

W. ALEF, M.P.I. fuer Radioastronomie, BONN, F.R.G.

S. ANANTHAKRISHNAN, Tata Inst.,OOTACAMUND, INDIA

H. ARP, Mount Wilson and Palomar Obs., PASADENA, U.S.A.

L.B. BAATH, Onsala Space Observatory, ONSALA, SWEDEN

D.C. BACKER, University of California, BERKELEY, U.S.A.

J. BALL, Harvard College Obs., CAMBRIDGE MASS., U.S.A.

N. BARTEL, C.F.A., CAMBRIDGE, U.S.A.

P.D. BARTHEL, Sterrewacht, LEIDEN, HOLLAND

A. BAUDRY, Universite'de Bordeaux, FLOIRAC, FRANCE

M.C. BEGELMAN, ILA, University of Colorado, BOULDER, U.S.A.

J.M. BENSON, N.R.A.O., CHARLOTTESVILLE, U.S.A.

R.V. BHONSLE, Physical Research Lab.,AHMEDABAD, INDIA

M. BIRKINSHAW, Cavendish Laboratory, CAMBRIDGE, U.K.

C.I. BJORNSSON, ESO, Garching bei Munchen, F.R.G.

R.D. BLANDFORD, CalTech., PASADENA, U.S.A.

A. BOISCHOT, Observatoire de Paris-Meudon, FRANCE

R. BOOTH, Onsala Space Observatory, ONSALA, SWEDEN

V. BORIAKOFF, Cornell University, ITHACA, U.S.A.

A. BRACCESI, Dipartimento di Astronomia, BOLOGNA, ITALY

W. van BREUGEL, Steward Observatory, TUCSON, U.S.A.

A.H. BRIDLE, N.R.A.O., CHARLOTTESVILLE, U.S.A.

F. BRIGGS, University of Pittsburgh, PITTSBURGH, U.S.A.

I. BROWNE, N.R.A.L., JODRELL BANK, U.K.,

G. de BRUYN, Radiosterrenwacht, DWINGELOO, HOLLAND

B. BURKE, M.I.T., CAMBRIDGE MASS., U.S.A.

J. CAMPBELL, Geodetic Institute, BONN, F.R.G.

A. CAPORALI, Telespazio S.p.A., ROMA, ITALY

A. CAVALIERE, Osservatorio Astronomico, PADOVA, ITALY

T.V. CAWTHORNE, Cavendish Lab., CAMBRIDGE, U.K.

C. CHIUDERI, Osservatorio di Arcetri, FIRENZE, ITALY

M.H. COHEN, O.V.R.O., CalTech., PASADENA, U.S.A.

J.M. CORDES, Cornell University, ITHACA, U.S.A.

W.D. COTTON, N.R.A.O., CHARLOTTESVILLE, U.S.A.

P. CRANE, N.R.A.O., SOCORRO, U.S.A.
N. D'AMICO, Istituto di Fisica, PALERMO, ITALY
B. DENNISON, Physics Dept. Virginia, BLACKSBURG, U.S.A.
P. DIAMOND, Onsala Space Observatory, ONSALA, SWEDEN
F. DRAGO, Osservatorio di Arcetri, FIRENZE, ITALY
A. ECKART, M.P.I. fuer Radioastronomie, BONN, F.R.G.
R. ESTALELLA, University of Barcelona, BARCELONA, SPAIN
C. FANTI, Dipartimento di Astronomia, BOLOGNA, ITALY
R. FANTI, Dipartimento di Astronomia, BOLOGNA, ITALY
I. FEJES, Geodetic Observatory, BUDAPEST, HUNGARY
M. FELLI, Osservatorio di Arcetri, FIRENZE, ITALY
L. FERETTI, Istituto di Radioastronomia, BOLOGNA, ITALY
A. FERRARI, Istituto di Fisica, TORINO, ITALY
A. FICARRA, Istituto di Radioastronomia, BOLOGNA, ITALY
P. FONTANELLI, Osservatorio di Arcetri, FIRENZE, ITALY
D.N. FORT, NRC of Canada, OTTAWA-ONTARIO, CANADA
Y. FUKUI, Nagoya University, NAGOYA, JAPAN
J.A. GARCIA BARRETO, Universidad de Mexico, MEXICO
B. GELDZAHLER, N.R.L., WASHINGTON, U.S.A.
R. GENZEL, University of California, BERKELEY, U.S.A.
G. GIOVANNINI, Istituto di Radioastronomia, BOLOGNA, ITALY
M.C. GORENSTEIN, C.F.A., CAMBRIDGE MASS, U.S.A.
S. GORGOLEWSKI, Torun Radio Astronomy Obs., TORUN, POLAND
D. GRAHAM, M.P.I. fuer Radioastronomie, BONN, F.R.G.
F. GRAHAM-SMITH, N.R.A.L., JODRELL BANK, U.K.
L. GREGORINI, Istituto di Radioastronomia, BOLOGNA, ITALY
G. GRUEFF, Istituto di Radioastronomia, BOLOGNA, ITALY
C.R. GWINN, Princeton University, PRINCETON, U.S.A.
P. HILL, Institut Geographique National, ST. MANDE, FRANCE
P.A. HUGHES, Cavendish Lab., CAMBRIDGE, U.K.
J.B. HUTCHINGS, Dominion Astrophysical Obs., VICTORIA, CANADA
M. INOUE, Nobeyama Radio Observatory, NOBEYAMA, JAPAN
D.L. JAUNCEY, CSIRO, CANBERRA, AUSTRALIA
K. JOHNSTON, N.R.L., WASHINGTON, U.S.A.
D. JONES, O.V.R.O., CalTech., PASADENA, U.S.A.
J.F. JORDAN, J.P.L., PASADENA, U.S.A.
K. KELLERMAN, N.R.A.O., GREEN BANK, U.S.A.
A. KONIGL, Institute for Advanced Study, PRINCETON, U.S.A.
V.K. KULKARNI, M.P.I. fuer Radioastronomie, BONN, F.R.G.
A. KUS, Torun Radio Astronomy Observatory, TORUN, POLAND
H. van der LAAN, Sterrewacht, LEIDEN, HOLLAND
A. LANE, N.R.A.O., CHARLOTTESVILLE, U.S.A.
C. LARI, Istituto di Radioastronomia del CNR, BOLOGNA, ITALY
T.H. LEGG, NRC of Canada, OTTAWA ONTARIO, CANADA

G. LEVY, J.P.L., PASADENA, U.S.A.
K.J. LO, Astronomy Dept., CalTech, PASADENA, U.S.A.
P. LONDRILLO, Dipartimento di Astronomia, BOLOGNA, ITALY
C.J. LONSDALE, N.R.A.L., JODRELL BANK, U.K.
J. MACHALSKI, Jagiellonian University, KRAKOW, POLAND
F. MANTOVANI, Istituto di Radioastronomia, BOLOGNA, ITALY
J. MARCAIDE, M.P.I. fuer Radioastronomie, BONN, F.R.G.
C. MATVEYENKO, Institute for Space Research, MOSCOW, U.S.S.R.
I.M. Mc HARDY, Leicester University, LEICESTER, U.K.
T.K. MENON, University of British Columbia, VANCOUVER, CANADA
A. MICHALEC, Jagiellonian University, KRAKOW, POLAND
L. MILLER, Royal Obs., EDINBURGH - SCOTLAND, U.K.
T.A. MOFFET, O.V.R.O., CalTech., PASADENA, U.S.A.
I.G. MOISEEV, Crimean Astrophysical Observatory, U.S.S.R.
R.L. MOORE, O.V.R.O., CalTech., PASADENA, U.S.A.
M. MORIMOTO, Nobeyama Radio Observatory, NOBEYAMA, JAPAN
R. MUTEL, University of Iowa, IOWA CITY, U.S.A.
T.W.B. MUXLOW, N.R.A.L., JODRELL BANK, U.K.
S. NEFF, Radiosterrenwacht, DWINGELOO, HOLLAND
G.D. NICOLSON, Radio Astronomy Obs., JOHANNESBURG, S. AFRICA
A.E. NIELL, J.P.L., PASADENA, U.S.A.
J.L. NIETO, Obs., du Pic du Midi, BAGNERES de BIGORRE, FRANCE
L. NOBILI, Osservatorio Astronomico, PADOVA, ITALY
R.P. NORRIS, N.R.A.L., JODRELL BANK, U.K.
H. OLTHOF, ESA, PARIS, FRANCE
F. PACINI, Osservatorio di Arcetri, FIRENZE, ITALY
L. PADRIELLI, Istituto di Radioastronomia, BOLOGNA, ITALY
N. PANAGIA, Istituto di Radioastronomia, BOLOGNA, ITALY
P. PARMA, Istituto di Radioastronomia, BOLOGNA, ITALY
I. PAULINY-TOTH, M.P.I. fuer Radioastronomie, BONN, F.R.G.
T.J. PEARSON, O.V.R.O., CalTech., PASADENA, U.S.A.
R. PERLEY, N.R.A.O., SOCORRO, U.S.A.
G.C. PEROLA, Istituto Astronomico, ROMA, ITALY
R.B. PHILLIPS, Haystack Observatory, WESTFORD, U.S.A.
G. PILBRATT, Onsala Space Observatory, ONSALA, SWEDEN
R.W. PORCAS, M.P.I. fuer Radioastrónomie, BONN, F.R.G.
R. PRESTON, J.P.L., PASADENA, U.S.A.
E. PREUSS, M.P.I. fuer Radioastronomie, BONN, F.R.G.
A.C.S. READHEAD, O.V.R.O., CalTech., PASADENA, U.S.A.
J.M. REES, Inst. of Astronomy, CAMBRIDGE, U.K.
M. REID, C.F.A., CAMBRIDGE MASS, U.S.A.
D.H. ROBERTS, Brandeis University, WALTHAM, U.S.A.
J.D. ROMNEY, M.P.I. fuer Radioastronomie, BONN, F.R.G.
B. RONNANG, Onsala Space Observatory, ONSALA, SWEDEN

H. de RUITER, Istituto di Radioastronomia, BOLOGNA, ITALY
R.D. RUSK, University of TORONTO, CANADA
M. SALVATI, ESO, Garching bei Munchen, F.R.G.
M. SANROMA, University of Barcelona, BARCELONA, SPAIN
R. SAUNDERS, Cavendish Lab., CAMBRIDGE, U.K.
P. SCHEUER, Cavendish Lab., CAMBRIDGE, U.K.
R.T. SCHILIZZI, Radiosterrenwacht, DWINGELOO, HOLLAND
M. SCHNEPS, C.F.A., CAMBRIDGE MASS, U.S.A.
G.C. SETTI, ESO, Garching bei Munchen, F.R.G.
D.B. SHAFFER ,ANNAPOLIS, U.S.A.
P.A. SHAVER, ESO, Garching bei Munchen, F.R.G.
R.S. SIMON, N.R.L., WASHINGTON, U.S.A.
R.E. SPENCER, N.R.A.L., JODRELL BANK, U.K.
K. SUBRAMANIAN, Tata Institute, BOMBAY, INDIA
S. TALLQUIST, Metsahovi Observatory, OTAKAARI, FINLAND
J. TAYLOR, Princeton University, PRINCETON, U.S.A.
G. TOFANI, Osservatorio di Arcetri, FIRENZE, ITALY
P. TOMASI, Istituto di Radioastronomia, BOLOGNA, ITALY
G. TORRICELLI, Osservatorio di Arcetri, FIRENZE, ITALY
R. TUROLLA, S.I.S.S.A., TRIESTE, ITALY
E. TRUSSONI, Istituto di Fisica, TORINO, ITALY
M.H. ULRICH, ESO, Garching bei Munchen, F.R.G.
S.C. UNWIN, O.V.R.O., CalTech., PASADENA, U.S.A.
F. VAGNETTI, Istituto Astronomico, ROMA, ITALY
C. de VEGT, Sternewarte, Universitat, HAMBURG,F.R.G
M. VERON, ESO, Garching bei Munchen, F.R.G.
P. VERON, ESO, Garching bei Munchen, F.R.G.
M. VIGOTTI, Istituto di Radioastronomia, BOLOGNA, ITALY
C.R. WALKER, N.R.A.O., CHARLOTTESVILLE, U.S.A.
WAN TONG-SHAN, Shanghai Obs., PEOPLE's REPUBLIC of CHINA
G.J. de WAARD, Sterrewacht, LEIDEN, HOLLAND
K. WEILER, N.S.F., WASHINGTON, U.S.A.
D. WEISTROP, G.S.F.C., GREENBELT, U.S.A.
N. WHYBORN, N.R.A.L., JODRELL BANK, U.K.
P.N. WILKINSON, N.R.A.L., JODRELL BANK, U.K.
A. WITZEL, M.P.I. fuer Radioastronomie, BONN, F.R.G.
D.M. WORRAL, University of California, LA JOLLA, U.S.A.
A.E. WRIGHT, A.N.R.A.O., AUSTRALIA
J.M. WROBEL, O.V.R.O., CalTech., PASADENA, U.S.A.
G. ZAMORANI, Istituto di Radioastronomia, BOLOGNA, ITALY
L. ZANINETTI, Istituto di Fisica, TORINO, ITALY
J.A. ZENSUS, M.P.I. fuer Radioastronomie, BONN, F.R.G.

INTERMEDIATE SCALE STRUCTURE [+]

I.W.A. Browne
Nuffield Radio Astronomy Laboratories,
Jodrell Bank.

What does intermediate (arcsecond) scale structure tell us about the activity deep in the VLBI cores of galaxies and quasars? There are certain key areas where a knowledge of this structure is particularly relevant. One is in defining the relationships between the various classes of objects leading to unified schemes. A second is in the understanding of jet physics. For the latter it is very important to know if the material in the jets supplied from the nucleus is heavy or light, slow or fast, ejected one side at a time or symmetrically. The optimum resolution to study jet material is on the arcsecond ($\equiv$ kilo-parsec) scale where very high dynamic range maps are possible and where it becomes feasible to find out something about the physical conditions in the surrounding medium from optical and X-ray observations.

UNIFIED SCHEMES

The justification for talking about unified schemes is that they simplify things and that they make testable predictions about a wide range of phenomena. Also one of the important inputs which prompted the idea of a unified scheme came from the study of the intermediate scale structure of compact VLBI sources. The basic proposition of such schemes is that the only difference between a flat and a steep spectrum source is one of aspect; projection and the Doppler boosting turn an extended double source into a compact core-dominated one.

Evidence that there exists some kind of unity between flat and steep radio spectrum quasars comes from several directions and has been discussed in some detail by Orr & Browne (1982). The situation concerning radio galaxies is not nearly so well defined. Only in one, 3C120, is there direct evidence of relativistic motion. In addition flat spectrum radio galaxies are rare, and most of these when studied optically, turn out to be BL Lac objects (e.g. Ap.Lib and 3C371). This latter point leads me to propose a 'unification' between elliptical radio galaxies and BL Lac objects.

[+] Discussion on page 415

1

R. Fanti et al. (eds.), VLBI and Compact Radio Sources, 1–5.
© *1984 by the IAU.*

If we assume that the characteristic optical continuum radiation
from a BL Lac object is relativistically beamed there must be a large
'parent population' not showing BL Lac behaviour. Such a parent population would have the following properties:-

1)   be extended radio sources, because BL Lacs themselves have diffuse
      radio emission (Stannard & McIlwrath 1982; Ulvestad et al. 1983)

2)   be elliptical galaxies since nearby BL Lac objects all lie in
      elliptical galaxies

3)   have very weak or non-existent emission lines

4)   have a significantly higher space density than BL Lac objects
      themselves.

The obvious conclusion is that the BL Lac 'parents' are elliptical
radio galaxies. Furthermore, by comparing the local space density of
BL Lac objects and elliptical radio galaxies of a given extended radio
luminosity we can deduce an upper limit to the average Lorentz factor
of the radiating material. The simplest possible situation is that it
is only when the angle to the line of sight, θ, is less than $1/\gamma$ does
the object display characteristic BL Lac properties. In this case the
proportion of objects expected to be BL Lacs is $[1 - \cos (1/\gamma)]$.
Comparing the local space densities of BL Lacs of extended radio
luminosity $\geq 10^{22}$ W Hz^{-1}sr^{-1}, with that of elliptical radio galaxies
of the same luminosity, gives $\gamma \lesssim 5.4$ (Browne, 1983).

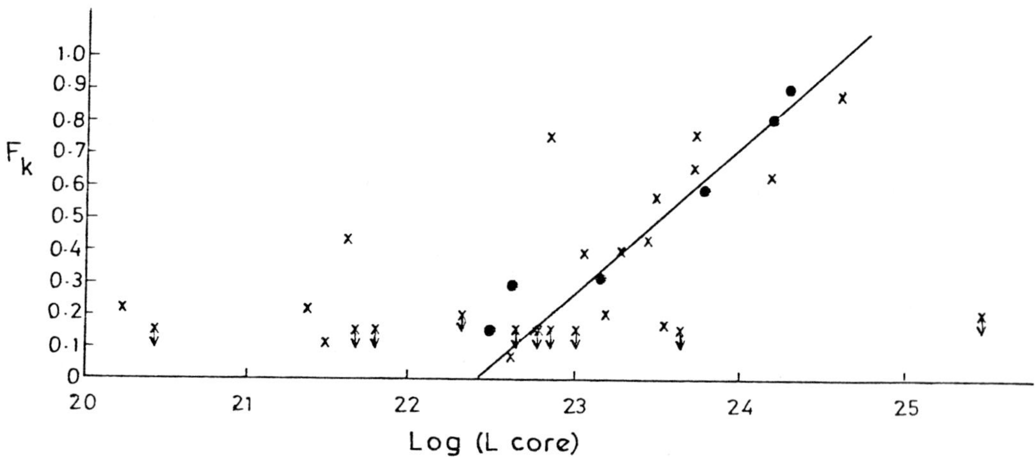

Fig.1   The fraction of galaxy flux in an infrared core as a function
        or core radio luminosity. F_K is the function of K band light
        in the galaxy core. L_{core} is the λ 6cm core luminosity in
        W Hz^{-1}sr^{-1}. Crosses denote radio galaxies; filled circles
        BL Lac objects. Data are taken from Heckman et al. (1983).

Before leaving the subject of BL Lacs and radio galaxies I would like to mention a consequence of the proposed unification. Such a scheme predicts the existence of optical BL Lac cores in all elliptical radio galaxies. How might we find these? In face-on objects ($\theta \sim 90°$) the cores would be too weak to detect, but there might be a chance in radio galaxies with fairly strong cores which we might expect to be only at moderate angles to the line-of-sight. Fortunately there exist some observations which at least partially check this prediction. Heckman et al.(1983) have looked for nonstellar optical/infrared nuclear continuum emission from 42 radio galaxies selected by their core strength. They find that most of these objects do have continuum emission and this is roughly proportional to the strength of the compact radio source. Fig.1 shows the fraction of the infrared flux contained in the core plotted against core radio luminosity for Heckman et al's sample. Whenever the core-radio luminosity is sufficient an infrared core is nearly always present. It is also worth pointing out that all objects appear to follow the same trends irrespective of whether they are classified as radio galaxies or BL Lacs. A careful study of all these cores in the optical to see if they have the characteristic variability and high polarization of BL Lacs is clearly a very important investigation.

JET PROPERTIES (ONE-SIDED OR TWO-SIDED; FAST OR SLOW?)

Do the central objects in radio galaxies and quasars always eject two beams or is the ejection sometimes one-sided? Are the jets relativistic? The study of kiloparsec scale jets offers some useful pointers, but not as yet any definitive answers to these questions. In the following I will try and review some of the evidence, concentrating on ways of estimating jet speeds.

Only if jets are relativistic on all scales can the idea of symmetrical ejection be reconciled with the observational evidence that jets appear very asymmetric. On the VLBI scale direct observation of superluminal behaviour indicates that jets start off relativistically. I will assume that there is relativistic motion in all VLBI cores and consider the question of the speed of Kiloparsec jets.

1. Jet Continuity

VLBI jets almost always look as if they will join up with the arcsecond jets. Misalignments are common but usually < 90° and, even in the case of large bends, the sense of curvature of the structure usually suggests continuity (Browne et al.1983). The obvious interpretation is that Doppler favouritism is responsible for the asymmetry on all scales. However, there is in fact no reason why the jets should not be intrinsically asymmetric, having their asymmetry further enhanced by Doppler boosting.

Could jets be one-sided, start off fast, but slow to subrelativistic

speed later? To answer this question the relative strengths of VLBI
cores + jets and arcsecond jets should be examined for a wide range of
angles to the line of sight. If there is slowing one would expect a
large dispersion in this ratio. Observationally, when we see an arcsecond jet it is usually associated with a VLBI core of roughly similar
flux density. This suggests that there is not much slowing as the jet
moves outwards from the nucleus. Moore (1983) has attempted to quantify
this argument, by trying to see if there is any evidence for 'disembodied
jets' (I.E. strong jets with no cores). From the numbers he concludes
that kiloparsec scale jets must be relativistic.

2.   Other evidence for jet speeds

a)   Precession models

Several groups have tried to model rotationally symmetric sources
using ballistic precessing beam models. Gower et al.(1982) have looked
at a wide range of sources and find that they can successfully reproduce
the main features of the structures with precession models. They usually
require mildly relativistic beams. Quasars require the fastest beams.

Muxlow (this Conference) has fitted a similar precession model to
the very twisted quasar 3C418. He ends up with a mild preference for
a beam with $v/c = 0.98$.

In contrast Hunstead et al.(1983) conclude that the beam speeds in
the strikingly rotationally symmetric quasar 2300-189 must be $< 0.3c$.
Their precession model reproduces the ridge line geometry of the quasar
very well, but not the intensity distribution which is much brighter on
one side of the core than the other.

b)   Bending by Ambient Medium

Like the results from precession models attempts to deduce jet
speeds from modelling jet bends produced by winds or atmospheres, give
varying answers. For the large scale bent tails in low luminosity
sources, velocities of a few thousand km s^{-1} are preferred. However,
for the small scale jets in the nucleus of 3C293 van Breugel et al.(1983)
deduce a beam velocity $\gtrsim 4 \times 10^4$ km s^{-1} .

To sum up the situation on beam velocities we can say the
following:-

1)   If beams start out at relativistic velocities they stay
     relativistic.

2)   Relativistic motion is sufficient but not necessary to
     explain the observed asymmetry.

3)   Evidence on beam velocities from modelling bends is
     inconclusive.

3. Jet Symmetry

So far I have discussed work primarily directed at deducing jet velocities. Is there any way in which the related question of jet symmetry can be addressed? One possibility is to investigate sources which show cores, jets and lobes. On the assumption that lobe emission is unbeamed and the observed jet asymmetry is solely due to Doppler beaming, the lobe to which the jet points should be statistically indistinguishable from the opposite lobe. Saika (1981) has attempted such an investigation for relatively low luminosity radio galaxies and finds that jets tend on average to point to the lobes containing the highest brightness features. Work on quasars (Shone, private communication) shows a similar, but less marked, trend. One additional and potentially very important point is that in the few cases we know of sources with multiple hotspots the jet points to the lobe which contains these hotspots. This evidence all supports the conclusion that there are some intrinsic differences in the beams and that Doppler beaming is not the sole cause of the observed asymmetry. The only other possibility is that some of the high brightness features in lobes may be Doppler boosted.

ACKNOWLEDGEMENTS

I thank David Shone and Peter Wilkinson for helpful discussions and Tim Heckman for permission to use unpublished data.

REFERENCES

Browne, I.W.A., 1983. Mon.Not.R.astr.Soc., 204, 23P.
Browne, I.W.A., Charlesworth, M., Muxlow, T.W.B., Tzanetakis, A., & Wilkinson, P.N. 1983. "Astrophysical Jets", eds. A. Ferrari & A.G. Pacholczyk, Reidel.
Gower, A.C., Gregory, P.C., Hutchings, J.B. & Unrah, W., 1982. Ap.J., 262, 478.
Heckman, T.M., Lebofsky, M.J., Rieke, G.H. & van Breugel, W., 1983. preprint.
Hunstead, R.W., Murdoch, H.S., Condon, J.J. & Phillips, M.M., 1983. Mon.Not.R.astr.Soc., submitted.
Moore, P.K., 1983. Mon.Not.R.astr.Soc., submitted.
Orr, M.J.L. & Browne, I.W.A., 1982. Mon.Not.R.astr.Soc., 200, 1067.
Saika, D.J., 1981. Mon.Not.R.astr.Soc., 197, 11P.
Stannard, D. & McIlwrath, B., 1982. Nature, 298, 140.
Ulvestad, J.S., Johnston, K.J. & Weiler, K.W., 1983. Ap.J., 266, 18.
van Breugel, W., Heckman, T.M., Butcher, H. & Miley, G.K., 1983. preprint.

HOW TO CHOOSE A RANDOMLY ORIENTED SAMPLE[+]

Tim Cawthorne
Mullard Radio Astronomy Observatory, Cavendish Laboratory,
Madingley Road, Cambridge CB3 OHE, U.K.

Efficient VLBI tests of relativistic models for superluminal motion require a complete sample of sources free from selection effects in orientation, coupled with an observational requirement for bright cores. Probably the best way to construct such a sample is to extract from a low frequency sample those sources with the highest radio luminosity, since these tend to have the strongest cores. One must then ask whether there are any sources included which would not be if they were differently oriented; that is, either the flux or the power falls below the limit for the sample when the core and jet flux is removed.

Taking the sample of 166 3C sources of Jenkins et al. (1977) and power limit $P_0 \geq 10^{28}$ WHz^{-1} sr^{-1} one finds 19 quasars and 8 galaxies with measured redshifts. (Unfortunately even the 3C sample contains a number of faint galaxies without measured redshifts. The number of these in our sample depends sensitively on the apparent magnitude-redshift relation used; but there are unlikely to be more than 7.)

Some of the powerful sources are famous and thoroughly investigated. However many remain obscure and have no useful map. Morrison, Muxlow and I have mapped some of these with the MERLIN interferometer which gives 0.3 arcsec resolution at 1.67 GHz. Combining our observation with those made elsewhere at higher frequencies enables us to estimate component spectral indices over a reasonable range of frequencies, and hence to guess the origin of the flux at the selection frequency.

We have confirmed that a number of sources (e.g. 3C181) are classical doubles. For some of the smaller sources further observation will be needed to fully elucidate their structures. For example 3C454 has a curious twisted shape : we cannot yet identify the components. For our high luminosity subset of the 166 sample we find :

(i)   Ten of the 34 sources have cores and there are a few "don't knows".

(ii) About 75% of the sample consists of classical doubles or triples.
The hotspots are quite symmetrically disposed about the nucleus

+ Discussion on page 416

R. Fanti et al. (eds.), VLBI and Compact Radio Sources, 7–8.

indicating that the hotspot advance speed is not highly
relativistic.

(iii)   A rough estimate of the hotspot advance speed for the Northern
        component of 3C9 (the most powerful member of the sample) can
        be made. It turns out to be $\sim 0.2c$, similar to that deduced
        for classical doubles as a whole. For ram pressure confinement
        present observations demand n $\sim 10^3$ m^{-3}, similar to that
        required for Cygnus A. Thus while question marks remain over a
        few sources, the majority are bonafide members of our sample.

REFERENCE

Jenkins, C.J., Pooley, G.G. & Riley, J.M., Mem. R.A.S., <u>84</u>, 61.

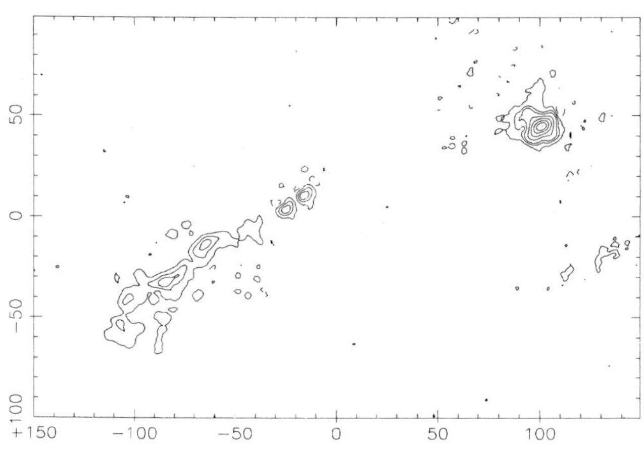

Fig. Caption: 3C280.1 : a candidate for exclusion?
                (1.67 GHz   HPBW 0.4" x 0.4"   coordinates are 0.1 arcsec)

                The jet has a surprisingly steep spectrum, $\alpha_{1.7}^{5} \sim 1$.
                If the jet were relativistic then this source might
                warrant exclusion from our sample. It seems unlikely
                however, that the absence of a Southern hotspot is an
                effect of orientation.

FAINT EXTENDED RADIO SOURCES SURROUNDING ACTIVE GALAXY NUCLEI [+]

H. van der Laan, S. Zieba[*]
Leiden Observatory

J.E. Noordam
Netherlands Foundation for Radio Astronomy

INTRODUCTION

As is clear from many papers in this volume, the phenomenology of
radio sources associated with active galaxy nuclei (AGNs) is very diverse.
It is attractive nevertheless to try and retain the belief that complex
effects may have simple causes. In this case the great variety of
structures seen in broad band radio surface photometry over a tremendous
range of angular resolution is preferably thought of as the result of
one basic process in which plasmas are ejected in an intrinsically
symmetric manner at bulk velocities which (at least initially) are
relativistic. The complexity is then attributable to many circumstances,
e.g.:
- the engine's time dependent output
- the engine's directionality which may
     a) be very steady
     b) precess
     c) jump eratically
- the plasma's opacity at radio wavelengths
- the geometry of the source-observer configuration
- the relativistic kinematic effects which so strongly depend on
  that geometry
- the circumnuclear environment of the engine, i.e. the gas and dust
  content of the AGN
- the circumgalactic environment of the parent galaxy
- the nearest neighbours of the parent galaxy
- the selection effects of our samples.

We report here some early results of a programme which searches for
faint radio emission and its morphology in the immediate vicinity of
very radio bright AGNs. This programme utilises a newly developped
capacity of the Synthesis Radio Telescope at Westerbork, based on the

[*] On leave from the Astronomical Observatory of the
    Jagiellonian University, Cracow, Poland.

[+] Discussion on page 417   9

R. Fanti et al. (eds.), VLBI and Compact Radio Sources, 9–13.
© 1984 by the IAU.

many redundant spacings inherent in the uniformity of this East-West
array (Noordam and De Bruyn, 1982). The method, its principles and power
are described by De Bruyn in this volume.

THE SAMPLE

From the catalogue of radio sources compiled by Kühr et al. (1981)
we culled the list of galaxies and BL Lacs satisfying the conditions:
(i) $S_{6\ cm}$ > 1 Jy, (ii) δ > 25°, (iii) z < 0.10. The latter two conditions
in order to have a beam no larger than 3.5 x 8 arc s FWHM and a linear
resolution of order 10 kpc or better (H_o = 50). From this list of 21
sources the four objects were selected which have the strongest core
flux and flat or inverted spectra. To these four we added, for the
purpose of this exploratory programme, the radio galaxy 3C 111, not in
the Kühr catalogue because of its low galactic latitude (b = -9°) but
otherwise meeting our criteria. Table 1 summarizes some pertinent facts
about this five source sample.

These sources were observed at a frequency of 4874 MHz with the
Westerbork SRT, in the configuration which yields 91 interferometers
with 38 different spacings from 72 to 2732 m in 72 m increments.

Taking advantage of the array's redundancy in the manner described
by De Bruyn (this volume), high dynamic range cleaned maps were obtained
from which, in the uv plane, 99 to 99.9% of the core source's peak flux
has been removed.

RESULTS

In *all cases* faint extended emission is detected surrounding the
ultracompact (more or less opaque) core source in the AGN.

For the three 'point sources' BL Lac, OQ 208 and 4C 39.49 the
extended emission is very weak but clearly visible as diffuse emission
enveloping the central source. Table 2 describes these sources.

The structure of the two remaining sources, 3C 371 and 3C 111, is
more complex. In addition to its core component 3C 371 has several other
features: A, a component unresolved in this observation, $\sim$ 3 arc s from
the core in p.a. $\sim$ -115°, known also from VLA (Perley et al., 1980) and
MERLIN (Browne et al., 1982) maps; B, an extended structure enveloping
the core *and* feature A, emanating from the core, elongated $\sim$ 12 arc s,
18 kpc in p.a. $\sim$ + 150° and $\sim$ 20 arc s, 30 kpc in p.a. $\sim$ - 30° at the
1% of peak flux contour level; finally feature C, an outer lobe of
diffuse emission about 25 arc s, 38 kpc from the central source along
p.a. $\sim$ - 100°, a direction also present on the VLBI scale as a one-
sided milliarcsecond jet according to Pearson and Readhead, 1981. Hence
there are three very different directions manifest in this source at
this level of brightness and dynamic range. It is particularly

TABLE 1

Source	Other Names	Radio position of the core α(1950)	δ(1950)	Approximate flux of central component at 6 cm (Jy)	z	Distance (Mpc) H_O=50 kms^{-1} Mpc^{-1}
0415+37	3C111	04^{h}15^{m}00^s.50	37°54'19.5	3.3 (variable)	0.0485	295
1404+28	0Q208, MK668	14 04 45.62	28 41 29.2	2.9	0.077	470
1652+39	4C39.49, MK501	16 52 11.75	39 50 24.6	1.4	0.033	200
1807+69	3C371	18 07 18.54	69 48 56.9	1.7	0.051	310
2200+42	BL Lac, OY401	22 00 39.36	42 02 08.5	4.8 (variable)	0.070	425

	Beam size (FWHM) angular size (arcs)	Beam size (FWHM) linear size kpc	Comments	
0415+37	3.5x5.7	5x8	Classical triple radio galaxy VLBI: one sided jet Inverted core spectrum	Laing, 1981 Linfield, 1981
1404+28	3.5x7.3	8x17	Markarian galaxy VLA unresolved Inverted spectrum	Ulvestad et al. 1981
1652+39	3.5x5.5	3x5	Markarian galaxy VLA unresolved Flat spectrum	Ulvestad et al. 1981
1807+69	3.5x3.7	5x6	N-galaxy Core + secondary component at $\sim$ 3 arc sec distance One side VLBI jet, misaligned with secondary comp. Complex flat spectrum	Browne et al. 1982 Perley et al. 1980 Pearson, Readhead, 1981
2200+42	3.5x5.2	7x11	VLA unresolved VLBI: two components Complex flat spectrum	Ulvestad et al. 1981 Philips, Mutel, 1982; Pearson, Readhead, 1981

Table 2

		Estimate of extended emission (% of total source flux $\quad$ WHz^{-1})		Approximate Dimensions of extended component (NS x EW)
BL	Lac	0.7	7.2×10^{23}	24 x 8 arc s 48 x 16 kpc
OQ	208	0.3	2.3×10^{23}	10 x 15 arc s 23 x 35 kpc
4 C	39.49	1.6	1.1×10^{23}	20 x 10 arc s 18 x 9 kpc

interesting that current activity, visible as VLBI structure in the form
of a one sided jet, points towards the diffuse, relatively distant and
presumably relatively old component C, not to the much brighter feature
A, which looks like a one sided jet at VLA and MERLIN resolutions and is
presumably rather younger than component C. The engine in 3C 371 could
be a multiple one, each turned on part of the time, with the current one
repeating a performance of $\sim 10^{7\pm1}$ years ago when the now diffuse
component C was formed.

The triple radio galaxy 3C 111, fifth source in our sample, did
not yield a reliable map of the active nucleus' immediate vicinity, due
to grating ring interference from the bright outer lobes. The map is
very interesting nevertheless, for it shows at very faint levels,
remarkable structure of the bridge connecting the bright active core to
the outer lobes. On the south side this bridge ~ 300 kpc in length, is
normal, i.e. sharply bounded. On the northern side the bridge has a
faint but remarkable veil extending up to 80 kpc away from the bridge
at the 0.15 % of peak flux brightness level. There are other examples
of such faint extensions, e.g. that of 3C 430 (Riley and Pooley, 1975)
but none of such great dimensions, either relative or absolute, are
known to us. We intend to pursue this phenomenon at other wavelengths.

CONCLUSIONS

This exploratory study of a small heterogeneous sample can only
lead to provisional conclusions.
- AGNs are not naked radio sources, they are at least enveloped by
  faint very extended radio sources.
- The extended radio sources are often amorphous, showing no
  particular structural features or symmetries.
- The extended radio sources have sizes and powers, intermediate
  between those of radio galaxies and normal galaxies (Hummel,
  1980), not unlike the radio properties of some Seyfert galaxies
  (Meurs, 1982; Meurs and Wilson, 1983).

- In 3C 371 the extended radio source has features which in relation
  to the milliarcsecond and subarcsecond structure suggest the
  presence of more than one basic direction which persist over long
  time scales.
- The radio galaxy 3C 111 has a radio bridge apparently eroded by
  an intergalactic wind or onesided diffusion.

ACKNOWLEDGEMENTS

We thank the staff of the Westerbork Observatory for their help
with the observations and Dr. A.G. De Bruyn for helpful discussions.
The Westerbork Radio Observatory is operated by The Netherlands
Foundation for Radio Astronomy, with the financial support of the
Netherlands Organization for the Advancement of Pure Research (ZWO).

REFERENCES

Browne, I.W.A., Orr, M.J.L., Davis, R.J., Foley, A., Muxlow, T.W.B. and
    Thomasson, P.: 1982, MNRAS 198, 673.
Hummel, E.: 1980, Ph.D. Thesis, Groningen University.
Kühr, H., Nauber, U., Pauliny-Toth, I.I.K. and Witzel, A.: 1981, A
    Catalogue of Radio Sources, Max-Planck-Institut für Radioastronomie,
    Bonn.
Laing, R.A.: 1981, MNRAS 195, 261.
Linfield, R.: 1981, Astrophys. J. 244, 436.
Meurs, E.J.A.: 1982, Ph.D. Thesis, Leiden University.
Meurs, E.J.A. and Wilson, A.S.: 1984, Astron. Astrophys. (in press).
Noordam, J.E. and de Bruyn, A.G.: 1982, Nature 299, 597.
Pearson, T.J. and Readhead, A.C.S.: 1981, Astrophys. J. 248, 61.
Perley, R.A., Fomalont, E.B. and Johnston, K.J.: 1980, Astron. J. 85, 649
Phillips, R.B. and Mutel, R.L.: 1982, Astrophys. J. 257, L19.
Riley, J.M. and Pooley, G.P.: 1975, Mem. Roy. Astron. Soc. 80, 105.
Ulvestad, I., Johnston, K., Perley, R. and Fomalont, E.: 1981, Astron.
    J. 86, 1010.

VLBI SURVEY OF A COMPLETE SAMPLE OF ACTIVE NUCLEI AND QUASARS [+]

T.J. Pearson and A.C.S. Readhead
Owens Valley Radio Observatory
California Institute of Technology

ABSTRACT

We have conducted a VLBI survey of a complete, flux-density limited
sample of 65 extragalactic radio sources, selected at 5 GHz. We have
made hybrid maps at 5 GHz of all of the sources accessible to the
Mark-II system. The sources can be divided provisionally into a number
of classes with different properties: central components of extended
double sources, steep-spectrum compact sources, very compact (almost
unresolved) sources, asymmetric sources (sometimes called "core-jet"
sources), and "compact double" sources. It is not yet clear whether
any of these classes is physically distinct from the others, or whether
there is a continuous range of properties.

1. INTRODUCTION

Since 1977 we have been engaged on a project to determine the structure
of a complete, unbiased sample of VLBI sources, and to measure changes
in the sources. The sample consists of 65 sources, of which 45 have
VLBI structure. Mapping 45 sources with our present instrumentation is
a very large project, but even so, 45 is a small number for statistical
studies. Thus, although statistical trends may become apparent, our
purpose is more to provide an overview of the range of morphologies
exhibited by the sources.

We cannot overemphasize the importance of studying a complete
sample of this type. There has been a tendency in VLBI to concentrate
on a small number of sources of special interest, such as the
superluminal sources, simply because VLBI is still very time-consuming;
although things are getting easier these days with the establishment of
the U.S. and European VLBI Networks. Studies of individual sources
can, of course, be very rewarding, but they leave many questions
unanswered, and may even bias our view of the VLBI universe. In
addition to exploring the menagerie of source morphologies, we can
address a number of specific questions by way of a complete survey, all

[+] Discussion on page 417.   15

R. Fanti et al. (eds.), VLBI and Compact Radio Sources, 15–24.
© 1984 by the IAU.

aimed ultimately at elucidating the fundamental cause of galactic
nuclear activity. For example, we may expect to find more superluminal
sources, and find out just how common they are - an important parameter
for the relativistic-beaming theories; and we can look for
correlations between VLBI and optical and X-ray properties of the
sources which may indicate physical connections and constrain
theoretical models.

2. THE COMPLETE SAMPLE

The sample we have chosen to study contains 65 sources, selected from
the NRAO-MPIfR 6-cm Strong Source Surveys S4 and S5 (Pauliny-Toth et
al. 1978, Kühr et al. 1981a). The selection criteria were: (a)
declination (1950) > 35 deg, for good u,v-coverage with existing
telescopes; (b) galactic latitude $|b|$ > 10 deg; and (c) total flux
density at 5 GHz $\geq$ 1.3 Jy (at the epoch of the original surveys). The
flux density limit was chosen to include a moderately large number of
sources strong enough to be mapped with the Mark-II system.

The sample thus defined contains 20 classical double radio sources
and 45 core-dominated sources (having > 95% of their flux density
arising in a region smaller than about 1 arcsec). Most of the
classical double sources were known from VLA and other observations to
contain no compact component strong enough to map with current Mark-II
VLB systems (i.e., > 0.25 Jy). All 65 sources are listed in Table 1,
which gives the source names, optical type (galaxy, quasar, BL Lac
object, or empty field), redshift where available, and large scale
radio structure (I or II indicates a classical double source of
Fanaroff-Riley class I or II; U indicates a core-dominated or
unresolved source); the remaining columns of Table 1 will be explained
below.

3. OBSERVATIONS

We made observations in three stages. All the observations were made
at 5 GHz. First, we made brief ("snapshot") observations to determine
which sources in the complete sample could be adequately mapped. A
detection is indicated by "Y" in the column headed "VLB?" in Table 1.
Then we made first-epoch maps of the sources detected in the finding
survey from full-track (10-11 hour) observations with three or four
antennas of the U.S. VLBI Network together with the 100-m antenna of
the MPIfR at Effelsberg (in most cases). The observations were
completed in seven sessions scheduled by the U.S. VLBI Network between
1978 and 1982, as indicated in Table 1, in which the antennas used are
shown by their conventional abbreviations. Finally, we have begun the
third stage: second-epoch maps of all the sources. We shall present
preliminary results of some of the second-epoch observations later in
this Symposium (Readhead et al. 1984).

TABLE 1
The Complete Sample

Source		Opt z	Radio	VLB?	First Epoch Mapped	First Epoch Antennas	Second Epoch Mapped	Second Epoch Antennas	Classification
0016+731		Q -	U	Y	1982 Apr	BKGFO			Very compact
0040+517	3C20	G 0.350	II	N	----				----
0108+388	OC314	EF -	U	Y	1979 Jul	GFOH	1982 Dec	BKGFO	Double
0133+476	OC457	Q 0.86?	U	Y	1978 Dec	GFOH			Compact
0153+744		Q -	U	Y	1982 Apr	B GFO			Double
0210+860	3C61.1	Q 0.184	II	N	----				----
0212+735		BL? -	U	Y	1980 Sep	BK FO			Asymmetric
0220+427	3C66B	G 0.0215	I	N	----				----
0314+416	3C83.1B	G 0.0181	I	N	----				----
0316+413	3C84	G 0.0172	U	Y	1978 Dec	GFOH			Unclassified
0404+768	4C76.03	G -	U	N	----				----
0454+844		BL -	U	Y	1981 Aug	BKGFO			Compact
0538+498	3C147	Q 0.545	U	Y	(see Simon et al.)				Steep-spect
0605+480	3C153	G 0.2769	II	N	----				----
0710+439	OI417	G -	U	Y	1980 Jul	BKGFO	1982 Dec	BKGFO	Double
0711+356	OI318	Q 1.620	U	Y	1980 Jul	BKGFO	1982 Dec	BKGFO	Double
0723+679	3C179	Q 0.846	II	Y	(see Porcas et al.)				Central cmpt
0804+499	OJ508	Q? -	U	Y	1979 Dec	BKGFO			Very compact
0809+483	3C196	Q 0.871	II	N	----				----
0814+425	OJ425	Q -	U	Y	1979 Jul	GFOH			Compact
0831+557	4C55.16	G 0.2420	U	Y	1979 Dec	KGFO			Steep-spect
0836+710	4C71.07	Q 2.17	U	Y	1980 Sep	BK FO			Asymmetric
0850+581	4C58.17	Q 1.322	U	Y	1980 Jul	BK FO			Unclassified
0859+470	4C47.29	Q 1.462	U	Y	1978 Dec	GFOH			Compact
0906+430	3C216	Q 0.670	U	Y	1979 Dec	BKGFO			Steep-spect
0917+458	3C219	G 0.1744	II	N	----				----
0923+392	4C39.25	Q 0.699	U	Y	1978 Dec	GFOH			Unclassified
0945+408	4C40.24	Q 1.252	U	Y	1979 Jul	GFOH			Steep-spect
0951+699	M82	G 0.0009	?	N	----				----
0954+556	4C55.17	Q 0.909	U	Y	----				Unclassified
0954+658		BL? -	U	Y	----				Unclassified
1003+351	3C236	G 0.0989	II	Y	(see Schilizzi et al.)				Central cmpt
1031+567	OL553	G? -	U	Y	----				Unclassified
1157+732	3C268.1	G -	II	N	----				----
1254+476	3C280	G 0.994	II	N	----				----
1358+624	4C62.22	G -	U	Y	----				Unclassified
1409+524	3C295	G 0.4614	II	N	----				----
1458+718	3C309.1	Q 0.905	U	Y	1981 Aug	BKGFO	1982 Dec	BKGFO	Steep-spect
1609+660	3C330	G 0.549	II	N	----				----
1624+416	4C41.32	G? -	U	Y	1980 Jul	BKCFO			Asymmetric
1633+382	4C38.41	Q 1.814	U	Y	1979 Apr	BKGFO			Asymmetric
1634+628	3C343	Q·0.988	U	N	----				----
1637+574	OS562	Q 0.745	U	Y	1979 Jul	GFOH			Compact
1641+399	3C345	Q 0.595	U	Y	(see Cohen et al.)				Asymmetric
1642+690	4C69.21	Q -	U	Y	1980 Jul	BKGFO			Asymmetric
1652+398	4C39.49	BL 0.0337	U	Y	1980 Jul	BKGFO			Unclassified
1739+522	4C51.37	Q 1.375	U	Y	1982 Apr	B GFO			Very compact
1749+701		BL -	U	Y	1982 Apr	BKGFO			Asymmetric
1803+784		BL? -	U	Y	1982 Apr	BKGFO			Asymmetric
1807+698	3C371	G 0.05	U	Y	1978 Dec	GFOH	1982 Dec	BKGFO	Asymmetric
1823+568	4C56.27	BL? -	U	Y	1979 Dec	BKGFO			Asymmetric
1828+487	3C380	Q 0.692	U	Y	1978 Dec	GFOH			Steep-spect
1842+455	3C388	G 0.0908	II	N	----				----
1845+797	3C390.3	G 0.0569	II	Y	1982 Apr	BKGFO			Central cmpt
1928+738	4C73.18	Q 0.36	U	Y	1980 Sep	BK FO			Asymmetric
1939+605	3C401	G 0.201	II	N	----				----
1954+513	OV591	Q 1.230	U	Y	1979 Jul	GFOH			Compact
2021+614	OW637	G 0.2266	U	Y	1979 Dec	KGFO	1982 Dec	BKGFO	Double
2153+377	3C438	G 0.292	II	N	----				----
2200+420	BL Lac	BL 0.07	U	Y	1978 Dec	GFOH			Asymmetric
2229+391	3C449	G 0.171	I	N	----				----
2243+394	3C452	G 0.0811	II	N	----				----
2342+821		EF -	U	N	----				----
2351+456	4C45.51	G -	U	Y	1980 Jul	BKGFO			Asymmetric
2352+495	OZ488	G 0.237	U	Y	1979 Dec	BKGFO			Double

Space allows us to present only a few of the maps here. They have
a resolution of about 1 mas (or 2 mas when Effelsberg was not
available). The dynamic range of the maps has improved considerably
since we presented the first results from this program (Pearson and
Readhead 1981): we can now usually achieve a dynamic range of between
100:1 and 200:1. The improvement over earlier results (which had a
dynamic range of about 20:1) is due primarily to the use of an
amplitude self-calibration scheme (based on CORTEL by Cornwell and
Wilkinson 1981). Even so, the reader should be aware that the quality
of the maps varies considerably from source to source, depending on the
exact u,v-coverage, the source structure, and the accuracy of the
calibration. The field of view of the maps is limited: we are likely
to have missed weak components more than about 25 mas from the
brightness peak, and the arrays are insensitive to structure on a scale
larger than about 10 mas.

4. RESULTS

In the initial finding survey, we detected 45 of the 65 sources. Of
the 20 undetected sources, 17 are classical double sources with central
components weaker than 0.25 Jy, and the other three are supposedly
"compact" sources (these should be studied at intermediate resolutions
~ 10-100 mas).

We have made first-epoch maps of 37 of the 45 detected sources.
The remainder include four sources which have been extensively studied
by other people (3C147, 3C179, 3C236, and 3C345, all described in other
contributions to this Symposium), and four which were too weak for
satisfactory mapping with Mark-II; we hope to map these later with
Mark-III.

The sample of 45 detected sources includes 30 quasars or BL Lac
objects, 14 galaxies, and one empty field. Unfortunately, only 27 have
measured redshifts. The redshifts range from 0.05 for the galaxy 3C371
to over 2 for some of the quasars. There is a correspondingly wide
range in the linear scales and luminosities of the sources.

5. CLASSIFICATION

In order to impose some order on the diversity of structures found in
the VLBI maps, we have attempted to separate the sources into a number
of groups with similar characteristics, based on the VLBI maps, the
radio spectra (obtained from the literature, primarily Kühr et al.
1981b), and the radio variability of the sources (monitored at 10.8 GHz
by Seielstad et al. 1983). The classification is necessarily somewhat
subjective and provisional, as there appears to be a wide range of
source types, with several sources showing characteristics of two or
more of the groups. Further observations may either clarify our
classification or show that the attempt is misguided. Essentially, we

have chosen a number of well-studied, "archetypal" sources, and looked
for similar characteristics amongst the others. Our assignment of
sources to classes is indicated in Table 1 and summarized in Table 2.
We have not attempted to classify eight of the sources in the sample:
four (0316+413 = 3C84, 0850+581, 0923+392 = 4C39.25, and 1652+398) do
not clearly belong in any of the classes we have described, and the
remainder are the four sources which we were unable to map with
Mark-II. We shall now discuss each class in turn.

TABLE 2 - Classification

Central components of classical double sources:	3
Steep-spectrum compact sources:	6
Very compact sources:	9
Asymmetric sources ("core-jets"):	13
"Compact double" sources:	6
Unclassified:	8
	--
Total:	45

6. CENTRAL COMPONENTS OF CLASSICAL DOUBLE SOURCES

This class is very easy to identify. We detected central components in
three of the 20 sources in the complete sample which are dominated by
extended emission. The results are clearly biased towards sources with
unusually strong central components. Two of the three sources (3C179
and 3C236) will be discussed by other participants in this Symposium;
the third is 3C390.3 which has also been mapped by Preuss et al. (1980)
and Linfield (1981). All three sources show good alignment between the
nuclear structure and the outer lobes. One (3C179) is definitely
superluminal (Porcas 1981), and one (3C390.3) is apparently not (v/c <
0.5: Preuss et al.). It is of interest that 3C390.3 shows evidence
for two-sided ejection: Linfield's 10.7-GHz map shows a "core-jet"
structure directed towards the bright, compact SE lobe, while both our
5-GHz map and that of Preuss et al. show another component 5 mas away
on the NW side of the "core".

7. STEEP-SPECTRUM COMPACT SOURCES

This class of source is the subject of several contributions to this
Symposium. The sources are characterized by a steep, straight
spectrum, usually with a low-frequency turnover near 100 MHz, and
sometimes flattening at high frequencies. They show structure on all
angular scales from 1 mas to 1 arcsec. The VLBI maps of three of these
sources are shown in Figure 1, together with VLA maps (Pearson, Perley,
and Readhead, in preparation). The VLBI structure is complex, showing
a compact, flat-spectrum core, and in some cases a central "jet" which
may be misaligned by as much as 90 deg from the large-scale structure.
All the sources, except 0831+557 which has a peculiar structure, are

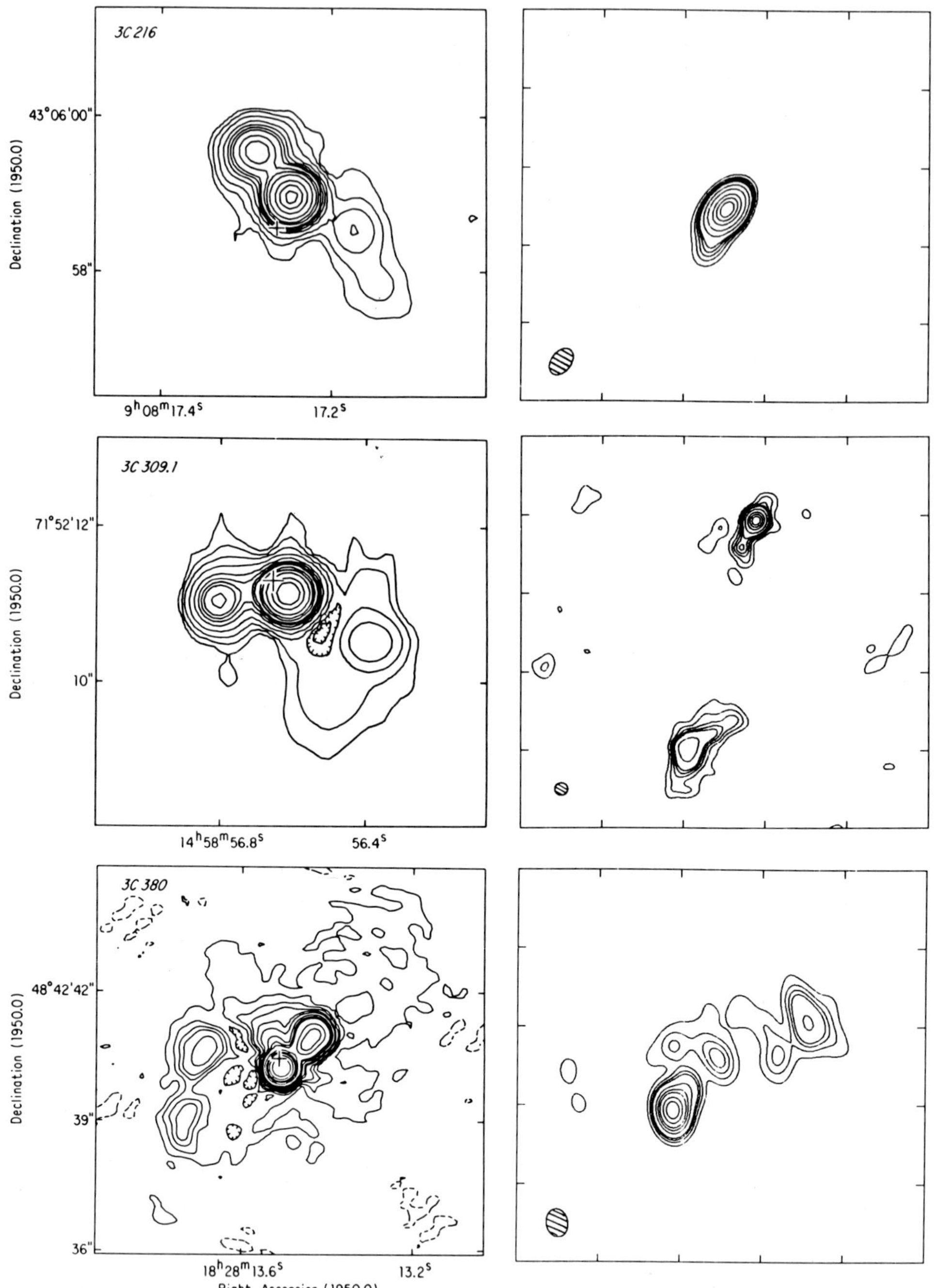

Fig. 1.   5 GHz VLA maps (left) and VLBI survey maps (right) of three "steep-spectrum compact" sources.   The VLBI map of 3C216 is 20 mas square;   the VLBI maps of 3C309.1 and 3C380 are 40 mas square.

quasars. These sources can perhaps be understood as normal extended double sources seen "end-on", but if this is the case, then relativistic beaming is required to account for the large number of these sources in the sample. There is some evidence from inverse-Compton arguments that relativistic bulk motion is occurring in two of these sources (3C147 and 3C390.3; Simon 1982).

8. VERY COMPACT SOURCES

A number of the sources were only barely resolved in our observations; for example, in 0016+731, 0804+499, and 1739+522, the visibility is > 95% even on transatlantic baselines. All the sources in this class are highly variable, changing by 50% or more in 4 yr or less (at 10.8 GHz). All are fairly strongly polarized (> 1%) and all are quasars or BL Lac objects. The hybrid maps show some evidence for low-brightness structure at a level of < 1% of the peak.

9. ASYMMETRIC SOURCES

A large number of the sources appear to be asymmetric sources like 3C345. The term "core-jet" is frequently used to describe these sources, although in only a few is there sufficient evidence to justify the word "jet". They are characterized by a strong, compact component at one end of a lower-brightness linear feature. When observations at a second frequency are available, the core is found to have a flat (self-absorbed) spectrum while the jet has a steeper, transparent spectrum. Two examples are shown in Figure 2. All these sources are variable, though not as violently so as the very compact sources. Two of the sources in this class are known to be superluminal (3C345 and BL Lac) and others may very possibly be (e.g., 3C371: Readhead et al. 1984). In some cases the VLBI structure is well aligned with outer arcsecond-scale structure (e.g., 3C371, 1642+690). Lack of alignment does not imply that there is no connection, however, as 3C345 illustrates (Browne et al. 1982).

10. COMPACT DOUBLE SOURCES

The existence of a distinct class of symmetric, compact double VLBI sources was first pointed out by Phillips and Mutel (1982). These objects are dominated by two components of almost equal brightness, separated by 5 - 20 mas. It turns out that several of them have additional weaker components, so the term "compact double" is a misnomer, although we shall continue to use it here. These sources appear to be distinctly different from the others in several respects: they have radio spectra with a single smooth "hump" peaking around 1 - 5 GHz, and no steep spectrum component; they have no structure on arcsecond scales brighter than 0.2% of the peak; they are only slightly variable (< 10% in 4 yr); and they have very low radio

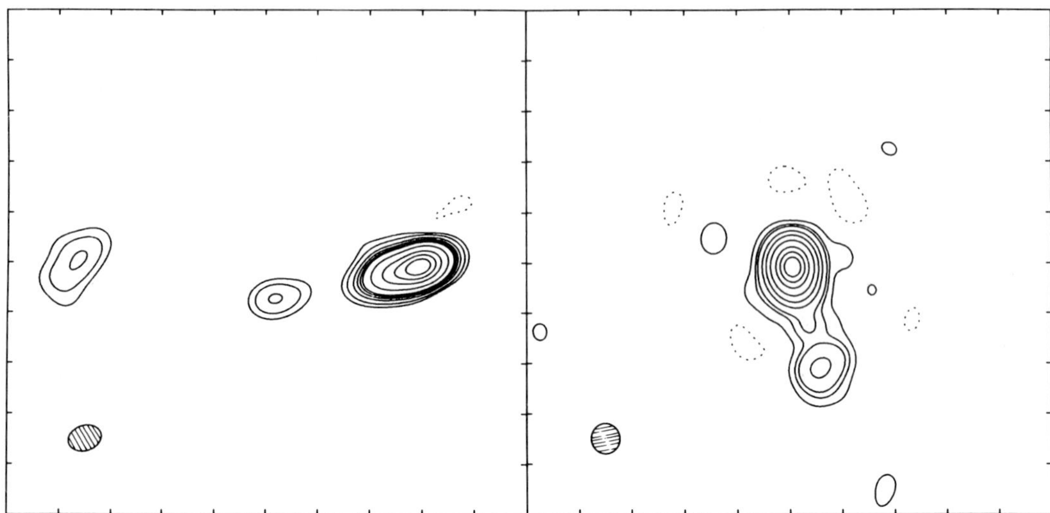

Fig. 2. 5 GHz VLBI survey maps of two "asymmetric" sources. Left: 0212+735; right: 1642+690. Both maps are 20 mas square; the restoring beams are indicated by the hatched ellipses.

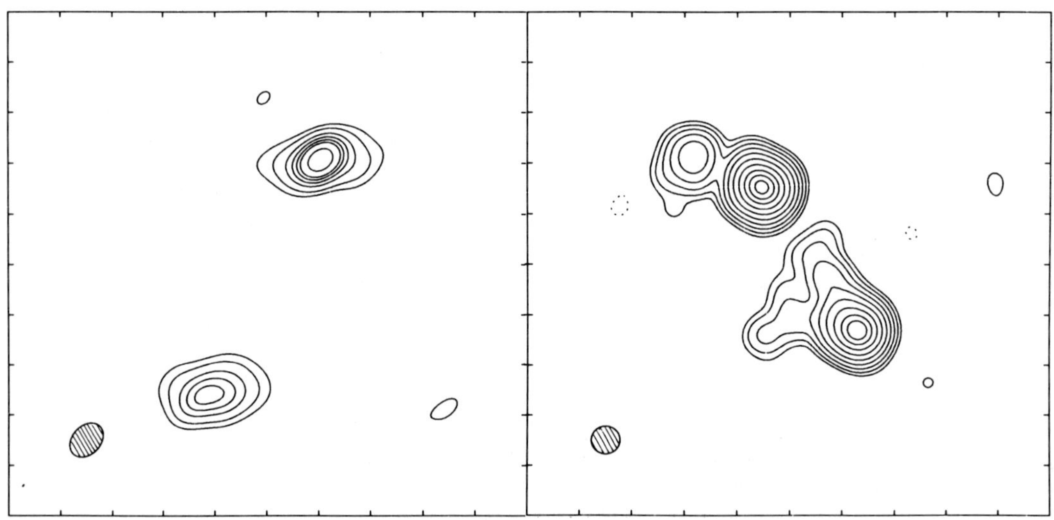

Fig. 3. 5 GHz VLBI survey maps of two "compact double" sources. Left: 0153+744; right: 2021+614. Both maps are 20 mas square; the restoring beams are indicated by the hatched ellipses.

polarization (< 0.2% in all except 0711+356). The correlation of double structure with humped spectrum was noticed in an independent sample by Phillips and Mutel (although not all sources with such spectra are compact doubles), and the correlation between humped spectrum, low variability, and low polarization was noticed by Rudnick and Jones (1982). Figure 3 shows VLBI maps of two of the sources in this class; further examples can be found in the contribution by Readhead et al. (1984). In some cases, observations at another frequency show that the sources are not as symmetric as they appear at 5 GHz, and some should perhaps be allocated to our "asymmetric" class instead (see Readhead et al. 1984). In particular 0711+356 looks out of place in the "compact double" class, and fits better into the "asymmetric" class.

11. TRENDS AND CORRELATIONS

Finally, we should like to summarize some of the trends which have already emerged from our work.

11.1. Radio spectrum

Most of the compact double sources have a distinctive "humped" spectrum. By contrast, most of the compact and very asymmetric sources have a flat or complex spectrum over a wide frequency range. Unfortunately, variability may corrupt the published spectra, and more single-epoch spectral measurements (like those of Owen et al. 1980) are needed to see if there is a real difference in spectrum between the classes.

11.2. Large-scale radio structure

A considerable body of information on the arcsec-scale structure of the sources is available from work with the VLA by Perley (1982). It is notable that none of the six "compact doubles" has any extended structure, while several of the very compact and asymmetric sources do (as do all the steep-spectrum compact sources). Of the seven asymmetric sources with large-scale structure, only three show a good agreement between mas and arcsec position angle, presumably owing to curvature in the sources.

11.3. Radio polarization

Perley (1982) and Rudnick and Jones (1982, 1983) have measured the integrated polarization of most of the sources at 5 GHz. The compact double sources all have low polarization (< 0.5%), while the very compact and asymmetric sources usually have high polarization (1 - 8%). Asymmetric sources generally have higher polarization than the very compact sources. There is no clear correlation of polarization position angle with structural position angle, perhaps because the former is corrupted by Faraday rotation.

11.4. Optical identification and spectrum

The optical data on the objects are very inhomogeneous, and in
particular, there has been no systematic attempt to distinguish between
quasars and BL Lac objects. We have begun a program to obtain
high-quality spectra of all the objects using the Palomar 5-m
telescope, which we hope will improve the situation. But with such
information as is available, we can again see a difference between the
compact double sources and the others: while only 12 of the 45 VLBI
sources are identified with galaxies, three of the six compact double
sources are galaxies (and thus fairly low-redshift objects).

VLBI research at the Owens Valley Radio Observatory is supported by NSF
grant AST 82-10259. We are very grateful to the U.S. Network, the
Max-Planck Institut für Radioastronomie, and the staffs of the
observatories, for making the observations possible.

REFERENCES

Browne, I.W.A., Clark, R.R., Moore, P.K., Muxlow, T.W.B., Wilkinson,
    P.N., Cohen, M.H., and Porcas, R.W.: 1982, Nature 299, pp. 788-793.
Cornwell, T.J., and Wilkinson, P.N.: 1981, Monthly Notices Roy. Astron.
    Soc. 196, pp. 1067-1086.
Kühr, H., Pauliny-Toth, I.I.K., Witzel, A., and Schmidt, J.: 1981a,
    Astron. J. 86, pp. 854-863.
Kühr, H., Witzel, A., Pauliny-Toth, I.I.K., and Nauber, U.: 1981b,
    Astron. Astrophys. Suppl. 45, pp. 367-430.
Linfield, R.: 1981, Astrophys. J. 244, pp. 436-446.
Owen, F.N., Spangler, S.R., and Cotton, W.D.: 1980, Astron. J. 85, pp.
    351-362.
Pauliny-Toth, I.I.K., Witzel, A., Preuss, E., Kühr, H., Kellermann,
    K.I., Fomalont, E.B., and Davis, M.M.: 1978, Astron J. 83, pp.
    451-474.
Pearson, T.J., and Readhead, A.C.S.: 1981, Astrophys. J. 248, pp. 61-81.
Perley, R.A.: 1982, Astron. J. 87, pp. 859-880.
Phillips, R.B., and Mutel, R.L.: 1982, Astron. Astrophys. 106, pp.
    21-24.
Porcas, R.W.: 1981, Nature, 294, pp. 47-49.
Preuss, E., Kellermann, K.I., Pauliny-Toth, I.I.K., and Shaffer, D.B.:
    1980, Astrophys. J. 240, pp. L7-L10.
Readhead, A.C.S., Pearson, T.J., and Unwin, S.C.: 1984, IAU Symp. No.110
    this volume p. 131.
Rudnick, L., and Jones, T.W.: 1982, Astrophys. J. 255, pp. 39-47.
Rudnick, L., and Jones, T.W.: 1983, Astron. J. 88, pp. 518-526.
Seielstad, G.A., Pearson, T.J., and Readhead, A.C.S.: 1983, Publ.
    Astron. Soc. Pacific, in press.
Simon, R.S.: 1982, Ph.D. dissertation, California Institute of
    Technology.

DISTORTIONS IN COMPACT STEEP-SPECTRUM RADIO SOURCES [+]

P.N. Wilkinson, R.E. Spencer.
Nuffield Radio Astronomy Laboratories, Jodrell Bank
A.C.S. Readhead and T.J. Pearson
Owens Valley Observatory, California Institute of Technology
R.S. Simon
Naval Research Laboratory, Washington, D.C.

The size, and therefore the importance, of the population of compact, steep-spectrum, radio sources has only recently been recognised[1,2]. While it is now clear that the extended steep-spectrum sources are powered by a pair of, originally anti-parallel, beams which transport energy to the outer lobes some 10^5-10^6 parcsec away, our understanding of the compact steep-spectrum sources is almost nil. This is largely because our radio maps have not had high enough resolution to show their structures in any detail. However 1.67 and 5 GHz MERLIN observations (resolutions 0".25 and 0".1) of the $\sim$20 steep-spectrum 3CR sources whose LAS is $\lesssim 2$" have now allowed us to classify their structures, at least in broad terms. These MERLIN maps, and recent VLBI maps, show that while there is a wide range of structures - from colinear doubles to amorphous "blobs" - the "peculiar" structures are strongly concentrated in the objects whose optical counterparts are called QSO's.

The 3CR QSO's which we include in this, as yet subjectively defined "peculiar" class, are 3C48 (see Fig.1) 3C147 (see Fig.2 and Preuss et al. and Simon et al.); 3C216 (no map available) 3C309.1 (see Fig.3); 3C343 (see Fanti et al.), 3C380 (see Fig.4); 3C454 (see Cawthorne) 3C380 is included although its LAS is >2" because the brightest components at 5 GHz do fall within 2" of each other. About half these 3CR sources are QSOs and about half are galaxies but among the galaxies only 3C299 clearly has a "peculiar" structure - most of the rest of them are rather colinear and undistorted (see Fanti et al.). While a few of these sources have not yet been mapped in detail the separation between QSOs and galaxies, in this specific morphological respect, is clear-cut.

What does this mean? First of all it is important to realise that the projected sizes of all these radio sources are of sub-galactic (i.e. a few tens of kpc) dimensions and thus the beams could well be propagating through interstellar gas throughout most of their paths. There is strong evidence, from observations of a few steep-spectrum radio cores in nearby galaxies (e.g. ref 3 and van Breugel et al.) that interstellar gas can interact strongly with beams. The pressures in the radio plasma and the emission line clouds in these galaxies are rather similar

[+] Discussion on page 418.   25

R. Fanti et al. (eds.), VLBI and Compact Radio Sources, 25–28.
© 1984 by the IAU.

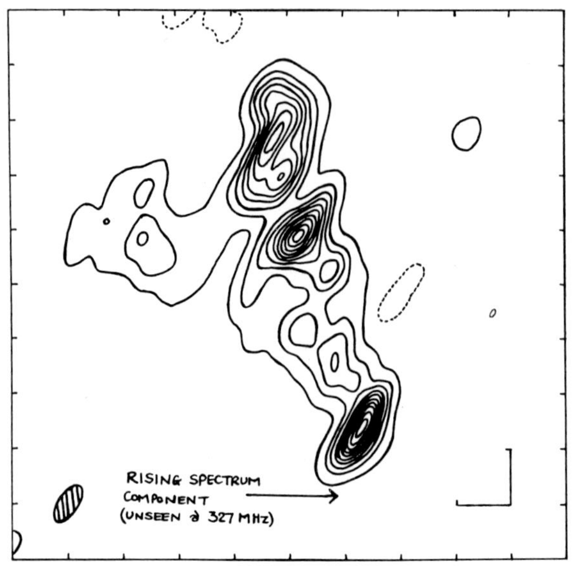

<u>Fig. 1</u> 3C 48 : 327 MHz VLBI (Simon et al., in preparation). Scale indicates $0.05'' = 150$ pc (for $H_0 = 100$ km s^{-1}Mpc^{-1} $q_0 = 0.5$)

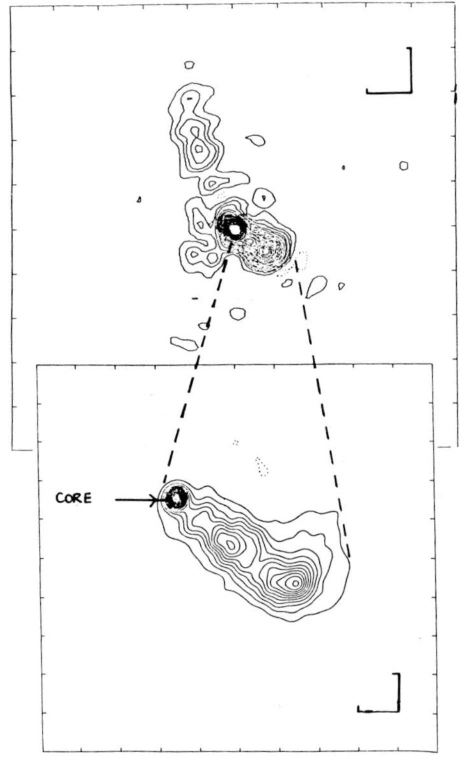

<u>Fig. 2</u> (top) 3C 147 5 GHz MERLIN (Wilkinson et al., in preparation) Scale indicates $0.2'' = 740$ pc.

(bottom) 3C 147 327 MHz VLBI (Simon et al., in preparation) Scale indicates $0.05'' = 185$ pc

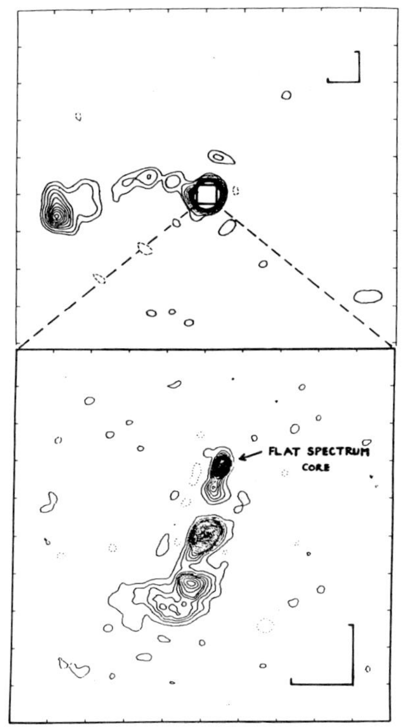

Fig. 3 (top) 3C 309.1 : 5 GHz
MERLIN (Wilkinson et al.,
in preparation. Scale indicates 0.2" = 840 pc.
N.B.: there is also low
brightness emission to the
west of the nucleus (see
Proc. IAU Symposium N_o.97,
p152,Fig.6)

(bottom) 3C 309.1 : 1.67 GHz
VLBI (Wilkinson et al., in
preparation). Scale indicates 0.02 = 84 pc.

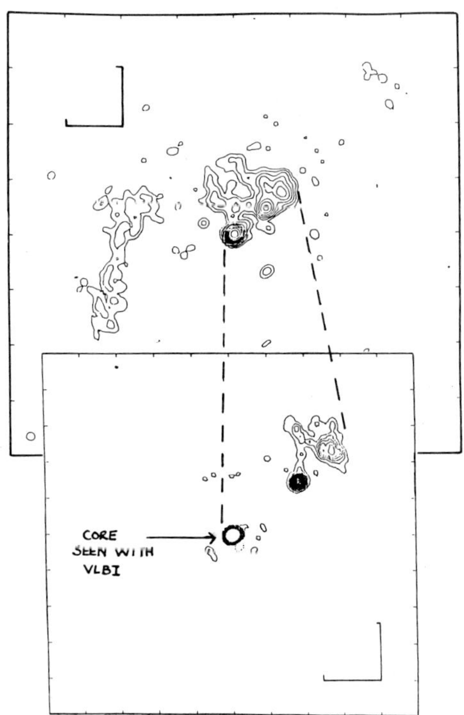

Fig. 4 (top) 3C 380 : 1.67 GHz
MERLIN+VLBI (Wilkinson et
al., in preparation).
Scale indicates 1" = 4 kpc.

(bottom) 3C 380 : 5 GHz
MERLIN (Wilkinson et al.,
in preparation). Scale indicates 0.5" = " kpc.

N.B.: there are other low
brightness features (see
Proc. IAU Symposium N_o.97,
p157 Fig.10)

$(10^{-8 \cdot 5}$dyne cm^{-2}), the optical line emission is usually brightest near radio knots or hotspots and the beams often appear to bend near bright emission-line clouds. If we are to ascribe the much greater distortions in these highly radio-luminous QSOs to similar interactions then clearly the energy density of the interstellar gas must be higher. Thus it is interesting that many QSO's seem to be associated with interacting galaxy systems and some of them are associated with spirals i.e. gas-rich objects[4].

The specific case of the low redshift QSO 0351+026 (ref 5) may be particularly important in this respect because it "strongly resembles" (see ref 6) one of our QSOs, 3C48. 0351+026 is clearly a pair of gas-rich galaxies undergoing a violent interaction and this gives us a clue that perhaps the other QSO's with peculiar radio structures are located in similar interacting/merging systems. Extended gas with a mean density of $>10^{-23 \cdot 5}$gm cm^{-3} and a bulk velocity relative to the radio plasma of $\gtrsim 500$ km s^{-1} would exert a ram pressure which is comparable with the radio plasma pressure of the kpc-scale components in these QSO's i.e. $(10^{-(8\pm1)}$ dyne cm^{-2}. Such energetic gas may well be common in merging systems. Thus, if the beams are very light, strong plasma/gas interactions may well be common in QSO's. Severe distortions or even disruption of the beams could occur in highly energetic gas and clearly computer simulations using 3-D hydrodynamic codes are needed to explore the range of circumstances under which these effects can happen.

The properties of the gas within $\sim$1 kpc of the nucleus in QSO's are very uncertain. However in nearby spiral galaxies the average gas density in this region is usually at least an order of magnitude higher than it is further out. Thus the morphological features in some of our VLBI maps e.g. the non-colinear feature in 3C48, the bend at the end of the 'jet' in 3C147, the sharp apparent bend in 3C309.1, may also be due to interactions with surrounding material on the 0.1-1 kpc scale. This is despite the fact that the radio plasma pressure in these components is typically two orders of magnitude higher than on the 1-10 kpc scale.

To summarise: QSOs preferentially contain very distorted radio sources compared with radio galaxies. By analogy with the compact steep-spectrum cores in a few nearby galaxies we suspect that this is due to an interaction with energetic interstellar gas. A circumstantial case can be made that this energetic gas is associated with mergers of gas-rich galaxies.

REFERENCES

N.B. Named references are to papers in this volume
1)   Kapahi, V.K. Astron.Astrophys.Suppl., 43, 381 (1981)
2)   Peacock, J. and Wall, J., M.N.R.A.S. 198, 843 (1982)
3)   van Breugel, W. and Heckman, T., Proc.IAU Symposium No.97, p.61
4)   Hutchings, J.B. and Campbell, B., Nature 303, 584 (1983)
5)   Bothun, G.D. et al. Astron.J. 87, 1621 (1982)
6)   Balick, B. and Heckman, T. Astrophys.J., 265, L1 (1983)

THE MILLIARCSECOND CORE OF 3C147 AT 6 CM [+]

E. Preuss, W. Alef
Max-Planck-Institut für Radioastronomie, Bonn, FRG
N. Whyborn, P.N. Wilkinson
Nuffield Radio Astronomy Laboratory, Jodrell Bank, Cheshire, UK
K.I. Kellermann
National Radio Astronomy Observatory, Green Bank, W.Va., USA

3C147 is a compact ($\lesssim 1''$), steep spectrum radio source identified with a quasar at z = 0.545 ($0''001$ = 7.4 pc; c/H_0 = 6000 Mpc and q_0 = 0.5). The radio structure shown by VLBI observations at 18 cm (Readhead & Wilkinson, 1980; Simon et al., this volume), at 50 cm (Wilkinson et al., 1977), and at 90 cm (Simon et al., 1980 and 1983) shows a bright 'core' $\lesssim 0''008$ (60 pc) at one end of a 'jet' $\sim 0''2$ (1.5 kpc) in length oriented in p.a. $\sim -130°$. In this sense 3C147 is typical of the one-sided 'core-jet' structures commonly found in the centres of other extragalactic radio sources. However, MERLIN observations at 6 cm (Wilkinson, this vol.) and VLA observations at 2 cm (Crane & Kellermann, unpubl.; Readhead et al., 1980) show a larger elongated feature extending $\sim 0''5$ (3.7 kpc) to the North East of the bright core in p.a. $\sim 25°$ or on the opposite side to the $0''2$ jet.

Higher resolution VLBI observations made at 6 cm wavelength (Preuss et al., 1982) showed that the core itself was elongated roughly along the same direction as the jet but that the lower surface brightness features appeared to point in the opposite direction from the $0''2$ VLBI jet. The extension of this core feature was also visible on high resolution 18 cm maps. We have now reobserved 3C147 at 6 cm in Dec. 1982 with an intercontinental 6 station array. The new data represent a considerable improvement in (u, v) coverage and as shown in Fig. 1, the nucleus appears very complex and is non-linear. The inner high brightness part appears somewhat S-shaped and two-sided with respect to the brightest component located in the middle. However, this bright feature is not very dominant compared with sources such as 3C273, and it is unclear where the actual core is located. The 6 cm map also shows clearly the apparent extension of 5 to 10 mas toward the north-west, which was apparent in the 18 cm and earlier 6 cm maps. On a large scale, the VLA and MERLIN maps show an extension of $\sim 0''3$ in the opposite direction toward the south-east. It is not obvious how to interpret this complex structure within the framework of relativistic beam models which typically predict asymmetric structure.

Simon et al. (1983) have observed an increase in the flux density of the 327 MHz core by about a factor of 2 (1 Jy) during the period 1975 to 1981. Using causality arguments together with the measured angular size of the core and time scale of variability, as well as the low measured

[+] Discussion on page 419

R. Fanti et al. (eds.), VLBI and Compact Radio Sources, 29–30.
© *1984 by the IAU.*

X-ray flux density, they conclude that bulk relativistic flow towards us
must occur in 3C147, and they 'predict' that superluminal motion will be
observed in the core structure with v/c $\sim$ 9. We have compared our new
6 cm map of the 3C147 core with those made in 1978.2 (3 stations) and
1981.3 (5 stations), and find no evidence for any relative component
motion with v/c $\gtrsim$ 0.5. Furthermore,the 6 cm core flux density has not
changed by more than 10% (30 mJy) in the 4.7 year span.

Relativistic flow may indeed be important in 3C147, but it is not
observed as superluminal motion,possibly because the highly specialized
geometric conditions required are not met. Higher resolution observations
with greater dynamic range will be necessary to clarify the complex geometry of 3C147; but considering the absence of the 'predicted' superluminal motion in 3C147 in spite of the general widespread observance of this
phenomenon (e.g. Cohen, this volume), it is not clear to what extent predictions of superluminal motion provide meaningful tests of theoretical
models.

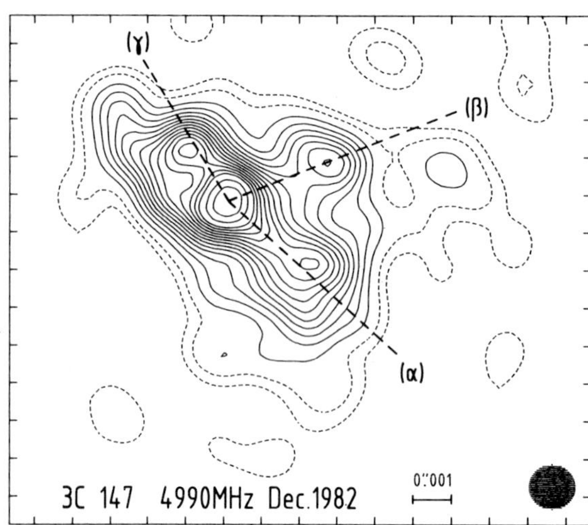

Fig. 1: Map of the core of 3C147
at 6 cm for epoch 1982.9. The
restoring beam is shown as a
shaded circle. Contour levels are
2, 5 (dashed), 10, 15 ... 70,
80, 90 percent of the peak brightness ($\sim$1.2 x 10^{10}K). The straight
lines indicate the main direc-
tions of elongated features in
maps on larger scales and/or at
longer wavelengths. (α is the
direction of the 200 milliarcsec
VLBI jet ($\lambda \gtrsim$ 18 cm); β is the
position angle of the extension
of the 18 cm core (Simon et al.,
this volume); γ is the direction
of the MERLIN/VLA North East
large scale extension.

REFERENCES

Preuss, E. et al.: 1982, IAU Symposium No. 97, p. 289
Readhead, A.C.S. & Wilkinson, P.N.: 1980, Astrophys. J. 235, 11
Readhead, A. et al.: 1980, Astrophys. J. 237, L55
Simon, R. et al.: 1980, Astrophysical J. 236, 707
Simon, R. et al.: 1983, Nature 302, 487
Wilkinson, P. et al.: 1977, Nature 269, 764

EXTRAGALACTIC VLBI AT 89 GHZ[+]

D. C. Backer
Radio Astronomy Laboratory
University of California, Berkeley

Radio interferometry at 89 GHz with transcontinental baselines
can probe the region of active galaxies and quasars where collimation
of relativistic outflows occur on length scales of 10^{17-18} cm. Three
experiments have been conducted recently to demonstrate the feasa-
bility and importance of this investigation. Highlights of the three
pioneering experiments are discussed below. Results obtained for the
luminous sources 3C84, 3C273 and 3C345 are then summarized.

The technical requirements for millimeter VLBI are demanding. The
apertures used are small, 6-11 m, and the system temperatures are
high, 350-1000 K SSB, in comparison with most cm-wavelength VLBI. Use
of the full 50 MHz bandwidth capability of the MkIII recording system
is essential. Coherence of independent oscillators requires
hydrogen maser frequency standards, the best available synthesizers,
and careful checking of local oscillator stability. Our three experiments are summarized below:

Date	Stations	Baseline Range	Sources
Oct'81	HCRK-OVRO	$0.7-1.4 \times 10^8 \lambda$	3C84
May'82	HCRK-OVRO-KTPK-QBBN-ONSL	$0.7-23 \times 10^8$	3C84,3C273,3C279,3C345
Apr'83	HCRK-OVRO-KTPK	$0.7-4 \times 10^8$	3C84,OJ287,3C273,3C345, OV-236

Where possible, we have included cm-wavelength VLBI, either sequen-
tially or simultaneously as a MkIII track pair, to determine the clock
at the various stations. Accurate clock determination removes one
source of ambiguity in coherent fringe searches over 50 MHz.

The coherence time for these experiments is typically 500 s;
coherent integrations for 700 s result in amplitude losses of 0.3-0.5.
While we expect the coherence to be limited by tropospheric fluctuations, the coherence time does not show marked variations with zenith
angle or with time of day.

+ Discussion on page 420

R. Fanti et al. (eds.), VLBI and Compact Radio Sources, 31–33.
© *1984 by the IAU.*

The technical highlights of these experiments were: first 89 GHz fringes on 3C84 from Oct'81 run (Readhead et al. 1983); long coherence time for '82 and '83 experiments; fringe detection in both RF sidebands in Apr'83; successful demonstration of double bandwidth mode, 224 Mbit/s in '83. We were unsuccessful in obtaining fringes from the West coast to either Quabbin or Onsala; this may result either from resolution or from hardware problems. We look forward to phasing the elements of the HCRK and OVRO interferometers for added sensitivity.

3C345. Observations at cm-wavelengths of this bright quasar (z=0.595) have shown striking evidence for superluminal motion. At 89 GHz the intensity of 3C345 has been declining from 13 Jy in May'82 to 10 Jy in Apr'83. The May'82 observations are consistent with a very compact elongated, or double, source with pa = -107° and size 0.65 mas (Fig. 1). The double model is consistent with the non-radial ejection description of recent 22 and 10 Ghz VLBI observations discussed elsewhere in this volume (Moore 1984). The source expansion and intensity decrease are probably responsible for the nondetection of 3C345 in the most recent experiment.

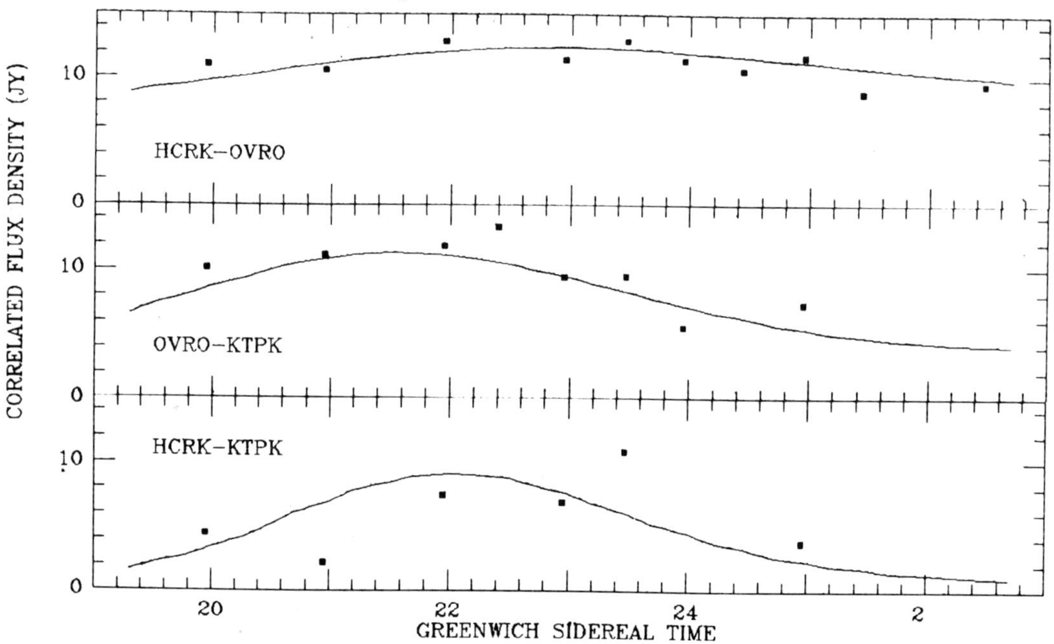

Fig.1 Observations of 3C345 in May'82 (squares) and elliptical gaussian model discussed in text (solid line).

3C84. Over the past decade VLBI observations of the nucleus of this nearby erupting galaxy have shown the presence of a non-collinear series of emission knots along a basically N-S axis. The most compact emission is located at the northern end of the source and is assumed to be associated with the energy production region. Our observations require the presence of at least three components: 14 Jy in structures

resolved by our minimum resolution of 5 mas; an unresolved core, < 0.2
mas, whose flux density has declined from 16 Jy in May'82 to 5 Jy in
Apr'83; a 0.7 mas halo surrounding the core which has remained constant. Closure phases indicate the presence of asymmetry in the source
which has not been modeled. Future transcontinental VLBI at 89 GHz
will be directed to resolution of the core whose physical size must be
less than 3 x 10 ^{17} cm.

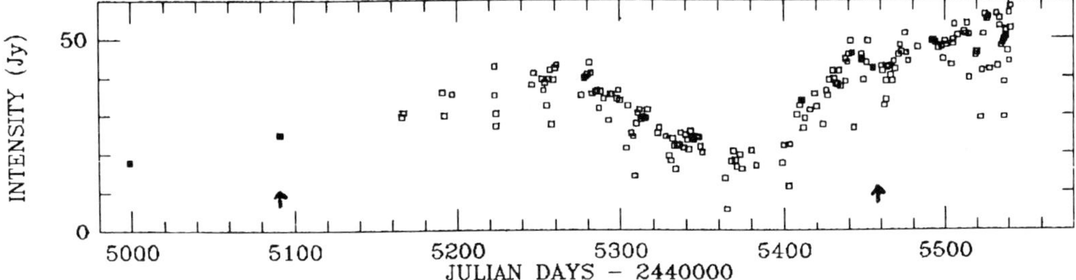

Fig.2 Intensity variations of 3C273 during 1982-83 at 89 Ghz. Open
symbols are from HCRK baseline data; filled squares are from other
measurements. Arrows mark May'82 and Apr'83 VLBI observations.

3C273. The intensity of 3C273 at 89 GHz has peaked several times
over the past several years. Our May'82 observation occurred at the
beginning of an increase that took place during 1982 (Fig.2). The
brief observation indicates that 3C273 was unresolved with a resolution of 0.8 mas. The Apr'83 observations coincided with the peak of a
rapid outburst which was first detected at infrared wavelengths in
Jan'83 (Ade et al. 1983). The 89-Ghz light curve in Figure 2
suggests that an 'old' component had decayed to 10 Jy by Apr'83 and
the 'new' component associated with the infrared outburst had risen by
35 Jy. The visibilities from the HCRK-OVRO-KTPK observations indicate
strong resolution of the source on all baselines. The old intensity is
heavily resolved on the short HCRK-OVRO baseline indicating a size of
0.7 mas (1 pc). If we assume a component age of 1 year, a superluminal
expansion with $\gamma= 3$ is required. The correlated flux density on the
HCRK-KTPK baseline is much less than the new component flux in the '83
outburst. The required source size of 0.2 mas for a source age of 120
days also requires a superluminal expansion with $\gamma= 3$. Future observations of 89 GHz outbursts will require frequent sampling.

These observations resulted from a large effort by a team including
Baath, Masson, Moffet, Moran, Pearson, Plambeck, Predmore, Readhead,
Rogers, Ronnang, Seielstad, Webber, and Wright with generous technical
assistance from the Jet Propulsion Laboratory, the Smithsonian Astrophysical Observatory, and the National Radio Astronomy Observatory.

References
Ade, P.A.R., et al. 1983:IAU Circular 3795.
Moore, R. 1984, IAU Symp. 110, this volume.
Readhead, A.C.S.R., et al. 1983:Nature,303,504.

MILLIARCSECOND POLARIZATION MEASUREMENTS [+]

David H. Roberts,[1] Robert I. Potash,[1] John F. C. Wardle,[1]
Alan E. E. Rogers,[2] and Bernard F. Burke[3]

[1]Brandeis University, [2]Haystack Observatory, [3]M. I. T.

Linear polarization measurements at milliarcsecond resolution have
been made at λ6 cm using four stations of the U. S. VLBI Network and the
Mark III recording system. Calibration of the cross-polarization contam-
ination of the feeds was done using the cross- to parallel-hand fringe
ratios of the essentially unpolarized source OQ 208, and is believed
accurate to 0.5%. Other sources, such as AO 0235+164 and 0106+013, which
are nearly unresolved in the unpolarized fringes, show clear signs of resolution in their polarized fringes. A preliminary map of the BL Lacertae
object OJ 287 is compatible with a weakly-polarized optically-thick core
and a highly-polarized optically-thin jet about 5 mas in length.

OBSERVATIONS

Data were taken for 24 hours in December 1981 with the 120-foot
Haystack, 140-foot Green Bank, and 130-foot Owens Valley antennas, and
the phased-up 27 antennas of the Very Large Array, using the Mark III
recording system. Seven 2-MHz tracks each of right and left circularly-
polarized radiation (RCP and LCP), centered at a sky frequency of 4983
MHz (λ6 cm), were recorded at Green Bank and the VLA, while only LCP was
available at Haystack and Owens Valley. All possible parallel- and cross-
hand correlations were performed on the Haystack Mark III correlator.

POLARIZATION CALIBRATION

Detection of the RCP and LCP components of the source radiation (the
electric fields E_R and E_L) in each of the two elements of an interferometer
enables one to measure the Stokes parameters (I, Q, U, V), where I is the
total intensity, Q and U are the linearly-polarized components, and V is
the circularly-polarized component. These are related to the complex
linear polarization P by $P = Q + i U = m I e^{2i\chi}$, where m is the fractional
linear polarization and χ is the linear polarization position angle. Any
real antenna and feed system will have arbitrary gain and some cross-
polarization response, and the voltage induced in given right- and left-

+ Discussion on page 420

R. Fanti et al. (eds.), VLBI and Compact Radio Sources, 35–38.
© *1984 by the IAU.*

circularly polarized feeds may be written approximately as

$$V_R = G_R(E_R\, e^{-i\phi_p} + D_R\, E_L\, e^{+i\phi_p}), \quad V_L = G_L(E_L\, e^{+i\phi_p} + D_L\, E_R\, e^{-i\phi_p}),$$

where the G's and D's are complex gains and cross-talks, and ϕ_p is the parallactic angle. If all cross-correlations of the voltages are formed, and terms second-order in m and D are dropped, the results are

$$L_1\, L_2^* = e^{i(\psi_1-\psi_2)}\, G_{1L}\, G_{2L}^*\, (I - V)\, e^{i(+\phi_1-\phi_2)} \quad,$$

$$R_1\, L_2^* = e^{i(\psi_1-\psi_2)}\, G_{1R}\, G_{2L}^*\, [\ (m\, I\, e^{+2i\chi})\, e^{i(-\phi_1-\phi_2)}$$

$$+ D_{2L}^*\, (I + V)\, e^{i(-\phi_1+\phi_2)} \quad + D_{1R}\, (I - V)\, e^{i(+\phi_1-\phi_2)}\ ] \quad,$$

plus similar equations for RR and LR. Here ψ_1 and ψ_2 are time-variable phases due to the differences in clock rate, atmosphere, etc., at the two stations. In order to calibrate the polarization response, it is necessary to put successive cross-hand fringes in the same phase-frame. Since the parallel-hand fringes on a given baseline at a given epoch are contaminated by exactly the same phases ψ_1 and ψ_2 as are the cross-hand fringes, their ratio is free of this phase ambiguity. Thus we form cross- to parallel-hand fringe ratios such as

$$\frac{R_1\, L_2^*}{L_1\, L_2^*} = \frac{G_{1R}}{G_{1L}}\, [\ m\, e^{2i(\chi-\phi_1)} + D_{2L}^*\, e^{2i(-\phi_1+\phi_2)} + D_{1R}]\quad,$$

neglecting terms of order V/I since the circular polarization of almost all compact radio sources is known to be small. These relations still contain the unknown (and possibly time-variable) ratios of right and left instrumental gains at a given stations, e.g., (G_{1R}/G_{1L}). Fortunately, the ratio of parallel-hand fringes (RR/LL) on the same baseline involves the same gain ratios, and may be used to determine how they change in time. Due to the method of phasing-up the array, there was substantial R-L phase drift for the VLA, but essentially no drift at Green Bank. The amplitude part of the R-L gain ratio proved to be quite stable (better than 1%). With this information, the cross- to parallel-hand fringe ratios contain one fixed vector and one or two rotating vectors, depending on the stations involved. Knowing the parallactic angles at each station for each scan, it is straightforward to solve for m and the D's. The single overall phase constant was determined by comparing the VLBI results with those obtained simultaneously at arcsecond resolution for sources which are substantially unresolved by VLBI.

Two independent sets of polarization calibration software were used to analyze the data. One is correlator-based, and determines the antenna cross-gains baseline-by-baseline; these routines use only the cross- to parallel-hand fringe ratios. The second set of routines does a full antenna-based solution, and is quite similar to that used to calibrate the VLA. Each set was applied to OQ 208 = 1404+286, which

TABLE 1:   INSTRUMENTAL CALIBRATION

| Station | Correlator-Based[#] | | Antenna-Based | |
	D_R^*(%,phase)	D_L(%,phase)	D_R^*(%,phase)	D_L(%,phase)
G	5.8 (−160)	11.0 (0)	6.2 (−163)	10.4 (0)
Y	2.0 (−135)	1.4 (− 39)	2.0 (−134)	1.2 (− 36)
K	−−	2.5 (+139)	−−	2.7 (+142)
O	−−	4.1 (+ 93)	−−	4.2 (+ 91)

[#]Average of baseline-by-baseline results.

is essentially unpolarized (< 0.3 %) at arcsecond resolution, yielding the
cross-gains given in Table 1. The root-mean-square deviation of the fits
to a individual correlators is typically about 0.4%; we estimate that the
overall polarization calibration is good to 0.5% . There are small but
measureable correlator-based differences in the cross-talks, probably due
to non-identical bandpasses. For example, the term D_{YR} has a value 1.8%
when determined on the GY base-line, but 2.3% when determined on the KY
baseline. Using the correlator-based calibration of the VLBI data, and
only the short baselines KG and OY, we find the source parameters given
in Table 2. Although they are nearly unresolved in total intensity, the
sources 0106+013, 0235+164, and 1749+049 showed clear signs of resolution
in their cross-hand fringes, indicating that their diffuse structure is
more highly polarized than their compact components.

TABLE 2:   SOURCE POLARIZATION

| Source | VLBI | | VLA | |
	m(%)	χ	m(%)	χ
0106+013[#]	3.4	−46	3.5	−52
0235+164[#]	1.3	+38	1.8	+44
OQ 208	<0.3	−−	[##]	[##]
1749+049[#]	4.5	+42	[##]	[##]

[#]Slightly resolved.   [##]Not observed.

POLARIZATION STRUCTURE OF OJ 287

A preliminary map of the polarized flux of OJ 287 = 0851+202 was
produced by the complex-polarization method. The cross- to parallel-hand
fringe ratios were calibrated using the correlator-by-correlator cross-
talks determined from OQ 208; the result of projecting the ratios onto the
u-axis is shown in Figure 1. These cross- to parallel-hand fringe ratios
were converted into polarization visibilities P(u,v) using a model $I_m(x,y)$
of the total intensity distribution of this nearly-unresolved source, by
multiplication with the corresponding visibility $I_m(u,v)$. A model consisting of a point source plus a 5 mas gaussian with flux ratio 5:1 and
5 mas separation was adopted, but the results are almost the same if a
single point source is used instead. After a u-v taper corresponding to
angular resolution of 3 mas, a one-dimensional Fourier transform and a

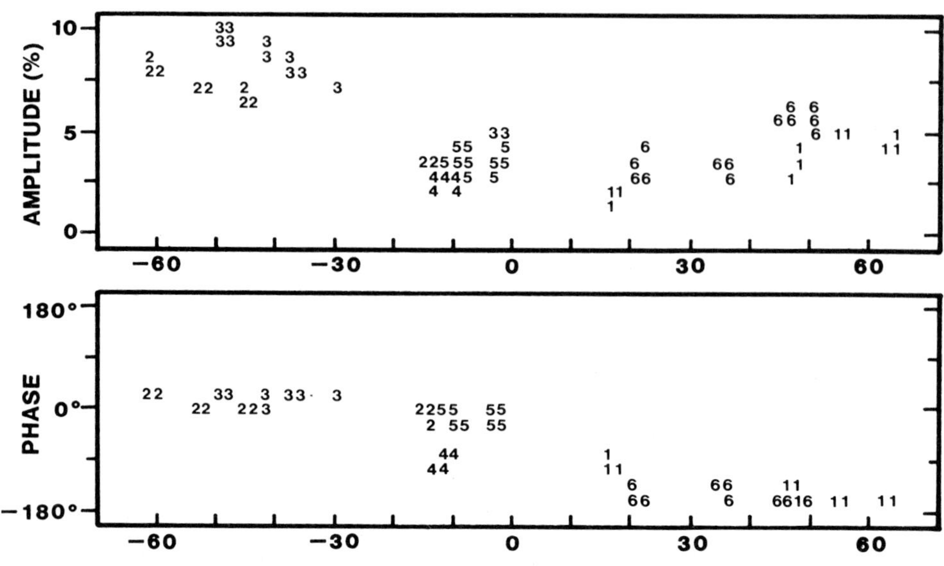

Figure 1. Cross- to parallel-hand fringe ratios for OJ 287, projected onto the u-axis. The various symbols distinguish the six correlators: 1 = RL(GY), 2 = LR(GY), 3 = LR(KY), 4 = LR(OY), 5 = LR(KG), 6 = LR(OG).

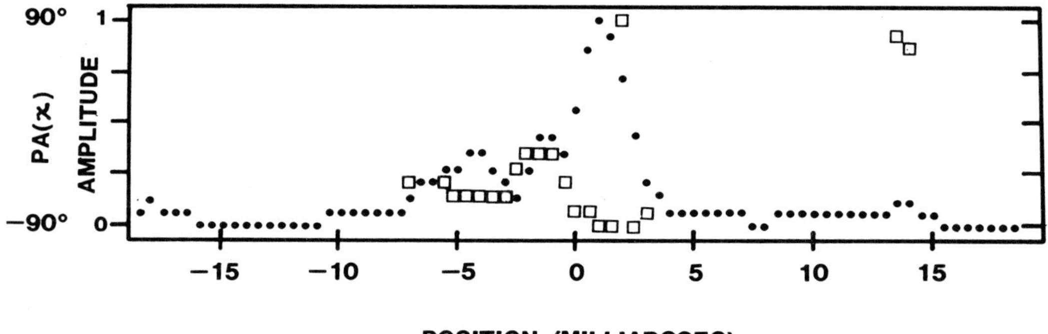

Figure 2. Complex polarization of OJ287, along PA = 90°. The dots and squares are amplitude and position angle; the amplitude scale is arbitrary

complex clean led to the map in Figure 2. The map is consistent with a weakly-polarized core and a highly-polarized jet, with significant changes in polarization position angle that may be related to changes in optical depth and/or orientation of magnetic field across the source.

The NRAO is operated by Associated Universities, Incorporated, under contract with the NSF. This research was supported in part by the NSF.

CORES IN EXTENDED QUASARS

P.D. Barthel and G.K. Miley
Leiden Observatory, Leiden, The Netherlands

R.T. Schilizzi
NFRA, Dwingeloo, The Netherlands

E. Preuss
MPIfR, Bonn, FRG

T.J. Cornwell
NRAO, Socorro, USA

The Nuclear Radio cores of several nearby extended radio galaxies (e.g. M87, 3C236) consist not only of optically thick (< 1 pc) components, but also of emission on somewhat larger scale. As extended radio sources associated with quasars have on average stronger and more luminous radio cores (see e.g. Miley, 1980, Ann. Rev. Astron. Astrophys. $\underline{18}$, 165), we have started a project to study the properties of these quasar cores. The most important reasons for this study are:
1. A comparison of the core structure with the overall morphology of the sources can give important information about the evolution of extended radio sources;
2. The statistics of the core sizes and their relation to other source parameters as radio, optical and X-ray luminosities, optical polarizations, emission line widths etc. can provide fundamental clues as to the nature of active galactic nuclei;
3. Observations of the core structures at different epochs may provide tests for the different models explaining superluminal motion.

Four large MkII VLBI experiments at 5 GHz have been carried out so far, and this contribution is meant as a progress report on the project.
Pilot surveys in January 1980 and April 1981 were followed by first and second epoch mapping programs in April 1982 and April 1983. The telescopes that we used were the Effelsberg 100-m, the Onsala 25-m, the Green Bank 43-m, the 27 antenna VLA and the Owens Valley 40-m telescopes, giving resolutions of about 1 mas.[*]

The source sample (1980) consisted of all quasi-stellar radio sources having (i) δ (1950) > 0°, (ii) known extended radio structure, and redshift, and (iii) 5 GHz core flux density greater than 100 mJy. This sample of 18 quasars is shown in Table 1.

[*]mas = milli arc second

R. Fanti et al. (eds.), VLBI and Compact Radio Sources, 39–40.
© *1984 by the IAU.*

For the pilot surveys we made typically 3-5 10-20 minute scans per source at different hour angles. Compact cores ($\lesssim$ 1 mas) were detected in 15 quasars, whereas in 8 sources significant variations in correlated flux density occurred, implying mas structure.

As follow-up, full tracks at two epochs have been made on the four (three) largest sources showing mas structure in their cores: 0610+260, 0742+318, 1137+660 (only one epoch) and 1721+343. Although the hybrid maps still have to be made, comparison of the visibility amplitudes on the longest baselines already showed that the core structures in these very large radio sources have not changed much in a year: μ < 0.15 mas/yr, implying v/c < 3. Comparison of the hybrid maps will yield better values or limits to the velocities involved.

Full account of this work will be given elsewhere.

Acknowledgements

PDB was supported by the Netherlands Foundation for Astronomical Research (ASTRON) with financial aid from the Netherlands Organization for the Advancement of Pure Research (ZWO). He also acknowledges travel support from the Leidsch Kerkhoven-Bosscha Fonds.

Table 1.

source	other name	redshift	proj. lin. size*
0003+158	4C15.01	0.450	140 kpc
0214+108	4C10.06	0.408	510
0610+260	3C154	0.580	250
0742+318	4C31.30	0.462	530
0836+195	4C19.31	1.691	70
0838+133	3C207	0.684	50
0855+143	3C212	1.048	50
0932+022	4C02.27	0.659	270
1040+123	3C245	1.029	30
1047+096	4C09.37	0.786	380
1055+201	4C20.24	1.110	130
1058+110	4C10.30	0.420	140
1137+660	3C263	0.652	240
1203+109	4C10.34	1.088	50
1222+216	4C21.35	0.435	100
1548+114A	4C11.50	0.436	300
1618+177	3C334	0.555	250
1721+343	4C34.47	0.206	1270

*calculated, assuming H_o = 75 km sec^{-1} Mpc^{-1} and q_o = 0.5

3C205 – A SOURCE WITH EXTRAORDINARY ALIGNMENT AND A VLBI HOTSPOT

C.J. Lonsdale
Nuffield Radio Astronomy Laboratories, Jodrell Bank
P.D. Barthel,
Sterrewacht te Leiden, Leiden

High resolution maps of 3C205 using MERLIN and the European VLBI network (EVN), some of which are shown in Fig.1, reveal unusual characteristics in this high redshift ($z = 1.534$) source. The most striking of these are as follows:

1)   The hotspots A and B, and the core C are aligned to within $0''.05$ (at the hotspot), as measured on the MERLIN 6 cm map.

2)   The compact feature in the southern hotspot (A_1) has projected dimensions of $\sim 0''.03 \times \leq 0''.015$ (see Fig.1d). The corresponding minimum internal energy density of this feature is $\gtrsim 2.3 \times 10^{-6}$ erg cm^{-3}.

3)   There is a continuous zig-zag ridge in the southern component, which starts with a definite spur of emission extending in a south-east direction from the compact feature A_1 (see Fig.1b). The magnetic field in A_1 is in approximately the same position angle as this spur and the VLBI elongation, which both point towards a bright secondary peak in the southern lobe (1a and 1b).

It is possible to put constraints on specific models, using these observations, which are more severe than hitherto. For example, briefly consider the standard beam model, in which the jet thrust must balance the hotspot internal pressure x area (in the hotspot frame of reference). Due to the high power of the hotspot, and the inferred small diameter of the beam, it is necessary that the beam be fast and light, and that the efficiency of energy conversion in the hotspot ε approach unity. Acceptable numbers would be $v_j \doteq 0.8c$, $\varepsilon = 0.7$. If the beam is required to power the entire lobe at its present luminosity, then $\varepsilon \sim 1$ otherwise the hotspot would be bigger.

A second important question concerns the confinement of the compact VLBI scale hotspot A_1. If it is unconfined, then expansion losses are presently running at $\sim 10^{47}$ erg s^{-1} (internal sound speed $c/\sqrt{3}$). Conversely, if dynamical (ram) pressure is confining the hotspot, external medium densities of $n > 10^{-2}$ cm^{-3} are required. The apparent morphological

R. Fanti et al. (eds.), VLBI and Compact Radio Sources, 41–42.
© *1984 by the IAU.*

link between A_1 and A_2 (item 3 above) further complicates this issue, since it suggests that A_2 is powered by outflow from A_1. If this is the case, the fact that A_2 contains $\sim 10^{59}$ ergs in visible energy sets a lower limit on the time taken for its formation ($\sim 10^6$ yrs), and an upper limit on the velocity of A_1 ($\sim 0.015c$). At such speeds, ram pressure balance requires an external medium density $n > 1$ cm^{-3}. The existence of filaments of such dense material >40 kpc from the nucleus is not ruled out by current data as far as we know, and if they are present, one might expect detectable optical emission to occur near A_1.

Many further constraints are possible for the beam model and other models, which are meaningful only because the parameters of 3C205 are so extreme.

FIG. 1

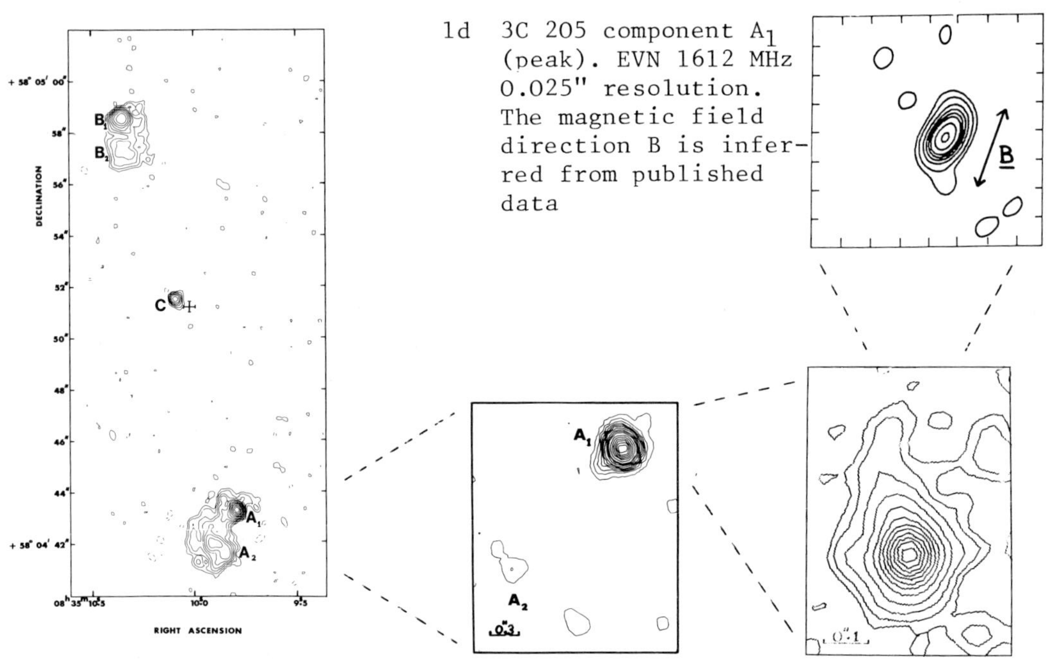

1a   3C 205 MERLIN 1666 MHz, 0.25" resolution (1" = 5 kpc for $H_o = 75$ km s^{-1} Mpc^{-1}, $q_o = 0.5$)

1b   3C 205 southern lobe, 5 GHz, 0.18" resolution.

1c   3C 205 component A_1; MERLIN 5 GHz, 0.1" resolution.

VLBI OBSERVATIONS OF 3C298[+]

D.A. Graham
Max-Planck-Institut für Radioastronomie, Bonn, F.R.G.

L.I. Matveyenko
Space Research Institute, Moscow, USSR

The quasar 3C298 (1416+06) has a complex radio-frequency spectrum, with a spectral bend below 100 MHz and a steep slope of -1.2 in the range 150 MHz - 3 GHz. At higher frequencies the spectrum flattens with some evidence of variability.

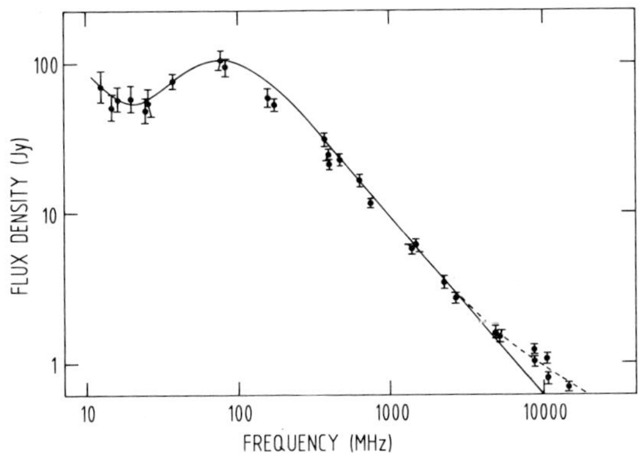

Fig. 1: Integrated radio spectrum of 3C298. The dashed line shows the sum of a straight spectrum of slope -1.2 and a flat spectrum component of 0.45 Jy.

Two VLBI experiments were performed: EVN+Crimea at 18 cm in April 1982 and US Network+Effelsberg in April 1983. The hybrid map constructed from the 18 cm observations is shown in Fig. 2. The overall extent of the structure is approximately 1.6 arcsec, corresponding to 5 kpc for $z = 1.44$ ($H_0 = 75$, $q_0 = 0.05$). The overall double structure was mapped by Anderson and Donaldson (1967), but our position angle differs from their value of 78 degrees, suggesting that one of the outer components has resolved structure not centered on the compact hotspots detected here.

Using the naming scheme of Fig. 2 the components A/B and E seem to be the outer hotspots of the source, probably responsible for the low-frequency turnover in the spectrum around 100 MHz. Results on the 2500 km Effelsberg-Crimea baseline at 18 cm show an unresolved 0.35 Jy source and

+ Discussion on page 421

R. Fanti et al. (eds.), VLBI and Compact Radio Sources, 43–44.
© *1984 by the IAU.*

a 0.2 Jy component of approximately 12 mas. The spacing of these
components identifies them as D and E of Fig. 2.

The low declination and the configuration of interferometers at 6 cm
give high resolution only along a line corresponding approximately to the
major axis of the outer structure. A fit of a single gaussian component
to the visibilities obtained gives a flux density of 0.45 Jy and size
along the source axis of 0.7 mas for the 6 cm core. The core size perpendicular to the source axis is less than 1.5 mas.

The high-frequency flattening of the integrated spectrum can be
explained by the combination of a flat-spectrum component (presumably D)
with a component of slope −1.2. The dashed portion of the fit in Fig. 1
shows a model with a 0.4 Jy flat-spectrum component.

In morphology and radio spectrum 3C298 shows similarities both to
the class of compact double sources (Phillips and Mutel 1982) and that of
large classical double radio sources, and is intermediate in physical
size between these two types. The asymmetrical placing of the core
relative to the outer components could have several explanations. The
most obvious is projection effects due to a small angle between source
axis and line of sight, which would enhance any small distortions present,
and incidentally enhance the brightness of the core by relativistic
beaming. It is also possible that the core illuminates first one outer
component and then the other, so that all the components A–C and E could
be blobs produced by sudden outbursts, the blobs then move outwards and
evolve. However, the present small size of the core suggests continuous
activity.

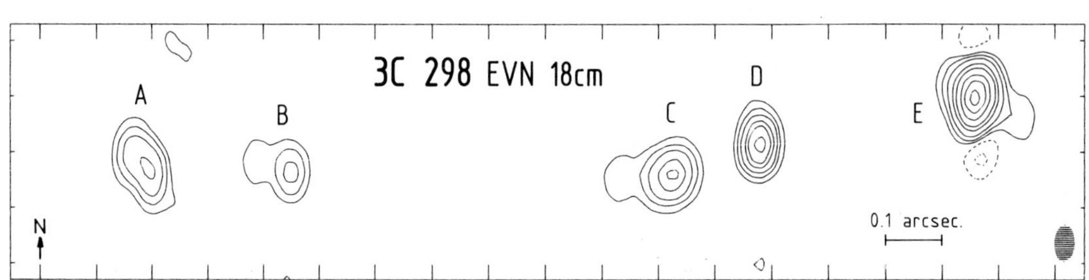

Fig. 2: Hybrid map of 3C298 at 18 cm. Contours at −5, −2, 2, 5, 10, 20,
30, 50, 70 and 90% of the peak of 0.65 Jy. A restoring beam of 65 x 35
mas has been used.

REFERENCES

Anderson, B., Donaldson, W.: 1967, Monthly Notices Roy. Astron. Soc. <u>137</u>,
   81
Phillips, R.B., Mutel, R.L.: 1982, Astron. Astrophys. <u>106</u>, 21

HOT SPOTS IN COMPACT SOURCES [+]

T.K. Menon
Department of Geophysics and Astronomy
University of British Columbia
Vancouver, B.C. V6T 1W5 Canada

The low-frequency turnover of a selected sample of quasars is found to be strongly correlated with their redshift (Menon 1983)). Assuming that the components which produce the low-frequency turnover are hot spots in the quasars I have derived the physical parameters of the hot spots and show that there is a continuity in the various parameters as the angular sizes of the hot spots vary from subarcseconds to milliarcseconds. Hence the low-frequency turnover can be used as a probe into the physical conditions at the earliest phases of quasar activity.

INTRODUCTION

I have recently shown (Menon (1983)) that for a selected sample of 19 quasars the frequency of the low-frequency turnover, in the rest frame of the sources, is strongly correlated with their redshift in the sense that sources at higher redshifts have higher turnover frequencies. The redshifts range from 0.367 to 3.53 and the turnover frequencies range from 64 MHz to 5 GHz. Assuming that the turnover is due to synchrotron self-absorption I have calculated the equipartition angular sizes for these sources for a $q_o = 0.5$, $H_o = 50$ km/s/Mpc cosmological model. The Fig. (2) of Menon (1983) shows that the angular sizes vary from subarc second to a few milliarcseconds and the variation is much faster than predicted by any reasonable cosmological model.

DISCUSSION

It has been suggested that the high surface brightness component which produces the low-frequency turnover in such sources is the outermost hot spot in the source seen at a small angle to the line of sight. Recent VLBI studies of sources such as 3C147 (Readhead and Wilkinson (1980)) appear to confirm such an interpretation. In such models a source may have a number of self-absorbed components the frequency of maximum flux of each decreasing as the components are further away from

+ Discussion on page 422

R. Fanti et al. (eds.), VLBI and Compact Radio Sources, 45–46.
© *1984 by the IAU.*

the nucleus. The linear sizes of these outermost components for the
present sample decrease from about 5.9 kpc to about 12 pcs. Since we
do not know the projection angles of the components we cannot directly
calculate the distance of the components from the centers of activity
of the sources. But it seems reasonable to assume that the observed
decrease of linear size is accompanied by a decrease in the distance of
the component from the center. As an approximation I shall assume that
the collimation angle is the same for all sources and I shall use a
standard hot spot of 3.5 kpc at a distance of 50 kpc from the center as
defining the collimation angle. Hence, knowing the linear size, we can
compute the distances of the hot spots from the center. The distances
vary from a few parsecs to a few kiloparsecs.

I have next calculated the equipartition energy density in the hot
spots using the known radio luminosity. Assuming a ram pressure confinement model for the hot spots and a velocity of motion of 0.1 c. the
necessary ambient density at the distance of the hot spots was calculated. A composite diagram of the densities as a function of distance
from the center for all quasars in the sample shows that the densities
vary continuously from a few hundred atoms per c.c. to 10^{-3} atoms per
c.c. These are average densities and as is known from optical studies
of emission line regions in quasars and galaxies extreme fluctuations
do occur in the density distribution with filling factors as small as
10^{-5} or 10^{-6}. If the collimation angle is smaller closer to the center
then the derived densities will be higher closer to the center than in
the sample model. In any case the derived data on the dimensions and
densities suggest that the energetics of the outermost hot spots as
delineated by their low-frequency turnover can be used as a probe to
study the physical conditions from the innermost regions to the boundary
of the parent galaxies. It is further noteworthy that the derived
physical parameters of hot spots show a remarkable continuity from
scales of the order of parsecs to tens of kiloparsecs.

The research reported here was supported by a grant from the
Natural Sciences and Engineering Research Council of Canada.

REFERENCES

Menon, T.K.: 1983, A.J. 88, 598.
Readhead, A.C.S., and Wilkinson, P.S.: 1980, Astrophys. J. 235, 11.

SUB-MILLIARCSECOND RADIO STRUCTURE OF AO 0235+164[+]

Dayton L. Jones
California Institute of Technology
Mike M. Davis
National Astronomy and Ionosphere Center
Steve C. Unwin
California Institute of Technology

AO 0235+164 is a very compact, flat-spectrum radio source. It is identified with a BL Lac object, and has optical absorption-line systems at $z = 0.524$ and $z = 0.852$. A complex set of HI absorption lines is seen at $z = 0.524$ (932 MHz), and several of these lines change significantly in depth over periods of less than a year. This is the only known case of variable extragalactic absorption lines. A faint nebulosity 2 arcsec south of AO 0235 + 164 has an emission-line redshift of $z = 0.52$ and may be an intervening galaxy. The radio spectrum of this source (between major outbursts) is remarkably flat, with the total flux density staying between about 1 and 3 Jy over a range of at least 1000 in frequency. Such a flat spectrum would lead one to expect a complex, wavelength-dependent structure consisting of several components with different self-absorption frequencies. However, the observed radio structure of 0235+164 is about as simple as one could imagine -- it is a nearly unresolved point source in VLBI experiments from 900 MHz to 22 GHz. Recent VLBI experiments at 6 and 13 cm have shown evidence for some elongation of the source in a generally NE-SW direction, but only at low contour levels (< 15% of the peak). The major portion of the flux density appears to come from a core which is unresolved in VLBI experiments over a range of $\sim$ 25 in frequency.

We observed 0235+164 in December 1982 at 22 GHz with a six-station VLBI array. The maximum baseline length was $\sim$ 600 million wavelengths, the minimum $\sim$ 40 million wavelengths. Our goal was to map the structure of this source with the highest possible resolution. The data were obtained with the Mk-III system in mode B, and correlated at Haystack. The Caltech VLBI programs were used to edit and calibrate the data, and to produce hybrid maps using both closure amplitudes and phases. These maps had dynamic ranges of about 50 to 1 near the central peak, and more than 100 to 1 several milliarcsecs away from the central peak. This is still a factor of 25 less than the dynamic range that could be achieved if the map noise level was set by the thermal noise from the receivers. The fact that the observed noise level is much higher than its theoretical value implies that there are uncorrected calibration errors in the data. These may be station-dependent errors which change more rapidly than the integration time used (60 seconds), or baseline-dependent errors. The usual

+ Discussion on page 422

47

R. Fanti et al. (eds.), VLBI and Compact Radio Sources, 47–48.
© *1984 by the IAU.*

self-calibration techniques can only adjust station gains once per integration interval, and thus can not correct for either type of potential error mentioned above.

The highest dynamic range map is one made with a 300 million wavelength taper in the U,V plane to search for extended, low surface brightness emission. Between 4 and 20 milliarcseconds from the core there are no peaks larger than ∿ 1 % of the central peak. However, the core is not completely unresolved. The untapered maps show a weak component approximately 1/2 milliarcsecond from the core component in position angle 210 degrees. It contains about 4 % of the flux of the core. The core itself appears to have some extended structure at low levels, but the resolution is not sufficient to determine the shape of this extension. The projected linear separation between the core and the component 1/2 milliarcsecond away is approximately 3 pc at z = 0.85, and the position angle is similar to that determined for larger-scale extensions at lower frequencies.

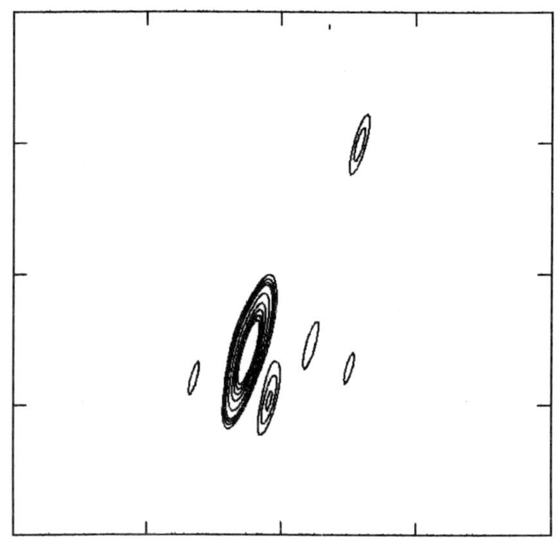

Figure 1. Hybrid map of AO 0235+164 at 1.3 cm. Contours are at 2, 3, 4, 5, 10, 20, 30, 40, and 50 percent of the peak, which equals 1.05 Jy per beam or a brightness temperature of 1.4 x 10**10 K. The clean beam is 1.00 by 0.18 milliarcsecs, with the major axis in position angle -15 degrees. The tick marks along the borders are 2 milliarcsecs apart.

The peak ∿ 3.5 milliarcseconds from the core in position angle -30 degrees may be real (it's about 5 sigma above the apparent noise level), but the most significant result is that this source continues to show a simple, nearly unresolved structure even at resolutions < 0.2 milliarcsec. As at lower frequencies, the secondary peaks are all very much weaker than the central peak.

AO 0235+164 has been detected in X-rays, and the observed X-ray flux can be combined with the measured radio parameters to determine a lower limit for the Doppler factor. This limit turns out to be > 3, which is independent evidence for bulk relativistic motion. This is consistent with the low-frequency variability which has been studied for many years and requires bulk relativistic motion to avoid severe Compton losses.

VLBI research at OVRO is supported by NSF grant AST82-10259.

THE VARIABLE HI ABSORPTION (z=0.524) IN THE SPECTRUM OF AO 0235+164[+]

F. H. Briggs
University of Pittsburgh

The spectrum of AO 0235+164 (z>0.85) reveals the presence of absorbing gas (z=0.524) which has been studied extensively at radio and optical wavelengths (See Wolfe, Davis and Briggs, 1982 for bibliography). A program by Wolfe et al. to monitor the radio absorption line at 932 MHz showed that the profile had changed during a 5 year period. The observation led to a tightly constrained model in which variations in the source continuum were capable of driving the HI hyperfine level populations (and hence the optical depths of the features in the absorption line) by virtue of the hypothesized close proximity of the source to the clouds. A continued monitoring program has not shown the predicted correspondence between the continuum variation and the behavior of the individual features, and this model has therefore been ruled out.

The violent variability of 0235+164 naturally leads to models where the radio source structure changes behind a non-uniform absorbing medium whose properties are time-invariant. Furthermore, while the individual features vary, the general character of the profile remains unchanged--as would be expected for a situation where radiating blobs die out in time, allowing the radio centroid to shift back toward the compact component. In this scheme, it is expected that certain patterns in the relative behavior of the fine depths will repeat, depending on the position of the source centroid. In fact a covariance analysis of the variations in the 4 principal features can be related to a model where a point source moves behind clouds with gradients in their properties (Briggs, 1983). The outcome of this model, which reproduces the observed behavior, are "pseudo-gradients" and "pseudo-motions" describing the clouds and the source motion overtime. The relation of these quantities to true motions and cloud structure requires high spatial resolution information obtainable only from VLBI. Knowledge of the milli-arc-second source structure or precise astrometric position of the centroid at the absorption frequency or VLBI spectra observations as a function of time for comparison with the single dish profile will help to define the free parameters of the model (angular size and orientation).

[+] Discussion on page 422

R. Fanti et al. (eds.), VLBI and Compact Radio Sources, 49–50.
© *1984 by the IAU.*

VLBI observations in the absorption line have been performed at
a few epochs (Wolfe et al., 1978; Johnston et al. 1979; Briggs,
Broderich, Condon, Davis, Johnston, Romney, Wolfe, in preparation) and
a systematic VLBI monitoring program has been initiated by Baath,
Benson, Briggs, Davis, Johnston, Jones, Ronnang, and Wolfe. One
result is that a comparison of spectral visibilities taken on a Bonn-
Arecibo baseline in 1977 and 1981 shows an apparent change in interferometer phase. Although the phase shift observed in feature B
corresponds to an angular displacement of only ~ 0.1 mas, its presence
implies that the source structure must extend at least ~ 2 mas and that
the cloud must vary in opacity over this extent (~ 20pc).

REFERENCES

Briggs, F. H., 1983, Ap. J., in press.

Johnston, K. J., Broderick, J. J., Condon, J. J., Wolfe, A. M.,
    Weiber, K., Genzel, R., Witzel, A., Booth, R. 1979, Ap. J.,
    234, 466.

Wolfe, A. M., Davis, M. M., Briggs, F. H. 1982, Ap. J., 259, 495.

Wolfe, A. M., Broderick, J. J. Condon, J. J., and Johnston, K. J.
    1978, Ap. J., 222, 752.

MULTIFREQUENCY OBSERVATIONS OF THE BL LAC OBJECT 1219+28 [+]

Donna Weistrop[*], David B. Shaffer[+] and Paul Hintzen[*]
[*]Lab. for Astronomy and Solar Physics, Goddard Space Flight Center
[+]Interferometrics, Inc.

The BL Lac object 1219+28 has been observed at visible wavelengths and at 2.3 GHz, 5.0 GHz and 8.3 GHz using the VLBI network.

The visible observations include spectra and direct images obtained at Kitt Peak National Obs. with the PF/CCD camera and the cryogenic camera, respectively. A single emission line at 5521 Å has been observed in the spectrum of 1219+28. No emission lines have been detected in this object previously (Strittmatter et al. 1972). The [OIII] 5007 Å line has been observed to be one of the strongest emission lines in several BL Lac spectra (Miller, French and Hawley 1978). If we assume this identification for the observed line, we find z = 0.103. At this redshift, the 4959 Å line is very close to a night sky line and could not be distinguished. There is a small bump where H_β would be expected, but the feature may not be significant.

Direct imaging in a broad red passband reveals the presence of an underlying elliptical nebulosity (Figure 1). The existence and approximate orientation of the nebulosity is confirmed by a much shorter integration in V. For several BL Lac objects, similar features have proved to have the characteristics of elliptical galaxies. There are two galaxies to the SW of 1219+28 (Figure 1). The presence of the closer one has been noted previously (Strittmatter et al. 1972). A spectrum obtained for this object is extremely noisy; thus far, we have not been able to determine its redshift from the data. The separations of these objects from the BL Lac are about 10" and 21". If the galaxies are at the same redshift as 1219+28, the distances correspond to 27 kpc and 53 kpc assuming H_o = 50 km sec^{-1} Mpc^{-1} and q_o = +0.1. Hutchings and Campbell (1983) have recently suggested that many QSO's appear to be interacting with other galaxies and that such interactions may activate the QSO. The presence of a galaxy apparently close to 1219+28 is consistent with this hypothesis. No definite conclusions can be drawn until redshift data have been obtained. There are other faint galaxies in the field, and the possibility that the two near 1219+28 are at different distances along the same line-of-sight cannot be dismissed.

+ Discussion on page 423

51

R. Fanti et al. (eds.), VLBI and Compact Radio Sources, 51–52.
© *1984 by the IAU.*

Radio observations were made at 2.3, 5.0 and 8.3 GHz. At all three frequencies there is a central unresolved source with a jet at p.a. ~ 110°. The jet has three components. Two are visible in the 5 GHz map, at distances of 3.4 mas and 6.2 mas from the core source (Figure 2). A third component is visible in the 8 GHz map, only 0.74 mas from the core. The component 4.0 mas from the core in the 2.3 GHz map is probably the same as the 3.4 mas component in the 5 GHz map. The jet shows some curvature which does not increase close to the core.

There is no obvious relationship between the orientation of the radio jet and the nebulosity underlying 1219+28. The p.a. of the apparent major axis of the nebulosity is about 65°. The difference in the position angles indicates the radio jet is not being emitted along the rotation axis of the nebulosity, assuming the rotation axis defines the minor axis. However, we cannot rule out the possibility that the jet is emitted in the plane of the nebulosity.

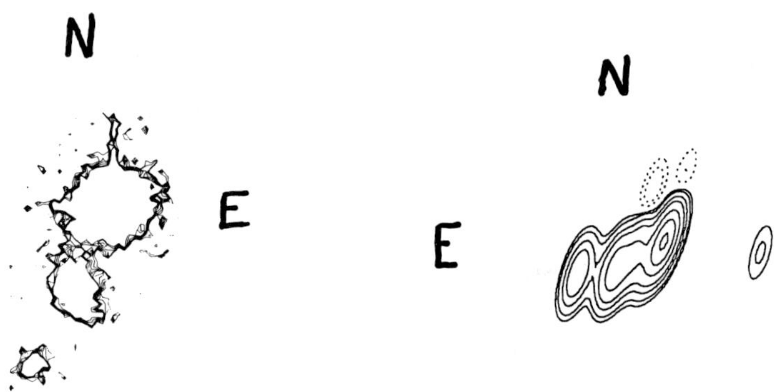

Figure 1. Low intensity contour map of the field of 1219+28 (northernmost object). The vertical feature at the top is due to saturation in the center of the image.

Figure 2. 5.0 GHZ map of 1219+28.

References

Hutchings, J.B. and Campbell, B. 1983, Nature (in press).

Miller,J.S., French, H.B., and Hawley, S.A. 1978, Pittsburgh Conference on BL Lac Objects, ed. A.M. Wolfe (Pittsburgh: University of Pittsburgh Press), pp.176-191.

Strittmatter, P.A., Serkowski, K.,Carswell, R., Stein, W.A., Merrill, K.M., and Burbidge, E.M. 1972, Ap.J. (Letters), 175, pp. L7-L14.

OPTICALLY QUIET QUASARS[+]

W. D. Cotton
National Radio Astronomy Observatory

1. OBSERVATIONS

The study of steep spectrum sources containing compact components
has become increasingly popular as the number of papers on the topic in
this volume indicates. I wish to discuss the properties of a number of
steep spectrum sources which are dominated by compact components whose
spectra remain steep to quite low frequencies. Most of the known
sources of this type do not have optical counterparts to the limit of
the Palomar Sky Survey prints and thus, with the implicit assumption
that they are extragalactic, have been referred to as "Optically Quiet
Quasars".

A number of the low frequency variable sources from Cotton 1976
have quite steep spectra to low frequencies and no optical counterpart
to the limit of the Palomar Sky Survey. The occurrence of low frequency
variations in these sources indicates the presence of compact structure
which has been confirmed by VLA and VLBI observations (Cotton 1983 and
Cotton et al. 1983). The observational material available for one of
these sources, 2147+145, is sufficiently better than the others that in
the following discussion it will be used to illustrate the properties of
these sources.

The spectrum of 2147+145 is shown in Figure 1. The values plotted
are from: Gower et al. 1967 (0.178 GHz); Sharp and Bash 1975
(0.365 GHz); Spangler and Cotton 1981 (0.430 GHz); Cotton 1983 (1.4,
2.7, 5.0, 8.1 and 14.9 GHz); Cotton unpublished 89.6 GHz upper limit;
Owen and Puschell, private communication (upper limit 2.2 microns);
upper limit in the visual due to the absence on the Palomar Sky Survey.

The spectrum of 2147+145 is steep from about 0.2 GHz up to optical
frequencies. There is no evidence in this spectrum for a flat spectrum
component at high frequencies as is frequently found in compact sources.

A VLBI map of 2147+145, given in Figure 2, shows a strong component
and a string of knots as is frequently seen in compact quasars and

+ Discussion on page 424

53

R. Fanti et al. (eds.), VLBI and Compact Radio Sources, 53–56.
© *1984 by the IAU.*

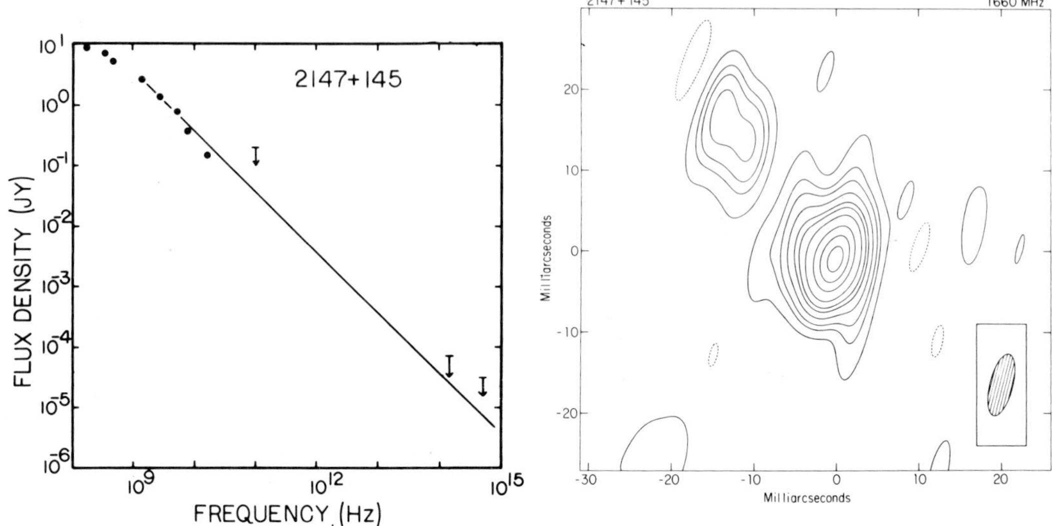

Figure 1. The spectrum of
2147+145. The straight line
is an extrapolation of the
1 GHz value with a spectral
index of 1.0

Figure 2. Structure of
2147+145 at 1.67 GHz
from Cotton et al. 1983.

active galactic nuclei. The size of the strongest component is quoted
by Cotton et al. 1983 as 3 milliarcseconds. Cotton 1983 gives a number
of other sources with steep spectra which, on the basis of VLA
observations, are dominated by compact structure.

2. INTERPRETATION

The observable properties of a synchrotron emitting source can be
used to constrain the magnetic field strength. Following Terrell 1966:

$$B \propto \nu_m^5 \theta^4 S_m^{-2} (1 + z)^{-1}$$

where ν_m is the turnover frequency in the spectrum, θ is the angular
size of the source, S_m is the flux density at ν_m, and z is the redshift
of the source. Most of the better studied sources with significant
structure on the milliarcsecond scale have turnover frequencies at 1 GHz
or above; i.e., have flat spectra. Since the magnetic field strength
derived from the observations is proportional to the fifth power of the
turnover frequency, sources such as 2147+145 must have magnetic fields
which are much weaker than those in the flat spectrum sources.

A weak magnetic field has other implications about the physical conditions in the source. In particular, the density of relativistic electrons must be significantly larger than in sources with "normal" magnetic field strengths because 1) the emissivity per electron is proportional to the magnetic field strength and 2) a given energy electron radiates at a lower frequency in a weaker magnetic field. The two effects just mentioned require more electrons at higher energies to produce comparable emission to·a source with a stronger magnetic field.

A high relativistic electron density in turn implies a high brightness temperature near the turnover frequency. The high brightness temperature will result in serious inverse Compton losses especially to the high energy electrons which radiate predominantly at the higher frequencies. Detailed modeling, as described in Cotton 1983, confirms the general statements made above. In order to produce a model with radio frequency properties similar to 2147+145 a very weak magnetic field and very high relativistic electron densities were required. The high peak brightness temperature produced in such a model resulted in relatively short lifetimes for electrons radiating synchrotron emission at higher frequencies and inverse Compton infrared and optical emission at levels in excess of the observational limits for 2147+145. If the relativistic electron density becomes too large the inverse Compton opacity can cause serious distortions to the synchrotron spectrum. If the strongest component of 2147+145 shown in Figure 2 still dominates the source at a few hundred megahertz then it is difficult to explain the source using a synchrotron model.

3. STEEP SPECTRUM SOURCE SURVEY

Recently a survey of steep spectrum sources was undertaken with the VLA at 5 GHz in collaboration with F. N. Owen. The results of this survey will help determine the frequency of compact, steep spectrum radio sources. The sample of sources observed were all those from the deep 5 GHz surveys of Condon and Ledden 1981 and Owen et al. 1983 with spectral indices steeper than 1.0. A preliminary analysis of the results indicate that as many as 25 percent of the 75 sources observed were dominated by a core less than about 0.2 arcseconds in size. The implication of this result is that as many as a few percent of all sources selected at 5 GHz may be dominated by steep spectrum, compact components.

REFERENCES

Condon J. J. and Ledden, J. E.: 1981, Astron. J. 86, pp. 643-652.
Cotton, W. D.: 1976, Astrophys. J. 204, pp. L63-L66.
Cotton, W. D.: 1983, Astrophys. J. in press.
Cotton, W. D., Owen, F. N., Geldzahler, B. J., Johnston, K.,
     Baath, L. B. and Romney, J.: 1983, submitted to Astrophys. J.
Gower, J.F.R., Scott, P. F. and Wills, D.: 1967, Mem. Roy. Astron.
     Soc. 71, pp. 49-144.

Owen. F. N., Condon, J. J. and Ledden, J. E.: 1983, Astron. J. 88,
    pp. 1-15.
Sharp, J. R. and Bash, F. N.: 1975, Astron. J. 80, pp. 335-352.
Spangler, S. R. and Cotton, W. D.: 1981, Astron. J. 86, pp 730-746.
Terrell, J.: 1966, Science 154, pp. 1281.

EVN OBSERVATIONS OF COMPACT STEEP SPECTRUM RADIO SOURCES AT 18 CM [+]

C.Fanti, R.Fanti, P.Parma
Istituto di Radioastronomia, Bologna, ITALY

R.T.Schilizzi
Radiosterrenwacht, Dwingeloo, THE NETHERLANDS

We consider the compact ($\lesssim$ 2-4 asec) radio sources in the 3CR catalogue with normal ($\alpha \gtrsim 0.5$) straight spectrum and low frequency($\nu_c \lesssim 200$ MHz) turnover. The curvature in the spectrum, if due to synchrotron self-absorption, implies structures of the order of tenths of an arcsec, while its straightness at high frequencies suggests that no dominant very compact ($\sim$1 mas) feature is present. The radio sources matching the above requirements are listed in the Table. The sample is probably not complete but it is representative of the class.

For a few radiosources VLB data existed. For the remaining we have undertaken an observing program with the European VLB Network (Onsala, Effelsberg, Dwingeloo, Jodrell Bank) at 18 cm. The status of VLB data is given in the Table. Here we present the maps of the 8 radio sources we have observed so far.

A large variety of structures is present: complex morphologies (3C287, 3C343) with no indication of symmetry at all, and core-jet like features (3C49, 3C138) have been found. Even double sources present some-time complex shapes (3C241,3C343.1). In a few cases collimation processes seem to be absent or occurring far from the possible core (3C303.1). No segregation in spectrum or morphology seems to be present between QSO's and Galaxies.

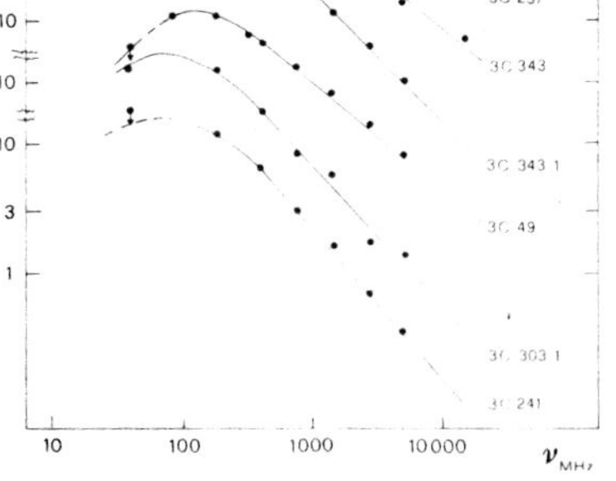

3C	Id	VLB	3C	Id	VLB	3C	Id	VLB	3C	Id	VLB
48	Q	3	147	Q	6	286	Q	4	303.1	G	1
49	G	1	237	G	1	287	Q	1	305.1	G	2
67	G	2	241	G	1	295	G	7	343	Q	1
119	Q	4	266	G	2	298	Q	8	343.1	G	1
138	Q	5	268.3	G	2	299	G	2			

1)Present paper; 2)Observations in progress; 3)Wilkinson et al,in prep.: 4)Pearson et al, Ap.J 1980,236,373; 5)Geldzahler et al,in press; 6)Readhead & Wilkinson, Ap.J 1980,235,11; 7)Spencer, in prep.; 8)Graham, this Symp.

+ Discussion on page 425

57

R. Fanti et al. (eds.), VLBI and Compact Radio Sources, 57–58.
© 1984 by the IAU.

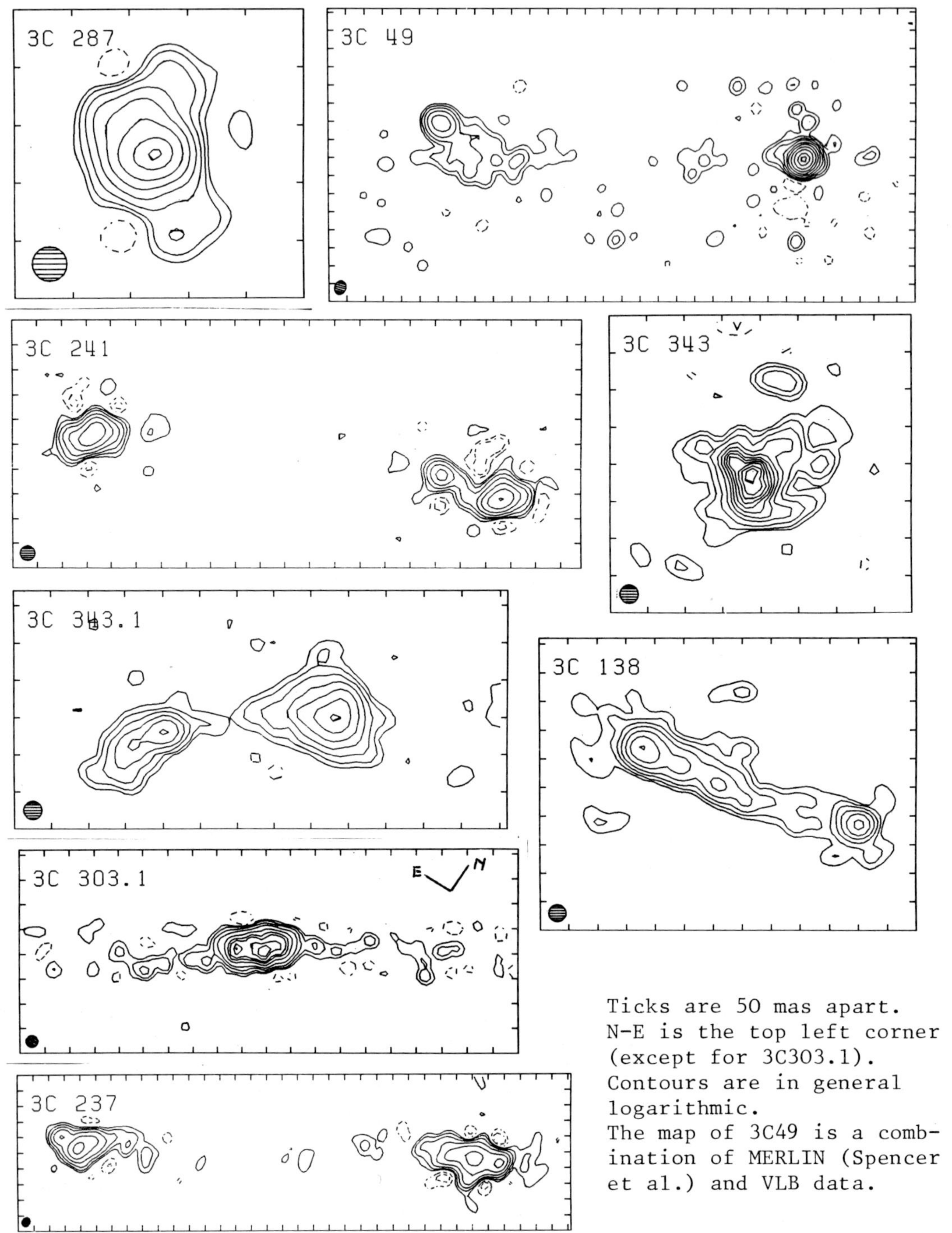

Ticks are 50 mas apart.
N-E is the top left corner
(except for 3C303.1).
Contours are in general
logarithmic.
The map of 3C49 is a combination of MERLIN (Spencer
et al.) and VLB data.

STEEP SPECTRUM RADIO CORES AND FAR-SIGHTEDNESS [+]

Wil van Breugel
Steward Observatory, University of Arizona

1. INTRODUCTION

Quite naturally so, VLBI'ers tend to be somewhat far-sighted: because of the very high resolutions, and poor UV-coverages, obtained with current VLBI networks only distant objects can be suitably studied. Unfortunately one thus tends to oversee some basic properties of a very popular class of objects, namely the compact, steep-spectrum radio cores. I would like to put these objects in their proper perspective.

2. STEEP SPECTRUM CORES (SSC's): A PHYSICAL DEFINITION

I define steep spectrum radio cores to be radio source components which have a spectral index $\alpha \geqslant 0.50$ ($S_\nu \sim \nu^{-\alpha}$) in a frequency range of at least 1 GHz – 15 GHz and which have linear sizes of approximately 0.1 kpc – 10 kpc, i.e. much smaller than a galaxy diameter but larger than the very compact (and often flat-spectrum) components associated with galaxy or quasar nuclei. Note that this definition is more general, and physically more meaningful, than the loose definition usually used by VLBI'ers (namely small [< 1"] steep spectrum sources, which can be mapped with VLBI).

3. PROPERTIES OF NEARBY SSC's

Defined as above, nearby and low luminosity SSC's comprise Seyferts such as NGC 1068 (Wilson and Ulvestad 1983). More powerful SSC's, which are somewhat further away, are 3C 305, 3C 293 and 4C 26.42 (e.g. van Breugel and Heckman 1982). Some objects have associated radio emission on a much larger scale (3C 293) and, in fact, even the largest radio galaxy known (3C236) has a SSC.

Nearby SSC's are accessible to spatially resolved optical studies. Presently known properties are (see for example, Miley 1981; van Breugel et al. 1983):
(1) SSC's have associated narrow-line regions of comparable size

+ Discussion on page 426

59

R. Fanti et al. (eds.), VLBI and Compact Radio Sources, 59–61.
© 1984 by the IAU.

and which are often spatially related.
  (2)   Their radio luminosities are correlated with both the luminosities and widths of the [OIII] $\lambda 5007$ lines.
  (3)   The asymmetries of the [OIII] $\lambda 5007$ lines are correlated with the Balmer line decrements which may be interpreted as gas and dust outflow, presumably caused by jets and expanding radio components.
  (4)   The radio emission has a very low degree of polarization, even at short wavelengths. This appears to be anti-correlated with the presence of optical line emission.
  (5)   The parent galaxies show all evidence for a dense interstellar medium: gaseous disks and dustlanes (3C 305, 3C 293), neutral hydrogen (3C 293, NGC 3801, NGC 1167), and a cooling X-ray core (4C 26.42).
  (6)   In Seyferts 3C 305 and 3C 293, the SSC's are embedded in rotating gaseous disks which are capable of breaking and bending the jets. This may account for their S-shaped morphologies, although precession of jets also remains a viable explanation.

4. DISTANT SSC's (QUASARS): MORE OF THE SAME

SSC's are also found in some quasars (such as 3C 48, 3C 147, 3C 286, etc.), which may be more powerful but otherwise similar objects. I mention the following circumstantial evidence for this:
  (1)   There may be some evidence that the radio power/narrow line width correlation for SSC's can be extrapolated to the more distant and more powerful objects (Heckman et al. 1981).
  (2)   A high resolution VLA survey of a sample of distant SSC's (van Breugel et al. 1983) suggests that they have similar properties as nearby SSC's: a low degree of polarization, typical sizes of 0.3 - 24 kpc, complex morphologies (i.e. large misalignments between central core components and more distant 'lobes').
  (3)   Optical imaging of quasars suggests that a fair fraction (> 30% ?) of them may reside in spiral galaxies and that the quasar phenomenon may be triggered by interactions with nearby companions (e.g. Hutchings and Campbell 1983). This would provide for the dense environments in quasar SSC's and might explain their complex morphologies (Section 3, point 6).
  (4)   Quasars are profuse optical line emitters, thus showing off these gas reserves. Also, in 3C 48 the narrow line region extends roughly in the same direction as the SSC.

5. LOW FREQUENCY TURNOVERS IN SSC's: FREE-FREE ABSORPTION

The radio spectra of many SSC's are turning over at low frequencies (Peacock and Wall 1982). Examples are, in order of radio luminosity: NGC 1068, 3C 293 and 3C 286. These turnovers are generally ascribed to synchrotron self-absorption. I believe, however, that free-free absorption of the radio emission by the line emitting gas (entrained?) may provide a more natural explanation.

The optical depth (τ_ν) to thermal free-free-absorption at a frequency ν is

$$\tau_\nu = T_e^{-1.5} \, \nu^{-2.1} \, \varepsilon \qquad\qquad (\text{c.g.s. units})$$

where T_e is the electron temperature, ν is the frequency and $\varepsilon = n_e^2 \, \phi \ell$ is the emission measure of the ionized gas (electron density n_e, volume filling factor ϕ, line of sight through the source ℓ).

Taking NGC 1068 as an example, I find that the spectrum turns over ($\tau_\nu \simeq 1$) near $\nu \simeq 100$ MHz (Kuhr et al. 1981). Assuming that the free-free absorption is due to the ionized gas of the line emitting clouds one further has $T_e \simeq 10^4$ K, so that the emission measure is $\varepsilon = 6\times10^{22}$ cm^{-5}. In NGC 1068, on average, $n_e \simeq 10^3$ cm^{-3} (Alloin et al. 1981) and the radio source 'lobes' have dimensions of ~ 0.4 kpc ($H_o = 75$; Wilson and Ulvestad 1983). The filling factor for the clouds must then be $\phi \simeq 10^{-4}$, which is entirely reasonable.

Similar order of magnitude estimates can be made for other SSC's and I conclude that free-free absorption by line emitting clouds is naturally to be expected. These clouds, which may have been entrained and are ionized by the source as its jets plough through the dense interstellar medium, will exhibit a range in densities, temperatures and velocities: small clouds will be accelerated and heated more by the jets than heavier clouds. Thus the absorption turnover will occur over a relatively broad region, rather than having a steep drop-off on the low frequency side which would occur for free-free absorption by a single, uniform cloud.

Because the emission measure for free-free absorption and optical line emission is the same, the turnover frequency (ν_T) and brightness of the optical line emission are correlated: $\varepsilon \sim \nu_T^{2.1}$. Thus brighter emission line objects (quasars) are expected to have higher turnover frequencies. This might also explain the redshift dependence of the low frequency turnovers in a sample of quasar SSC's found by Menon (1983).

REFERENCES

Alloin, Lacques, Pelat, Despian: 1981, A & A Suppl, 43, 231.
Heckman, Miley, Balick, van Breugel, Butcher: 1981, Ap. J., 262, 529.
Hutchings and Campbell: 1983, preprint.
Kuhr, Witzel, Pauliny-Toth, Nauber: 1981, A & A Suppl., 45, 367.
Menon: 1983, A. J., 88, 598.
Miley: 1981, ESO/ESA Workshop on Optical Jets in Galaxies, ESA SP-162, p. 9.
Peacock and Wall: 1982, M.N.R.A.S., 198, 843.
van Breugel and Heckman: 1982, Proceedings of IAU Symposium 97, p. 67, (Reidel Publishing Company).
van Breugel, Heckman, Miley: 1983, A. J. (submitted for publication).
Wilson and Ulvestad: 1983, preprint.

COMPACT DOUBLES: A GENUINE OR ILLUSORY CLASS? [+]

R. B. Phillips
Haystack Observatory

M. W. Hodges and R. L. Mutel
University of Iowa

Symmetric, two-sided morphology seems to argue against relativistic effects dominating compact radio emission. This kind of structure has been reported for a number of sources (Phillips and Mutel 1982; Pearson 1983) based on maps made at one frequency. Various arguments, all indirect, can be made for these sources being (1) Twin regions formed at the ends of jets which emerge from an invisible core, or (2) misidentified core-jet sources wherein the core and an unusually bright knot are wrongly taken to be a "double." A telling test of both hypotheses is to map the sources in question over an octave or so of frequency. Proponents of view (1) would predict that the two double components will show nearly identical spectral indices and that weak central cores with flat or rising spectra might even be revealed. Champions of view (2) would predict that one end or the other will dominate at high frequencies (the core!) or that complex bridges of emission (the jet!) will be revealed between the components at low frequencies. We have followed our initial discovery of 5 symmetric compact doubles by (A) attempting to enlarge the sample of symmetric sources available to study, and (B) by investigating at 5 GHz those doubles for which the best maps exist at 1.7 GHz.

A. RESULTS OF THE SEARCH FOR DOUBLES

Two Mark II VLB sessions were carried out at 1.7 GHz, observing the next 10 strongest sources showing self-absorption between 0.5 and 1.5 GHz and an optically thin spectral index $\alpha \leq -0.6$. We chose these empirical criteria to eliminate "flat" spectrum sources and because they are an amalgam of spectral properties seen in the original 5 compact doubles. The results can now be summarized. Six of the ten are single component "core" objects slightly resolved with $0".006$ resolution; three are complex. Only 1225+36 shows a double morphology: it _is_ an unequal (2:1 brightness ratio) double, $\sim0".020$ in size. It appears that many simple spectra are formed by sources that _are_ simple! Our earlier choices seem to have fortuitiously been doubles.

+ Discussion on page 427

63

R. Fanti et al. (eds.), VLBI and Compact Radio Sources, 63–64.
© _1984 by the IAU._

B.   NEW HIGH FREQUENCY MAPS FOR SOME ORIGINAL DOUBLES

We have made maps at 5 GHz for 3 of the original compact doubles. Two, 1518+047 and 2050+363, show detail within <u>one</u> end of the "double." 2050+363 (shown, Fig. 1) shows that the east end of the double, which appeared slightly larger at 1.7 GHz, is actually three bright spots which form a small hook. Though spectral differences are slight, one might now argue that 2050+363 is in fact an asymmetric system with an invisible jet.

A third, CTD 93, shows no particular structure within its major double components. There seems to be a significant <u>spectral</u> gradient across one or both components (outside edge "flatter"). We also find evidence (Fig.2) for a previously unnoticed, third component roughly midway between the two others. While this component is not impressively strong (11% of peak), we have applied gain-correcting algorithms in the mapping process and believe 5-stations maps should be able to attain 9:1 fidelity. The best confirmation of the central object would be corroborative discovery at a higher frequency, at which, presumably, its rising or flat spectrum would make it even more prominent between the two double components.

SUMMARY

Between 20% (Pearson "complete" sample) and 30% (our old and new <u>spectral</u> sample) of compact sources have the "compact double" morphology. While some sources show mild asymmetry when mapped at two or more frequencies, CTD 93 shows no evidence for asymmetry and tentative evidence for a flat spectrum central component. If such a third component can be confirmed, we will be faced with - for the first time - a compact source wherein the apparent core of activity appears not at one end, but in its midst. We conclude that all doubles may not ultimately be symmetric, but that further study has so far been quite illuminating.

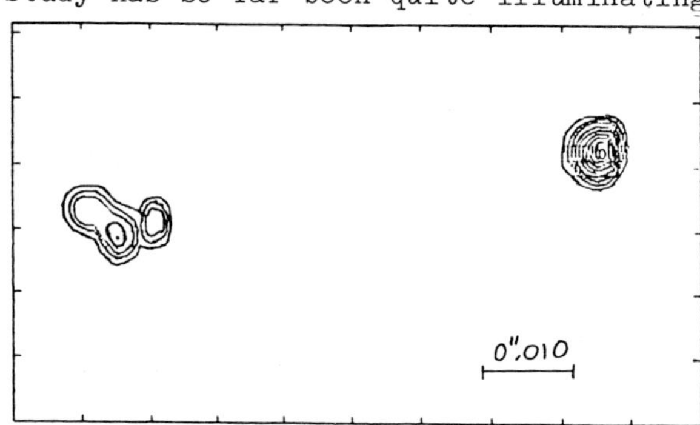

Figure 1.   2050+363

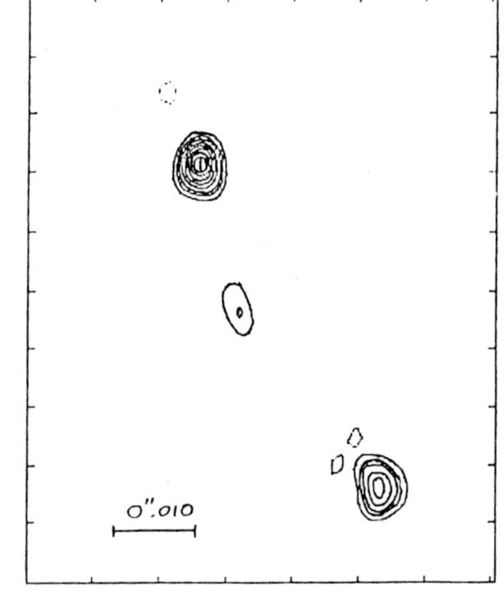

Figure 2.   CTD 93

REFERENCES

1.   Phillips, R. B., and Mutel, R. L., 1982 Astron. and Astrop., <u>106</u>, 21

2.   Pearson, T. J., 1983, in preparation.

SPATIAL STRUCTURES OF A COMPLETE SAMPLE OF EXTRAGALACTIC RADIO SOURCES NORTH OF DECLINATION 70°

A. Eckart, A. Witzel

Max-Planck-Institut für Radioastronomie, Bonn, W.-Germany

With the aim of studying the spatial structures of a complete sample of radio sources found at 6 cm wavelength we selected 13 sources from the "S5-survey" (Kühr et al., 1981) which fulfil the following criteria

1) $\delta > 70°$, $|b^{II}| > 10°$

2) $S_{5\ GHz} > 1$ Jy (at the epoch of the S5-survey)

3) $\alpha_{5\ GHz}^{2.7\ GHz} > -0.5$ $(S \propto \nu^{\alpha})$

6 of these sources are identified by means of optical spectroscopy with QSOs, 6 are BL Lac-type objects, and the identification of 1 source is presently unknown. Observations at frequencies ranging from radio to X-ray have been reported (Biermann et al., 1981; Biermann et al., 1982; Eckart et al., 1982). Table 1 lists the present status of the survey.

Table 1

Source name	Ident.	Z	VLA 20 cm	MERLIN 18 cm	V L B I 18 cm	6 cm	2.8 cm	1.3 cm
0016+73	QSO	X	X	X	map	model	model	
0153+74	QSO	X	X	X	map	model		
0212+73	BL		X	X	map	model	model	proc. struct.
0454+84	BL		X	X	map	model		proc. struct.
0615+82	?		X	X	map	model	model	
0716+71	BL		X	X	map	2 x map		proc. struct.
0836+71	QSO	X	X	X	map	2 x map	map	proc. struct.
1039+81	QSO	X		X	map	map		
1150+81	QSO	X	X	X	map	model map		
1749+70	BL		X	X	map	model		
1803+78	BL		X	X	map	map		
1928+73	QSO	X	X	X	map	2 x map		
2007+77	BL		X	X	map	2 x map		

"proc. struct." means: processed, structure

4 of the 13 sources display extended arcsec structures based on VLA and MERLIN observations.

VLBI-measurements at different frequencies showed source structures ranging from point-like to complex (see table 2). At 18 cm about half of the objects are point-like or just slightly resolved. Although the information is still incomplete at 2.8 cm and 1.3 cm it is obvious that the number of resolved and complex sources increases with rising frequency.

R. Fanti et al. (eds.), VLBI and Compact Radio Sources, 65–66.
© *1984 by the IAU.*

Table 2

Structure	λ18 cm	λ6 cm	λ2.8 cm	λ1.3 cm
pointlike	2 (1)	–	–	–
slightly resolved	5 (2)	3 (1)	2	–
core-jet (incl. complex jets)	6 (3)	10 (5)	2 (1)	4 (3)

(The numbers in parentheses refer to BL Lac objects)

Although the investigation of the spectral indices of individual components, the analysis of the total spectra, and the statistical investigations of apparently superluminal motions in the sample sources is not yet finished, we will briefly comment on four sources of the sample for which we have presently analyzed two epochs of 5 GHz VLBI observations.

The experiments at 6 cm were performed in 1979.9/1980.1 and 1982.9/1983.3 and therefore cover an interval of about three years.

0716+71: No significant changes in the source structure at 5 GHz have been observed for this BL Lac object. At both epochs the source is well described by a close (separation 1.3 mas) double source with PA 45°. The 18 cm map displays extended emission in PA –150° and a separation of about 4 mas. Therefore this object is a candidate for VLB structure on both sides of the compact emission. Furthermore the source has a two sided arcsecond structure in PA –60° and 125°. This may represent a bending of the jet of about 70° to 100°.

0836+71: This high redshifted quasar (z = 2.16) is well described by three compact sources located along PA –144°. A comparison with the 18 cm VLB map suggests that the northern component has an inverted spectrum and can therefore be identified with the core. Between the two 5 GHz epochs no significant structural changes have been observed.

1928+73: This source is a quasar with a redshift of 0.36 and consists of at least 3 bright components which are part of a large jet. This jet has also been detected on the arcsecond scale. Our data does not reveal any significant changes of separations. The decrease of total flux density between the two epochs of about 0.5 Jy can be attributed to the core. The two dominant secondary components in the 1982.9 map are located at a distance from the core at 3.3 mas (PA 165°) and about 9 mas (PA 175°). A 18 cm MERLIN map shows an extension along PA 180°.

2007+77: This BL Lac source consists of 3 components along PA 95°. The object is a good candidate for superluminal expansion. The separations between the core and the first secondary component at the different epochs is 1 mas and 1.8 mas respectively.

REFERENCES
Biermann, P. et al.: 1981, Astrophys. J. 247, pp. L53–L56.
Biermann, P. et al.: 1982, Astrophys. J. 252, pp. L1–L5.
Eckart, A. et al.: 1982, Astron. Astrophys. 108, pp. 157–160.
Kühr, H. et al.: 1981, Astron. Astrophys. Suppl. 45, pp. 367–430.

THE SOUTHERN HEMISPHERE VLBI EXPERIMENT (SHEVE)[+]

Preston, R.A.[1], Jauncey, D.L.[2], Tsioumis, A.[3], Meier, D.L.[1],
Skjerve, L.J.[1], Ables, J.[2], Batchelor, R.[2], Faulkner, J.[1],
Hamilton, P.M.[4], Haynes, R.[2], Johnson, B.[5], Louie, A.[1],
McCulloch, P.[4], Moorey, G.[2], Morabito, D.D.[1], Nicolson, G.[6],
Niell, A.E.[1], Robertson, J.G.[7], Royle, G.W.R.[8], Slade, M.A.[1],
Slee, O.B.[2], Watkinson, A.[3], Wehrle, A.E.[1], Wright, A.[2]

[1]Jet Propulsion Laboratory, Pasadena, California
[2]CSIRO, Sydney, Australia
[3]Sydney University, Sydney, Australia
[4]University of Tasmania, Hobart, Australia
[5]Ford Aerospace and Communications Corporation, Goldstone,
   California
[6]Hartebeesthoek Radio Astronomy Observatory, Johannesburg,
   South Africa
[7]Anglo-Australian Observatory, Sydney, Australia
[8]Division of National Mapping, Canberra, Australia

The Southern Hemisphere VLBI Experiment (or SHEVE) was a joint
US-Australian-South African venture with both astronomy and geodesy
goals. The principle astronomy goal was to make models or maps of the
following sources: at 2.3 GHz (with six antennas and 9 usable baselines)
-- Centaurus A (the nearest active galaxy), Circinus X-1 (a flaring
binary), the VELA pulsar, and 26 other active galactic nuclei and
quasars; at 8.4 GHz (only one baseline) -- Centaurus A and the galactic
center.

The SHEVE array consisted of six antennas -- five in Australia and
one in South Africa. JPL's 64m Deep Space Network antenna at Tidbinbilla,
the CSIRO 64m antenna at Parkes, and one of the 14m antennas at Fleurs
form an equilateral triangle of about 250 km on a side. About 1000 km
away from the triangle is the 14m antenna at Hobart and about 2000 km
away is the 9m Landsat antenna at Alice Springs. The 26m antenna at
Hartebeesthoek, South Africa provides an 8000 km baseline.

Of the 30 sources observed, only three will be discussed. Centaurus
A is a dust-lane and shell galaxy only 5 Mpc distant. Early radio maps
showed very extended (10 degree) north-south double structure plus a
smaller (8 arc minute) double at a position angle of 50 degrees. Recent
VLA maps show a northeast jet emanating from the nucleus and aligned
with the inner double. Earlier one baseline VLBI observations (Preston
et al. 1983) showed an elongated central component about 50 milliarcseconds long ($\sim$1 pc) consistent with the VLA jet position angle. The

+ Discussion on page 427   67

R. Fanti et al. (eds.), VLBI and Compact Radio Sources, 67–69.
© *1984 by the IAU.*

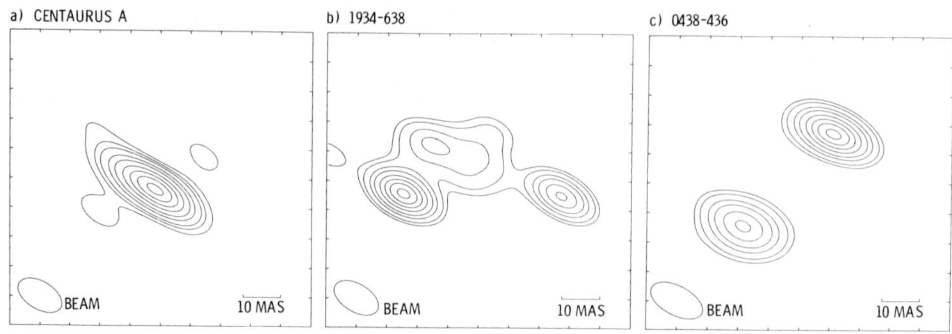

Figure 1. Hybrid Maps

SHEVE data pin down this angle much more precisely to about 51° and indicate the core is slightly extended in the northeast direction.

A hybrid map of Centaurus A generated from the SHEVE data is shown in Figure 1a. Model fitting was also performed on the Centaurus A data. It is less reliable than hybrid mapping, but tends to bring out more features. All components were assumed to have Gaussian shapes and to lie at a position angle of 50° on the northeast side of the nucleus. The data were fitted with four components: the central source splits into two -- an elongated 12 milliarcsecond core at position angle 51° and a blob about 40 milliarcseconds away. There is also a very narrow component at position angle 49° whose distance from the center is uncertain and a component about 0".25 away which may be observable with the VLA.

Parkes 1934-638 is also a dust-lane galaxy, but at z = 0.138. Data from a 1970 one-baseline VLBI experiment at 2.3 GHz were modeled with a 41.9±0.2 milliarcsecond (or 87 pc) double plus another weaker component (Gubbay et al. 1971). The 1982 SHEVE map (Figure 1b) at 2.3 GHz also shows a double with almost exactly the same separation (42.0±0.06 milliarcseconds) yielding a limit of 0.06c±0.2c on the expansion velocity. We also find a third component between the two, but offset from their line of centers. The 1 GHz spectral peak of 1934-638 is similar to that of other compact double sources (Phillips and Mutel 1981).

The final object is Parkes 0438-436. At a redshift of 2.863 it is the most radio-luminous quasar known ($L_{Radio} \sim 10^{46}$ erg s^{-1}). Several other high redshift quasars appear generally unresolved with VLBI. The SHEVE map (Figure 1c) reveals an equal double separated by about 35 mas (about 250 pc). The southeast component is resolved in our map while the northwest one is unresolved.

In summary, these three sources show some similarities and some differences among their optical and radio properties. Centaurus A and 1934-638 are both dust-lane galaxies, but display quite different radio structure, a jet and a compact double. 1934-638 and 0438-436 have similar radio structure, but one is a galaxy and the other a quasar. Theories of compact radio sources which employ a common central energy mechanism will have to account for such variations of form.

REFERENCES:

Preston et al. 1983, Ap. J. (Letters), 266, 93.
Phillips and Mutel 1981, Ap. J., 244, 19.
Gubbay, J. S. et al. 1971, A. J., 76, 965.

STATISTICAL PROPERTIES OF COMPACT SOURCES

J. Machalski
Jagiellonian University, Cracow, Poland.

The analysis is based on VLA 1465-MHz observations of GB and GB2 sources (Machalski et al. 1982; Machalski and Condon 1983a, 1983b). From those flux-density limited samples compact sources were selected. The following definition of source "compactness" has been used

$$C \equiv S_{1465}(\Theta < 1'') / S_{1400} \geqq 0.85$$

Subjects of analysis were: (a) fraction of compact sources among radio galaxies, quasars, and "empty field" sources, (b) identification rate vs. high-frequency spectral index for compact sources, (c) spectral-index distributions for compact identified and/or EF sources, (d) flux-density ratio for double compact sources, and (e) correlation between radio and optical luminosities for compact sources.

Some conclusions from this work are:

Table 1. The fraction of compact sources vs. optical type.

Type	Flat-spectrum	Steep-spectrum	All
comp. Gal / Gal	55±15(%)	7±2(%)	12±3(%)
comp. QSO / QSO	93± 4(%)	42±8(%)	67±5(%)
comp. EF / EF	91± 9(%)	38±4(%)	41±4(%)

(a) The observed difference between the abundance of compact galaxies and quasars (Table 1) is partly due to the distance effect, however, the great differences between Gal/QSO ratios for flat- and steep-spectrum sources confirm that quasars are intrinsically much more compact. The similar fractions suggest that compact EFs are mostly quasars.

(b) The identification rate for compact steep-spectrum

R. Fanti et al. (eds.), VLBI and Compact Radio Sources, 71–72.
© *1984 by the IAU.*

sources is lower than that for extended sources suggesting
that those sources have higher radio-to-optical luminosity
ratio than an average source {see also par.(e)}. There is not
any significant change of the identification rate vs. angular
size of sources in the size range 0.1 - 100 arcsec. This confirms that statistically small angular size reflects intrinsic compactness of those sources.

(c) High-frequency, $\alpha(1400,5000)$, spectral-index distributions for both identified and EF compact sources indicate
that there is a distinct correlation between the average size
and spectral index. Compact sources statistically have flatter spectra.

(d) The flux-density ratio S_1/S_2 distribution for compact sources is much broader than that for extended sources.
The mean from our analysis is $<S_1/S_2>=6.6\pm1.5$ and confirms
that compact sources are indeed more asymmetric than extended sources.

(e) The dependence of radio-to-optical luminosity ratio
on high-frequency spectral index has been noted by Condon et
al. (1983). Authors conclude that steep-spectrum objects
(mainly QSOs) have this ratio intrinsically higher than those
with flat spectra, and suggest that the optical luminosities
of radio-selected QSOs decline some time after the last ejection of relativistic plasma - then these objects become EFs.
Our analysis show that (1) The effect for compact galaxies
is evidently due to limited radio luminosity, thus the observed correlation is explained by the selection effect. (2)The
correlation for QSOs seems to be real. The small scatter of
the radio-optical spectral index in this case is a strong
evidence on the correlation between radio and optical luminosities in the core region of a QSO. (3) Our data do not indicate any rapid cut-off of identification for the steep-
spectrum objects. Oppositely, there are very compact EF sources with flat high-frequency spectra. These sources might be
"EF" because of the thirty-year period between radio and optical observations. Until a series of simultaneous radio and
optical observations of these sources is made there is no
observational evidence for any recognizable decline of the
optical luminosity for steep-spectrum QSOs.

REFERENCES:

Condon, J.J., Condon, M.A., Broderick, J.J., and Davis, M.M.
        1983, Astron. J. (in press).
Machalski, J., Maslowski, J., Condon, J.J., and Condon, M.A.
        1982, Astron. J. 87, 1150.
Machalski, J., and Condon, J.J. 1983a, Astron. J. 88, 143.
ibid   1983b, Astron. J. (in press).

THE INFRARED, OPTICAL AND ULTRAVIOLET PROPERTIES OF ACTIVE NUCLEI

Marie-Helene Ulrich
European Southern Observatory
Garching bei München

ABSTRACT

We review the important properties of active nuclei, in particular
(i) the optical polarization and its relation to the jets found by VLBI
(ii) the energy distribution and the temporal variations of the
continuum spectrum and (iii) the distribution of the matter in the broad
line region.

I. INTRODUCTION

The highest angular resolution which can be achieved with the
existing large ground based telescopes is obtained by speckle
interferometry and corresponds to the diffraction limit of the
telescopes: 2×10^{-2} arc sec for telescopes of 5m diameter. This
technique has yielded results on relatively bright objects ($m_V < 18$) of
simple structure. With the Space Telescope it is expected that a similar
resolution will be obtained by deconvolution of direct images. Thus the
best angular resolution achievable now or in the near future in the
optical/UV range is coarser than the 10^{-3} arc sec currently achieved
with radio VLBI. It is however possible to obtain important information
on the linear dimensions of the most compact components of the sources
emitting the optical/UV radiation from the time scale of their temporal
variations and through theoretical arguments.

The components of active nuclei which have dimensions comparable to
or smaller than the VLBI jets are the sources of optical and X-ray
radiations, and the broad emission line regions. As for the broad UV
absorption lines observed in some quasars and Seyfert galaxies, it can
be shown that they are formed by clouds which cover a large fraction of
the broad emission line region as viewed from the observer and which are
outside the broad line regions. The absorbing region is thus much larger
than the VLBI jets and for this reason it is not discussed here.

In NGC 4151 the linear dimensions of the UV continuum source and of

R. Fanti et al. (eds.), VLBI and Compact Radio Sources, 73–83.
© *1984 by the IAU.*

TABLE 1

Sources with VLBI maps and measured Optical Polarization

Source Name	VLBI	Optical Polarization p%	Polarization P.A.	Ref. - VLBI - Optical Polarization
3C 84	10°	1-6	usually in the range 100-160°	Pearson and Readhead, 1981 A.S. 1980
3C 371	83°	0-12	65° to 100°	Pearson and Readhead, 1981 A.S., 1980
3C 273	63°	0.2	52°	Readhead et al., 1979 Stockman et al., 1979
3C 345	101°	3.6	10° to 173° but preferred angle 80°	Unwin et al., 1983 Stockman et al., 1979 Impey et al. 1982
CTA 102	146°	1-11	100-170	Pearson et al., 1980 A.S., 1980
3C 390.3	139°	1-4	149-165	Linfield, 1981 A.S., 1980 Martin et al., 1983
3C 120	63°	0.9-1.2	99-103	Readhead et al., 1979 A.S., 1980 Martin et al., 1983
BL Lac	10°	2-23	0-180, no preferred angle	Kellermann et al., 1977 Pearson and Readhead, 1981 A.S., 1980 Puschell and Stein, 1980 Moore et al., 1982
3C 454.3	130°	0-16	0-170, no preferred angle	Pearson et al., 1980 A.S., 1980
0235+164	21±21	6-25	15-175, no preferred angle	Bååth et al., 1981 Impey et al., 1982
0735+178	45±13	3-31	0-175, no preferred angle	Bååth et al., 1981 Impey et al., 1982

Footnote to Table 1:

The values of the preferred angles (Angel and Stockman, 1980 - referred to above as A.S. 1980) are provisional and could be found to be different or even non-existent with further monitoring.

the broad emission line region as obtained from the observations of the temporal variations are of the order of 0.25 and 0.5 light-months respectively. At the distance of NGC 4151, ~15 Mpc, this corresponds to an angular dimension of approximately 10^{-4} arc sec. Note that if one makes the not unreasonable assumption that in first approximation the characteristic linear dimension of the broad line region of an active nucleus or quasar scales with $L^{1/2}$ (where L is the absolute luminosity of the UV central component), then all the active objects of comparable apparent luminosities have broad line regions of comparable angular dimensions.

II. POLARIZATION PROPERTIES

A) The Highly Polarized Objects

Among the seven compact radio sources where superluminal velocities have been observed (Cohen and Unwin, 1982) three have large and highly variable optical polarization which suggests a relationship between the two phenomena. The variations of the polarization in percentage and angle occur on time scales of days, i.e., much shorter than the variations occurring in the radio structure observed with VLBI. The variable optical polarization may come from a region which is smaller than the VLBI jet, perhaps the base of the relativistic jet. It must be noted that the relationship between highly variable optical polarization and superluminal sources is not a simple one since there are superluminal sources such as 3C 273, where the optical polarization is very weak, 0.2%. Table 1 gives the optical polarization properties of the radio sources which show jets on the VLBI scale and for which VLBI maps were published in June 1983. The table shows that when the optical polarization angle covers a relatively small range of position angle ($\lesssim 45°$) then the direction of the VLBI jet falls in this range.

In BL Lacertae, which has been studied extensively, the spectrum of the time variations of the optical polarization angle and of the total optical flux is consistent with a random walk spectrum (Moore et al., 1982). In some BL Lac objects and highly polarized QSOs there is transient evidence for wavelength dependent polarization (with the percentage polarization increasing with decreasing wavelength) while at other times and in other QSOs the polarization is wavelength independent (Bailey, Hough and Axon, 1983; Impey et al., 1982; Puschell et al., 1983). A simple - but not unique - model fitting all these observations is one in which the energy is emitted by a small number of independent sources (Moore et al., 1982), each emitting a fully polarized (70%) synchrotron radiation of random orientation and varying on a time scale of a few days. The "independent sources" could be inhomogeneities or sub-units in a beamed jet. In this model, the dependence of polarization on wavelength is easily explained if the spectral domains of the sources are different, as would happen if the spectral domains were limited by different amounts of radiative energy losses. The polarization properties are also consistent with non-uniform Faraday effect, and

since electron radiative lifetimes are probably very short, there must
be a large number of non-relativistic electrons (Puschell et al., 1983).

A satisfactory model must also be able to account for extreme
variations such as the occasional large outbursts where the total energy
increases by a factor up to 100 (Rieke et al., 1976; Moore et al., 1982)
and the polarization outbursts like the one which occurred in 0235+164
during which the polarization reached 43% without much change in total
flux (Impey, Brand and Tapia, 1982). Moore et al. (1982) speculate that
the extreme total flux outbursts might be caused by an increase in the
average luminosity of the components. Regarding the polarization
outbursts, Impey et al. (1982) argue that these changes are caused by a
large scale physical reordering of the magnetic field. Alternatively, if
the emitting material is beamed towards us, then aberration exaggerates
the effects, and small changes in the source geometry are amplified by
relativistic effects and cause the observed large changes in flux, in
percentage polarization, and in angle of polarization.

B) <u>Seyfert Galaxies and Weakly Polarized Quasars</u>

In the majority of quasars (20 out of 24 in the sample of Stockman,
Angel and Miley, 1979) and of Seyfert 1 galaxies (5 out of 5 in the list
of Antonucci, 1982) the optical polarization angle is rather well
aligned with the angle of the associated radio structure on the VLA
scale, i.e. the difference between the two angles, $\Delta\Theta$, is less
than $\sim 20°$. In the objects with $\Delta\Theta \sim 0°$ the polarization could be
intrinsic to the radiation mechanism. The fact that in NGC 4151 neither
the broad nor the narrow emission lines have the same polarization as
the continuum (Schmidt and Miller, 1980) strongly suggests that this is
the case in at least some objects. More abundant data on spectral
polarization and time variations should clarify this point. It must be
pointed out that the discovery of variability of the angle of
polarization in Seyfert 1 galaxies, e.g. NGC 4151, would be the best
proof that some of the continuum is of synchrotron origin.

In Seyfert 2 galaxies and in Broad Line Radio Galaxies (BLRG) the
percentage polarization can be very large and furthermore in many of
these objects $\Delta\Theta$ is close to $90°$.

It has been suggested that in Seyfert 2 galaxies and in BLRG the
polarization is produced by scattering with, in some cases, intrinsic
polarization of the continuum contributing to the total polarization.
Rudy et al. (1983) argue that in BLRG, the polarization is due to
scattering by dust because they find a correlation between the
percentage polarization and the value of the Balmer decrements of the
broad emission lines while Antonucci (1983) proposes that polarization
is caused by electron scattering, with $\Delta\Theta \sim 90°$ being produced by a
thick scattering disk, and $\Delta\Theta \sim 0°$ by a thin scattering disk.

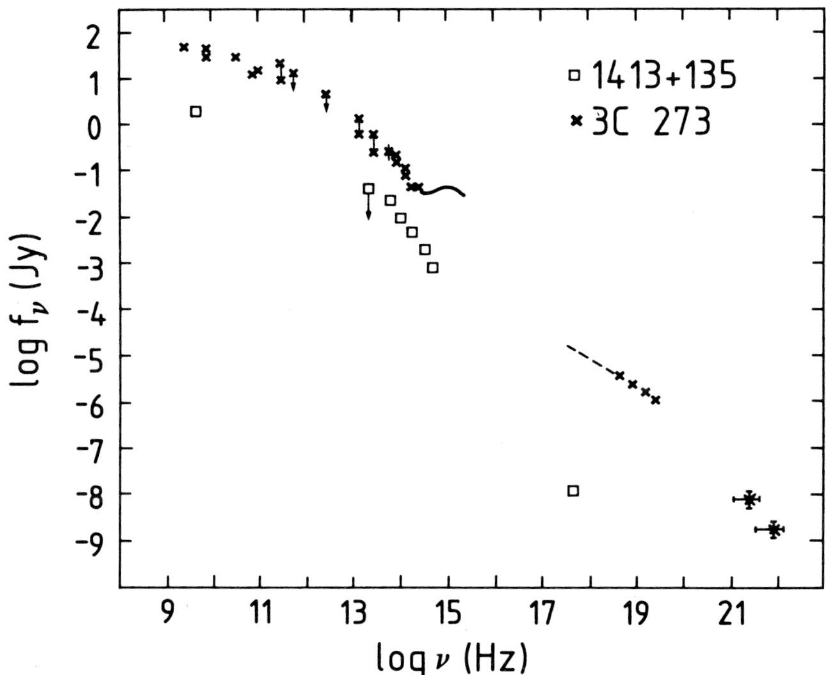

Figure 1. The electromagnetic spectrum of the quasar 3C 273 (Ulrich et al. 1980) and of the red object 1413+135 which is believed to be a BL Lac object (Bregman et al., 1981). Note the 3000 Å bump in the spectrum of 3C 273. This bump almost always occurs in conjunction with broad emission lines and is thought to be emitted in part by a $\sim 30000°$K accretion disk. In contrast, the spectra of BL Lac objects show a downward bend between 10^{14} and 10^{15} hz; this bend is particularly strong in 1413+135.

III. THE CONTINUUM SPECTRUM

A characteristic feature of the optical/UV continuum spectrum of quasars and Seyfert 1 galaxies is the presence of an excess of radiation in the range 3000 - 4000 Å over the extrapolation of the power law which represents the spectrum between 10μ and 5000 Å. (Figure 1). This excess, "the 3000 Å bump", seems always to be present in the spectrum <u>in conjunction</u> with broad emission lines and is generally believed to be a combination of black body radiation, optically thick Balmer continuum, 2-photon continuum, and blends of Fe II lines (Baldwin, 1977; Grandi, 1982; Malkan and Sargent, 1982).

In contrast the IR/optical/UV spectrum of BL Lac objects steepens between 10^{14} and 10^{15} hz, the exact frequency of the bend depending on the object (Maraschi, Tanzi and Treves, 1983). A similar bend might be present in the spectrum of Seyfert galaxies but remain undetected because of the 3000 Å bump.

It is not clear, at least to the reviewer, whether there is a difference between the nonthermal radiation of BL Lac objects and the nonthermal radiation of quasars and Seyfert galaxies. The IR/optical/UV spectrum of the latter could be nonthermal radiation as in BL Lac objects to which is added re-emission by dust in the IR and the 3000 Å bump in the optical/UV range.

Seyfert galaxies and BL Lac objects differ also by the temporal variations of their UV spectrum. In Seyfert galaxies the UV spectrum becomes harder when the object brightens (e.g. Perola et al., 1982; Boisson and Ulrich, 1982) while in the BL Lac object Mrk 421 (Figure 2) and in intrinsically bright BL Lac objects there is no correlation between spectral slope and intensity (Ulrich et al., 1984). The difference in temporal behavior between the UV continuum of Seyfert galaxies and BL Lac objects could be due to real differences in the properties of the non-thermal continuum related to differences in the accretion regime in Seyfert nuclei and in BL Lac objects (e.g. Maraschi, Perola and Treves, 1980). On the other hand, the correlation observed in Seyfert galaxies may result from the variations of the different components of the continuum.

IV. THE BROAD EMISSION LINE REGION

The broad emission line region is formed by many dense gas clouds distributed in the region with a very small filling factor. Some basic properties of the broad emission line gas in quasars and Seyfert galaxies have been known for nearly a decade and others are being explored. They are reviewed below in order of decreasing certainty.

(i) The particle density is in the range 10^9-10^{11} cm^{-3}. The line width is $\sim 10^4$ km s^{-1} and caused by gas motions. The degree of ionization and the particle density in the gas clouds are similar in objects differing

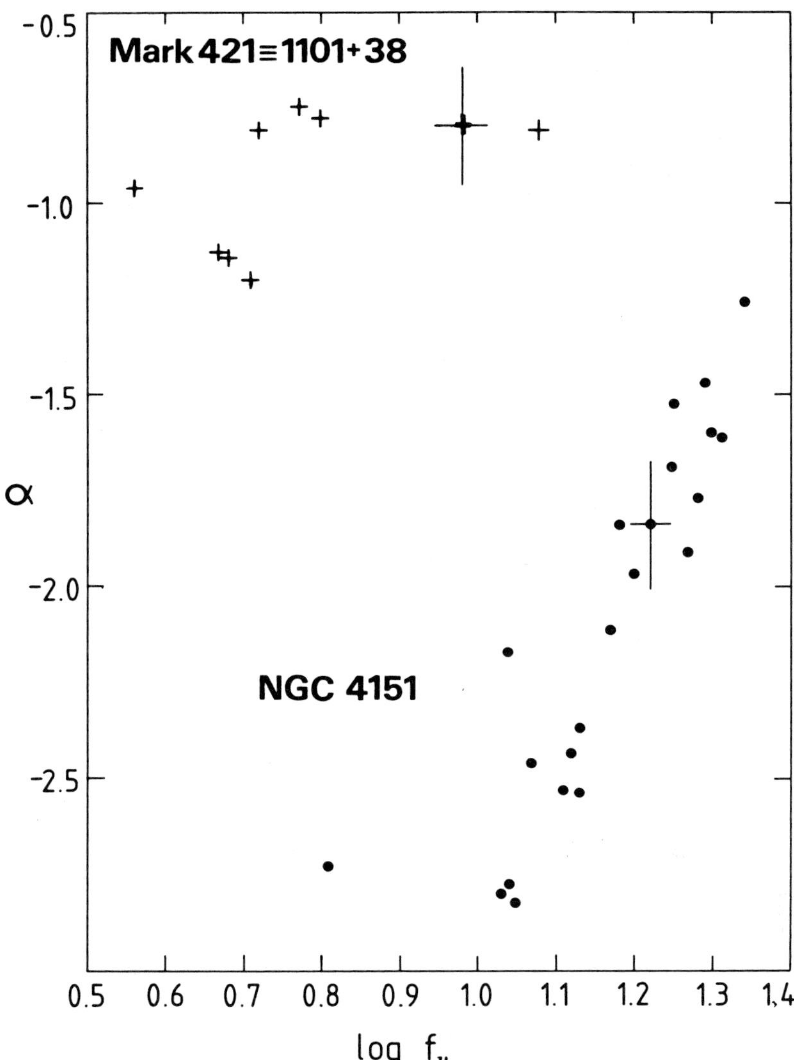

Figure 2. The spectral index, α, plotted as a function of the flux
level at 2500 Å in NGC 4151 (Perola et al. 1982) and in Mrk 421 (Ulrich
et al. 1984). In NGC 4151 the spectral index was measured between 2000
and 3000 Å. The measurements plotted here come from 21 different epochs
of observations. In Mrk 421 the spectral index was measured between 1300
and 3000 Å at 9 epochs. In NGC 4151, and in other Seyfert galaxies,
there is a strong correlation between α and log f_ν at 2500 Å. No such
correlation has yet been detected in Mrk 421 nor in intrinsically bright
BL Lac ojbects. Two typical error bars are marked.

by 10^4 in absolute luminosity; this suggests that the linear dimension
of the broad line region must be roughly proportional to the square root
of the UV continuum luminosity. In quasars similar to 3C 273 in
luminosity the linear dimension of the broad line region is $\sim$ 1 to 5 pc
i.e. comparable to the dimension of the VLBI jet. In NGC 4151, it is
approximately 5×10^{16} cm.

(ii)   More recently it has become clear that the large $H\beta/Ly\alpha$ intensity
ratio observed in the spectra of quasars and Seyfert galaxies is due to
collisional enhancement of $H\beta$ and implies a very large optical depth. In
such conditions the clouds emitting the bulk of $H\beta$ must emit the
$Ly\alpha$ line in a very anisotropic way which imposes strong constraints on
models for the broad line region. Also it has been realized that the
column density in each cloud is $\leq 10^{23}$ cm^{-2}, otherwise the intensity of
CII$]\lambda2326$ would be larger than observed (Kwan and Krolik, 1981). This,
together with the high value of n_e, indicates a very small linear
dimension for the cloud, $\sim 10^{13}$ cm, at least in the direction to the
continuum source.

(iii)   In the last two years, it has been recognized that the analysis
of the variations of the emission line intensity and profile could give
information on the <u>structure of the broad line region</u>. In the case of
NGC 4151 which is the best studied object, this analysis leads to a
model for the broad line region in which the velocity of the clouds,
their optical thickness and their distance from the central continuum
source are correlated, the clouds having the largest velocities and
smallest optical thickness being closer to the continuum source (Ulrich
et al., 1983; Ulrich, 1983). The mass of the central object deduced from
the line width and variability time-scales is 0.5 to 1.0×10^8 M$_\odot$. The
size of a 30000°K accretion disc surrounding a black hole of this mass
(Lynden-Bell, 1969) is consistent (perhaps by mere coincidence) with the
excess of radiation observed at 1455 Å over the power law component
fitting the rest of the UV/optical spectrum (Ulrich et al., 1983).

(iv)   Recently several arguments have been put forward suggesting that
the broad line region in quasars and Seyfert galaxies is made up of two
types of clouds differing by their physical conditions and also by their
location with respect to the central continuum source:

- The high degree of symmetry of $Ly\alpha$ and the similarity of the $Ly\alpha$ and
  CIV profiles in the few well studied high redshift quasars (Wilkes
  and Carswell, 1982) indicate that most of $Ly\alpha$ and CIV is likely to
  come from small optically thin clouds emitting isotropically.

- The lack of correlation in Seyfert galaxies between the variations of
  $Ly\alpha$ and the variations of MgII $\lambda2800$ suggests that MgII, which is
  emitted by optically thick, partially ionized gas, does not come from
  the same clouds which emit the bulk of $Ly\alpha$ (in preparation).

- The extreme difficulty in reproducing in one type of clouds the
  observed line intensity ratios when the transfer is properly treated

(Collin-Souffrin, Dumont and Tully, 1980) is a strong argument in favor of two types of clouds.

These arguments lead to models of the broad line regions where most of Lyα and CIV is emitted by small optically thin clouds distributed more or less spherically around the central continuum source. As for the optically thick clouds emitting the Balmer lines and the FeII and MgII lines, Gordon, Collin-Souffrin and Dultin-Halcyan (1981) suggest that they could be mechanically heated and possibly form the outer part of an accretion disk, while Norman and Miley (1983) propose that the optically thick clouds could form the cocoon of a relativistic radio jet.

(v) The broad line clouds are so small (10^{13} cm in at least one dimension) that if they are not constrained by some external pressure their linear dimension doubles in a time much shorter than the crossing time of the broad line region. There must therefore exist an <u>intercloud medium</u> exercising a high thermal pressure, without causing a strong drag which would prevent the acceleration of the clouds at the observed high velocities. A mildly relativistic medium at $T > 10^8$ $^{\circ}$K would fulfil these requirements and Weymann et al. (1982) have recently discussed the properties of such a medium. Alternatively, the broad line clouds could be instabilities in a cold medium (M. Begelman, this Volume).

Finally, a review article should not end without a list of as yet unanswered questions:

- What is the nature of the IR/optical/UV continuum in Seyfert galaxies and quasars? Is there a difference between the non-thermal continuum of BL Lac objects and that of other objects, a difference which would be related to differences in the accretion regimes?

- Is there circulation of gas in and out of the central region with some gas being accreted and a fraction of it being ejected by mechanical or radiation pressure?

- What is the nature of the very dense gas clouds? Are they formed as such by evaporation from a disk or star disruption or are they cool condensations in a hot medium?

- Is the 30000 Å bump caused by black body radiation from an accretion disk, and how are the dense gas clouds related to this disk?

- What is the role of dust in changing the apparent shape of the spectrum in the infrared, optical and UV ranges, as well as in changing the line intensities?

- Why is there no dense gas observed in BL Lac objects?

REFERENCES

Angel, J.R.P. and Stockman, H.S., 1980, Ann. Rev. Astr. Ap., 8, 321.
Antonucci, R.R.J;, 1982, Nature 303, 158
Antonucci, R.R.J., 1983, preprint.
Bååth, L.B., Elgered, G., Lundqvist, G., Graham, D., Weiler, K.W.,
    Seielstad, G.A., Tallqvist, S. and Schilizzi, R.T. 1981, Astron.
    Astrophys. 96, 316.
Bailey, J., Hough, J.H. and Axon D.J., 1983, Monthly Not. Royal Astr.
    Soc., 203, 339.
Baldwin, J.A., 1977, Monthly Not. Royal Astr. Soc. 178, 678.
Boisson, C. and Ulrich, M.H., 1982, Third European IUE Conference, ESA
    SP-176, p. 549, Madrid, 10-13 May 1982.
Bregman, J.N., Lebofsky, M.J., Aller, M.F., Rieke, G.H., Aller, A.D.,
    Hodge, P.E., Glassgold, A.E. and Huggins, P.J., 1981, Nature 293,
    714.
Cohen, M.H. and Unwin, S.C., 1982, in "IAU Symposium 97, Extragalactic
    Radio Sources", D.S. Heeschen and C.M. Wade eds. (Dordrecht Reidel),
    p. 345.
Collin-Souffrin, S., Dumont, S. and Tully, J., 1982, Astron. Astrophys.
    106, 362.
Gordon, C., Collin-Souffrin, S. and Dultin-Hacyan, D., 1981, Astron.
    Astrophys. 103, 69.
Grandi, S.A., 1982, Ap.J;, 255, 25.
Impey, C.D., Brand, P.W.J.L. and Tapia, S., 1982, Monthly Not. Royal
    Astr. Soc., 198, 1.
Impey, C.D., Brand, P.W.J.L., Wolstencroft, R.D. and Williams, P.M.,
    1982, Monthly Not. Royal Astr. Soc. 200, 19.
Kellermann, K.I., Shaffer, D.B., Purcell, G.H., Pauliny-Toth, I.I.K.,
    Preuss, E., Witzel, A., Graham, D., Schilizzi, R.T., Cohen, M.H.,
    Moffet, A.T., Romney, J.D., and Niell, A.E., 1977, Ap.J. 211, 658.
Kwan, J. and Krolik, J.H., 1981, Ap.J. 250, 478
Linfield, R., 1981, Ap.J. 244, 436.
Lynden-Bell, D., 1969, Nature, 223, 690.
Malkan, M.A. and Sargent, W.K.W., 1982, Ap.J. 254, 22.
Maraschi, L., Perola, G.C. and Treves, A., 1980, Ap.J. 241, 910.
Maraschi, L., Tanzi, E.G. and Treves, A., 1983, preprint.
Martin, P.G., Stockman, H.S., Angel, J.R.P., Maza, J. and Beaver, E.A.
    1983, Ap.J.255, 65.
Moore, R.L., McGraw, J.T., Angel, J.R.P., Duerr, R., Lebofsky, M.J.,
    Rieke, G.H., Wisniewski, W.Z., Axon, D.J., Bailey, J., Hough, J.M.,
    Thompson, I., Breger, M., Schulz, H., Clayton, G.C., Martin, P.G.,
    Miller, J.S., Schmidt, G.D., Africano, J. and Miller, H.R., 1982,
    Ap.J. 260, 415.
Norman, C. and Miley G.K., 1983, preprint.
Pearson, T.J. and Readhead, A.C.S., 1981, Ap.J. 248, 61.
Pearson, T.J., Readhead, A.C.S. and Wilkinson, P.N., 1980, Ap.J. 236,
    714.
Perola, G.C., Boksenberg, A., Bromage, G.E., Clavel, J., Elvis, M.,
    Elvius, A., Gondhalekar, P.M., Lind, J., Lloyd, C., Penston, M.V.,
    Pettini, M., Snijders, M.A.J., Tanzi, E.G., Tarenghi, M., Ulrich,

M.H. and Warwick, R.S., 1982, Monthly Not.Royal Astr. Soc. 200, 293.
Puschell, J.J. and Stein, W.A., 1980, Ap.J. 237, 331.
Puschell, J.J., Jones, T.W., Phillips, A.C., Rudnick, L., Simpson, E.,
    Sitko, M., Stein, W.A. and Moneti, A., 1983, Ap.J. 265, 625.
Readhead, A.C.S., Pearson, T.J., Cohen, M.H., Ewing, M.S. and Moffet,
    A.T., 1979, Ap.J.. 231, 299.
Rieke, G.H., Grasdalen, G.L., Kinman, T.D., Hinzen, P., Wills, B.J. and
    Wills, D., 1976, Nature 260, 754.
Rudy, R.J., Schmidt, G.D., Stockman, H.S. and Moore, R.L., 1983, Ap.J.
    271, 59.
Schmidt, G.D. and Miller, J.S., 1980, Ap.J. 240, 759.
Stockman, H.S., Angel, J.R.P. and Miley, G.K., 1979, Ap.J.227, L55.
Ulrich, M.H., Boksenberg, A. Bromage, G., Carswell, R., Elvius, A.,
    Gabriel, A., Gondhalekar, P., Lind, J., Lindegren, L., Longair, M.S.,
    Penston, M.V., Perryman, M.A.C., Pettini, M., Perola, G.C., Rees,
    M.J., Sciama, D., Snijders, M.A.J., Tanzi, E., Tarenghi, M. and
    Wilson, R., 1980, Monthly Not. Royal Astr. Soc. 192, 561.
Ulrich, M.H., Boksenberg, A., Bromage, G.E., Clavel, J., Elvius, A.,
    Penston, M.V., Perola, G.C., Pettini, M., Snijders, M.A.J., Tanzi,
    E.G. and Tarenghi, M., 1983, Monthly Not. Royal Astr. Soc., in press.
Ulrich, M.H., Hackney, K.R.H., Hackney, R.L. and Kondo, Y., 1983, Ap.J.,
    15 January in press.
Ulrich, M.H., 1983, XI Texas Symposium on Relativistic Astrophysics,
    Austin, 12-17 December 1982, D.S. Evans ed. New York academy Press,
    New York.
Unwin, S.C., Cohen, M.H., Pearson, T.J., Seielstad, G.A., Simon, R.S.,
    Linfield, R.P. and Walker, R.C., 1982, preprint.
Weymann, R.J., Scott, J.S., Schiano, A.V.R. and Christiansen, W.A.,
    1982, Ap.J. 262, 497.
Wilkes, B.J. and Carswell, R.F., 1982, Monthly Not. Royal Astr. Soc.
    201, 645.

X-RAY PROPERTIES OF EXTRAGALACTIC COMPACT OBJECTS [+]

G.Zamorani

Istituto di Radioastronomia, C.N.R,

Bologna, Italy.

1. INTRODUCTION

The compact objects which are the subject of this talk are essentially quasars and Seyfert galaxies; I will only briefly mention a couple of results about BL Lac objects. Before describing the X-ray properties of these objects, it is useful to introduce a "working" definition of radio-quiet and radio-loud quasars: I will call radio-loud quasars all the objects which have been detected at radio frequencies and have a spectral index between radio (5 GHz) and optical frequencies (2500 A) greater than 0.35 (Zamorani et al. 1981); all the other objects will be considered radio-quiet. Note that this definition is independent of distance and is a function only of the relative importance of radio and optical emission.

Before discussing in detail the X-ray properties of radio-loud quasars (Section 3), and their correlations with radio characteristics (i.e. spectral index, morphology and radio luminosity), it is necessary to know which are the X-ray properties of radio-quiet quasars (Section 2). The need of this approach is clearly shown by two very interesting sets of observations due to Owen and collaborators. In the first one (Owen, Helfand and Spangler 1981), they observed with the Einstein Observatory a sample of 25 extragalactic sources selected on the basis of a high millimeter flux (S > 1 Jy at 90 GHz). From the high rate of X-ray detections and the small dispersion in the ratio of the millimeter to X-ray flux density they concluded that there is some physical relationship between the emission in the millimeter and X-ray regimes, at least for sources selected for their strong millimeter emission. In order to check about the possible extension of this correlation to other classes of quasars, Owen and Puschell (1982) observed at 90 GHz a sample of about 50 quasars which were already known to be X-ray sources (Zamorani et al. 1981). About 60% of these

[+] Discussion on page 428   85

R. Fanti et al. (eds.), VLBI and Compact Radio Sources, 85–94.

objects were not detected at 90 GHz, showing a significantly wider distribution of the ratio of millimeter to X-ray fluxes. The problem is then to understand which mechanism can give rise to a one-way correlation such that, while the X-ray flux can be predicted within a factor of two for a sample of millimeter-loud QSOs, the knowledge of the X-ray flux does not allow any prediction about the millimeter flux.

A qualitative explanation for these effects can be given if one accepts the following working hypothesis: The X-ray emission is due to at least two different mechanisms. The first one, present in all classes of quasars, has a broad α_{ox} distribution, peaked at about 1.45, which is the typical value observed in radio-quiet quasars. The second one, which connects millimeter and X-ray frequencies, has a narrow α_{mx} distribution around a typical value of about one and is likely to be Inverse Compton emission on radio photons, as suggested by Owen, Helfand and Spangler (1981). The relative importance of the two mechanisms is visualized in Fig. 1 a) and b) where the data for the two samples (millimeter selected and X-ray detected respectively) are plotted in the plane $\alpha_{mo} - \alpha_{ox}$. In this plane the loci of constant α_{mx} are straight lines parallel to each other; in both figures the straight line corresponds to the average value of α_{mx} found by Owen, Helfand and Spangler (1981). Looking at Fig.1, it is clear that, as far as $\alpha_{mo} > 0.75$, the X-ray emission is dominated by the second mechanism (for $\alpha_{mo} > 0.75$ and α_{mx} about one, the expected α_{ox} is < 1.45), and the correlation between millimeter and X-ray emission becomes evident. When $\alpha_{mo} < 0.75$, the α_{ox} expected on the basis of the second mechanism is >1.45, so that the correlation with millimeter emission disappears and we see the intrinsic dispersion in α_{ox} due to the first

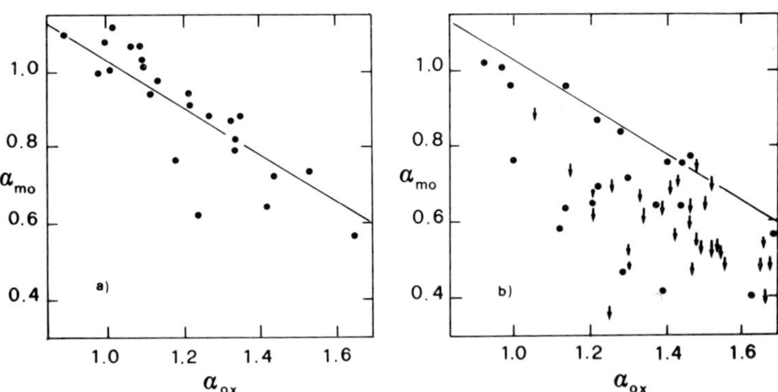

Fig. 1 Spectral index between millimeter and optical frequencies (α_{mo}) versus spectral index between optical and X-ray frequencies (α_{ox}) for: a) millimeter selected sources; b) X-ray detected quasars. The straight line corresponds to $\alpha_{mx} = 1.0$.

mechanism. From this example, it follows the necessity of studying first the correlation between optical and X-ray luminosities in radio-quiet quasars if one wants to understand the different properties which are seen in radio-loud quasars.

2. THE RADIO-QUIET QUASARS

During the lifetime of the Einstein Observatory a few hundred quasars, both optically and radio selected, have been observed. In this Section, I will give a summary of the X-ray properties of optically selected quasars. The characteristics of the samples which will be used (range of optical magnitudes, number of X-ray detections and X-ray upper limits) are given in Table 1.

T A B L E   1

X-ray Observations of Optically Selected Quasars.

Sample	m_b	Detections	Upper limits	X-ray References
Schmidt and Green	15-16	57	9	Tananbaum et al.(1983b)
Braccesi	18-20	16	19	Marshall et al.(1983)
Zamorani et al.	17-19	20	24	Zamorani et al.(1981)
Kriss et al.	14-16	11	0	Kriss et al.(1980)
Total	14-20	104	52	

The first two samples consist of X-ray observations of either a complete (Braccesi sample; see Marshall et al. 1983) or an unbiased subset of a complete optically selected sample (Schmidt and Green 1983); the other two samples consist of X-ray observations of "miscellaneous" optically selected quasars (Zamorani et al. 1981) and Seyfert 1 galaxies (Kriss, Canizares and Ricker 1980).

Figure 2 a) and b) show the monochromatic X-ray luminosity at 2 KeV versus the monochromatic optical luminosity at 2500 A for the X-ray detections and upper limits, respectively. The best fit regression of X-ray versus optical luminosity (drawn in both panels of Fig. 2) gives:

$$\log L_x = 4.95 + 0.71 \, (\pm 0.04) \, \log L_o \tag{1}$$

This fit has been obtained using the Maximum Likelihood Method of Avni et al.(1980) (see also Avni and Tananbaum 1982) which makes full use of all the available data, including the upper limits. The fundamental

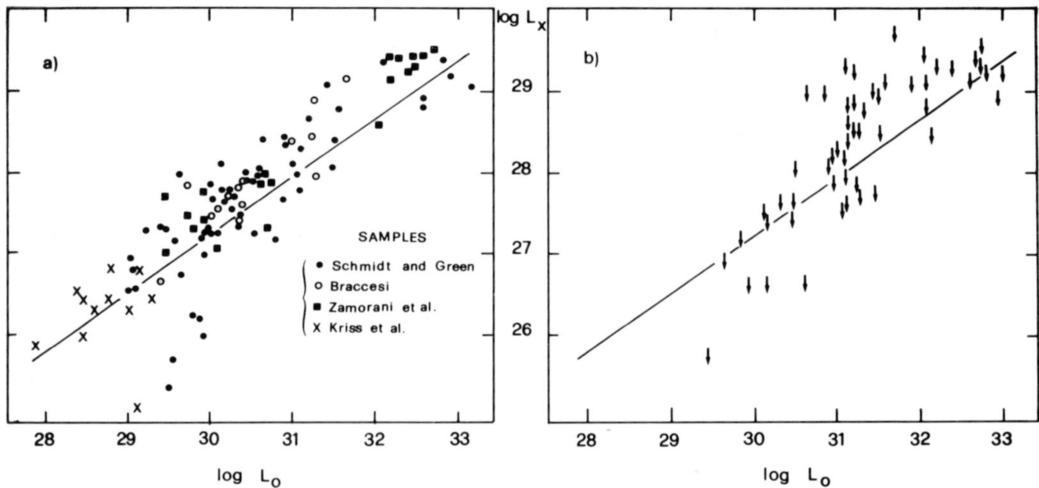

Fig. 2 Monochromatic X-ray luminosity at 2 Kev versus monochromatic optical luminosity at 2500 A for optically selected quasars and Seyfert galaxies. Panel a) shows the data for 104 detections; panel b) shows the data for 52 upper limits. In both panels the straight line represents the best fit regression of X-ray versus optical luminosity.

assumption which assures the applicability of this method is that all the objects in the sample (both the detections and the upper limits) are drawn from the same parent population with respect to the property which is being analyzed. Even if this assumption is not warranted by the existing data, the distribution of the upper limits (much more numerous at high optical luminosity) is such that the slope obtained with a fit which uses only the detections (0.77 $\pm$ 0.04) can be considered an overestimate of the real slope. This result confirms that in a sample of optically selected quasars the average ratio of X-ray to optical luminosity decreases with increasing optical luminosity (Zamorani et al. 1981, Avni and Tananbaum 1982).

From Fig. 2 three rather firm conclusions can be derived:
a) There is a definite correlation between X-ray and optical luminosities over a range of at least four orders of magnitude; the dispersion around the best fit line corresponds to a factor of about 3.2;
b) There is a continuity in the X-ray properties of Seyfert 1 galaxies and optically selected quasars;
c) There is no difference in the X-ray properties of quasars as a function of apparent magnitude in the range m_b = 16 - 20.

3. THE RADIO-LOUD QUASARS

It was shown by Zamorani et al. (1981) that radio-loud quasars are stronger X-ray emitters than radio-quiet quasars. Now, the increased amount of available data allows a more detailed study of the X-ray properties of radio-loud quasars dividing them into different classes. An obvious classification is in terms of the radio spectral index. We will call flat (steep) spectrum quasars the quasars with a high frequency radio spectrum flatter (steeper) than 0.5. This classification corresponds largely to a morphological separation between compact and extended radio sources. Using all the data in Zamorani et al. (1981), Ku, Helfand and Lucy (1980) and Giommi et al. (1983) we have 57 flat spectrum quasars (54 detections and 3 upper limits in X-ray) and 48 steep spectrum quasars (43 detections and 5 upper limits).

Fig. 3 shows the X-ray luminosity versus the optical and radio luminosities for both samples. In both cases there is a definite correlation between the X-ray and optical luminosities, but well above the correlation which was found for the sample of radio-quiet quasars (dashed straight line). For a given optical luminosity, the X-ray luminosity of a radio flat (steep) spectrum quasar is about four (two) times higher than that of a radio-quiet quasar. This difference

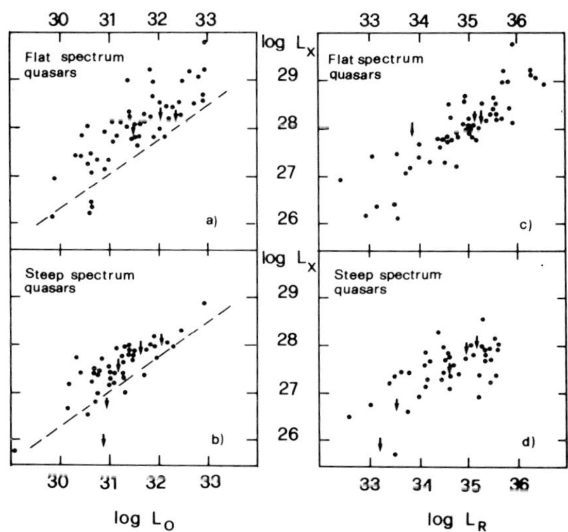

Fig. 3   X-ray luminosity versus optical and radio luminosity for radio loud quasars (flat and steep spectrum objects, respectively). The dashed line in panels a) and b) represents the best fit obtained for optically selected quasars (see Fig. 2).

suggests that the presence of radio emission is somehow connected with an enhanced X-ray emission. However, while for radio flat quasars the correlation between radio and X-ray luminosity appears to be significantly better and with smaller dispersion than the correlation between optical and X-ray luminosity, the opposite is true for radio steep quasars.

In order to study in more detail the correlations between X-ray, optical and radio luminosities we have performed a multi-variate, non-linear fit of the type:

$$L_x = a*L_o^{\alpha} + b*L_r^{\beta} \tag{2}$$

Physically, this representation is equivalent to assuming that the total X-ray emission from these objects is the sum of two mechanisms, related to the optical and radio emission, respectively. Note that this fit is different from that used by Tananbaum et al. (1983a) in studying the X-ray properties of the complete sample of 3CR quasars ($\log L_x = a*\log L_o + b*\log L_r + c$). Table 2 gives the values of the slopes (and relative errors) obtained for the various samples.

T A B L E   2

Slopes of various samples

Sample	α	β
Radio-Quiet	0.70 ± 0.05	
Radio-Loud Flat	0.75 ± 0.10	0.95 ± 0.15
Radio-Loud Steep	0.63 ± 0.10	0.75 ± 2.18

A few interesting points resulting from this analysis are:
 a) The slope α in the correlation between X-ray and optical luminosities is consistent with being the same for all the three samples;
 b) For both samples of radio-loud quasars the X-ray luminosity produced by the "optical mechanism" (i.e. the first term of Eq. 2) is about a factor of two higher than for radio-quiet quasars;
 c) There is a strong correlation between X-ray and radio luminosities for radio flat quasars, with a slope of the order of one;
 d) The correlation between X-ray and radio luminosities in radio steep quasars is, at best, weak: the slope β is consistent with zero. This implies that on the average the magnetic field in the extended lobes (which are the main contributors to the radio flux for these objects) is larger than about 0.05 times the equipartition field. Otherwise, a significant and detectable Inverse Compton X-ray emission would be expected from these regions.

As for flat spectrum radio-loud quasars, the direct proportionality between radio and X-ray luminosities (see Table 2) suggests a non-thermal origin for the X-ray emission. Moreover, the fact that the spectral index between the radio and X-ray frequencies is of the order of unity (Owen et al. 1981, Biermann et al. 1981 and this paper) is typical of a self-regulated Synchro-Compton mechanism (Pacini and Salvati 1978), in which approximately the same amount of energy is emitted at radio (Synchrotron) and X-ray (Compton) frequencies. Using the formulae given in Burbidge et al. (1974), the ratio of Compton to Synchrotron emission can be written as:

$$L_c/L_s = 1/(1-\alpha)i_\alpha^4 \, (\nu_b/10^{11})(\nu_n/\nu_b)^\alpha \, (T_n/10^{12})^5 \qquad (3)$$

where i_α is a function (of the order of unity) of the radio spectral index α, ν_n is the frequency at which the optical depth to synchrotron self-absorption is unity and ν_b is the high frequency cutoff. With typical values of α of the order of zero (flat spectrum sources) and ν_b of the order of 10^{11} Hz, our data imply the existence of an upper limit for the maximum brightness temperature (T_n) at about 10^{12} K. This is consistent with the fact that so far no measured brightness temperature exceeds this limit (Preuss 1981). Moreover, the observed dispersion in the ratio between the radio and X-ray fluxes would imply, if taken at the face value, an implausibly small dispersion in the value of the brightness temperatures ($\pm 30\%$). This is in contrast with existing VLBI observations. For example, Preston et al. (1983) report a wide distribution of brightness temperatures between 3×10^{10} and 1×10^{12} for a sample of about 100 sources observed at 2.3 GHz. This discrepancy can easily be explained if the contributions to the X-ray emission from the two parts of Eq. 2 (optical and radio luminosities) are of the same order of magnitude for an average object of our sample of radio flat quasars. As a consequence, it is difficult to separate the two mechanisms and almost impossible to sample low values of X-ray to radio fluxes. The best available sample of objects in which this separation can be done, at least in principle, is the sample of optically quiet, compact radio sources recently published by Ledden and O'Dell(1983). In fact, for these objects very little X-ray emission is expected as a contribution of the "optical mechanism". By comparing a sample of ten such sources with a sample of ten "control" sources, Ledden and O'Dell find that in these objects the X-ray luminosity is significantly weaker, relative to the radio, than in optically identified objects with comparable radio properties; at least four objects are still undetected in X-ray at values of X-ray to radio fluxes five times smaller than the average value observed in our sample. This result suggests that the real dispersion in the X-ray to radio luminosities is higher than what observed in our sample of radio flat quasars.

With this caveat in mind, we can still derive order of magnitude estimates of the physical parameters for the typical objects in our sample. Assuming a simple Synchro-Compton mechanism in a homogeneous and uniform spherical source, we obtain typical sizes of the order of $(10^{19} - 10^{20})$ cm, corresponding to angular sizes of $(0.3 - 3.0)$ milliarcsec at $z = 1$, and magnetic fields of the order of $(10^{-3} - 10^{-2})$ gauss. These values are in reasonable agreement with existing VLBI measurements (Kellermann and Pauliny-Toth 1981); similar values for the magnetic field and the size of the radio and X-ray emitting region have been recently derived by Bregman et al.(1983) from an analysis of multifrequency observations of the BL Lac object 0735+178.

Despite these apparently promising results, there are a few problems far from being understood within this extremely simple model:
1.) With the sizes and magnetic fields derived here and the typical overall spectrum of these objects (the optical emission lying well above the straight line connecting radio and X-ray frequencies), the energy density in the inner $10^{19} - 10^{20}$ cm would be dominated by the optical photons coming from the nucleus. If this is the case, the relativistic electrons would tend to lose their energy through Inverse Compton emission on the optical photons, instead of through Synchro-tron-Self Compton mechanism, and no obvious correlation between X-ray and radio emission should be expected. A possible way out is to assume that the relativistic electrons are accelerated far away from the nucleus, at distances of 10^{19} cm at least (see, for example, Protheroe and Kazanas 1983);
2.) Some radio-loud, flat spectrum sources show X-ray variability on timescale of a few days (Zamorani et al. 1983) or a few months (Schwartz, Madejski and Ku 1983, Zamorani et al. 1983) implying sizes of the emitting region smaller than those derived here. A possibility is that in these cases the observed variability is not due to the Synchro-Compton mechanism, but to the X-ray emission mechanism which is present also in radio-quiet quasars and is likely to originate in the inner regions of the accretion disk around the central black hole (Tucker 1983);
3.) After subtracting the contribution of the "radio mechanism", the normalization in the correlation between X-ray and optical luminosities for radio-loud quasars (both flat and steep spectrum) is still about a factor of two higher than in radio-quiet quasars. No obvious explanation for this difference is apparent. A possibility would be Inverse Compton emission of low energy relativistic electrons ($\gamma \sim 30$)

on optical photons. Also in this case sizes of the order of a few times 10^{19} cm are required in order not to give too high an X-ray emission. This effect would appear as a correlation of X-ray luminosity with optical luminosity (rather than radio), because these electrons would emit at very low frequencies (1-10 MHz) through synchrotron emission in the typical magnetic field.

4. CONCLUSION

From the analysis of a large sample of radio-quiet quasars we have shown that there is a well defined correlation between X-ray and optical luminosities over a range of at least four orders of magnitude, with a continuity in the X-ray properties from low luminosity Seyfert galaxies to the most luminous quasars. The dispersion around the best fit line corresponds to about a factor of three. This value has been obtained assuming that the upper limits (about one third of the total number of data points for radio-quiet quasars) have been drawn from the same parent population as the detections. In order to test this assumption, deeper X-ray observations of some of the objects which have not been detected with the EINSTEIN Observatory are clearly called for. The future German X-ray Observatory ROSAT (Trümper 1983) has all the required capabilities to make such observations.

As for radio-loud quasars, there appears to be a difference in the X-ray properties of objects with steep and flat radio spectrum. While quasars of both classes are stronger X-ray emitters than radio-quiet quasars with the same optical luminosity, only in flat spectrum quasars there is a clear correlation between X-ray and radio luminosities. This suggests that in these objects the observed "excess" X-ray luminosity is somehow related to the presence of compact radio emission. A simple Synchrotron-Self Compton model requires magnetic fields and sizes of the radio and X-ray emitting regions consistent with the existing data. Multifrequency (radio through X-ray) and VLBI simultaneous observations of a selected sample of "representative" objects would be of great value to clarify the physical processes in these sources.

REFERENCES

Avni,Y., Soltan,A., Tananbaum,H., and Zamorani,G.: 1980, Ap.J., 238, 800.
Avni,Y., and Tananbaum,H.: 1982, Ap.J.Letters, 262, L17.

Biermann,P., Duerbeck,H., Eckart,A., Fricke,K., Johnston,K.J., Kühr,H., Liebert,J., Pauliny-Toth,I.I.K., Schleicher,H., Stockman,H., Strittmatter, P.A., and Witzel,A.: 1981, Ap.J.Letters, 247, L53.

Bregman,J.N., Glassgold,A.E., Huggins,P.J., Aller,H.D., Aller,M.F., Hodge,P.E., Rieke,G.H., Lebofsky,M.J., Pollock,J.T., Pica,A.J., Leacock,R.J., Smith,A.G., Webb,J., Balonek,T.J., Dent,W.A., O'Dea, C.P., Ku,W.H.M., Schwartz,D.A., Miller,J.S., Rudy,R.J., and LeVan, P.D.: 1983, Ap.J., in press.

Burbidge,G.R., Jones,T.W., and O'Dell,S.L.: 1974, Ap.J., 193, 43.

Giommi,P. et al.: 1983, in preparation.

Kellermann,K.I., and Pauliny-Toth,I.I.K.: 1981, Ann. Rev. of Astron. Astrophys., Vol. 19, Pag. 373.

Ledden,J.E., and O'Dell,S.L.: 1983, Ap.J., 270, 434.

Marshall,H.L., Avni,Y., Braccesi,A., Huchra,J.P., Tananbaum,H., Zamorani,G., and Zitelli,V.: 1983, Ap.J., submitted.

Kriss,G.A., Canizares,C.R., and Ricker,G.R.: 1980, Ap.J., 242, 492.

Ku,W.H.M., Helfand,D.J., and Lucy,L.B.: 1980, Nature, 288, 323.

Owen,F.N., Helfand,D.J., and Spangler,S.R.: 1981, Ap.J.Letters, 250, L55.

Owen,F.N., and Puschell,J.J.: 1982, Astron.J., 87, 595.

Pacini,F., and Salvati,M.: 1978, Ap.J.Letters, 225, L99.

Preston,R.A., Morabito,D.D., and Jauncey,D.L.: 1983, Ap.J., 269, 387.

Preuss,E. : 1981, Proceedings of the Second ESO/ESA Workshop on "Optical Jets in Galaxies", Pag. 97.

Protheroe,R.J., and Kazanas,D.: 1983, Ap.J., 265, 620.

Schmidt,M., and Green,R.F.: 1983, Ap.J., 269, 352.

Schwartz,D.A., Madejski,G., and Ku,W.H.M.: 1983, Highlights of Astronomy, Vol. 6, Pag. 449.

Tananbaum,H., Wardle,J.F.C., Zamorani,G., and Avni,Y.: 1983a, Ap.J., 268, 60.

Tananbaum,H. et al.: 1983b, in preparation.

Trümper,J.: 1983, Physica Scripta, in press.

Tucker, W.: 1983, Ap.J., 271, 531.

Zamorani,G., Henry,J.P., Maccacaro,T., Tananbaum,H., Soltan,A., Avni, Y., Liebert,J., Stocke,J., Strittmatter,P.A., Weymann,R.J., Smith, M.G., Condon,J.J.: 1981, Ap.J., 245, 357.

Zamorani,G., Giommi,P., Maccacaro,T., and Tananbaum,H.: 1983, Ap.J., in press.

SUPERLUMINAL EFFECTS AND BULK RELATIVISTIC MOTION

M.H. Cohen and S.C. Unwin
Owens Valley Radio Observatory
California Institute of Technology

ABSTRACT

The observational status of superluminal radio sources is reviewed, and
the results for one well-studied source (3C 345) are discussed in detail.
We discuss the relativistic-beam model of superluminal sources, and
concentrate on one aspect, namely the geometric constraints provided by
combined X-ray and proper-motion measurements.

1. OBSERVATIONAL SUMMARY

A superluminal source is one which shows internal motions with an
apparent transverse velocity greater than the speed of light. There has
been a marked improvement in our knowledge of these sources since the
last VLBI symposium in Heidelberg, in 1978, and even since we wrote a
review in 1981 (ref. 1). Whereas five years ago one could still question
whether the motions were artifacts of analysis or of inadequate sampling,
today we have accurate tracks of several components in each of the
better-studied sources, and are beginning to get data on the evolution of
the moving components.

The eight known superluminal sources are listed in Table 1, with the
redshift z in column 3, proper motion relative to the core in column 5,
and apparent transverse velocity in column 6. We assume that the sources
are at cosmological distances, with $q_o = 0.05$ and $\underline{h} = H_o/100$, with $H_o =$
Hubble constant in km/sec/Mpc. References in column 7 are only to recent
articles; earlier references may be found in Ref. 1 and other reviews.
Each of the sources (except NRAO 140) is discussed in separate articles
in this volume.

Sequences of maps made from multi-station VLBI observations have
been shown for 3C 120 (Ref. 2), BL Lac (5), 3C 273 (8,25), and 3C 345
(10,13,16,17,18). These clearly show the characteristic core-jet
structure, with an optically-thick core and optically-thin jet components
separating superluminally from the core. 3C 120 is highly variable and

R. Fanti et al. (eds.), VLBI and Compact Radio Sources, 95–103.
© *1984 by the IAU.*

TABLE 1

Superluminal Radio Sources

Name	IAU	z	Component	μ (mas/yr)	v/c	References
3C 120	0430+052	0.033	A	1.35	2.1 h^{-1}	1,2,3,4
			B	2.57	4.1	
			C	2.34	3.7	
			D	1.65	2.6	
BL Lac	2200+420	0.070		0.6	2	5,6,7
3C 273	1226+023	0.158	C3	0.79	5.5	1,8
			C4	0.99	6.9	
			C5	0.79	5.5	
3C 279	1253-055	0.538		0.17	3.5	1,8
3C 345	1641+399	0.595	C2	0.42	9.5	1,9 - 15
			C3	0.31	7.0	
			C4*	0.17	3.9	16,17,18
				0.27	6.1	
4C 39.25	0923+392	0.698		(-0.1)**	(-2.6)**	1,19
3C 179	0723+679	0.846		0.14	4.2	1,20,21,22
NRAO 140	0333+321	1.258		0.13	5.4	1,23,24

Notes: * This component shows an acceleration.
**Tentative result; negative sign means contraction.

has the largest proper motions in Table 1 (consistent with the lowest
redshift) and a consistent picture has only been obtained in the last two
years, with frequent observations by Walker et al. (2). Four individual
components have been tracked in 3C 120, since 1979. Their proper motions
differ by a factor 2 (Table 1) but the errors on the slow components (A
and D) are perhaps large enough that all four have nearly the same speed.
3C 120 shows no evidence for acceleration, non-radial motion, or ejection
at different position angles; however, it is at $\delta = 5°$ and the N-S
resolution is very poor. 3C 345 ($\delta = 40°$) shows more complicated
behavior, as discussed below and in Ref. 16. In 3C 120, 3C 273, and
3C 345 the individual components decay with a time constant of one or two
years.

BL Lac is also very active and must be observed several times per
year to track the components. It is the only superluminal source for
which there is evidence of a "counterjet" (5). The other sources (except
4C 39.25) are all one-sided on a small scale (to a level of $\sim$ 20:1 of the
peak flux). 3C 273 is the only source which remains one-sided on a large
scale, when looked at with very high dynamic range (26,27). 4C 39.25 is
particularly interesting because in 1980 it began to contract, after a
decade of serving as a "standard" stationary simple double source (19).

NRAO 140 originally came to attention because of its low X-ray flux density (23), and the prediction of superluminal motion was confirmed by Marscher and Broderick (24). A similar prediction for 3C 147, however, has not been confirmed (28,29,44).

Bartel (ref. 9) has made the important measurement of the proper motion between 3C 345 and a nearby quasar, NRAO 512. They show that the core of 3C 345 is stationary with respect to NRAO 512, to a limit of 200 micro-arcsec in 9 years. This is an order of magnitude less than the internal proper motions in 3C 345 (Table 1).

Recent maps of 3C 279 (ref.8) show a new epoch of superluminal expansion at only one third of the rate found in 1971. The apparent velocity is now comparable with the other sources in Table 1. Other sources which are currently suspected of showing superluminal behavior include 0735+178, 2007+776, 2134+004 (ref. 30), 3C 454.3 (30), and 3C 446 (31).

2. BEAMING AND BULK RELATIVISTIC MOTION

Scheuer (32) discusses various models which can explain the superluminal effect, and we shall confine our remarks to the relativistic beam model. The beam lies at a small angle Θ to the line of sight, and the jet components are luminous "blobs" traveling with Lorentz factor γ along the beam. The core is stationary or moving very slowly (9); in one model it is the throat or region of the jet where the optical depth is unity (33,34). The Doppler factor δ and the apparent transverse velocity v/c are fixed by Θ and γ:

$$\delta = \gamma^{-1} (1 - \beta \cos \Theta)^{-1} \text{ and } v/c = \beta \sin \Theta (1 - \beta \cos \Theta)^{-1} \qquad [1]$$

where βc is the actual bulk velocity. Conversely, if δ and v/c can be measured, Θ and γ can be determined. Figure 1 shows a convenient graphical display of these quantities (13). VLBI gives v/c, and X-ray fluxes yield limits to δ; thus the kinematic parameters Θ and γ can be determined for the moving components. These parameters are derived below for 3C 345. If this procedure could be followed accurately for many sources, then in principle the Hubble constant could be determined (35).

Three independent types of observations suggest bulk relativistic motion in quasars and active galactic nuclei. Taken together, they provide strong evidence for bulk relativistic motion, and thus that the beam model or a variant thereof is most likely the correct explanation for the superluminal effect. A further virtue of these beams is that they lead fairly naturally into the jets seen on a kpc scale, which also are needed to power hot spots and the outer lobes of radio sources. However, the question of whether the extended jets are relativistic is by no means settled.

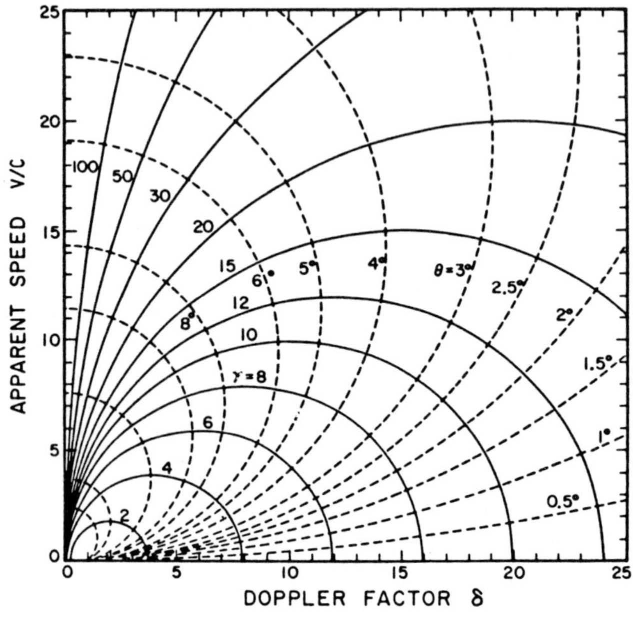

Figure 1

We now sketch these three lines of evidence, but omit discussion of beam collimation, the primary source of energy, and statistics.

a) Superluminal Motion

The measured quantities are proper motion μ and redshift z, which is converted to distance with a standard cosmological model and an assumed value of the Hubble constant. These give v/c, which in turn yields an upper limit to Θ, namely $\Theta_{max} = 2 \cot^{-1} (v/c)$, and a lower limit to γ, $\gamma_{min} = \sqrt{1 + (v/c)^2}$].   The Doppler factor δ is not fixed by v/c alone, but if $\gamma = \gamma_{min}$, then $\sin \Theta = 1/\gamma$ and $\delta = \gamma$ (see Fig. 1). From Table 1, values of γ_{min} range from 2 to 20 for $\underline{h}$ between 0.5 and 1.

b) Inverse-Compton X-rays

Standard synchrotron theory predicts a definite ratio between the X-ray and radio fluxes from a cloud of high-energy particles, with the X-rays arising in the Inverse-Compton process. Gould (36) has made the calculation for a series of spherical models, and the application to determine the Doppler factor of radio sources was stressed by Burbidge, Jones, and O'Dell (37) and more recently by Marscher and Broderick (23). The principal result is that very weak X-rays imply bulk relativistic motion with $\delta > 1$.

X-rays can also be produced by other processes: e.g. thermal bremsstrahlung, so that the measured X-ray flux density gives a lower limit to δ via the self-Compton process. The formula for δ_{min} may be written (13,38):

$$\delta_{min} = a(\alpha)\,(1+z)\,F_m \nu_m^{-p}\,\phi^{-q}\,S_x^{-r} \qquad [2]$$

where F_m is the peak radio flux density (Jy) at the turnover frequency ν_m (GHz), ϕ is the angular diameter (in milli-arcseconds, or mas) of the equivalent spherical radio source, and S_x is the X-ray flux density in units of $[10^{-11}(1+z)^{\alpha-1}]$ erg/cm^2/sec, in the band $(0.5 - 4.5)/(1+z)$ keV. The coefficient $a(\alpha)$ is of order unity, and the exponents p, q, and r are tabulated in ref. (13); for spectral index $\alpha = 0.75$, $a \simeq 0.65$, $p = 1.32$, $q = 1.64$, and $r = 0.18$.

The following points may be made about Eqn. [2]. (a) The calculation is based on the ratio of X-ray to radio brightness and is distance-independent, if the angular size is measured directly with VLBI. (b) r is small, so the calculation of δ is insensitive to S_x. (c) The calculation is most sensitive to ϕ, since typically $q \gtrsim 1.5$. (d) The spectral turnover (F_m, ν_m), must be known before δ_{min} can be calculated. If only an upper limit is known for ν_m, then a limit can still be calculated for δ but it is weak because $p > 1$. (e) The diameter ϕ is very difficult to determine, even with VLBI, and is usually the cause of most uncertainty in Eqn. [2].

If a source consists of several discrete homogeneous components then each has its own values of F_m, ν_m, ϕ, and δ_{min} (see the discussion below on 3C 345). If a component is inhomogeneous Eqn. [2] does not apply in a strict sense, but it should give a reasonable answer if $\phi = \phi_m$ is the effective diameter near the turnover frequency (38).

In summary, measurements of the radio synchrotron spectrum, angular diameter, and X-ray flux density give δ_{min}, a lower limit on the Doppler factor. This immediately gives an upper limit on Θ, $\sin \Theta'_{max} = 1/\delta_{min}$, and a lower limit on γ, $\gamma'_{min} = (\delta_{min} + 1/\delta_{min})/2$, which occurs at $\Theta = 0°$ (Fig. 1). If Θ takes its largest value (Θ'_{max}), then $\gamma = \delta_{min}$. Tighter limits can be set on both γ and Θ if v/c is also measured (Section 3 below).

c) <u>Variability</u>

Compact radio sources often show variability with a time scale $\tau = S\,|dS/dt|^{-1}$, where S is the flux density of the variable component. Velocity-of-light arguments suggest that the radius is less than $c\tau^*$, where $\tau^* = \tau\,\delta/(1+z)$ is the time scale in a co-moving frame. If the distance is known this gives ϕ_v, a limit to angular diameter which is often useful, although less so than a direct VLBI measurement. The apparent brightness temperature T_o, based on τ rather than τ^*, can be as high as 10^{15} K for the low-frequency variables (39), in apparent violation of the self-Compton limit $\simeq 10^{12}$ K. However, the limit applies only to the co-moving brightness $T_b^* \sim T_o\, \delta^{-3}$. Thus the lower limit to δ from low-frequency variability is about 10, in rough agreement with the largest values of δ_{min} deduced from the X-ray observations.

If there also are X-ray observations then ϕ_v can be used in Eqn. [2] to estimate δ_{min}. However, this must be done with caution as the variability diameter appropriate to the turnover frequency should be used. In particular, it is incorrect to use the X-ray variability time scale, since the rapidly variable X-rays presumably come from a region which is much smaller than the optically-thick radio components (38). Simon et al. (28) used radio variability data and X-ray flux data for 3C 147 to deduce $\delta_{min} \simeq 4$ (for $\underline{h} = 1$).

The need to assume $T_b^* < 10^{12}$ K can be dropped if the diameter of a variable component can be measured with VLBI, or with interstellar scintillations (ISS). In this case one can set $\phi = \phi_v$ and directly solve for δ_{min}. Unwin et al. (13) showed that in 3C 345 the VLBI, X-ray, and variability data were all self-consistent with $\delta_{min} \simeq 10$. Dennison and Condon (40) have discussed the lack of ISS and showed that bulk relativistic motion is most likely responsible for the high brightness temperatures.

The superluminal motions, weak X-rays, and variability combine to provide powerful evidence for bulk relativistic motion with values of γ and δ up to 10. The strongest case comes from NRAO 140 and 3C 345, where VLBI measurements of angular size and spectrum, and X-ray flux measurements, give estimates of $\delta_{min} \simeq 10$, and $\gamma_{min} \simeq 5\ \underline{h}^{-1}$ (see next Section.) The proper motion, variability measurements, and X-ray flux are all self-consistent if $\underline{h} \simeq 1$. Bulk relativistic motions cannot be dismissed by assuming non-cosmological distances, or gravitational lenses (41). The X-ray argument depends only on the (invariant) ratio of X-ray to radio brightness.

3.   OBSERVATIONS OF 3C 345

We now describe in more detail the state of our understanding of one particular superluminal source. 3C 345 is probably the best-studied of the superluminals, and appears to be prototypical, in that it exhibits most of the common properties of the group. The following is a summary of the results reported in refs. 9-18.

Hybrid maps of 3C 345 made at intervals of about 6 months show a "core-jet" structure at every epoch, with a bright "core" at one end of the source, and a "lumpy" jet extending ~ 5 mas in PA $\simeq -80°$. No "counter-jet" has been seen, although the limits are rather weak ($\lesssim 2\%$ of the core, or $\lesssim 10\%$ of the jet brightness). Superluminal motion at $v/c \simeq 8\ \underline{h}^{-1}$ is seen in two "knots" in the jet at 5 and 10 GHz and in a third close-in component seen at 10 and 22 GHz (Table 1). The two outer components move with the same speed (to within the errors), and show no acceleration; they are very compact ($\phi \lesssim 0.5$ mas) and do not appear to expand. As they separate from the core, they decay with half-lives of about 2 years, and their spectra steepen to $\alpha \gtrsim 1$. This is rather steeper than the mean spectral index of the 2-arcsecond jet seen by MERLIN and the VLA (26,42). Presumably, particle reacceleration must occur in the large-scale jet.

The new jet component (C4 in Table 1) shows acceleration and does not move radially from the core. The data (ref. 16) are roughly consistent with constant acceleration at 3 c $\underline{h}^{-1}$ yr^{-1}. The track is consistent with straight-line motion which does not intersect the core. or with a curved trajectory starting at the core. In the latter case projection can greatly amplify the intrinsic curvature, which could be as small as $2°$ (ref. 16). The tracks of components C2 and C3 lie at different position angles, and the outer one (C2) has the same PA as reported for the "double" in 1971-74. This has given rise to the suggestion that the motion is rectilinear (i.e. ballistic), at least beyond a radius of 1 mas (ref. 12).

Observations at 2.3 GHz of 3C 345 and 3C 273 (43) show that the cores of both sources are strongly self-absorbed at this frequency and below. Combining fluxes from maps made in the frequency range 2.3 - 22 GHz enables the spectral turnovers (F_m, ν_m) to be made for the core and moving components. The X-rays from 3C 345 are very weak for a source with such a high radio brightness temperature. When the X-rays are combined with VLBI determinations of spectrum and diameter, Eqn. [2] yields $\delta_{min} \gg 1$. Estimates from ref. 13 for v/c, δ_{min}, Θ_{max}, and γ_{min} are shown in Table 2. The core (D) is assumed to be stationary (ref. 9). An adequate spectrum could not be determined for the outer jet component (C2), so it has no limit for δ.

Component D in Table 2 has values of Θ_{max} and γ_{min} which are fixed geometrically by δ_{min} (see Fig. 1). For C2 they are fixed by v/c and thus involve $\underline{h}$. For C3 the limits are stronger because both v/c and δ_{min} are known. It seems likely that δ is near δ_{min} because this gives the largest allowed range of Θ and thus the most solid angle in which the jet direction must lie. If we take $\delta = \delta_{min}$ then the largest range of solid angle is obtained for $\delta =$ v/c. Thus $\underline{h} \simeq 1$ is consistent both with the

TABLE 2

Geometric Constraints for 3C 345

		D	C_3	C_2
v/c		-	7 h^{-1}	9.5 h^{-1}
δ_{min}		18	8	-
Θ_{max}	(deg)	3	7 h	12 h
γ_{min}		9	7 h^{-1}	9.5 h^{-1}

solid angle argument, and with the observed values of proper motion, redshift, radio spectrum, diameter and X-ray flux density.

The X-ray limit applies to the core as well as to the moving components, and $\delta_{min} \simeq 18$ for component D. Thus the radiating material in the core is moving relativistically towards us, even though the core is assumed to be stationary. This is in accord with the beam model in which the core is a stationary region where the diverging relativistic beam has optical depth unity (33). We note again that this conclusion depends on brightness ratios and is independent of distance or lensing effects.

VLBI Research at the Owens Valley Radio Observatory is supported by NSF grant AST 82-10259.

REFERENCES

1. Cohen, M.H., and Unwin, S.C.: 1982, "Extragalactic Radio Sources", IAU Symp. No. 97, pp. 345-354.
2. Walker, R.C., Benson, J.M., Seielstad, G.A., and Unwin, S.C.: 1984, IAU Symp. No. 110, this volume.
3. Benson, J.M.: 1984, IAU Symp. 110, this volume.
4. Walker, R.C., Seielstad, G.A., Simon, R.S., Unwin, S.C., Cohen, M.H., Pearson, T.J., and Linfield, R.P.: 1982, Astrophys. J. 257, pp. 56-62.
5. Mutel, R.L.: 1984, IAU Symp. 110, this volume.
6. Aller, H.D.: 1984, IAU Symp. 110, this volume.
7. Mutel, R.L., and Phillips, R.B.: 1982, "Extragalactic Radio Sources", IAU Symp. No. 97, pp. 385-386.
8. Unwin, S.C., and Biretta, J.A.: 1984, IAU Symp. 110, this volume.
9. Bartel, N.: 1984, IAU Symp. 110, this volume.
10. Cohen, M.H., Unwin, S.C., Simon, R.S., Seielstad, G.A., Pearson, T.J., Linfield, R.P., and Walker, R.C.: 1981, Astrophys. J. 247, pp. 774-779.
11. Readhead, A.C.S., Hough, D.H., Ewing, M.S., Walker, R.C., and Romney, J.D.: 1983, Astrophys. J. 265, pp. 107-131.
12. Cohen, M.H., Unwin, S.C., Pearson, T.J., Seielstad, G.A., Simon, R.S., Linfield, R.P., and Walker, R.C.: 1983, Astrophys. J. Letters 269, pp. L1-L4.
13. Unwin, S.C., Cohen, M.H., Pearson, T.J., Seielstad, G.A., Simon, R.S., Linfield, R.P., and Walker, R.C.: 1983, Astrophys. J. 271,
14. Schraml, J., Pauliny-Toth, I.I.K., Witzel, A., Kellermann, K.I., Johnston, K.J., and Spencer, J.H.: 1981, Astrophys. J. Letters 251, pp. L57-L60.
15. Spencer, J.H., Johnston, K.J., Pauliny-Toth, I.I.K., and Witzel, A.: 1981, Astrophys. J. Letters 251, pp. L61-L63.
16. Moore, R.L., Biretta, J.A., Readhead, A.C.S., and Bååth, L.B.: 1984, IAU Symp. No. 110, this volume.
17. Biretta, J.A., Cohen, M.H., Unwin, S.C., and Pauliny-Toth, I.I.K.:

    1983, Nature (in press).
18. Moore, R.L., Readhead, A.C.S., and Bååth, L.B.: 1983, Nature (in press).
19. Shaffer, D.B.: 1984, IAU Symp. 110, this volume.
20. Porcas, R.W.: 1984, IAU Symp. 110, this volume.
21. Porcas, R.W.: 1981, Nature 294, pp. 47-49.
22. Porcas, R.W.: 1983, Proceedings Bangalore Winter School on Energetic Extragalactic Sources (in press).
23. Marscher, A.P., and Broderick, J.J.: 1981 Astrophys. J., 249, pp. 406-414.
24. Marscher, A.P., and Broderick, J.J.: 1982 Astrophys. J. Letters, 255, pp. L11-L15.
25. Pearson, T.J., et al.: 1981, Nature 290, pp. 365-368.
26. Browne, I.W.A., Clark, R.R., Moore, P.K., Muxlow, T.W.B., Wilkinson, P.N., Cohen, M.H., and Porcas, R.W.: 1982, Nature 299, pp. 788-793.
27. Schilizzi, R.T., and de Bruyn, A.G.: 1983, Nature 303, pp. 26-31.
28. Simon, R.S., Readhead, A.C.S., Moffet, A.T., Wilkinson, P.N., Allen, B., and Burke, B.F.: 1983, Nature 302, pp. 487-490.
29. Whyborn, N.: 1984, IAU Symp. No. 110, this volume.
30. Pauliny-Toth, I.I.K.: 1984, IAU Symp. 110, this volume.
31. Brown, R.L., Johnston, K.J., Briggs, F.H., Wolfe, A.M., Neff, S.G., and Walker, R.C.: 1981, Astrophys. Letters 21, pp. 105-110.
32. Scheuer, P.A.G.: 1984, IAU Symp. 110, this volume.
33. Blandford, R.D., and Königl, A.: 1979, Astrophys. J. 232, pp. 34-48.
34. Scheuer, P.A.G., and Readhead, A.C.S.: 1979, Nature 277, pp. 182-185.
35. Marscher, A.P., and Broderick, J.J.: 1982, "Extragalactic Radio Sources", IAU Symp. No. 97, pp. 359-360.
36. Gould, R.J.: 1979, Astron. Astrophys. 76, pp. 306-311.
37. Burbidge, G.R., Jones, T.W., and O'Dell, S.L.: 1974, Astrophys. J. 193, pp. 43-54.
38. Cohen, M.H.: 1983, Proceedings Bangalore Winter School on Energetic Extragalactic Sources (in press).
39. Fanti, R., Padrielli, L., and Salvati, M.: 1982, "Extragalactic Radio Sources", IAU Symp. No. 97, pp. 317-324.
40. Dennison, B., and Condon, J.J.: 1981, Astrophys. J. 246, pp. 91-99.
41. Subramanian, K.: 1984, IAU Symp. 110, this volume.
42. Perley, R.A., and Johnston, K.J.: 1979, Astron. J. 84, pp. 1247-1252.
43. Cohen, M.H., et al.: 1983, Astrophys. J. 272 (in press).
44. Simon, R.S.: 1984, IAU Symp. No. 110, this volume.

VLBI OBSERVATIONS OF THE SUPERLUMINAL SOURCES 3C 273 AND 3C 279 [+]

S.C. Unwin and J.A. Biretta
Owens Valley Radio Observatory
California Institute of Technology

Monitoring of the radio structure in the core of 3C 273 has yielded a series of hybrid maps in which substantial source evolution can be seen. We discuss the superluminal motion of components in the jet of 3C 273, and the evolution of their flux densities and spectra. For 3C 279 we identify a new epoch of superluminal expansion with $v/c = 6.5$.

<u>3C 273</u>. We have observed the quasar 3C 273 at intervals of about 6 months, alternating between 5.0 and 10.7 GHz, using a VLBI network of at least 5 telescopes. Since 1979 the Effelsberg telescope has been included, effectively doubling the angular resolution available. Figure 1(a) shows the sequence of maps obtained at 5.0 GHz. The 1981.26 map was made from more extensive observations than the others and is of higher quality. A weak extended outer component was detected on this map (C_2 in the convention defined on the 2.3-GHz map of Cohen <u>et al</u>. 1983). Most striking is the 'core-jet' morphology, which is seen in all our maps of 3C 273, and by Pearson <u>et al</u>. (1981) at 10.7 GHz. The superluminal motion of C_3 away from the core (D), which can be seen in Figure 1(a), is a continuation of that seen by Pearson <u>et al</u>. Note the bending of the jet: the source axis curves through $\gtrsim 30^o$, but a line joining C_3 and C_2 lies within about 1° of the direction of the optical jet. Thus the bend occupies only $\sim 0.05\,\%$ of the total jet length.

Figure 1(b) shows the component motions as measured from the complete sequence of 5.0 and 10.7-GHz maps. C_3 shows no evidence for acceleration or frequency-dependence in its motion, and has an apparent speed $v = 10.0 \pm 0.3\ c$ ($H_o = 55$ km/s/Mpc, $q_o = 0.05$). C_4 has the same speed, to within the errors, and there is evidence in these and more recent data for motion with a similar speed in the knot C_5. Because of poor NS resolution, no change in PA has been seen for an individual knot; such changes should become measurable if components follow the jet curvature.

[+] <u>Discussion on page 430</u>

R. Fanti et al. (eds.), VLBI and Compact Radio Sources, 105–107.
© *1984 by the IAU.*

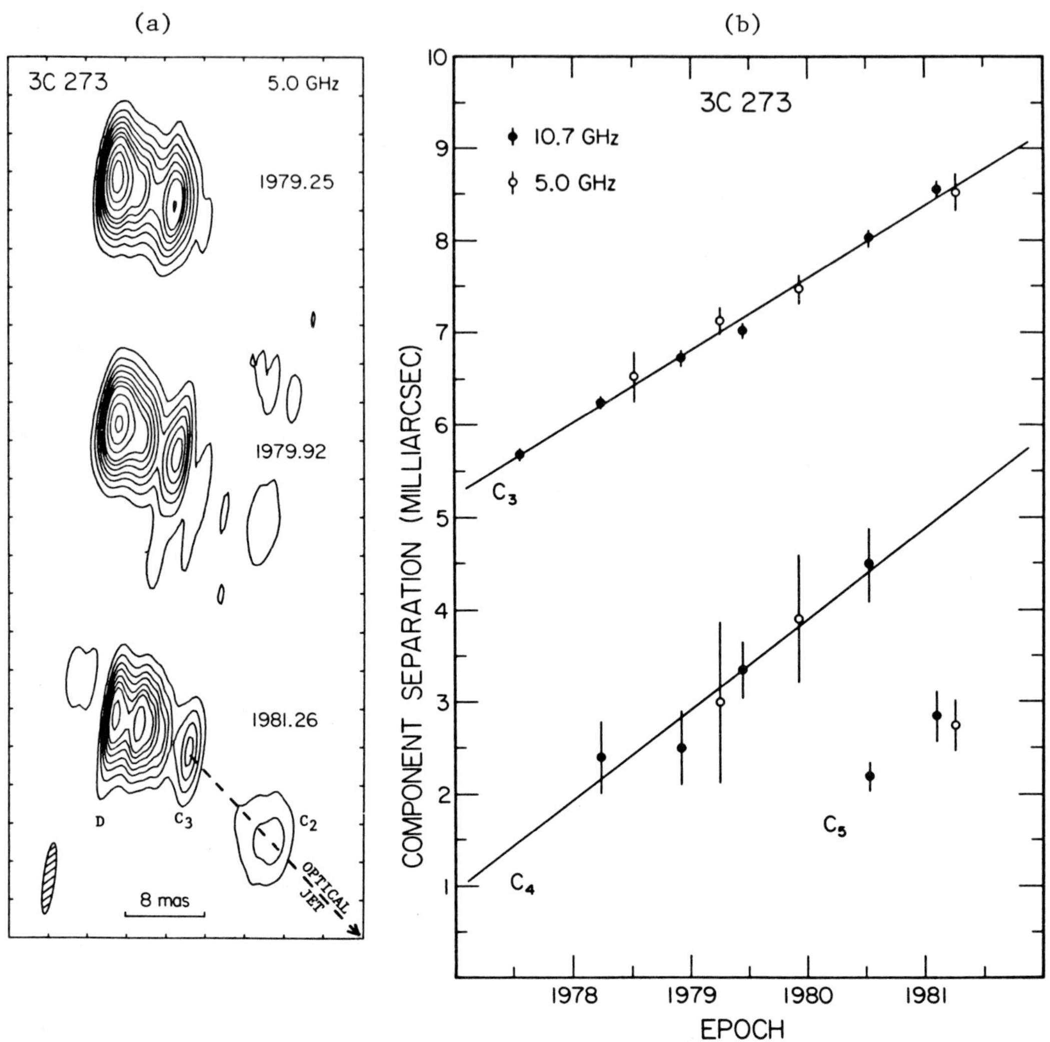

Figure 1. (<u>a</u>) Hybrid maps of 3C 273 at 5.0 GHz from three epochs. The
7.5 x 1.15 mas beam is shown as a shaded ellipse. Contours at -0.3,
0.3, 0.6, 0.9, 1.2, 1.5, 2, 2.5, 3, 4, 5, and 6 Jy/beam. (<u>b</u>) Component
separations from the core of 3C 273, measured from the 5.0 and 10.7 GHz
hybrid maps. Straight lines show linear least-squares fits to the
motion of components C_3 and C_4.

    The basic evolution seen in our data is one of ´birth´ of new
components, their motion away from the core, and decay; such behaviour
is expected if the core is the ´central engine´ which eventually
supplies energy to the large-scale structure via a jet. We find that

the knot flux densities decay with half-lives of 1-2 years, but with no obvious change in spectra (Unwin et al. 1983 found a slight spectral steepening of the knots in the jet of 3C 345). Several of the knots appear resolved, especially at 10.7 GHz, but we are not able to measure the presumed expansion directly from the maps.

3C 279. This low-declination ($\delta = -5^{\circ}$) quasar has had relatively few VLBI observations since its discovery as the first superluminal source (Whitney et al. 1971). We now have four hybrid maps made since 1981. The core is very compact, and only slightly resolved on the 5.0-GHz map. The two 10.7-GHz maps and the 22.2-GHz map all show the source as a close double, with separation $\approx$ 1.2 mas in PA $\approx 225^{\circ}$, and with roughly comparable component flux densities. Between 1981.1 and 1983.1 the separation increased at 0.17 mas/yr, (corresponding to v = 6.5 c), only one third of the rate found by Cotton et al. (1979) for epoch 1971. The spectral index of the NE component is flatter than that of the SW component, which is the expected sense for a core-jet picture of the VLBI structure: the arcsec-scale jet seen with the VLA (de Pater and Perley 1983) is in PA 206°, about 20° different from the VLBI structure.

Many people have helped in collecting and reducing the data discussed here, including M.H. Cohen, K.R. Lind, R.P. Linfield, T.J. Pearson, A.C.S. Readhead, G.A. Seielstad, R.S. Simon, and R.C. Walker. VLBI research at OVRO is supported by NSF grant AST 82-10259.

REFERENCES

Cohen, M.H., et al.: 1983, Astrophys. J. 272 (in press).
Cotton, W.D., et al.: 1979, Astrophys. J. 229, pp. L115-L117.
de Pater, I., and Perley, R.A.: 1983, Astrophys. J. (in press).
Pearson, T.J., Unwin, S.C., Cohen, M.H., Linfield, R.P., Readhead, A.C.S., Seielstad, G.A., Simon, R.S., and Walker, R.C.: 1981, Nature 290, pp. 365-368.
Unwin, S.C., Cohen, M.H., Pearson, T.J., Seielstad, G.A., Simon, R.S., Linfield, R.P., and Walker, R.C.: 1983, Astrophys. J. 271 (in press).
Whitney, A.R., et al.: 1971, Science 73, pp. 225-230.

SUPERLUMINAL ACCELERATION OF THE NEW COMPONENT IN 3C 345 [+]

R. L. Moore, J. A. Biretta, A. C. S. Readhead
California Institute of Technology, Pasadena, CA, USA

L. Bååth
Onsala Space Observatory, Onsala, Sweden

VLBI observations of 3C 345 at 10.8 GHz and 22.2 GHz show that the
position angle of the new component is increasing as it separates from
the core. Also, the apparent velocity of the component is increasing.
This is the first clear evidence for non-radial motion and acceleration
of an individual component in an extragalactic radio source.

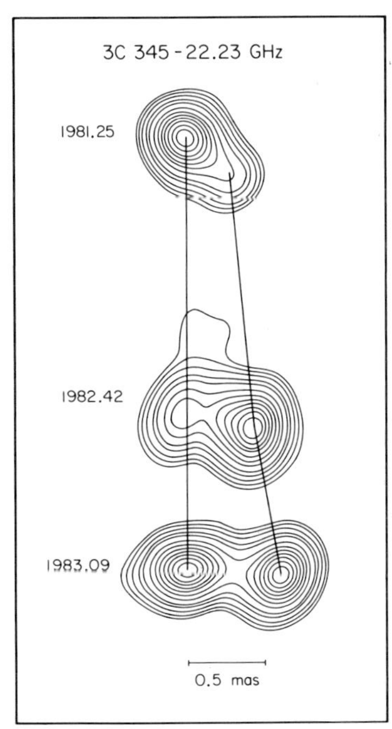

Figure 1

The superluminal quasar 3C 345 has a curved, one-sided, jet-like radio structure (refs. 1, 2). VLBI components have been observed moving at apparent speeds of 13c to 17c (ref. 3). An extension of the core has been observed in 1977 and 1978 at 22.2 GHz in position angle -135° (ref. 4). A well-defined component was clearly detected in early 1981 at a similar position angle and a radial separation of ~0.3 mas (ref. 5). We report here further observations of this component.

During the period 1981.1 to 1983.1, we have obtained three epochs of VLBI observations at 22.2 GHz (Figure 1) and four epochs at 10.8 GHz. The observations are described in detail elsewhere (refs. 5, 6, 7, 8). The radial separations and position angles of the western component relative to the core have been derived from model-fitting (Table 1). It is clear that the position angle of the western component has rotated ~35° as its radial separation has increased from 0.3 to 0.6 mas. The radial separation at 22.2 GHz appears to be larger by ~0.06 mas than at 10.8 GHz; this effect may be due to opacity gradients in the core or western component.

+ Discussion on page 430

R. Fanti et al. (eds.), VLBI and Compact Radio Sources, 109–110.
© *1984 by the IAU.*

TABLE 1 - Relative Positions and Velocities of Western Component

Freq. (GHz)	Epoch	r (mas)*	θ*	β_{app}*
22.23	1981.25	0.37 ± .01	-128° ± 5°	6.8 ± 1.2
22.23	1982.42	0.45 .02	-103° 2°	11.2 2.4
22.23	1983.09	0.61 .02	-93° 2°	
10.86	1981.10	0.32 ± .04	-132° ± 6°	4.1 ± 1.6
10.86	1982.10	0.35 .02	-115° 2°	9.9 1.6
10.86	1982.86	0.49 .06	-97° 6°	
10.86	1983.10	0.53 .02	-94° 4°	

* Errors represent ±3σ.

Apparent transverse velocities (relative to c) derived from the proper motion of the western component are also tabulated. (H_o = 55 km s^{-1} Mpc^{-1}, q_o = 0.05). The apparent velocity definitely increases with separation from the core. This result is significant at the 5σ level for both frequencies.

The observed non-radial trajectory of the component can be interpreted as a straight line which does not intersect the core, or as a curved path which originates in the core. The first case might occur if there were more than one supermassive body in 3C 345 (ref. 9). This explanation appears contrived because of the near equality of flux densities of the strongly-beamed core and western component.

In the latter explanation, projection effects will strongly amplify intrinsic curvature of the path. We have reproduced the observed trajectory with a simple curved path in a plane with an intrinsic curvature of <2° (ref. 8). However, we note that for such a path, the change in angle to the line-of-sight is too small to account for the large observed increase in the apparent velocity; intrinsic acceleration would be necessary. In order to explain both the trajectory and acceleration, a model with additional degrees of freedom will probably be required (e.g. non-radial shocks in a broad ejection cone).

The work at Caltech was supported by NSF grant 82-10259.

REFERENCES
1 Cohen, M. H. et al.: 1977, Nature, 268, pp 405-409.
2 Readhead, A. C. S. et al.: 1978, Nature, 276, pp. 768-771.
3 Unwin, S. C. et al.: 1983, Ap. J., in press.
4 Bååth, L. B. et al.: 1981, Ap. J. (Letters), 243, pp. L123-126.
5 Readhead, A. C. S. et al.: 1983, Ap. J., 265, pp. 107-131.
6 Bååth, L. et al.: 1983, in preparation.
7 Biretta, J. A. et al.: 1983, Nature, submitted.
8 Moore, R. L. et al.: 1983, Nature, submitted.
9 Begelman, M. C. et al.: 1980, Nature, 287, pp. 307-309.

A 10 Station VLBI Map of 3C147 at 1661 MHz

R. S. Simon[1,4], A. C. S. Readhead[2], and P. N. Wilkinson[3]
[1]E. O. Hulburt Center for Space Research, Naval Research
Laboratory, Washington D. C. 20375
[2]Owens Valley Radio Observatory, California Institute of
Technology, Pasadena, California 91125
[3]Nuffield Radio Astronomy Laboratories, Jodrell Bank,
Macclesfield, Cheshire, SK11 9DL United Kingdom
[4]NRC/NRL Cooperative Research Associate

ABSTRACT

The quasar 3C147 was observed in October 1982 with a ten-station
VLBI interferometer in an effort to search for possible superluminal
motion (Simon et al. 1983) in the core. Previous observations, in 1976,
had revealed that the core consisted of two compact components embedded
in a larger halo. The new high quality map reveals that this double
structure has disappeared and it is not possible to interpret the structural changes unambiguously in terms of an expansion or contraction.

DISCUSSION

We have made a 10-station VLBI map of 3C147 at 1661 MHz (see Fig.
1). The details of the observations and data reduction will be given
elsewhere (Simon, Readhead, and Wilkinson, in preparation). The global
fringe fitting technique of Schwab and Cotton (1983) was used to extract
the maximum possible information from the video tapes, and the resulting
map is shown in Fig. 1. The excellent (u,v) coverage afforded by these
observations has enabled us, for the first time, to map the 0".2 jet with
the full resolution at this frequency. We believe that this is the best
VLBI map, in terms of dynamic range and number of pixels, produced thus
far.

In Figs. 2a and 2b we show maps of the core from the 1976.74
(Readhead and Wilkinson 1980) and 1982.77 observations, respectively.
At the lowest contour levels the core is slightly larger at the later
epoch, especially in position angles -70 degrees and -150 degrees. We
note that the maps of Preuss et al. (in preparation) at 5 GHz suggest an
increase in size of the core in pa -90 degrees. It is not possible on
the basis of these observations to discriminate between an expansion of
the outer regions of the core and a brightening of stationary components. In 1976 the core could be modelled as a simple 2.6 milliarcsecond
separation double, embedded in a halo. This structure can be seen in
Fig. 2a at the higher contour levels as a double, or long ridge of
emission. This structure has been replaced in the 1982 map by a simple
symmetric broad peak. The change could be due to a contraction or
simply to variations in brightness of different parts of the core.

R. Fanti et al. (eds.), VLBI and Compact Radio Sources, 111–112.
© 1984 by the IAU.

REFERENCES
Readhead, A. C. S. and Wilkinson, P. N.: 1980, Ap.J. 235, pp. 11.
Schwab, F. R. and Cotton, W.D.: 1983, A.J. 88. pp. 487.
Simon, R. S. et al.: 1983, Nature 302, pp. 487.

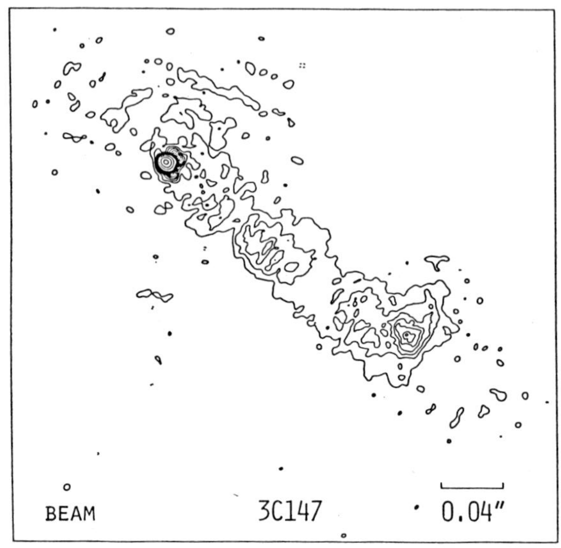

1.   Map of 3C147 at 1661 MHz. Contour levels are –5, 5, 15, 25, ...,
115, 230, 460, 900 mJy/beam. The restoring beam used is an elliptical
gaussian-shaped beam, with major axis 4.2 milli arc sec, minor axis 3.7
milli arc sec, with the major axis lying along position angle –45
degrees.

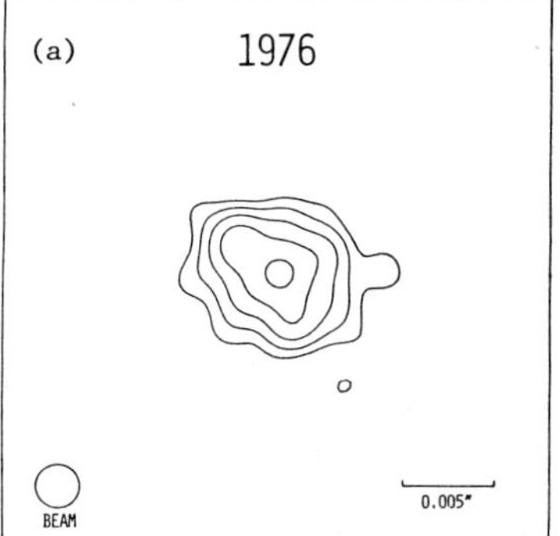

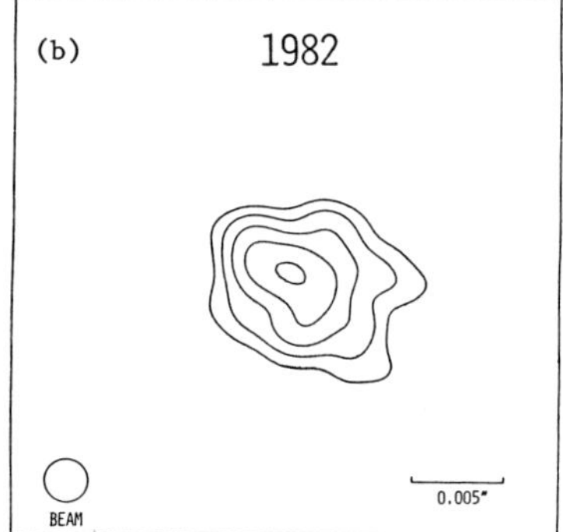

2.   Maps of the core of 3C147.   Contour levels are 0.02, 0.04, 0.08,
0.16, 0.32, and 0.64 Jy/beam in each map. The restoring beam used was a
circular gaussian of 2.4 mas FWHM. (a) The 1976.74 core map of Readhead
and Wilkinson.   The peak brightness in this map is 0.78 Jy/beam. (b)
The core from the 1982.77 map (the full map is shown in fig. 1); the
peak is 0.72 Jy/beam.

PROPER MOTION OF COMPONENTS OF THE QUASAR 3C 345[+]

N. Bartel, M.I. Ratner, I.I. Shapiro, T.A. Herring
Harvard-Smithsonian Center for Astrophysics

B.E. Corey
NEROC, Haystack Observatory

ABSTRACT
From five sets of VLBI observations spaced between 1972 and 1981, we
estimated the positions of components of the superluminal quasar 3C 345
relative to the position of the single component of the quasar NRAO 512.
The relative proper motion of the easternmost component of 3C 345,
believed to be the "core", was found to be 0.02±0.02 mas/yr. This
result is consistent with the "core" being stationary and the "jet"
components moving with respect to NRAO 512.

INTRODUCTION
The radio brightness distribution of the quasar 3C 345 (1641+399) has
been reported to have several components of milliarcsecond size.
Since 1971 several VLBI groups have measured increases in the separation
between these components of up to 0.4 mas/yr (e.g., Wittels et al. 1976,
Unwin et al. 1983). If the quasar lies near its redshift distance
(z=0.595), some of these increases require apparent superluminal
transverse motion of the components relative to each other. Our goal
was to determine the motion of these components with respect to an
external reference.

METHOD
The right ascension and declination of a radio source can now be
determined via VLBI with an uncertainty of a few milliarcseconds (see,
for example, Ma et al. 1981), likely too high to detect the proper
motion of any component of 3C 345 within a few years. Much higher
accuracy can be achieved in the determination of the relative position
of two sources, if their separation on the sky is sufficiently small:
most systematic errors tend to cancel and, further, the fringe phase
can be used as the observable (see Shapiro et al. 1979). For two
sources separated by, say, half a degree or less, one can interleave
observations of them and determine the difference in their sky positions
with an uncertainty below 1 milliarcsecond (mas). With present inter-
continental VLBI arrays, to achieve a precision of ∼0.1 mas in the
measurement of the relative position of two such sources, certain
geodetic and astrometric parameters have to be known with sufficiently

+ Discussion on page 432

113

R. Fanti et al. (eds.), VLBI and Compact Radio Sources, 113–116.
© *1984 by the IAU.*

high accuracy. Approximate accuracies (within a factor 3) for such
parameters are given in Table 1; for comparison we show the accuracies
now obtainable from "geodetic" VLBI experiments.

TABLE 1

Required Accuracy of Astrometric and Geodetic Parameters
to Determine Relative Positions to within 0.1 mas for Two Sources
Separated by 0.5 deg on the Sky

Parameters	Required Accuracy (approximate)	Obtainable Accuracy (approximate)	References
1. Baseline lengths	10 cm	5 cm	Rogers et al. 1983
2. Position on sky of reference source	5 mas	5 mas	Ma et al. 1981
3. Earth orientation (equivalent length on surface)	50 cm	40 cm	Robertson et al. 1983
4. Ionospheric delay	10 cm	1 cm	Herring 1983
5. Tropospheric delay	10 cm	10 cm	educated guess

To make use of such precision when at least one of the sources is
partially resolved, the position must refer to a well-defined point in
each source. Our estimates of position refer to the center of brightness
of the easternmost component of 3C 345. The positions of the other
compact components are determined from knowledge of the components'
separation, obtained, for example, from hybrid maps.

OBSERVATIONS
Interleaved observations of 3C 345 and the nearby source NRAO 512
(1638+398), with redshift z=1.67 and a single compact component, were
made at λ3.8 cm in the early and mid 1970's using the Mark I VLBI system,
and at λ3.6 cm and λ13 cm (simultaneously) since 1980 using the Mark III
VLBI system. Data from five epochs are discussed here.

RESULTS
The difference in right ascension between the easternmost compact
component (the putative "core" component) in 3C 345 and the single
compact component in NRAO 512 is given in Figure 1 for five epochs.
(We omit the declination because changes in declination are nearly
orthogonal to the motion of interest, and because our relative position
estimates are subject to larger errors in declination than in right
ascension.) If the individual observations are assigned standard errors
equal to the root-mean-square of the postfit residuals, the statistical
standard error in the estimate of the relative position of the two
sources is about 0.1 mas for the first two epochs and 0.02-0.03 mas for
the last three; in all cases the fractional uncertainties are less than
1 part in 10^7.

To display the positions of the "jet" components, a ten times larger scale for the right ascension is used in Figure 2. This figure includes relative positions of "jet" components previously reported by other VLBI groups; estimated errors are also shown. The positions are plotted with respect to the mean from all five epochs of the separation between the easternmost component of 3C 345 and the single component of NRAO 512. Although it is not clear whether specific "jet" components can be traced for the whole decade from 1972 to 1982 (see, e.g., Schraml 1981), one component can be traced from 1972 to 1976 and two from 1979 onwards. Within at least these periods proper motions of the "jet" components are apparent.

CONCLUSION
Our results show that the "core" of the superluminal quasar 3C 345 is stationary (0.02±0.02 mas/yr) and the "jet" components are moving with respect to the external reference, NRAO 512.

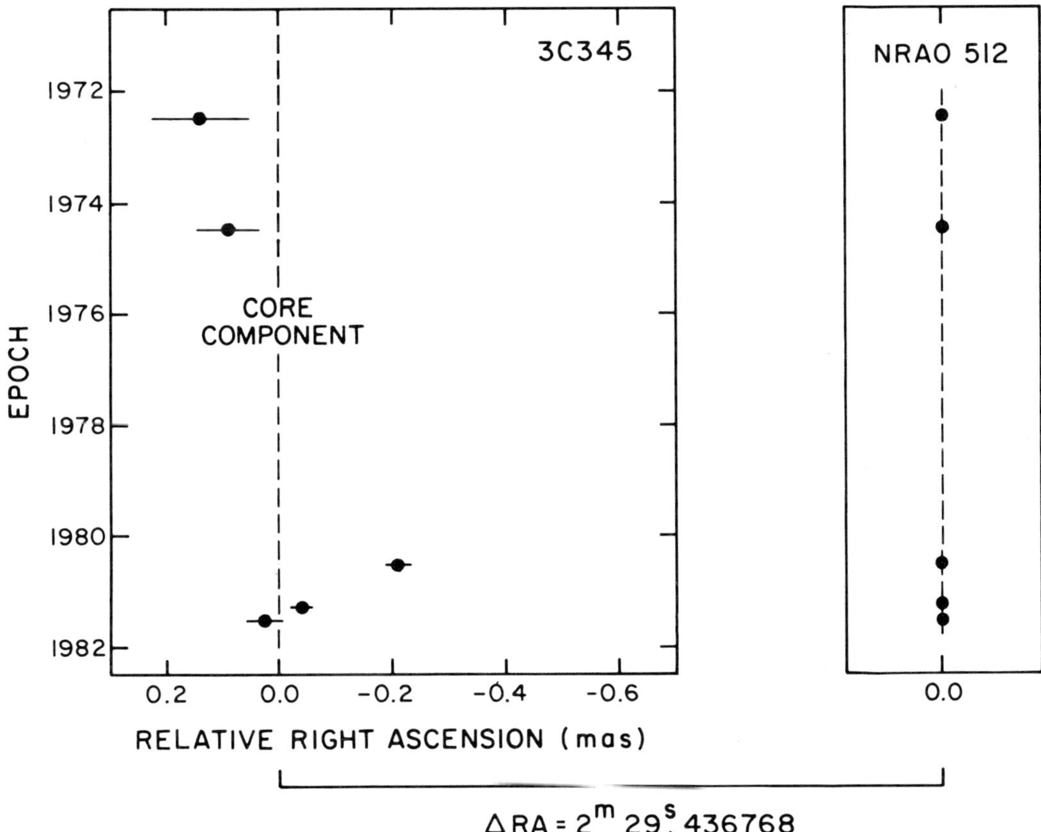

$$\Delta RA = 2^m 29^s.436768$$

Figure 1: The differences in right ascension between the "core" component in 3C 345 and the single compact component in NRAO 512 for each of five epochs. The left dashed line marks the mean of these differences. Statistical standard errors are shown.

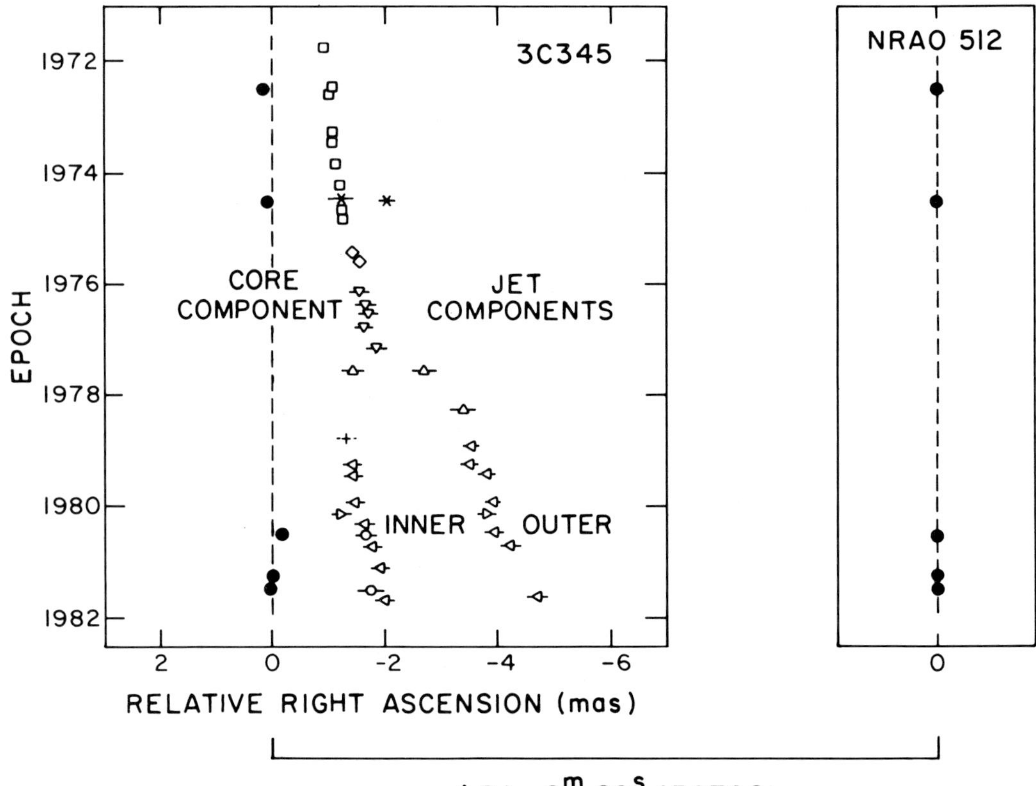

$$\Delta RA = 2^{m}\ 29^{s}.436768$$

Figure 2: Same as Figure 1, but a ten times larger scale. Additional symbols mark positions of inner and outer jet components relative to the dashed line. Data are from: +Baath et al. 1981, ◊Cohen et al. 1977, △ Cohen et al. 1981, x Cotton 1980, ▽ Seielstad et al. 1979, ▷ Spencer et al. 1981, ◁Unwin et al. 1983, □Wittels et al. 1976, ●and ○this paper.

REFERENCES

Bååth,L.B. et al.: 1981, Ap.J.(Lett.) 243, p.L123.
Cohen,M.H. et al.: 1977, Nature 268, p.405.
Cohen,M.H. et al.: 1981, Ap.J. 247, p.774.
Cotton,W.D.: 1980, Radio Interferometry Techniques for Geodesy,
     NASA Conference Publication No. 2115, p.193.
Herring,T.A.: 1983, Ph.D. Thesis, M.I.T., Cambridge, MA.
Ma,C., Clark,T.A., and Shaffer,D.B.: 1981, BAAS 13, p.899.
Robertson,D.S. et al.: 1983, Nature 302, p.509.
Rogers,A.E.E. et al.: 1983, Science 219, p.51.
Schraml,J. et al.: 1981, Ap.J.(Lett.) 251, p.L57.
Seielstad,G.A. et al.: 1979, Ap.J. 229, p.53.
Shapiro,I.I. et al.: 1979, A.J. 84, p.1459.
Spencer,J.H. et al.: 1981, Ap.J.(Lett.) 251, p.L61.
Unwin,S.C. et al.: 1983, Ap.J., in press.
Wittels,J.J. et al.: 1976, A.J. 81, p.933.

SUPERLUMINAL MOTION IN BL LACERTAE: 10.65 GHz VLBI OBSERVATIONS FROM
1981.7 to 1982.7 +

R. L. Mutel
Dept. Physics & Astron., Univ. of Iowa, Iowa City, IA
R. B. Phillips
Haystack Observatory, Westford, MA

We report here VLBI observations at four epochs of the source BL
Lacertae made with the U.S. VLBI Network (plus Bonn) at 10.65 GHz which
show clear evidence of uniform superluminal expansion at v/c = 4.4 $\pm$
0.2 (H_0 = 55 km s^{-1} mpc^{-1}). The maps, which were made every three
months from 1981.76 to 1982.75 (Figures 1, 2), were part of an ongoing
program of VLBI observations of BL Lacertae at 5 and 10.65 GHz starting
in 1980.4, near the beginning of a series of violent flux outbursts
(Aller, Hodge, and Aller, 1983; Aller and Aller, 1984). Previous
results from earlier epochs have been reported by Mutel, Aller, and
Phillips (1981) and Phillips and Mutel (1982).

All four maps are dominated by a nearly unresolved component at
the northern end of an extended "jet" of emission along p.a. 10° $\pm$ 2°.
The northern component ("A") is very likely responsible for the highly
polarized flux event reported by Aller, Hodge, and Aller (1983) since
the flux of the jet is not large enough to account for the observed
increase in total flux. As pointed out by Aller and Aller in these
proceedings, the intrinsic polarization angle of the linear polarization is nearly the same as the VLBI jet, so that the magnetic field is
nearly perpendicular. This is consistent with a picture of the radio
emission arising from a shock which causes axial compression of the
magnetic field. The radio emission would arise from electron density
enhancement behind the shock which then radiates by ordinary synchrotron radiation.

If this scenario is at least qualitatively correct, it seems
unlikely that component "A" is coincident with the "core" source, as is
commonly assumed for the brightest end component in the well-known
core-jet superluminal sources. Indeed, even though we have detected
relative motion between two components, both of them are more likely to
be "knots" or other temporary features along a longer-lived channel.
The core for this source is not obviously visible, so that the relative
velocity of the core and the jet components is unknown.

+ Discussion on page 433   117

R. Fanti et al. (eds.), VLBI and Compact Radio Sources, 117–118.
© *1984 by the IAU.*

The uniform motion between components "A" and "B" reported here results in an almost identical value of γ as that derived from the 1980.93 map (Mutel, Aller, and Phillips, 1981). In that case, however, the evidence was less direct since it depended on comparison of the flux and separation of two components made at a single epoch with the time scale of the flux event prior to the observation. The data are consistent with bulk relativistic motion of material with a constant Lorentz factor γ, moving along a long-lived channel in p.a. 10°, at least for the series of flux outbursts between 1980.2 and 1982.7. Inspection of maps for the latest two epochs (1982.44 and 1982.75) shows a tendency for the more diffuse southern end of the jet to bend to the east. This is more pronounced in our 5 GHz maps during the same epochs and can be clearly seen in other 5 GHz maps by Pearson and Readhead (1984) and by Bäath (1983). The total amount of bending appears to be approximately 25°.

This research was supported by NSF grants AST81-18337 and AST82-16890 to The University of Iowa. We are grateful to L. Bäath and H. Aller for data in advance of publication.

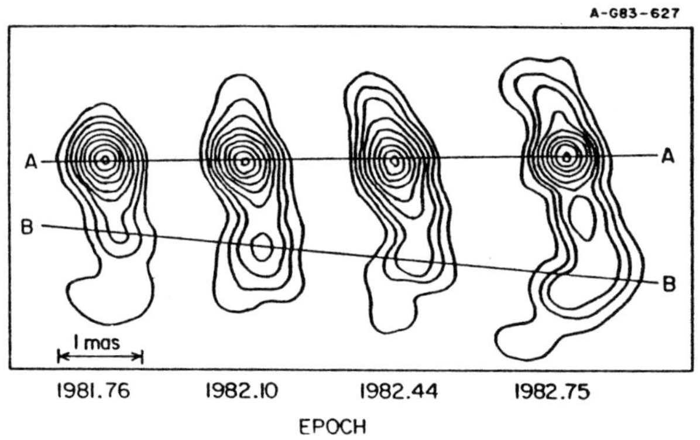

Figure 1. Contour maps (levels: 10, 20, ... , 90% of maximum) of four epochs of BL Lac at 10.65 GHz.

Figure 2. One-dimensional cuts along p.a. 10°.

REFERENCES

Aller, H. D. and Aller, M. F.: 1984, this volume, p. 119.
Aller, H. D., Hodge, P. E., and Aller, M. F.: 1983, Astrophys. J., in press.
Bäath, L.: 1983, private communication.
Mutel, R. L., Aller, H. D., and Phillips, R. B.: 1981, Nature 294, p. 236.
Pearson, I. J. and Readhead, A. E. E.: 1984, this volume, p. 15.
Phillips R.B. and Mutel R.L., 1982, Ap.J. (Letters), 257, L19

THE RADIO POLARIZATION OF BL LACERTAE: SHOCKS IN A JET [+]

Hugh D. Aller and Margo F. Aller
Radio Astronomy Observatory
University of Michigan

Abstract: The behavior of the linear polarization of BL Lacertae during
the recent series of radio bursts indicates the presence of Faraday
rotation in the vicinity of the source and the formation of an axial
compression in the radio jet.

In the past four years BL Lacertae has exhibited a series of radio
outbursts (Figure 1a) which were preceded by a three year period of
relative inactivity. During these bursts the source has exhibited
apparent superluminal motions along a position angle of 10° (Phillips
and Mutel 1982, Phillips 1983); and the polarization characteristics
have changed dramatically. This paper describes the evidence for
Faraday rotation and the behavior of the intrinsic polarization position
angle since 1980 based upon data obtained with the automated University
of Michigan 26-meter radio telescope.

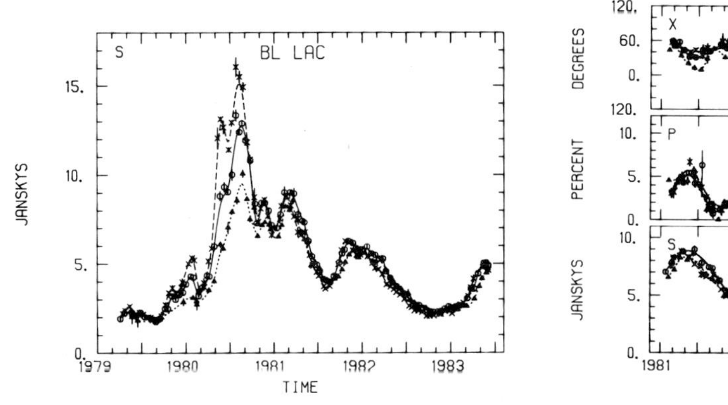
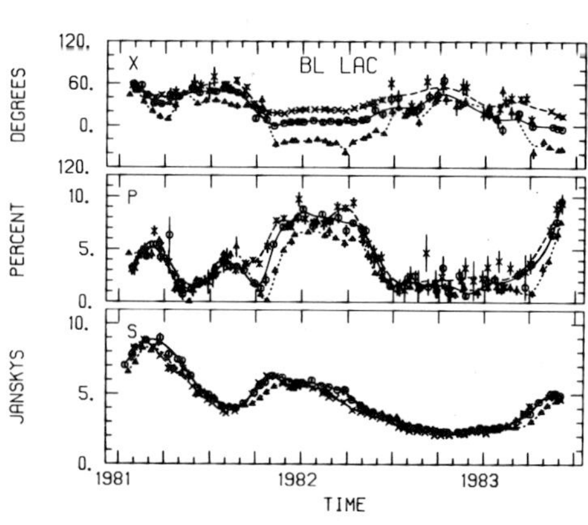

a)   b)

Figure 1: Flux density (a) and polarization (b) of BL Lacertae: Two-week
averages of the data at 4.8,8.0 and 14.5 are indicated by Δ, o, and x.

+ Discussion on page 434   119

R. Fanti et al. (eds.), VLBI and Compact Radio Sources, 119–120.
© 1984 by the IAU.

Throughout the bursts in 1980 the degree of polarization was relatively low (less than 5 percent) and the source exhibited rapid polarization variations including the position-angle rotation phenomenon (Aller, Hodge, and Aller 1981). In the bursts since 1980 the degree of polarization has been systematically higher and the position angles have been relatively stable (Figure 1b). We interpret the increased degree of polarization in the later bursts as due to an increased degree of order of the magnetic field as the emitting region has evolved. In the burst which started in late 1981, the degree of polarization increased dramatically, and the polarization position angles at the three frequencies separated into a pattern characteristic of Faraday rotation. The derived rotation measures before and during the burst were -103 and -234 rad.m^{-2} respectively. The rapid change of -131 rad.m^{-2} in the rotation measure is most easily accounted for by a small cloud ($N_e = 1$ cm^{-3} and $B = 2 \times 10^{-4}$ Gauss) in the vicinity of BL Lacertae's radio emitting region. Placing the rotation in our galaxy or within the emitting region gives problems with the small angular size of the source or with the constant value of the rotation measure during the evolution of the burst. As the burst died out, the polarizations appear to have resumed their previous values; but with the appearance of a new burst in 1983, the pattern is repeating. The rotation measure during the 1983 burst is derived to be -255 rad.m^{-2}.

The intrinsic polarization position angles during the late 1981 burst were in the range of 23 to 28 degrees and subtraction of the pre-burst polarized flux places these values in the range of 15 to 20 degrees. For the 1983 burst the observed IPA is 21 degrees, and removal of previous polarized emission yields a value of 18 degrees. In both bursts the polarization data indicate an orientation of the magnetic field in the emitting region which is nearly perpendicular to the "jet" observed by VLBI. The orientation of the magnetic field and the high degrees of polarization during the bursts suggest the formation of an axial compression (or shock) down the jet which results in the reacceleration of the emitting particles and the appearance of a new source component. VLBI observations (Mutel et al. 1984) show the appearance of a new source component coincident with the late 1981 flux density event, and we predict that future VLBI observations will detect a new source component appearing in 1983. The polarization data, together with the VLBI observations, furnish strong evidence for substantial reacceleration of emitting particles within the jet-like emitting regions of extragalactic variable sources. This has important implications for not only the centimeter variability phenomenon but may also have application to low frequency variable sources. This work was supported in part by the National Science Foundation Grant No. AST 8021250 A01.

REFERENCES

Aller, H.D., Hodge, P.H., and Aller, M.F. 1981, Ap.J.(Letters), <u>248</u>, L5.
Mutel R.L. and Phillips, R.B. 1984, this volume, p. 117.
Phillips, R.B. and Mutel, R.L. 1982, Ap.J.(Letters), <u>257</u>, L19.

OBSERVATIONS OF SUPERLUMINAL MOTIONS IN 3C 120 [+]

R. C. Walker and J. M. Benson
National Radio Astronomy Observatory[*]
G. A. Seielstad and S. C. Unwin
Owens Valley Radio Observatory

Superluminal motions were first seen in 3C 120 between 1972.5 and
1974.4 (Seielstad et al. 1979 and references therein). Between 1975
and 1980, the source was monitored along with 3C 273 and 3C 345 by the
Caltech group. One reasonably clear episode of expansion was seen in
1979 (Walker et al. 1982) but, for most of the time, the source
evolution was so rapid that it was difficult to relate the structures
seen at successive epochs.

In 1981, the monitoring of 3C 120 was separated from that of the
other sources so that more frequent observations could be made at one
frequency and so that more stations could be used in order to improve
the dynamic range of the maps. The dates, frequencies, beam sizes,
stations, and total flux densities for all of the observations made
since those described in Walker et al. (1982) are given in Table 1. The
maps from the first 4 epochs in which the new observing style was used
are shown in Figure 1. More complete details and the maps from the
other epochs will be published elsewhere. The new observations provide
a much clearer picture of the evolution of 3C 120 than was available
before. It is apparent that the evolution is rapid and that the
structure is complex, so it is not surprising that the earlier, more
limited data produced ambiguous results.

The new maps show three components (labeled B, C, and D; A is the
component seen in 1979) that are moving away from the eastern
component. Component B was also seen in the maps from 1980 and early
1981. The eastern component is at a sharp end to the structures seen
in the maps and displays more erratic flux density behavior than the
other features. In the terminology of the core-jet models (e.g.,
Blandford and Konigl, 1979), it is assumed to be the core and it serves
as the reference for measuring the positions of the other features.
The separation of each component from the core as a function of time is

*The National Radio Astronomy Observatory is operated by Associated
 Universities, Inc., under contract with the National Science
 Foundation.

+ Discussion on page 434

R. Fanti et al. (eds.), VLBI and Compact Radio Sources, 121–124.
© 1984 by the IAU.

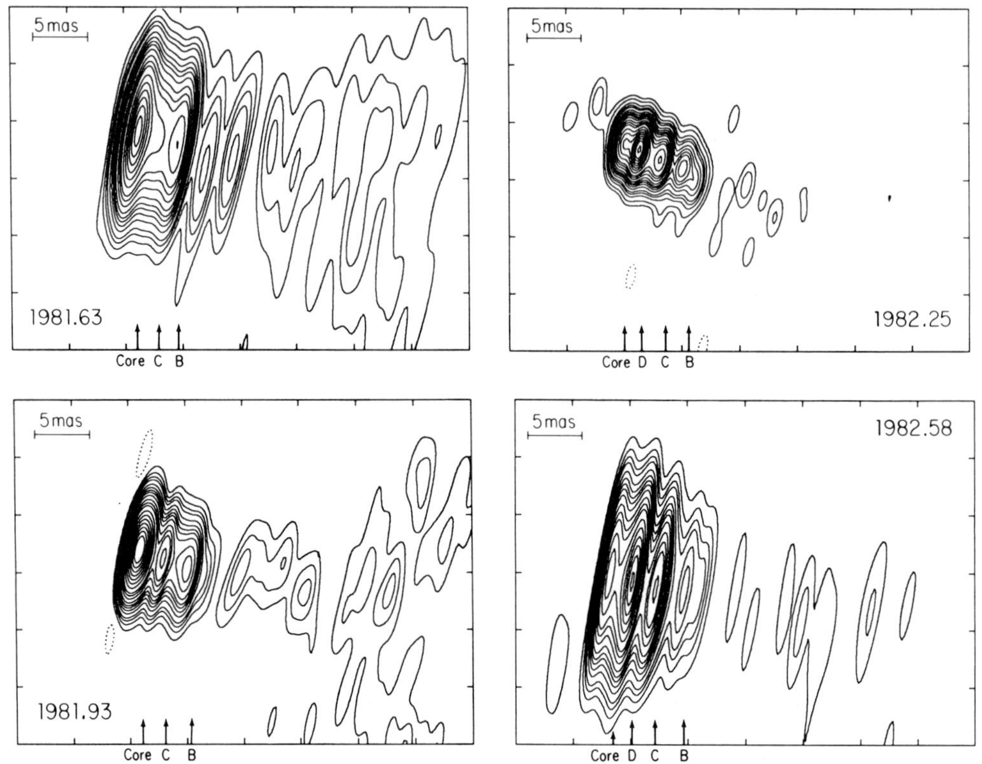

Figure 1. Hybrid maps of 3C 120 based on observations at 5.0 GHz at
4 epochs during 1981 and 1982 (See Table 1 for details). The feature
labeled "core" is at the eastern end of the source and is the feature
that is used as a position reference. The features labeled B, C, and D
are those whose separations from the core are shown in Figure 2. The
contour levels are -30, -20, -10, 10, 20, 30, 40, 60, 80, 100, 120,
150, 200, 250, 300, 400, 500, 600, 700, 800, 900, 1000, 1150, 1300, and
1600 mJy/beam. The 1982.25 map has additional contours at -5 and
5 mJy/beam. Because of missing short spacings, that observation was
not sensitive to the low brightness features seen at the other epochs.

shown in Figure 2. The error bars reflect estimates of the uncertainty
in the position of the component and the uncertainty in the position of
the core. At the times of the birth of a new component, the core and
the new component are not resolved, and the position of the core is
especially uncertain. This accounts for the large error bars in
1981.25 and 1981.92.

The two important results from the separation measurements are
that the velocity of a feature is constant and that different features
in the same source can have different velocities. The first result is

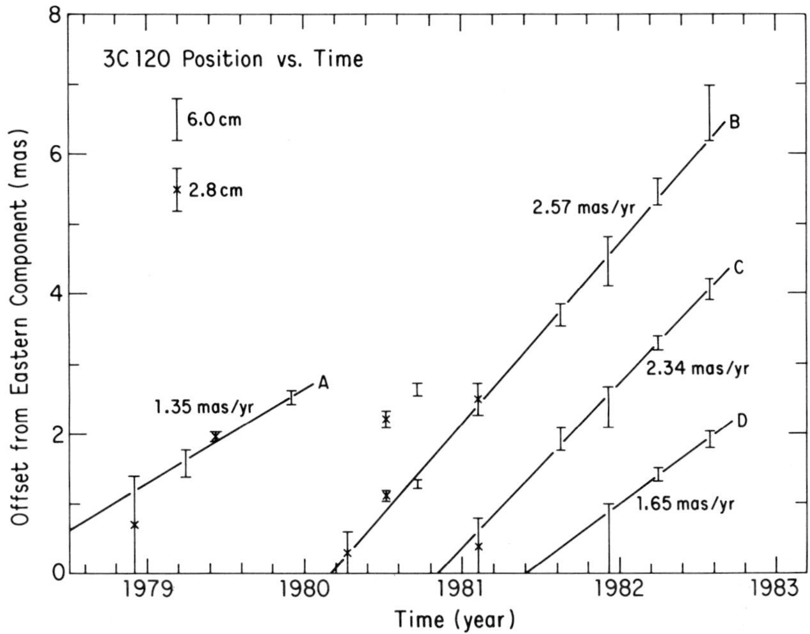

Figure 2. Separation of features from the eastern feature (core) as a function of time. Angular rates are shown for the feature seen in 1979 (Walker et al. 1982) and for the three features seen in the recent observations. The rates are the result of fits to the data. The rate of component D is poorly determined. For z = 0.033 and a Hubble constant of 100 km s^{-1} Mpc^{-1} one milliarc-second per year corresponds to 1.52 times the speed of light.

Table 1: Experiments

Epoch (yr)	Wavelength (cm)	Beam Axes (mas)	PA (deg)	Stations[1]	Total Flux Den. (Jy)
1980.52	2.8	0.5 x 5.5	− 9	BKGFO	3.3
1980.72	6.0	1.0 x 8.1	−10	BKFO	3.8
1981.10	2.8	0.8 x 3.5	−10	KGFO	3.2
1981.63	6.0	1.3 x 8.9	−11	BKGFVO	3.1
1981.93	6.0	1.0 x 5.6	−13	BSKGFVOH	3.7
1982.25	6.0	1.0 x 3.2	−13	BSAGFVOH	3.9
1982.58	6.0	1.0 x 9.7	−10	BKGFVO	3.5

[1]Abbreviations: B: Bonn, S: Onsala, A: Arecibo, K: Haystack, N: NRL, G: Green Bank, F: Fort Davis, V: VLA, O: Owens Valley, H: Hat Creek.

based on component B which was observed from when it was blended with
the core to when it was over 6 milliarc-seconds from the core. The
second result is based on the variations by nearly a factor of two in
the velocities of the various components. Note that the velocity of
the 1972-1974 component was very close to the velocity of component A.
It is not yet clear whether components seen at the same time have the
same velocity.

The observations show that the flux density of the core has varied
significantly but not in an obvious systematic manner, other than
contributing to the over all decrease in the total flux density of
3C 120 seen over the last several years. The moving components either
decreased systematically in flux density with time (features A, B and
maybe D) or were constant over the period of the observations analyzed
so far (feature C). New components were seen first as a brightening of
the core, usually at times when there were indications that the core
was slightly resolved. By the time the observations resolved a new
component from the core, the rise in flux density was over.

The high dynamic range maps also revealed the presence of
structures on scales of tens of milliarc-seconds. These structures are
clearly real but are very poorly constrained by the 5 GHz data.
Subsequent observations at 1.6 GHz show a jet extending to 0.2
arc-seconds. They are discussed elsewhere in these proceedings by
Benson et al.

REFERENCES

Blandford, R. D., and Konigl, A. 1979, Ap. J., 232, pp. 34-48.
Seielstad, G. A., Cohen, M. H., Linfield, R. P., Moffet, A. T.,
     Romney, J. D., Schilizzi, R. T., and Shaffer, D. B. 1979,
     Ap. J., 229, pp. 53-72.
Walker, R. C., Seielstad, G. A., Simon, R. S., Unwin, S. C.,
     Cohen, M. H., Pearson, T. J., and Linfield, R. P. 1982,
     Ap. J., 257, pp. 56-62.

200 MILLIARCSECOND STRUCTURE IN 3C 120 [+]

J. M. Benson and R. C. Walker
National Radio Astronomy Observatory[*]
G. A. Seielstad and S. C. Unwin
Owens Valley Radio Observatory

We present VLBI and VLA maps of the superluminal radio source 3C 120 (z = 0.03). The 18 cm VLBI maps shown in Figures 1 and 2c were constructed from a 14 station VLBI observation on 10, 11 October 1982. The map in Figure 1 is at full spatial resolution. The Figure 2c map was made from a uv-tapered data set and shows the extended emission. The VLBI maps show a jet ~175 pc long with distinct knots and bends (H_O = 55 kms^{-1} Mpc^{-1}). The 18 cm jet smoothly connects to the 6 cm superluminal jet (Figure 2d). The VLBI jet remains well collimated through several bends and only begins to spread out 100 pc beyond the core. At 175 pc from the core, the jet is wide and weak, and disappears into the under-sampled region of the uv-plane ($< 2(10)^6$ λ's).

The four maps in Figure 2 show structure ranging in size from the < 5 pc superluminal components to the 130 kpc extended emission regions. The 3C 120 jet is nearly continuously connected from the superluminal jet (p.a. = -110°, Figure 2c) through the larger 175 pc jet with a p.a. = -90° (Figure 2d) and into the 17 kpc jet in the VLA "B" array map (Figure 2b). Figure 2 shows the first clear observational evidence of a jet that is <u>relativistic</u> near the core, and is smoothly connected to a classical extended radio lobe. In the largest scale map (Figure 2a) an extended lobe is present south-east of the core. There is, however, no clear evidence for an eastern jet in the VLBI or VLA "B" array maps. In fact, for a γ = 5, the receding jet would be doppler suppressed by a factor of ~10^{-6} below the approaching jet.

The VLA "B" array jet trajectory may be represented by a simple relativistic precessing jet model where γ = 5, θ_{INCL} = 12° (consistent with the observed superluminal motions), ϕ_{PREC} = 9°, and a precession period ~5 (10)^4 years. The periodicity in the bends in the 175 pc VLBI jet may be represented by a 20 year nutation (ϕ_{CONE} = 1° -2°) on top of the precession period.

*The National Radio Astronomy Observatory is operated by Associated Universities, Inc., under contract with the National Science Foundation.

<u>+ Discussion on page 435</u>   125

R. Fanti et al. (eds.), VLBI and Compact Radio Sources, 125–126.
© 1984 by the IAU.

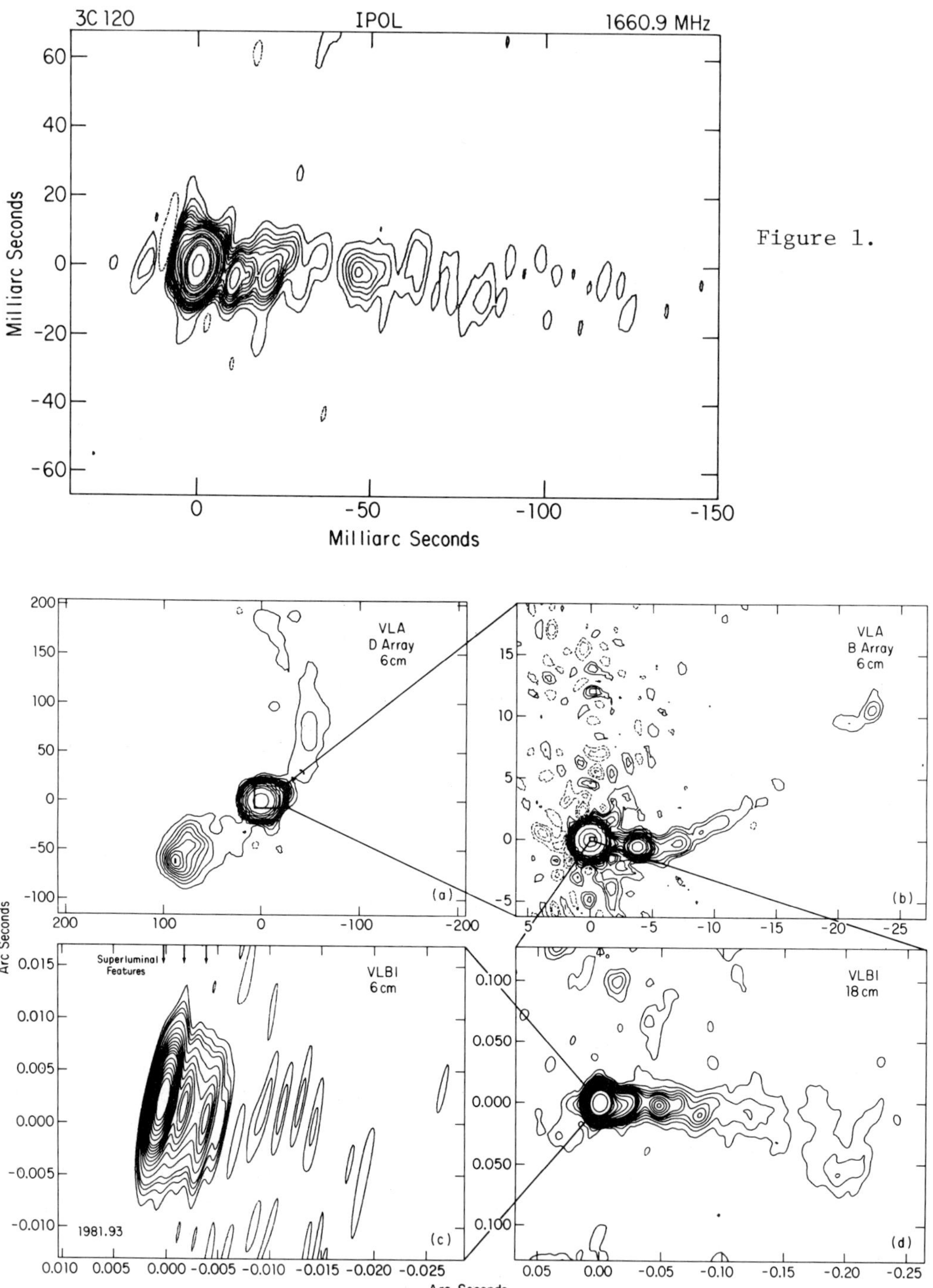

Figure 1.

Figure 2.

VLBI MONITORING OF BL LAC-OBJECTS [+]

Lars B. Bååth
Onsala Space Observatory, S-439 00 Onsala, Sweden

This is a report on an ongoing project designed to study structural variations on the milliarcsecond scale in four BL Lac-type objects. The observations are made at λ6 cm with a global VLBI network consisting of the combined US and European networks and with some additional antennae like one in South Africa and one in Finland. A typical experiment involves 7-9 stations and produces maps with a resolution (FWHM) of 1-1.5 m.a.s depending on the declination of the source.Seven experiments have been made since the start in 1978, resulting in hybrid maps at three epochs for each of the four main sources: AO 0235+164,0735+178,Mk 421, and 1749+701. The same restoring beam has been used for each source at all epochs. The epoch March 1979 is an exception though, the maps from then were restored with a larger beam corresponding to the resolution on the European baselines only. The results from the epochs 1978-1979 have already been published (Bååth et al. 1981), and papers describing the results from the later epochs are in preparation.

All the sources varied in flux density over the period 1978-1981.The most varying source was AO 0235+164;0735+178 decreased steadily in flux; Mk 421 and 1749+701 both had at least two outbursts. The four sources will be presented here in order of increasing right ascension, thus avoiding to group them together into any artificial subclasses.

AO 0235+164

AO 0235+164 has been previously discussed at this meeting by D.Jones and F.Briggs.The source is particularly interesting since its optical spectrum shows absorption lines corresponding to two redshift systems:z=0.52 and z=0.85. It is also one of the most violently varying extragalactic sources known with several large outbursts since the flux monitoring started in 1975 (G.Nicholson, private comm.). We observed the source with VLBI close to the peak of the two outbursts in March 1979 and October 1980, and also in December 1981 just at the beginning of the 1982 outburst (Fig. 1).The only component in each map which is intense enough to house the outburst is the core,and therefore the outer component should, in analogy with outbursts followed in other core-dominated radio sources like 3C 345 and 3C 273,be a remnant from a previous outburst.Doing this

[+] Discussion on page 435   127

R. Fanti et al. (eds.), VLBI and Compact Radio Sources, 127-130.
© *1984 by the IAU.*

analogy we would identify the northeast extension in the first map as
emanating from the 1975-outburst, in the second map from the 1979-out-
burst, and in the third map from the 1980-outburst.This results in an
apparent speed of angular separation, consistent for all three events,
of 1-1.5 m.a.s. per year or 45c (z=0.85,H_O=55 km sec^{-1} Mpc^{-1},q_O=0.05).
Caution must be taken though since each component is observed only once.
There is some evidence in our data from the second epoch of a component
that is either far away or too large to be more than marginally observed
with our interferometer,a possible remnant from the previous epoch. The
components must evolve faster than in other well studied varying VLBI
sources,and it is therefore necessary to observe the source more often
than once a year.The position angle of the jet is also different in the
three maps.Blandford and Königl (1979) concluded from the change in polarization angle during the 1975-outburst that the new component should
move northward from the core.Our map from 1979 does not show any extension in this direction, but the next epoch map, observed closer in time
to the event, does. The first two epoch maps are therefore consistent
with a scenario in which the new components move northward to start with
and then turn towards east. The third map does not fit into this scenario.
The position angle of the jet is around 45° though the size is about the
same as in the map at the previous epoch.Therefore the outburst probably
started in a direction different from that of the previous ones.

0735+178

This radio source has been called "the cosmic conspiracy" because of the
way a number of partly selfabsorbed radio components conspire to form a
flat spectrum (Cotton et al. 1980;Marscher 1980). 0735+178 was observed
at the same epochs as AO 0235+164, and the maps are shown in Fig. 2. The
VLBI structure is dominated by two components.The northeast (NE) component decreased in flux and the position angle relative the stronger (core)
component changed significantly between the first two epochs.This continued over the third epoch,the peak brightness of the NE component decreased whereas its southeast part increased in intensity.Gaussian fittings
to the 1980 and 1981 maps show that a component had moved from the core-
component towards the northeast with an apparent angular speed of 0.2
m.a.s. per year. The new component can be seen in the maps as an increasing extension of the core towards the northeast. It is also observed in
the spectrum as an increase of the flux density at frequencies higher
than 10 GHz during 1980 followed by a decrease during 1981.Also,a component with a high-frequency turnover had to be added to the radiospectrum
of 1979 in order to fit this as the superposition of the components observed with VLBI (Cotton et al. 1980;Bååth 1980). The new component remains optically thick at 5 GHz during the period we observed it and is
therefore not a strong feature in any of the maps.The only way to definitely observe the component is to increase the resolution, and therefore
we observed the source with a global VLBI network at λ1.35 cm,resulting
in a resolution of 200 microarcsec. The component was detected, and the
apparent speed of angular separation, based on three epochs, was 0.18
m.a.s. per year, or 2.6c. 0735+178 is thus a member of the superluminal
family of radio sources as has been predicted (Marscher 1980;Bååth et al.
1981).

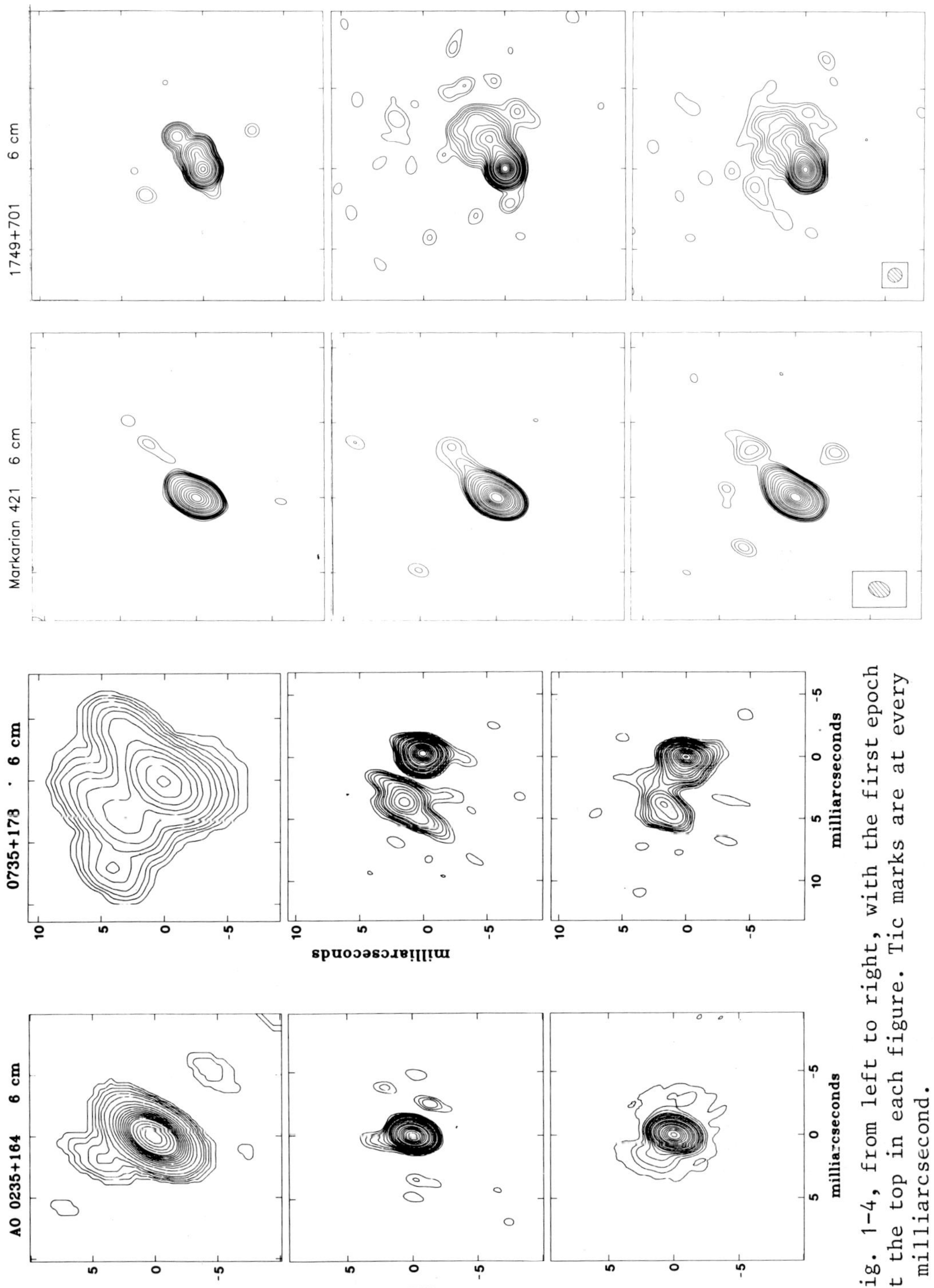

Fig. 1-4, from left to right, with the first epoch at the top in each figure. Tic marks are at every 5 milliarcsecond.

Mk 421(=B2 1101+38)

Mk 421 is one of the handful of BL Lac-objects proven to have underlying
stellar components (Margon et al. 1978). The variation in flux density
and the X-ray flux (Rickets et al. 1976) both indicate a size <0.1 m.a.s.
(Margon et al. 1978), if the X-ray radiation is assumed to be caused by
inverse Compton scattering of radio photons. The source was observed
with VLBI at the epochs May 1980, April 1981, and December 1981 (Fig.3).
Almost all of the variation in integrated flux density can be accounted
for by a change in the flux of a single component <0.2 m.a.s. in diameter, corresponding to a linear size <150 mpc. The size of this component is even more constrained by VLBI observations at λ1.35 cm, where
most of the flux was concentrated to a point source <0.1 m.a.s. in diameter, consistent with the Compton size. The maps from the three epochs
also show a thin 5 m.a.s. long jet, which brightened and formed a rather
distinct component in December 1981. More observations are needed in
order to investigate how the events in the jet are related to the flux
variation of the core.

1749+701

This source was observed at the same epochs as Mk 421. The hybrid maps
are shown in Fig. 4. As for Mk 421 there are two, possibly unrelated,
events. The core became more intense at the second epoch, and remained
so during the third epoch. This is consistent with the flux density
monitoring at λ3.8 cm done at NRL (B.Geldzahler, private communication)
which shows a distincts increase in flux density during the second half
of 1980. The other event observed by us is that the northwest part of
the VLBI source grew more intense and then expanded. This event coincides with a bump in the flux density curve at λ3.8 cm which occurred
shortly after the large increase of flux in 1980. As for Mk 421 it is
not yet clear how the two events are related.

REFERENCES

Blandford,R.D.,Königl,A.:1979, Astrophys. J., 232, 34
Bååth,L.B.: 1980, PH.D.Thesis, Chalmers University of Technology
Bååth,L.B.,Elgered,G.,Lundqvist,G.,Graham,D.,Weiler,K.W.,Seielstad,G.A.,
    Tallqvist,S.,Schilizzi,R.T.:1981, Astron.Astrophys., 96, 316
Cotton,W.D.,Wittels,J.J.,Shapiro,I.I.,Marcaide,J.,Owen,F.,N.,Spangler,
    S.R.,Rius,A.,Angulo,C.,Clark,T.A.,Knight,C.A.:1980, Astrophys.J.,
    238, L123
Margon,B.,Jones,T.W.,Wardle,J.F.C.:1978, Astron.J., 83, 1021
Marscher,A.P.:1980, Nature, 288, 12
Rickets,M.J.,Cooke,B.A.,Pounds,K.A.:1976, Nature, 259, 546

MULTI-EPOCH OBSERVATIONS OF SURVEY SOURCES

A.C.S. Readhead, T.J. Pearson, and S.C. Unwin
Owens Valley Radio Observatory
California Institute of Technology

ABSTRACT

Five objects mapped as part of a VLBI survey have been re-observed at
5 GHz, and four of them have also been observed at 10 GHz. Three of
the objects show no substantial structural variations: an upper limit
of 2c can be placed on the apparent relative velocities of the
components. One object (0711+356) shows structural variations which
are mostly simply described in terms of a superluminal <u>contraction</u>.
The remaining object (3C371, 1807+698) shows substantial structural
variations which suggest that it probably is a superluminal source.
The source 0710+439 is especially interesting as it consists of a
central flat-spectrum core component straddled by two compact
steep-spectrum components.

SECOND-EPOCH MEASUREMENTS AT 5 GHz

In an earlier contribution to this Symposium, the results of first
epoch mapping of a complete sample of sources were presented (Pearson
and Readhead, this volume). Second-epoch 5-GHz maps of four of the
objects are shown in Figure 1. The separations of the major components
(three components in the case of 0710+439) are given in Table 1
for all the objects except 3C371, for which it was not possible to
identify unambigously the knots observed in the jet at the two epochs.
It is clear that 3C371 is a likely candidate for superluminal motion
and more frequent observations of it are needed. Of the remaining four
sources, three (0108+388, 0710+439, and 2021+614) show no significant
relative motion; but in 0711+356 the separation of the two major
components has <u>decreased</u> by 0.13±0.02 mas. Results obtained from
model-fitting with a three-component model agree well with measurements
from the maps. Furthermore, the change in structure can be seen
directly in the visibility amplitudes and closure phases. Thus there
is no doubt that the separation of these components has decreased
significantly. If the change is due to real motion of the components,
the source has contracted superluminally with an apparent speed of 5c,
but more observations are needed to confirm this.

R. Fanti et al. (eds.), VLBI and Compact Radio Sources, 131–134.
© *1984 by the IAU.*

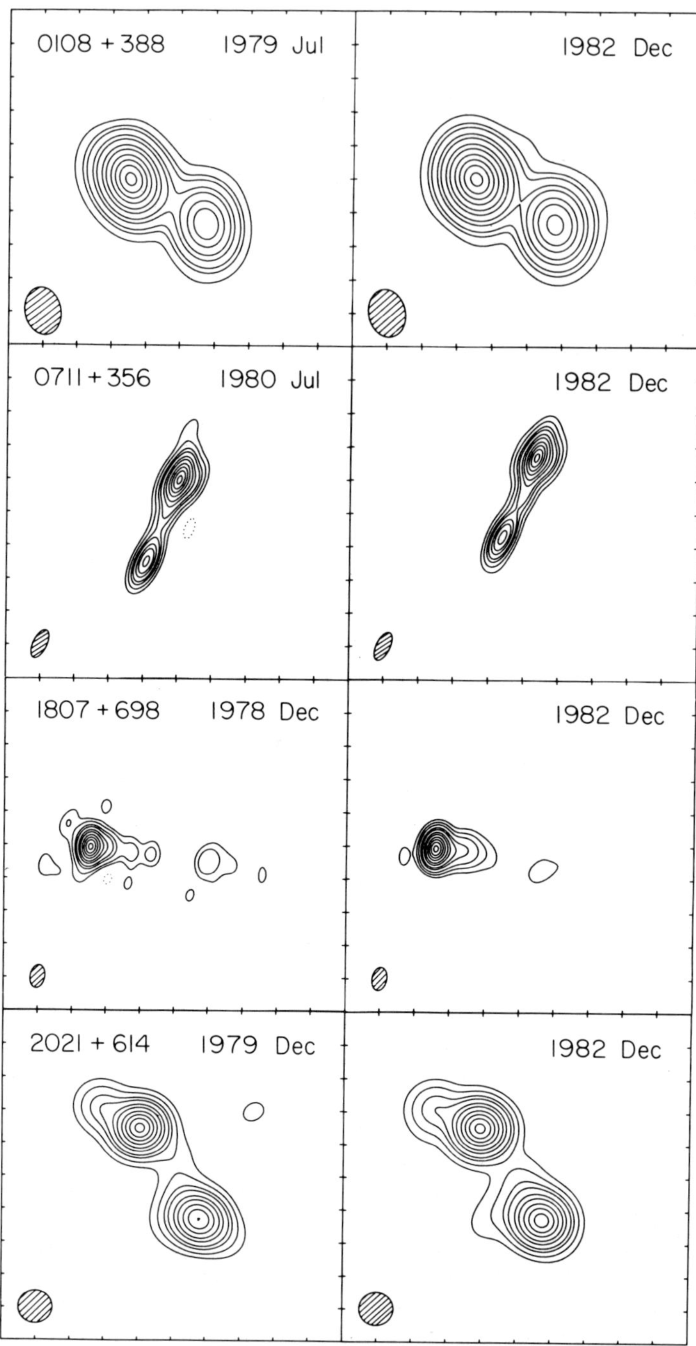

Fig. 1.   5-GHz maps of four sources at each of two epochs;   the same restoring beam (indicated by the shaded ellipse) has been used for both epochs. All the boxes are 20 mas square, except those for 1807+698 which are 40 mas square.

TABLE 1 - LIMITS ON RELATIVE MOTION

Object	ID	z	Epoch	Separation (mas)	Δ (σ) (mas)	v/c (σ) (apparent)
0108+388	EF	–	1979 Jul	5.13	0.08 (0.06)	1.5 (1.0)*
			1982 Dec	5.21		
0710+439	G (21m)	–	1980 Jul	8.28	0.05 (0.04)	1.3 (1.0)*
			1982 Dec	8.33		
			1980 Jul	23.78	0.12 (0.06)	3.0 (1.5)*
			1982 Dec	23.90		
2021+614	G (20m)	0.23	1979 Dec	6.47	0.05 (0.04)	0.20 (0.15)
			1982 Dec	6.52		
0711+356	Q (18m)	1.62	1980 Jul	5.30	-0.13 (0.02)	-5 (1)
			1982 Dec	5.17		

Velocities calculated assuming H_o = 55 km/(s.Mpc), q_o = 0.05, and using measured redshifts where available; (*) indicates z = 1 assumed.

10-GHz OBSERVATIONS AND SOURCE CLASSIFICATION

The objects mapped in the VLBI survey were divided into different morphological classes. Two of the objects (0711+356 and 2021+614) appeared at first to belong to the class of double objects, and were classified as such, but the observations at 10 GHz revealed that they are probably asymmetric, since one of the dominant components has a flat spectrum ($\alpha \geq -0.5$) while the other has a steep spectrum ($\alpha \leq -0.5$). The full-resolution 10-GHz maps also show that the more compact features have the flatter spectra.

0710+439

Perhaps the most interesting structure is that of 0710+439. Maps of this object at both 5 GHz and 10 GHz are shown in Figure 2. Figure 2 also shows profiles along the long axis of the source at 5 GHz and 10 GHz. It is not possible to align the maps so that the peaks of all three components coincide at both frequencies. If the two outer components are aligned (as in Fig. 2), the central component becomes an asymmetric one-sided jet, with a compact flat-spectrum core at the southern end and a steep-spectrum jet pointing north towards the stronger of the two outer components. An alternative is to align the central component, in which case the northern edges of the other two components must have flat spectra. This alternative appears less likely, because for the southern component it would contradict our observation that flat-spectrum components are generally more compact than steep-spectrum components. The maps therefore suggest that in 0710+439 the center of activity coincides with the southern end of the

central component, and that the other two components are enhanced
features in two oppositely directed jets of emitted material. This
indicates that the basic ejection mechanism is two sided in at least
some sources; it was pointed out in an earlier contribution that a
similar structure is found in 3C390.3.

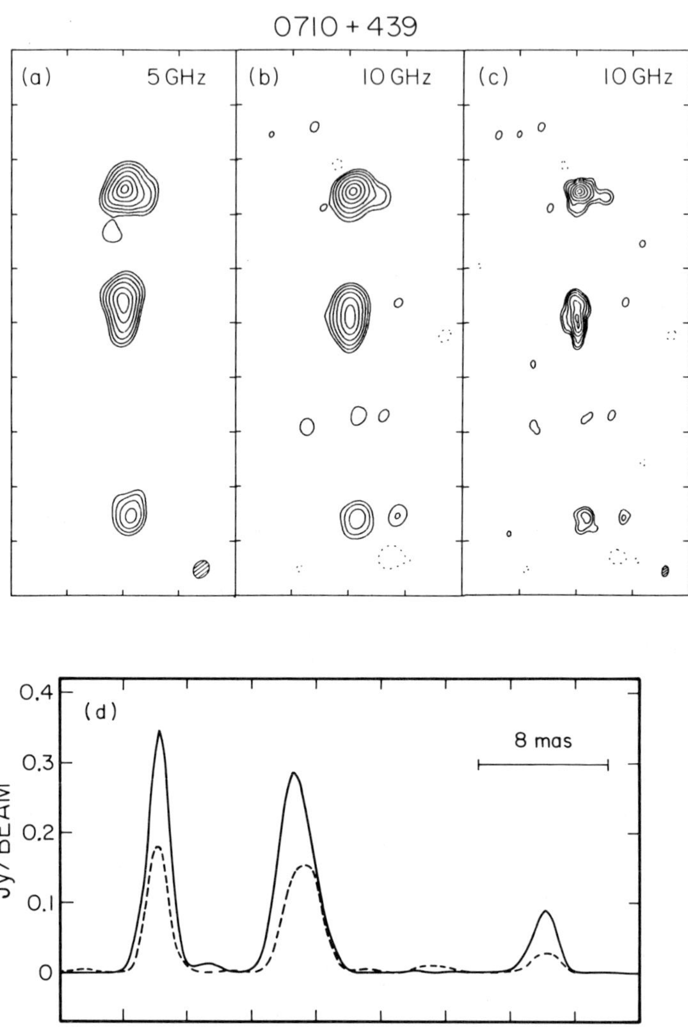

Fig. 2. Maps of 0710+439 (a) at 5 GHz, 1982 December; (b) at 10 GHz,
1982 February, displayed with the same resolution as (a); (c) at
10 GHz, with full resolution; each box is 16 x 40 mas. (d) Profiles
along the north-south axis of the maps in (a) and (b); solid line: 5
GHz, dashed line: 10 GHz.

4C39.25 - A CONTRACTING SOURCE?

David B. Shaffer
Interferometrics Inc.

The quasar 4C39.25, whose nuclear radio structure was stable for many years, now shows a contraction in the brightness distribution. Possible explanations are discussed.

From 1972 until early 1979, at frequencies from 5 GHz to 15 GHz, the radio nucleus of the quasar 4C39.25 could be characterized as a simple double source, with a separation of 2.0 mas (Shaffer _et al._ 1977, and unpublished; Pauliny-Toth _et al._ 1981; Bååth _et al._ 1980). During these years, the strengths of the components changed in a manner consistent with evolving synchrotron-self-absorbed sources. The relative constancy of separation put 4C39.25 in marked contrast to the superluminal sources which it resembles in many ways, such as radio spectrum and variability.

4C39.25 has been observed two or three times per year at 8.4 GHz since late 1979 for NASA Crustal Dynamics Project Mark III VLBI geodesy experiments. These observations provide enough data to make low dynamic range (about 20:1) maps. The first map from 1979 November showed that 4C39.25 had contracted, with a component separation of about 1.8 mas. By the latest epoch for which I have a map, 1982 December, the peaks in the brightness distribution were only 1.6 mas apart. The apparent contraction rate is between 0.05 and 0.1 mas per year, or v/c between 1.4 and 2.8 for Ho=55 km $\sec^{-1}Mpc^{-1}$(z=0.698). The contraction is confirmed by observations at 10.7 GHz in 1982 June (Shaffer, Romney, and Marcaide: private communication), which show that the first minimum in the visibility function has moved out from the origin. The equivalent separation at 10.7 GHz was 1.68 mas.

In 1979, the source was still very much the simple double source that was seen throughout the '70's, although at a somewhat smaller separation. By 1982, the source structure had become more complex and distinctly elongated along the almost east/west line of separation between the components. The probable core (the western component based on size and evolution) was relatively weaker at all wavelengths in the early '70's, but it strengthened continuously until it was the stronger component in the 1979 map. From 1979 to 1982, the western component weakened somewhat with respect to the eastern component. The total 8 GHz flux density of 4C39.25 reached a peak of about 11 Jy in 1972 and decreased almost linearly to about 6 Jy by the end of 1982 (Aller and Aller, private communication).

R. Fanti et al. (eds.), VLBI and Compact Radio Sources, 135–136.

This decline must have occurred almost totally in the eastern component. Its strength decreased by about a factor of three in the 1972-1982 interval, while the western component roughly doubled in strength, peaking about 1980, and then also starting to decline.

The decrease in the separation between the peaks in the brightness distribution of the two compact components in 4C39.25 is certainly real. Does this change represent a true contraction, though? If so, is the change consistent with the beaming models so in vogue at this time? Can the observations be explained by other than actual component motion?

The beaming model can give rise to (superluminal) contractions if the line of sight to the source lies in, or very close to, the plane which contains the curved trajectory of moving components. Curved trajectories are seen directly in transverse or oblique views of the sources 3C309.1 and 3C418 as shown in maps at this Symposium, and are postulated to explain the 1.3 cm observations of 3C345 (Moore, this Symposium). In this orientation, the constant separation from 1972 to 1979 occurs while the moving component is seen head-on, and the contraction occurs after the component "turns the corner" and starts to cut across our line of sight, in front of the quasar core. This explanation predicts that the components should eventually appear to merge. Later, the moving component should be seen on the other side of the core, giving the appearance of an expanding source. This model may explain the elongation seen in the 1982 map as radiation from material which trails the head of a beam that has just turned the corner. Unfortunately for this model, it is the supposed core (the western component) which seems to be most elongated, making it the beam rather than the core. Perhaps the rapid variation of this component and its smaller size are artifacts of our almost exactly head-on viewing of it.

Motion in a source can be mimicked by the appearance of new components. Stationary components of varying strength were postulated to explain superluminal motion until 3C273 and 3C345 were convincingly shown to be expanding (Pearson et al. 1981; Cohen et al. 1981). In 4C39.25, a new stationary component appearing between the two existing components can not be ruled out by current data. Neither can the ejection of a new component from the western core. In either case, a rather delicate balancing act in flux density between the old and new components must have occurred, since no rapid variations larger than 1 Jy have been seen, whereas the new component would have to be 2 to 3 Jy in late 1982 to cause the observed shift in the brightness peak. If a new component has been ejected from the core, 4C39.25 would be different from all other expanding sources. In those objects, all the non-core components move, whereas in 4C39.25 the eastern component is stationary. If a new component has been ejected, it may eventually run into the eastern component.

Bååth, L. B. et al.: 1980, Astron & Astrophys 86, pp. 364-372.
Cohen, M. H. et al.: 1981, Ap. J. 247, pp. 774-779.
Pauliny-Toth, I. I. K. et al.: 1981, A. J. 86, pp. 371-385.
Pearson, T. J. et al.: 1981, Nature 290, pp. 365-368.
Shaffer, D. B. et al.: 1977, Ap. J. 218, pp. 353-360.

THE NUCLEUS OF NGC1275

J.D. Romney, W. Alef, I.I.K. Pauliny-Toth, and E. Preuss
Max-Planck-Institut für Radioastronomie, Bonn, FRG

K.I. Kellermann
National Radio Astronomy Observatory, Green Bank, USA

The bright radio nucleus 3C84 of the peculiar galaxy NGC1275 was among the first sources studied by VLBI techniques sixteen years ago (Clark et al., 1968). Early observations at short centimeter wavelengths using arrays of up to four telescopes (Schilizzi et al., 1975, and references therein) demonstrated the presence of structure significantly more complex than that seen in other strong compact objects, but were insufficient to determine the brightness distribution with the relatively primitive methods then in use. By the time of the most recent international VLBI symposium (Aug. 1978, in Heidelberg, FRG), Pauliny-Toth et al. (1976) had succeeded in modelling the structure seen by a 4-station transatlantic array, revealing three distinct bright emission regions aligned in position angle $-9°$, and connected by non-colinear, more diffuse features. Further observations and a reanalysis of earlier data by Preuss et al. (1979) confirmed previous inferences that the three major condensations remained stationary but varied in brightness.

The predominant impression arising from these early observations at or near 2.8 cm wavelength was one of extreme, even "pathological" structural complexity, much in contrast to the relatively simpler morphology seen in other sources. More recently, the spectral index study by Unwin et al. (1982) and especially the 1.3 cm map of Readhead et al. (1983) have evinced a structural affinity between 3C84 and many other active extragalactic objects, where a compact variable flat-spectrum "core" and one or more relatively diffuse steeper-spectrum "jet" components are typical. The failure to recognize this connection ·in the earlier observations of 1972-76 is attributable largely to the lack of activity in the core during this period, and to the coincidentally equal intensities of the major components at 2.8 cm, the only wavelength then affording both sufficient angular resolution and acceptable calibration.

3C84 remains an unusually complex source. To some extent at least this is a selection effect arising from its very near distance, although the extreme radio luminosity and anomalous optical morphology suggest an intrinsic complexity. The recognition that 3C84 can be classified as a core-jet structure (e.g., Readhead et al.) provides an opportunity to study in detail the processes occurring in core-jet sources.

137

R. Fanti et al. (eds.), VLBI and Compact Radio Sources, 137–140.
© *1984 by the IAU.*

Since 1979 we have obtained 2.8 cm maps of 3C84 at three epochs using transatlantic arrays of seven (1979.1) and five (1981.1, 1982.1) stations. This observing wavelength is optimal for studies connecting the very compact and more extended structures; the major bright features are well resolved, while instrumental factors and the component spectra combine favorably to allow mapping with high dynamic range. The maps, presented in Fig. 1 together with the 1974.5 model of Pauliny-Toth et al., were derived using an adaptive calibration technique similar to that of Schwab (1980); earlier versions of the 1979.1 and 1981.1 maps were presented by Romney et al. (1982) before application of the amplitude corrections. We describe below the general structural evolution observed, and then consider in more detail the component motions so strikingly evident in Fig. 1.

The observations at epoch 1979.1 yielded by far the most complete aperture coverage, allowing a dynamic range of about 50 to be achieved. The core-jet configuration pointed out by Romney et al. is evident: the northern core component is resolved at this epoch into two intensity peaks separated by 0.9 milliarcsec (1.5 ly) along position angle $\sim 45^{\circ}$, the southwestern peak coinciding with a sharp, nearly right-angle bend leading into the diffuse 6 milliarcsec region which itself exhibits internal curvature. Contained within this diffuse structure are the relatively weak central component and the southern condensation which is the brightest part of the source at this epoch. There are no corresponding features to the north of the core greater than a few percent of this peak brightness.

The 1981.1 map has relatively limited dynamic range because only abbreviated tracks were observed, but the strengthening of the core is evident. This is presumably linked to the outburst seen to begin about 1980 at millimeter wavelengths (O'Dea, private communication), although the fading southern component limits the 2.8 cm flux density increase to 13% over the 1979.1 measurement. The marked increase in separation between the core and the southern feature, previously described by Romney et al., is discussed below. At epoch 1982.1 all these trends are seen to continue. The brightening core dominates the diffuse region; the southern component continues to fade as it recedes farther from the core, although it has become so extended that its location is increasingly indistinct. The double structure in the core does not appear at either of these later epochs.

In quantifying the motion of the southern component we have adopted the conventional registration in which the core is presumed stationary. The double northern component in the 1979.1 map presents an additional problem; we have identified the northeastern peak as the core, as seen in Fig. 1, on the basis of its spectral index and compactness. With this registration a linear fit yields an angular velocity of 0.24 ± 0.06 milliarcsec y^{-1}, corresponding to $v = 0.4c$ if $H_0 = 50$ km s^{-1} Mpc^{-1}. 3C84 is unique among active radio nuclei in manifesting this subluminal but finite velocity.

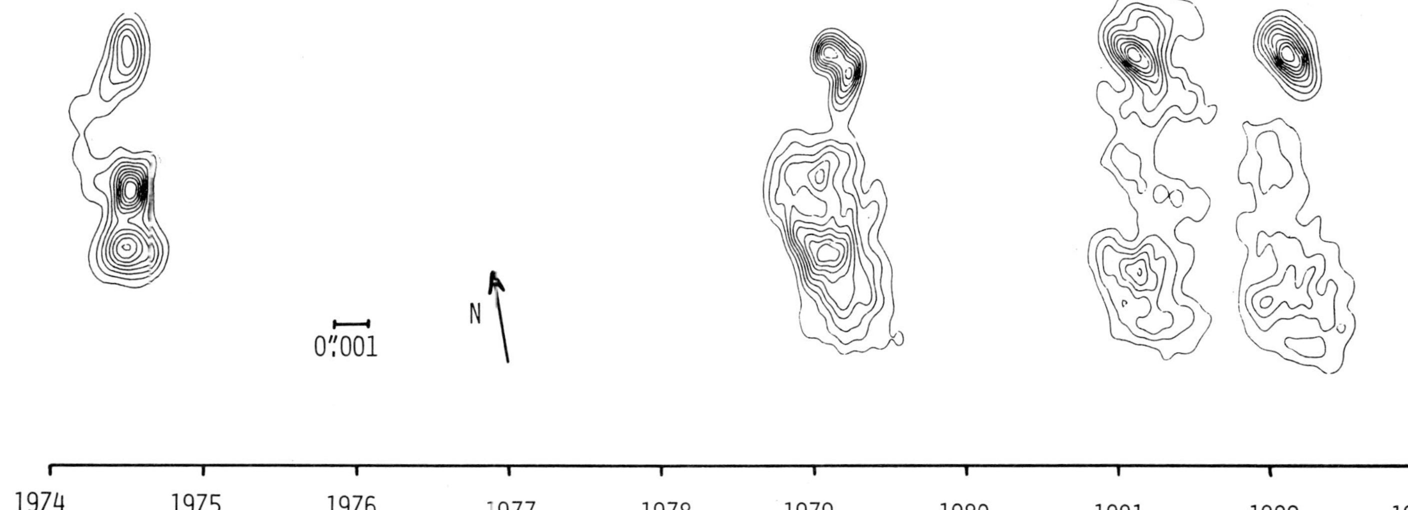

Fig. 1: Evolution of the compact structure in 3C84. The abscissa scale is observing epoch; the ordinate is angular distance along position angle $-9°$. Registration of the individual maps is discussed in the text. The contour unit is 10% of the peak brightness in each map. A uniform restoring beam was used, 0.85 by 0.43 milliarcsec in position angle $-20°$.

It is not clear how best to interpret the central component and the earlier observations. The best fit to the southern component for the three recent epochs passes quite near the central region of the 1974.5 model, although Preuss et al. reported no measurable positional changes prior to 1978, and the southern peak in the 1979.1 map is consistent with its location at the earlier epochs. The central feature is essentially stationary over the entire 10-year interval.

Although the observed transverse velocity of the one-sided structure is subluminal, it is instructive to consider an interpretation in the context of the relativistic-beaming model commonly invoked to explain both the apparent transverse motion and the one-sided morphology in superluminal sources. An intrinsically two-sided jet is assumed, with an orientation near the line of sight and velocity such that the counter-jet is suppressed (i.e., the approaching jet enhanced) by the necessary factor of ~ 50 and an apparent transverse motion at $\sim 0.4c$ is observed. This configuration can also explain naturally the sharp observed bend in the jet as a magnification of a minor intrinsic bend passing close to the line of sight; Allan (1984) has analyzed both the geometric and Doppler effects. The required values of the jet parameters offer an interesting contrast to the superluminal case: for example, at a typical inclination to the line of sight of 5°, the jet is only mildly relativistic, with Lorentz factor $\gamma \sim 2$, and a de-projected length of ~ 40 pc. For a more typical superluminal $\gamma \sim 7$ the jet must lie within a few tenths of a degree of the line of sight, and is more than an order of magnitude longer. Of course in the case of 3C84 it is also possible to postulate an intrinsic-ally one-sided jet, containing knots moving at $0.4c$ in or near the plane of the sky.

REFERENCES

Allan, P.: 1984, this volume, p. 195.
Clark, B.G., Kellermann, K.I., Bare, C.C., Cohen, M.H., and Jauncey, D. L.: 1968, Astrophys. J. 153, pp. 705-714.
Pauliny-Toth, I.I.K., Preuss, E., Witzel, A., Kellermann, K.I., Shaffer, D.B., Purcell, G.H., Grove, G.W., Jones, D.L., Cohen, M.H., Moffet, A.T., Romney, J.D., Schilizzi, R.T., and Rinehart, R.: 1976, Nature 259, pp. 17-20
Preuss, E., Kellermann, K.I., Pauliny-Toth, I.I.K., Witzel, A., and Shaffer, D.B.: 1979, Astron. Astrophys. 79, pp. 268-273.
Readhead, A.C.S., Hough, D.H., Ewing, M.S., Walker, R.C., and Romney, J.D.: 1983, Astrophys. J. 265, pp. 107-131.
Romney, J.D., Alef, W., Pauliny-Toth, I.I.K., Preuss, E., and Kellermann, K.I.: 1982, I.A.U. Symposium 97, "Extragalactic Radio Sources", pp. 291-292.
Schilizzi, R.T., Cohen, M.H., Romney, J.D., Shaffer, D.B., Kellermann, K.I., Swenson, G.W.Jr., Yen, J.L., and Rinehart, R.: 1975, Astro-phys. J. 201, pp. 263-274.
Schwab, F.R.: 1980, Proc. Soc. Photo-Opt. Instrum. Eng. 231, pp. 18-25.
Unwin, S.C., Mutel, R.L., Phillips, R.B., and Linfield, R.P.: 1982, Astrophys. J. 256, pp. 83-91.

THE RADIO JET IN 3C418 [+]

T.W.B. Muxlow[1], M.Jullian[1], R. Linfield[2]
[1] NRAL, Jodrell Bank, Macclesfield, Cheshire, U.K.
[2] University of California, Berkeley, U.S.A.

ABSTRACT

Models are here presented interpreting the arcsecond radio structure found in the quasar 3C418 as a precessing jet. No simple solution exists and it has been found necessary to complicate the model geometry by allowing the precession cone-angle to increase with time. The timescales suggest the presence of a close binary black-hole system within the quasar core. The modelled behaviour of the precession cone close to the core is explained either by interaction of the jet with the surrounding galactic material or by the observation of the source during the capture of one black-hole by the other.

INTRODUCTION

Much recent interest has arisen regarding precessing jets in radio-galaxies and quasars following the work of Gower <u>et al</u>. (1982) who suggest that a number of extragalactic radio sources may possess ballistic precessing jets similar to those found in the galactic source SS433. For the quasars in their sample, they derive jet velocities near 0.7c together with precession timescales of order 10^4 years. Begelman, Blandford and Rees (1980) have shown that such periods can be produced by a close binary black-hole system within the quasar core; and that a system undergoing mutual capture could account for those cases where the precession cone-angle increases on timescales comparable with the precession period.

MODELLING

We have attempted to account for the highly curved arcsecond radio structure seen in the quasar 3C418 (z=1.69) with a twin-sided precessing jet model: assuming that the jet velocity remains constant throughout its length and that the precession period and angle between the precession axis and the line of sight do not vary. Since the jet may not be intrinsically smooth, we have concentrated on explaining the observed geometry, the detailed radio brightness distribution being considered to be of secondary importance.

+ Discussion on page 436

R. Fanti et al. (eds.), VLBI and Compact Radio Sources, 141–143.
© *1984 by the IAU.*

No satisfactory geometrical solution was found for a simple precessing jet, since the extreme curvature of the observed structure within 1 arcsec of the core could not be reconciled with the small degree of curvature in its outer regions. We have thus found it necessary to introduce an additional model parameter: we have allowed the precession cone-angle to increase close to the quasar core (i.e. increase with time). This allows the line of sight to lie within the precession cone for the inner jet region which appears highly curved, whilst lying outside it for the outer jet where the apparent bending is much less pronounced. With the increased complexity of our revised model, no unique solution exists. Many satisfactory fits to the geometry were found. Four representative solutions are shown here (γ= 1.05, 1.41, 2.10, 5.00).

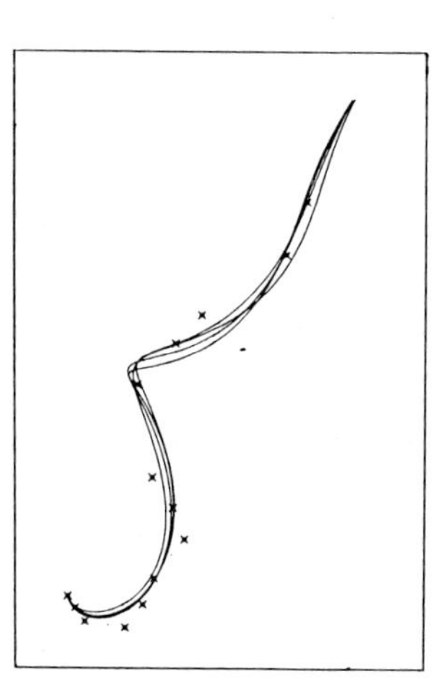

Left:
Ridge-line geometries for the illustrated four solutions overlayed upon component positions taken from the MERLIN/EVN map.

Right:
MERLIN/EVN map at 1.7GHz.

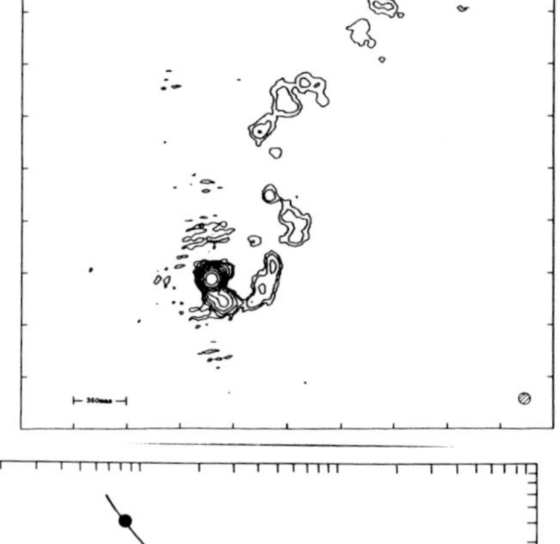

Right:
Model parameters for the illustrated solutions.

Below:
MERLIN map at 1.7GHz.

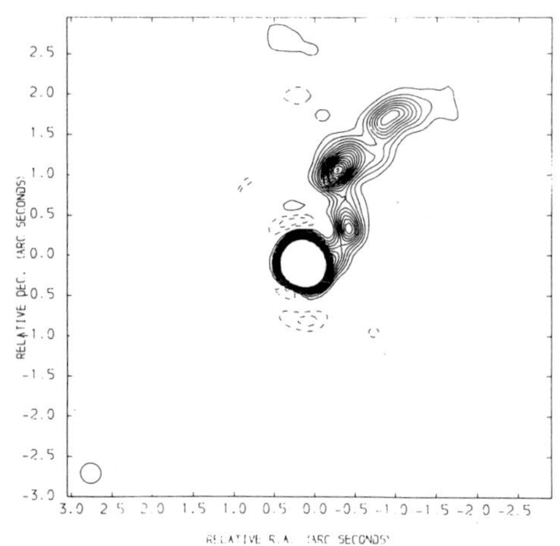

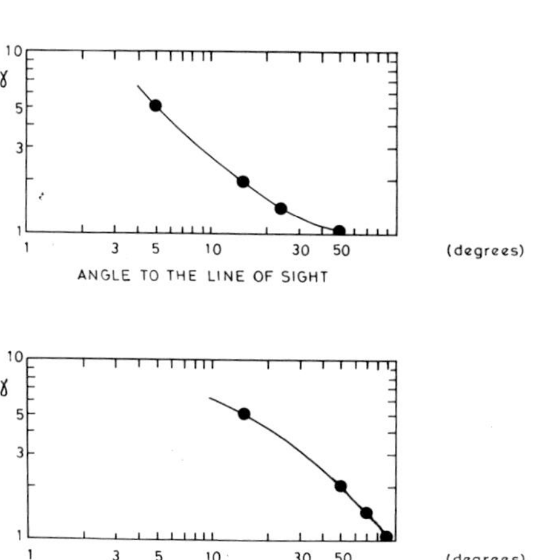

DISCUSSION

The precession angle opens rapidly within $\simeq 0.2$ cycles of the core for all the solutions. At greater distances this angle drops to a few degrees. The slow solutions lie substantially in the plane of the sky, must be intrinsically one-sided, and constitute a short jet with a large precession angle at the core. (The $\gamma = 1.05$ jet is some 15 kpc in length and has an opening half-angle of precession at the core of 90°), whereas the fastest solutions have jets of length in excess of 100kpc with a much reduced cone-angle at the core. Both the line of sight and precession cone-angles decrease with increasing jet speed. The geometry of the fastest ($\gamma = 5.00$) solution is compatible with the scheme of Orr and Browne (1982) to unify steep and flat spectrum radio quasars. The precession periods in these solutions are such that they could be produced by a close massive binary black-hole pair within the quasar core.

If the jet motion is purely ballistic, then the behaviour of the precession cone close to the core requires that the cone-angle be changing on timescales of order or shorter than the precession period. This could occur if the black-hole pair were undergoing mutual capture. This, however, places our observations at a very specific epoch in the source history which would seem somewhat contrived, although the probable promotion to a short lived high luminosity state which may be associated with such an event could account for our detection. Alternatively, we may identify the cone-angle behaviour as evidence for non-ballistic motion in the surrounding galactic material. Such bending might arise from fast winds or buoyant refraction. We can estimate the required pressure gradients from the radius of curvature and the observed internal energy density of the jet in this region ($\simeq 100$mas from the core). Assuming that the jet contains no thermal material we find that in all cases we require that $\nabla p \simeq 10^{-29} Nm^{-3}$. This is insensitive to γ because the faster solutions have lower values of internal energy density and larger radii of curvature. It is possible to envisage circumstances where such a high value of pressure gradient could exist; they do however lie outside the range of environments found in known galaxies.

REFERENCES

Begelman,M.C., Blandford, R.D. and Rees, M.J., 1980, Nature 287, 307.
Gower, A.C., Gregory, P.C., Hutchings, J.B. and Unruh, W.G., 1982,
          Ap.J. 262, 478.
Orr, M.J. and Browne, I.W.A., 1982, Mon.Not.R.astr.Soc., 200, 1067.

SUB-LUMINAL MOTIONS IN THE NUCLEUS OF M 87 [+]

M. J. Reid and J. H. M. M. Schmitt
Harvard-Smithsonian Center for Astrophysics

P. N. Wilkinson
Nuffield Radio Astronomy Laboratories

K. J. Johnston
U. S. Naval Research Laboratory

We have observed the nucleus of the elliptical galaxy M87 with VLBI arrays at 18 cm wavelength at two epochs separated by about two years. The results of the first epoch observations taken at 1980.12 are described in detail by Reid <u>et al</u> (1982, Ap. J. 263, pp 615-623). They indicated that the nucleus of M87 exhibited a highly asymmetric emission pattern with a bright "core" and long, thin "jet" extending for more than 0.05 arcsec towards the 20 arcsec jet seen in radio, optical, and X-ray radiation. This morphology is remarkably similar to that seen in the quasar 3C273. Asymmetric core-jet structures, possibly coupled with super-luminal motions as in 3C273, have been interpreted in the framework of relativistic beaming models (cf. Blandford, Mckee, and Rees: 1977, Nature 267, pp 211-216). Therefore, we conducted second epoch observations of M87 at 1982.27 to look for such motions.

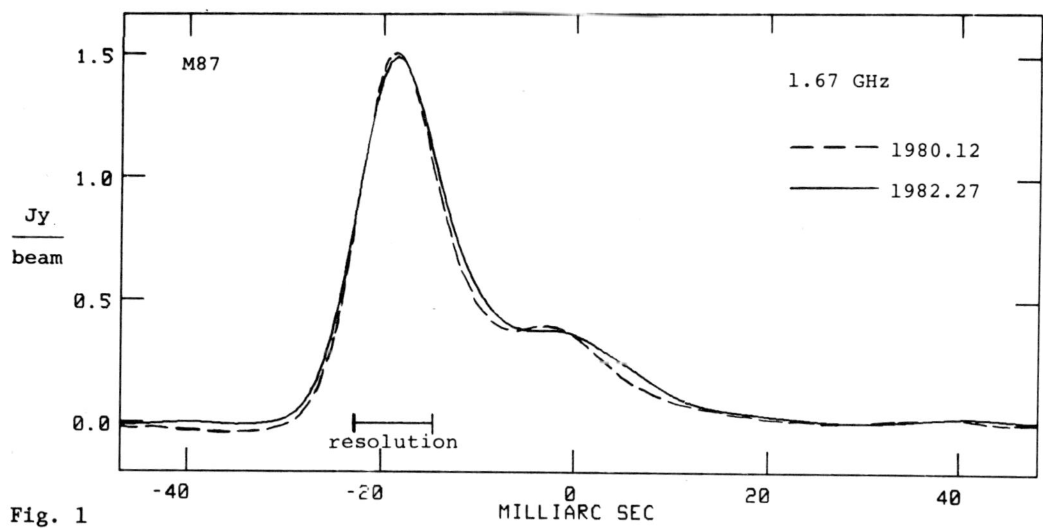

Fig. 1

+ <u>Discussion on page 437</u>

145

R. Fanti et al. (eds.), VLBI and Compact Radio Sources, 145-146.
© *1984 by the IAU.*

Our maps of the nucleus of M87 for the two epochs are remarkably
similar. In Figure 1 we present the intensity of the emission seen in
our maps along the position angle of the jet (288° E of N). While the
emission from the second epoch appears slightly more extended than from
the first epoch, the differences are probably not significant at the
two-sigma level. We estimate any component expansion to be
< 0.5 milli-arcsec/year, which corresponds to < 0.2 c. Under the
simplest assumptions, the absence of a detectable counter-jet requires
that the ratio, R, of the observed intensity of the jet and counter-jet
is given by

$$R > [\ (1 + \beta \cos \theta) \ / \ (1 - \beta \cos \theta) \ ]^{2.5} \tag{1},$$

where β is the flow speed of the jet in units of c and θ is the angle
the jet makes to our line-of-sight. Our limit on relative proper
motions, β_{ob} , of components in the jet requires that

$$\beta_{ob} < \beta \sin \theta \ / \ (1 - \beta \cos \theta) \tag{2}.$$

As shown in Figure 2, most of the parameter space for θ (plotted in equal
solid angle intervals) and β are excluded by relations (1) and (2).

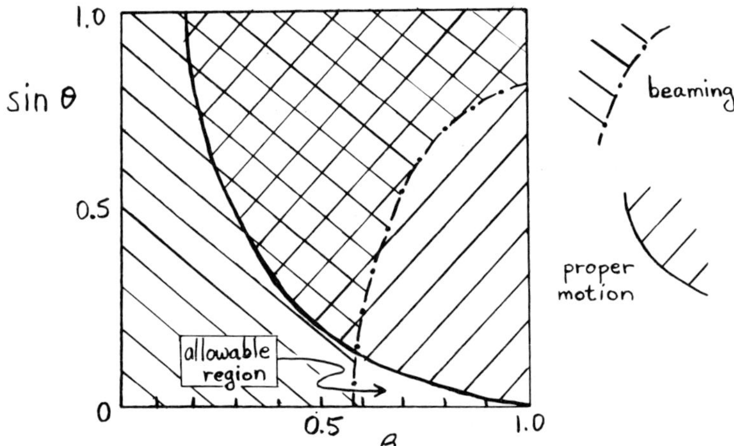

Fig. 2

Therefore, the most straight-forward implementation of a beaming model
appears unlikely to hold for M87. However, some reasonable combinations
of β and θ are still allowed and we cannot exclude these possibilities
at this time.

Alternatively, it may be that the simple beaming model requires
alteration to allow for smooth jets with relativistic flows but
stationary apparent images. This would require jets which do not form a
small number of discrete condensations, and whose central energy source
is very stable. Finally, were jets intrinsically one-sided, there would
be much greater freedom in choosing β and θ . This possibility
deserves further study.

INNER OSCILLATIONS IN THE M87 JET [+]

J.-L. Nieto (1)
G. Lelièvre (2)
J.-L. Heudier (3)
A. Maury (3)

(1) Observatoire du Pic-du-Midi et de Toulouse, France
(2) CFH Telescope Corporation, U.S.A
(3) Telescope de Schmidt, CERGA, France

Continuing a long project of investigations of the optical morphology of the M87 jet (see review in Nieto, 1983), we have obtained at the Cassegrainan focus of the CFH telescope in March 1983 three UV exposures with the wide-field electronographic camera (Lallemand et al., 1970). The aim was to achieve the same order of resolution as that of the photographic plates obtained by Nieto and Lelièvre (1982), but with a much higher signal to noise ratio.

This last characteristics of these observations allowed the detection of :
i) a very faint wave at the end of the jet,
ii) oscillations in the inner jet,
iii) a continuous component between the inner complexes,
iv) a faint jet-like nuclear extension elongated along the jet ($\sim 1''$) and bent southwards. This curvature is curiously similar to that found by the VLBI observations (Reid et al., 1982) at a scale 20 times smaller, suggesting that the VLBI jet and this small optical jet are the same feature.

This is evidence for wiggles starting right at the nucleus and continuing for a few hundreds of parsecs in the inner jet, supporting the hypothesis of a precessing nucleus. If we assume that the regular spacing of the inner knots found by Nieto and Lelièvre (1982) and confirmed by Biretta et al. (1983) is related to the precession of the nucleus, we derive, for an ejection velocity on the sky plane $\beta \sim 0.02\text{-}0.1$, a period of precession $P = 5 \times 10^3 - 2 \times 10^4$ years.

A more detailed description of the M87 and the 3C 273 jets (from data in U, B, V obtained in the same conditions), as well as a comparison of the two jets is presented in submitted papers (Lelièvre et al., 1983 a and b).

+ Discussion on page 437   147

R. Fanti et al. (eds.), VLBI and Compact Radio Sources, 147–148.
© *1984 by the IAU.*

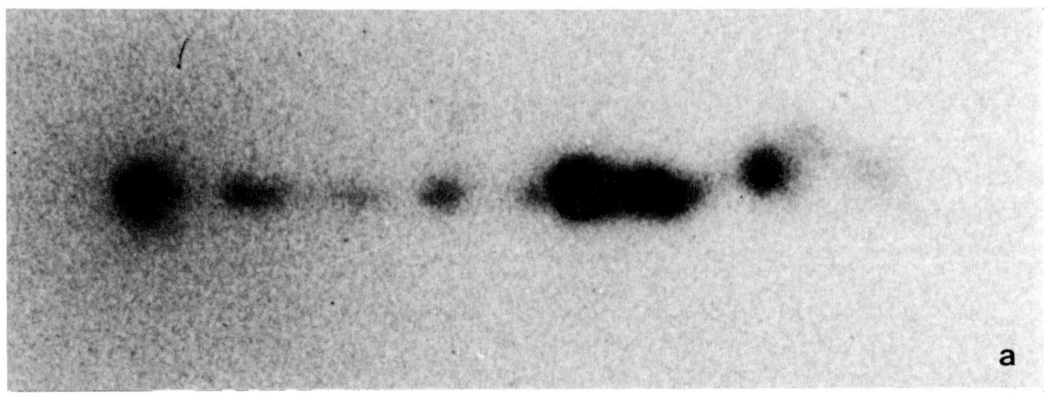

Figure : (a) 10 mn UV exposure of the M87 jet
         (b) 30 mn UV exposure after an image treatment showing the
             small inner jet (arrow).

REFERENCES :

Biretta, J.A., Owen, F.N., Hardee, P.E., 1983, Astrophys. J., in press
Lallemand, A., Renard, L., Servan, B., 1970, Comptes rendus Acad. Sci.,
     270, pp. 385
Lelièvre, G., Nieto, J.-L., Wlérick, G., Servan, B., Renard, L., Horville,
     D., 1983 a, in press
Lelièvre, G., Nieto, J.-L., Horville, D., Renard, L., Servan, B., 1983 b,
     submitted
Nieto, J.-L., 1983, Torino workshop on extragalactic jets, Ferrari, A.,
     and Pacholczyk, A.G. ed., Reidel : Dordrecht, p. 113.
Nieto, J.-L., Lelièvre, G., 1982, Astron. Astrophys., 109, pp. 95
Reid, M.J., Schmitt, J.H.M.M., Owen, F.N., Booth, R.S., Wilkinson, P.N.,
     Shaffer, D.B., Johnston, K.J., Hardee, P.E., 1982, Astrophys. J.,
     263, pp. 615

THE STRUCTURAL VARIATIONS OF 3C454.3 AND 2134+004

I.I.K. Pauliny-Toth, R.W. Porcas, A. Zensus
Max-Planck-Institut für Radioastronomie, Bonn, F.R.G.

K.I. Kellermann
National Radio Astronomy Observatory, Green Bank, W.Va., U.S.A.

ABSTRACT

3C454.3 has a core-jet structure on the milliarcsec scale. Following a flux density outburst, the core showed a "superluminal brightening". Comparison with older observations suggests superluminal motion in the jet. 2134+004, in contrast, is a double source, with no significant ($v/c < 1$) relative motion of the components.

INTRODUCTION

The quasars 3C454.3 ($z = 0.859$) and 2134+004 ($z = 1.936$) are both strong, variable radio sources. 2134+004 has shown only slow variations, while 3C454.3 has had large changes in flux density at cm wavelengths, as well as significant variations at metre wavelengths. The time scale of the latter has led Jones and Burbidge (1973) to suggest that bulk relativistic motion with $\gamma \sim 50$ should be present. A large outburst in 3C454.3 occurred in mid-1981.

We have observed both sources with an array consisting of the 100 m telescope in Effelsberg and 4 US antennas at wavelengths of 2.8 cm (1981.4) and 6 cm (1981.6), and reobserved 3C454.3 at 2.8 cm with the same array at epoch 1982.1.

3C454.3

The hybrid maps of the source at 6 cm wavelength (Fig. 1) show a bright core and a narrow, elongated "jet" extending to about 10 milli arcsec (~ 90 pc, for $H_0 = 55$ km s^{-1} Mpc^{-1}, $q_0 = 0.05$) from the core in P.A. -65°. The peak in the brightness of the jet lies about 6.5 mas (milliarcsec) from the core. The arcsec jet (Browne et al. 1982) is in P.A. -48°, so that the curvature typical of superluminal sources is present. At 2.8 cm, the jet component is weaker, with a peak brightness of ~ 1 percent of that of the core, which, however, shows an extension

R. Fanti et al. (eds.), VLBI and Compact Radio Sources, 149–152.
© *1984 by the IAU.*

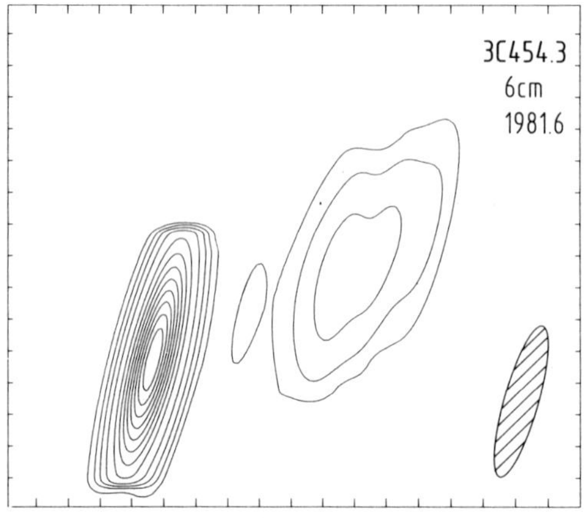

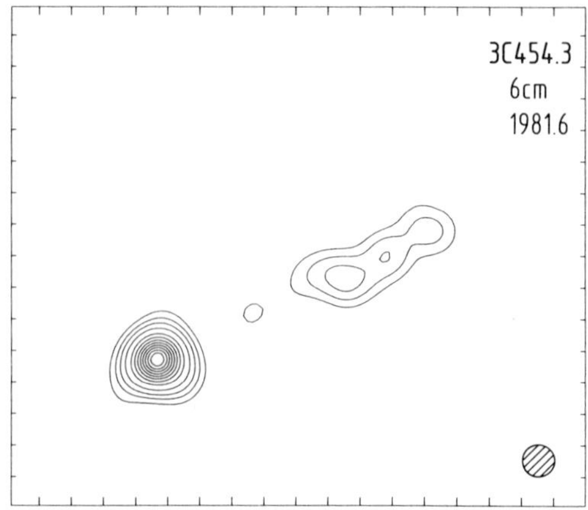

Fig. 1: Hybrid maps of 3C454.3 at 6 cm wavelength. Left: restored with clean beam of 5 x 1 mas; right: restored with a 1 mas circular beam. Contours are 1, 2.5, 5, 10, 20 ... 90 percent of the maximum.

in the same P.A. of $-65°$.

The core shows no evidence for any double structure on a scale > 0.3 mas. The 2.8 cm data from the long, transatlantic baselines for each of the two epochs are well fitted by an "inner core" ($\lesssim$ 0.25 mas) and an "outer core" with a diameter of about 0.5 mas. The flux density of the source increased from 16.3 at epoch 1981.4 to 18.9 Jy at epoch 1982.1. This increase occurred in the outer core, with little change in the flux density of the inner core:

Epoch	Inner core		Outer core		Total core
	Flux(Jy)	Size(mas)	Flux(Jy)	Size(mas)	flux(Jy)
1981.4	9.4	$\lesssim$ 0.2	5.4	0.55	14.8
1982.1	8.7	0.25	8.2	0.55	16.9

The core of 3C454.3 has thus shown a "superluminal brightening": the increase in brightness of a region $\sim$ 0.3 mas in radius in 0.7 years requires v/c $\sim$ 25.

Earlier 6 cm data, obtained at epoch 1974.1 (Pauliny-Toth et al. 1981) show a similar core-jet structure, but with the peak in the brightness of the jet component about 4 mas from the core, rather than the 6.5 mas seen at epoch 1981.6. The earlier data were sparser, but taken at face value, suggest a motion of 0.33 mas per year, or v/c $\sim$ 18, a value similar to that required to account for the superluminal brightening of the core.

The available data on 3C454.3 thus support the predicted relativistic bulk motion, both in the core and in the jet components.

2134+004

Hybrid maps of the source at 6 and 2.8 cm are shown in Fig. 2. At
both wavelengths, the source is double, with a component separation of
1.7 mas (19 pc) in P.A. 62°. The separation is identical with that
obtained at epoch 1973.5 (6 cm; Pauliny-Toth et al. 1981). A similar
separation, of 1.8 mas, was found by Schilizzi et al. (1975) for epochs
between 1972.3 and 1974.1, if we identify two of their three components
(only 0.4 mas apart) with one of the components given here.

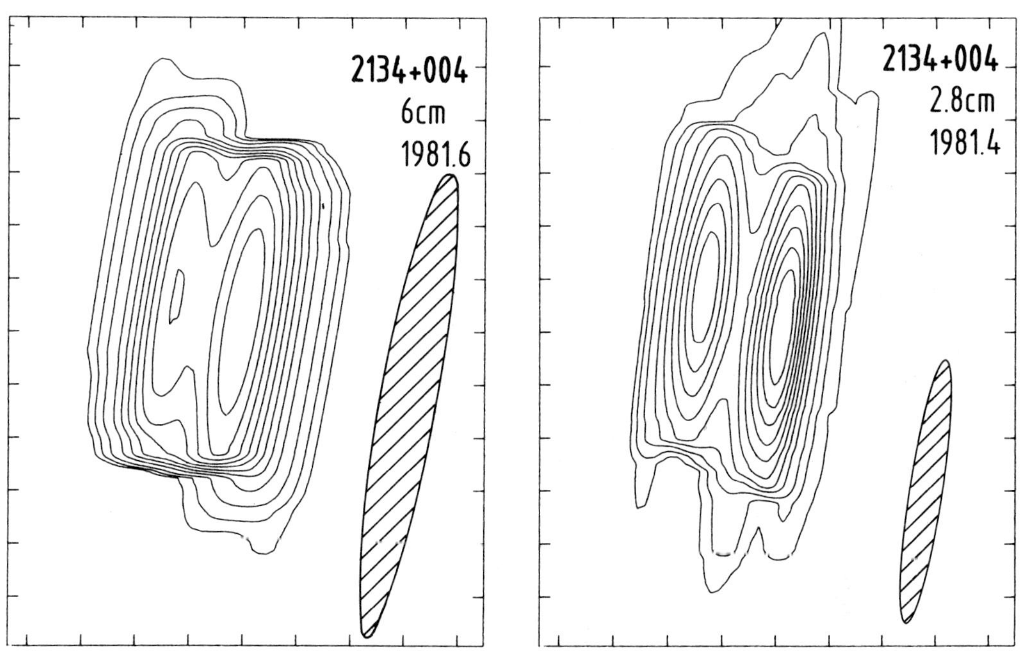

Fig. 2: Hybrid maps of 2134+004 at 6 (left) and 2.8 (right) cm wavelength, restored with the clean beams of 9 x 1 mas and 5 x 0.6 mas
respectively. Contours are 2.5, 5, 10, 20 ... 90 percent of the maximum.

We place a limit of 0.1 mas on any change in the component separation over the 9.3 year interval covered by the observations. The limit
on the relative rate of separation is 0.01 mas per year, or v/c < 1.

Nevertheless, the source appears more resolved than at earlier
epochs: the models of Schilizzi et al. (1975) give half the total flux
density in a component < 0.3 mas in size, whereas the present data give
equivalent gaussian sizes of ∿ 0.9 mas for both components. The behaviour
of the source appears similar to that of 3C84, which shows relative

component motion of $\sim$ 0.6 c and component expansion velocities of $\sim$ 0.1 c
(Preuss et al. 1979, Romney et al. 1982). In the case of 2134+004, however, if the early data are correct, the required velocity of expansion
of the component(s) is at least 3 c.

The frequency of the spectral peak in 2134+004 is about 5 GHz,
compared with about 8 GHz in 1972-1973, reflecting the component expansion. The spectral indices between 6 and 2.8 cm are α = -0.4 and -0.1
for the W and E components, respectively $_.$ (S $\propto \nu^{\alpha}$). Assuming the same
turnover frequency of 5 GHz for both components and taking the component
sizes of 0.9 mas gives a magnetic field in the emitting regions of
$\sim$ 5 x 10^{-3} gauss and a lifetime for relativistic electrons radiating at
10 GHz of $\sim$ 10^3 years.

REFERENCES

Browne, I.W.A., Clark, R.R., Moore, P.K., Muxlow, T.W.B., Wilkinson, P.
    N., Cohen, M.H., and Porcas, R.W.: 1982, Nature 299, pp. 788-793.
Jones, T.W. and Burbidge, G.R.: 1973, Astrophys. J. 186, pp. 791-799.
Pauliny-Toth, I.I.K., Preuss, E., Witzel, A., Graham, D., Kellermann,
    K.I. and Rönnäng, B.: 1981, Astron. J. 86, pp. 371-385.
Preuss, E., Kellermann, K.I., Pauliny-Toth, I.I.K., Witzel, A. and
    Shaffer, D.B.: 1979, Astron. Astrophys. 79, pp. 268-273.
Romney, J.D., Alef, W., Pauliny-Toth, I.I.K., Preuss, E. and Kellermann,
    K.I.: 1982, IAU Symp. No. 97, *"Extragalactic Radio Sources"*, publ.
    D. Reidel, Dordrecht, pp. 291-292.
Schilizzi, R.T., Cohen, M.H., Romney, J.D., Shaffer, D.B., Kellermann,
    K.I., Swenson, G.W., Yen, J.L. and Rinehart, R.: 1975, Astrophys.
    J. 201, pp. 263-274.

EVIDENCE FOR RELATIVISTIC MOTION BASED ON ARCSECOND STRUCTURE [+]

R. A. Perley
National Radio Astronomy Observatory, Socorro, U.S.A.

ABSTRACT   Observations with arcsecond-resolution interferometers can be used to detect motion or changes of the large scale structure associated with core dominated objects. If these arcsecond structures are expanding at relativistic velocities, detectable changes should be measureable within a few years. Indirect estimates of jet velocities can be made by a number of arguments, but none are model independent or free of dependence upon poorly established parameters. Application of some of these methods are given on sources with established VLB structure.

1.   <u>Direct Measurement of Proper Motion</u>.   Nearly all the known superluminal sources contain arcsecond structures located on the same side as the relativistic jet, an observation which suggests that the arcsecond structure is being continually fed by the milliarcsecond jet. Generally, the arcsecond structures are not smooth, but contain bright knots of considerable brightness gradient. Are the knots moving or changing in a detectable way? Since the cores of these sources provide a reliable phase and amplitude reference, this question may be answered in a reasonable time scale through observations with instruments such as the VLA. Such a detection will not necessarily give the jet velocity - the feature may be a slower moving shock or density enhancement (e.g. Blandford and Konigl, 1979). Noting that it is in regions of strong brightness contrast that the most sensitive measurements of change of structure can be made, a simple analysis shows that the 3-σ angular change, Δd, that can be measured is:

$$\Delta d = 2\sigma/s,$$

where σ is the rms noise on the map, and s is the maximum brightness <u>slope</u> (seen by the instrument) of the knots whose motion or change we wish to measure.

Ron Ekers and I have initiated an experiment on the VLA to perform this measurement on the sources 3C273 and 3C279, both of which contain arcsecond jets with considerable brightness enhancements. At this time, only first epoch observations have been made, so no results are yet available. From these first-epoch maps, we estimate that for 3C279 at 2cm, a motion as small as 8 milliarcseconds can be detected. With careful

[+] <u>Discussion on page 438</u>   153

R. Fanti et al. (eds.), VLBI and Compact Radio Sources, 153–156.
© *1984 by the IAU.*

data editing and map analysis, this limit should drop considerably. The
theoretical limit (using thermal noise in 12 hours) predicts than an
angular discrimination of only 0.4 marc should be possible. For 3C273,
the current maps indicate an angular change discrimination of better than
4 marc, or 2.5 marc using the polarized flux (where dynamic range problems
are less severe). The theoretical limit for this source is as low as 0.1
marc. Note however, that better than 60dB dynamic range will be required
to reach this limit. Current best dynamic ranges are 40 to 45 dB - we do
not fully understand why we cannot reach higher dynamic ranges.

This method should be applicable to any source with a strong,
unresolved core and arcsecond structure with good brightness contrast.

2. <u>Periodicities in structure.</u> The curvature seen in VLB maps of the
jets in core-dominated sources (e.g. 3C345, 3C418, 3C273, VLB maps of
which have been shown in this conference) lends credence to the idea that
a precessing collimator is operating in the core. If, in fact, this is
occuring, we might expect the arcsecond structures to reflect the
periodicity set by the collimator. Fig 1. shows a 6cm VLA map with 0.6"
resolution of the arcsecond component of 3C273.

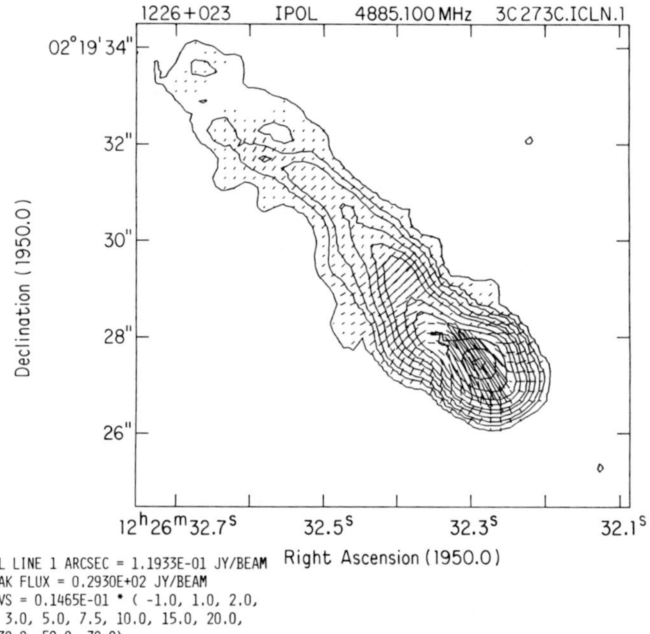

Fig. 1. The arcsecond jet of 3C273 at 6 cm with 0."6 resolution. The
intensity and direction of the polarized flux are shown by the dashes.

Note the sinusoidal shape of the ridge of maximum brightness. If this is
due to the collimator (rather than a hydrodynamic instability as suggested

by Conway et al., 1981), a direct measure of the jet velocity, v_j, is possible:

$$v_j = L_o(1-\beta\cos\theta)\csc\theta/T_c.$$

Here, L_o is the apparent spatial period, T_c the precession period, $\beta = v_j/c$, and θ is the angle of jet motion from the line of sight. Noting in Fig 1 that the jet ridge line is remarkably sinusoidal for a jet which is probably greatly foreshortened, a geometrical argument can be used from which the following equation arises:

$$v_j > 40R(1-\beta\cos\theta)\cotan\theta/T_c.$$

Here, R is the radius of the channel into which the jet is wound, and other symbols are as before. Inserting measured values into this latter equation yields:

$$\beta > (T_{p5}\tan\theta + \cos\theta)^{-1}.$$

Here, T_{p5} is the precession period in units of 10^5 years. If $\theta < 10$ degrees, as is probable, then β is relativistic if the precession period is less than 10^5 years.

3. <u>Brightness Ratios</u> This argument is applicable if jets are oppositely directed and equal in their respective frames (i.e. each evolves and radiates equally). Then, the measured flux ratio between knots identified as having been ejected at the same time is:

$$F^+/F^- = [(1+\beta\cos\theta)/(1-\beta\cos\theta)]^{3-\alpha}.$$

For intensity ratios of extended objects, the index should be $2+\alpha$ (e.g. Blandford and Konigl, 1979). The problem with this approach, besides the assumption of symmetry, is that the ratio is such a strong function of the index that mildly relativistic motion will lead to flux ratios beyond the dynamic range of current instruments. For example, in 3C273, the flux ratio is >500:1 - leading to $\beta\cos\theta>0.67$ - not an impressive limit.

4. <u>Jet/Counterjet Symmetry.</u> This argument again requires the jets to be intrinsically identical and oppositely directed. Then, for ballistic motion (or non-ballistic motion through identical media with no disruptive instabilities), any feature on the approaching jet should be mirrored on the receding jet, but due to time-delay effects, the location of the feature is closer to the nucleus. The ratio between approaching and receding separations is:

$$\theta^+/\theta^- = (1+\beta\cos\theta)/(1-\beta\cos\theta).$$

For the velocity limit given above for 3C273, this equation predicts the counter-blob (if it exists at all) should be within 4.5" of the nucleus.

The spectacular jet in NGC6251 provides an interesting test of these last two relationships. The jet/counter-jet brightness ratio leads to the limit $\beta\cos\theta > 0.6$. However, new VLA maps show that the ratio of angular separations for prominent bends in the jets is less than 1.1, leading to $\beta\cos\theta < 0.05$. Clearly, both velocity limits cannot apply, so one of our initial assumptions is wrong, most likely that the jets are radiating equally in their own frames. This provides important evidence that other factors are of great importance in the determination of jet brightness.

    5. <u>Global Lobe Properties</u>. In this approach, one attempts to estimate the properties of the jet from known properties of the lobes which the jets feed with energy, momentum and mass. The transport equations for a relativistic jet of cross-section A, mass density ρ, internal energy density μ, velocity $\beta = v/c$ and Lorentz factor γ are:

Kinetic Energy: $\gamma^2 Av^3 \rho [\gamma^3/(\gamma+1) + \alpha/\beta^2]$, where $\alpha = \mu/(\rho c^2)$.

Momentum $A\rho v^2 \gamma^2$.

Mass Flux $A\rho v\gamma$.

We can estimate the K.E. flux from the (roughly) known total luminosity of the lobe which the jet is supposed to feed, the momentum flux can be estimated from the familiar argument of the required thrust for the jet to enter the hot spot where the jet is supposed to terminate (using the usual equipartition method), while, at least in principle, the mass flux can be derived from the total mass (from depolarization), and source age (from spectral steepening, or some other way.)

    Unfortunately, none of the above parameters can be estimated with any degree of accuracy. The K.E. flux of the jet goes to other sinks besides the radiated luminosity (see Saunders et al., 1981 for a discussion), the thrust argument surely provides a lower limit to the momentum, while interpretation of depolarization data (if it exists at all) is fraught with problems (see Perley et al., 1983, for a discussion). Even if any one of these global quantities could be estimated accurately, the problem of the jet density remains. This vital quantity is completely unknown, and dependable measures of it remain a long way off. Independence from this parameter can be gained by taking ratios amongst the equations - however, this requires ratios of the poorly determined global parameters as well, with the usual accompanying deterioration of accuracy. Thus, it appears this approach to measuring jet velocities is not a fruitful one.

    Recently, a jet in Cygnus A has been discovered (Perley and Cowan, in preparation). Application of these ideas to Cygnus A (a source whose global prameters are known better than most) leads to only one firm conclusion - that the jet in Cygnus A cannot have an internal energy density exceeding the dynamic pressure ρv^2 and be relativistic - for this case cannot provide enough thrust for the jet to terminate in the hot spots found at the ends of the source. To make use of the global approach, much better measures of the energy, momentum and mass of the lobes will be needed.

REFERENCES

Blandford, R.D., and Konigl, A. Ap.J. <u>232</u>, 34 (1979)
Conway, R.G., Davis, R.J., Foley, A.R., and Ray, T.P. Nature <u>294</u>, 540 (1981)
Perley, R.A., Bridle, A.H., and Willis, A.G. Submitted to Ap. J.
Saunders, R., Baldwin, J.E., Pooley, G.G., and Warner, P.J. MNRAS, <u>197</u>, 253 (1981).

SUPERLUMINAL MOTION IN WEAK QUASAR CORES

R.W. Porcas
Max-Planck-Institut für Radioastronomie, Bonn, F.R.G.

ABSTRACT

The importance of studying the statistics of the occurrence of
superluminal motion in an unbiased sample of radio sources is stressed,
and methods for selecting such samples are reviewed. The weak radio
cores of quasars exhibiting extended, double-lobed emission can be used
for this purpose. Recent studies of the superluminal motion in the core
of 3C 179, a quasar selected from such a sample, are described.

INTRODUCTION

The observation of apparent superluminal motion by VLBI in a number
of strong, compact radio sources poses an intriguing problem for astronomers. While many explanations of the phenomenon have been proposed, the
most popular ones invoke highly relativistic motion of the radio components, at a small angle to the observer's line of sight. In the relativistic jet model (Blandford and Königl, 1979; Scheuer and Readhead, 1979),
this motion is in a collimated beam of energetic particles emanating
from the central energy source, which energises the outer, extended
radio-emitting regions. The ingredients of the model are the angle, θ,
with respect to the line of sight, and the Lorentz factor, $\gamma (=(1-\beta)^{-1/2}$,
$\beta = v/c)$. To produce an observed faster-than-light velocity, $\beta' = v_{obs}/c$,
then $\theta \simeq 1/\beta'$ and $\gamma \simeq \beta'$.

Whilst this mechanism explains many of the observed properties of
the superluminal sources, it is obviously important to try and find independent evidence for the high values of γ and small values of θ.
Marscher and Broderick (1981, 1982) have drawn attention to the (red-
shift-independent) evidence for high γs from the lack of inverse-Compton
X-ray emission from some radio sources. Using this argument they suggested, and indeed found, superluminal motion in the quasar NRAO 140.
If superluminal sources are really at small θ, one would expect
i) the size of any outer structure to be smaller, on average, by a factor of sin θ (but see Schilizzi and de Bruyn, 1983), and

R. Fanti et al. (eds.), VLBI and Compact Radio Sources, 157–161.
© *1984 by the IAU.*

ii) the superluminal effect in samples of sources with unbiased θ should be rare, about 1 in every γ^2 sources.

DEFINING AN UNBIASED SAMPLE

For the purpose of such statistical studies, it is clear that one cannot use a sample of quasars selected on the basis of the core flux density, because of the associated "Doppler Boosting" of the flux density when θ is small. Optically selected quasars could, in principle, provide a suitable sample free of orientation bias, but the radio emission from such quasars is typically very weak (e.g. Strittmatter et al., 1980; Kellermann et al., 1983) and is below the sensitivity of present VLBI. X-ray selected quasars are not suitable because if the X-ray emission is produced by the inverse Compton effect in the jet, it, too, will be Doppler Boosted. Selecting quasars on the basis of the flux density in their extended radio emission seems to be an ideal method. Firstly, for the case of double-lobed sources, Longair and Riley (1979) have

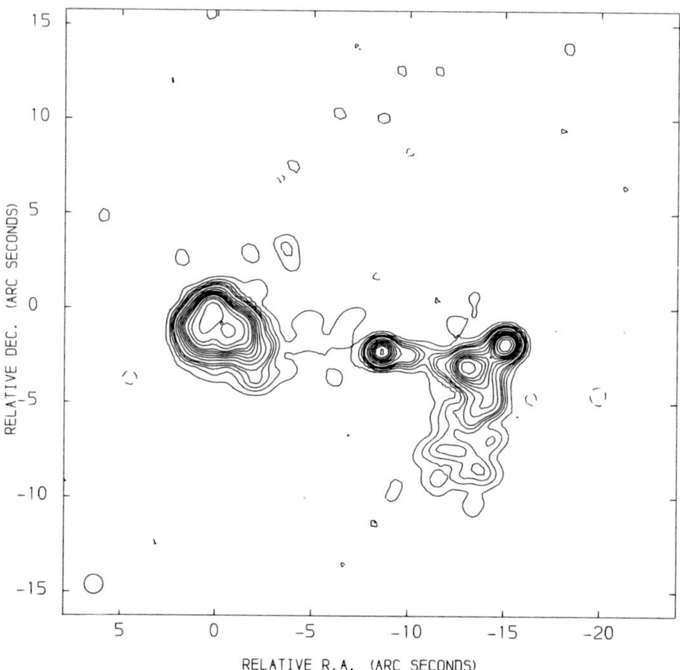

Figure 1: 408 MHz MERLIN map of the arc second scale structure of 3C 179, June, 1981. Restoring beam is 1.0 x 1.0 arcsec. Contour levels 2, 4, 6, 8, 10, 15, 20, 25, 30, 35, 40, 50, 60, 80 percent of the peak.

argued that the expansion speed of the lobes is low, because of the symmetric way they straddle the central object. Thus the lobe flux dens-

ity cannot be boosted, and flux density limited samples of classical
double quasars should have an unbiased distribution of θ. Secondly, nearly all such quasars have a detectable radio core, many in the range 50 –
300 mJy (Owen et al., 1978; Owen et al., 1982) and hence are observable
by VLBI.

OBSERVATIONS OF 3C 179

We have undertaken VLBI observations of such a sample of quasars,
selected from the Jodrell Bank 966 MHz survey. The selection criteria
are described by Porcas (1981, 1982). Zensus and Porcas (this volume)
report on first epoch Mark III observations of a number of the sources.
One quasar from the sample, 3C 179 (Fig. 1), has been observed since
1979 at 10.7 GHz using Mark II VLBI, and superluminal motion was found
from the first two observing epochs (Porcas, 1981). A number of subsequent observations have been made in order to confirm and to investigate
further,the superluminal behaviour. Fig. 2 shows an "expansion graph" of
the motion, and it is clear that the separation of the milli-arc-second
components has continued to increase in a roughly linear fashion. Most
of the values of the separation are derived from model-fitting to the

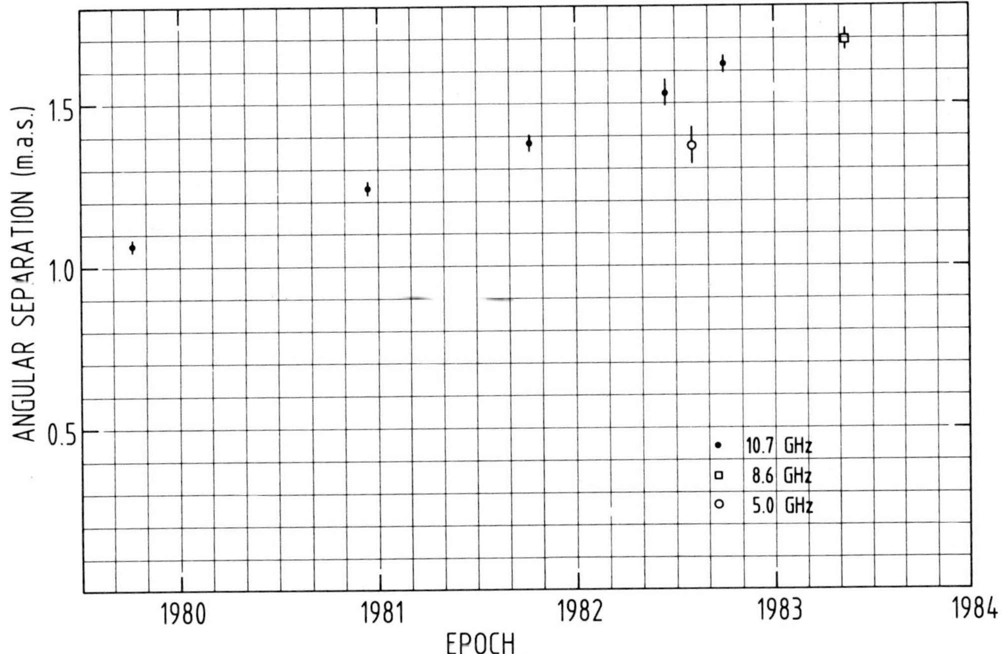

Figure 2: Expansion graph of the superluminal motion in the core of
         3C 179.

visibility functions, since the radio core is very weak, and, indeed,
has decreased in flux density since 1980. However, it has been possible
to make hybrid maps for two epochs: for December 1980 (when the 10.7 GHz
flux density was at a maximum) and for May 1983, from 8.6 GHz data, obtained using a sensitive VLBI array consisting of the 100 m Effelsberg
antenna and the two 64 m DSN dishes at Madrid and Goldstone (in collaboration with A. Rius). These maps are shown in Fig. 3, plotted on the same
angular scale and with the same CLEAN restoring beams. At both epochs the

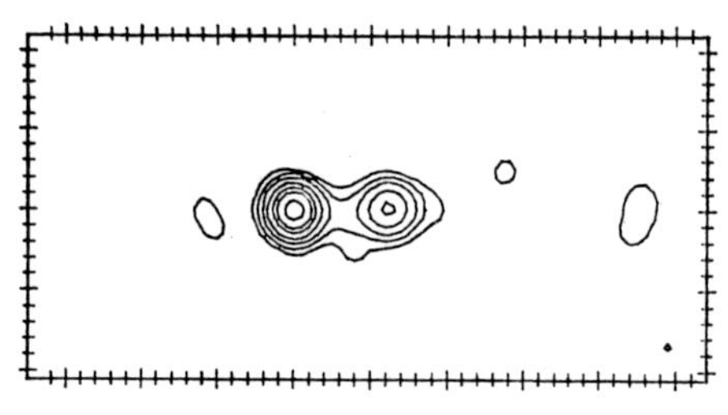

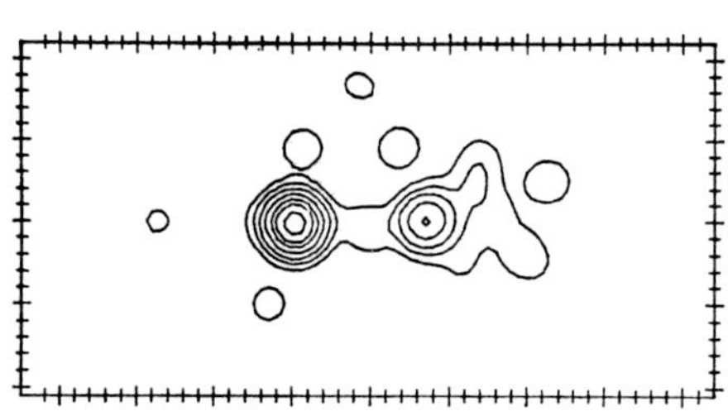

Figure 3: Hybrid maps of the core of 3C179 at December 1980 (10.7 GHz)
          and May 1983 (8.6 GHz). Restoring beam is 0.45 x 0.45 mas.
          Contour levels 2, 5, 10, 20, 35, 50, 80 percent of the peak.
          Tick interval is 0.2 mas.

source structure is dominated by two components, whose separation has
changed from 1.24 mas to 1.70 mas over the 2.4 year interval. Taking
H_o = 55 km s^{-1} Mpc^{-1} and q_o = 0.05, this corresponds to an apparent
expansion velocity of 10 c.

REFERENCES

Blandford, R.D. and Königl, A.: 1979, Astrophys. J. 232, 34
Kellermann, K.I., Sramek, R., Shaffer, D., Schmidt, M. and Green, R.:
    1983, Proceedings of 24th Liège International Astrophysical Symp.
    "Quasars and Gravitational Lenses" (to be published)

Longair, M.S. and Riley, J.M.: 1979, Monthly Notices Roy. Astron. Soc.
    188, 625
Marscher, A.P. and Broderick, J.J.: 1981, Astrophys. J. Lett. 247, L49
Marscher, A.P. and Broderick, J.J.: 1982, Astrophys. J. 255, L11
Owen, F.N., Porcas, R.W. and Neff, S.G.: 1978, Astron. J. 83, 1009
Owen, F.N., Puschell, J.J. and Laing, R.A.: 1982, IAU Symp. No. 97,
    p. 435
Porcas, R.W.: 1981, Nature 294, 47
Porcas, R.W.: 1982, IAU Symp. No. 97, p. 361
Scheuer, P.A.G. and Readhead, A.C.S.: 1979, Nature 277, 182
Schilizzi, R.T. and de Bruyn, A.G.: 1983, Nature 303, 26
Strittmatter, P.A., Hill, P., Pauliny-Toth, I.I.K., Steppe, A. and
    Witzel, A.: 1980, Astron. Astrophys. 88, L12

VLBI OBSERVATIONS OF WEAK CORES IN EXTENDED QUASARS

J.A. Zensus, R.W. Porcas
Max-Planck-Institut für Radioastronomie, Bonn, F.R.G.

ABSTRACT

We present Mark III VLBI results for a sample of weak cores associated with quasars with extended double lobes. The m.a.s. scale structure is found to be well aligned with the outer structure.

The relativistic beaming models invoked to explain superluminal motion predict that the effect should be seen only rarely in the cores of double radio sources selected without orientation bias (e.g. Scheuer and Readhead 1979; Porcas, this volume). Such a sample is provided by the 30 quasars from the Jodrell Bank 966 MHz survey mapped by Owen et al., 1978 ($S_{966} \geq 0.7$ Jy, $m_B \leq 19$ and angular sizes ≥ 10 arcsec, Porcas, 1981). We have chosen a small group of 'classical double' sources from this sample (see Table below) for high resolution VLBI observations at $\lambda 2.8$ cm, and include 3C 179, for which Porcas (1981) found the superluminal effect. MERLIN maps made at 1.6 GHz are presented in Fig. 1 for the 2 sources for which arc-second resolution maps have not previously been published. Our goal is to test whether the m.a.s. structure of the weak cores is consistent with the assumption of large viewing angles of a relativistic jet and, as a second step, to check for possible motions.

The low core flux densities of the sources require the use of the Mark III VLBI system. So far we have made first epoch observations of 4 sources, in addition to the Mark II observations of 3C 179. The correlation of the data for 2 of the sources was done with the new MPIfR Mark III processor in Bonn which is a replica of the Haystack processor.

Source	z	$S_{966\ \mathrm{MHz}}$	$S(core)_{2.7\ \mathrm{GHz}}$
0723+67 (3C 179)	0.846	2.87	$\sim$540 mJy
0833+65 (3C 204)	1.112	2.09	$\sim$80
1137+66 (3C 263)	0.652	4.72	$\sim$90
1206+43 (3C 268.4)	1.400	2.64	$\sim$40
1732+65	0.856	1.07	$\sim$60
1951+49	0.466	1.10	$\sim$120

R. Fanti et al. (eds.), VLBI and Compact Radio Sources, 163–164.
© *1984 by the IAU.*

In three of the four sources observed we were able to detect compact structure on the m.a.s. scale. Maps of 1137+66 (Zensus et al., 1983) show a double structure, of separation ~0.5 m.a.s. in p.a. 110°, which is closely aligned with the p.a. of the outer double lobes (Pooley and Henbest, 1974; Owen et al., 1978). Inspection of the visibility data for 1951+49 shows that the structure is elongated in p.a. ~90°, roughly that of the weaker, Eastern lobe. 3C 179 also shows reasonable alignment between the m.a.s. double structure and the arc-second scale jet (Browne et al., 1982). Thus, a first result from our study is that these sources do not show the frequent structural misalignments which have been found in some of the core-dominated sources (Readhead et al., 1978; Browne et al., 1982). The double structure found in the core of 1137+66 makes this source a suitable candidate for measuring possible component motions with second epoch observations.

Figure 1: 1.6 GHz MERLIN maps

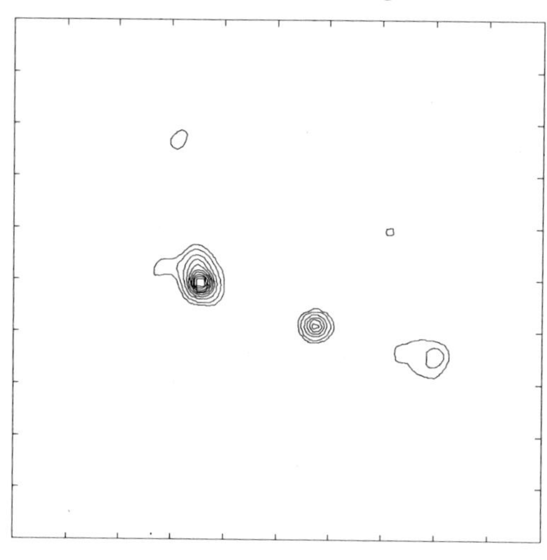

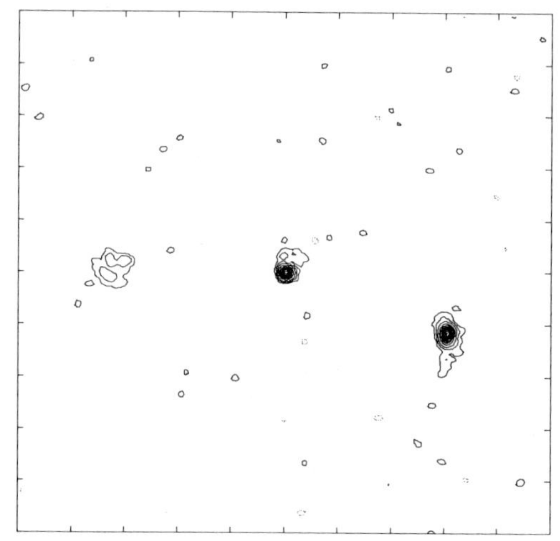

1732+65
Tick interval = 4 arcsec
Restoring beam = 1.3 x 1.3 arcsec

1951+49
Tick interval = 2 arcsec
Restoring beam = 0.35 x 0.35 arcsec

REFERENCES

Browne, I.W.A., Clark, R.R., Moore, P.K., Muxlow, T.W.B., Wilkinson, P.N., Cohen, M.H., Porcas, R.W.: 1982, Nature 299, 788
Owen, F.N., Porcas, R.W. and Neff, S.G.: 1978, Astron. J. 83, 1009
Pooley, G.G. and Henbest, S.N.: 1974, Monthly Notices Roy. Astron. Soc. 169, 494
Porcas, R.W.: 1981, Nature 294, 47
Readhead, A.C.S., Cohen, M.H., Pearson, T.J. and Wilkinson, P.N.: 1978, Nature 276, 768
Scheuer, P.A.G. and Readhead, A.C.S.: 1979, Nature 277, 182
Zensus, J.A., Hough, D.H. and Porcas, R.W.: 1983 (in preparation)

EXTENDED STRUCTURE AROUND SUPERLUMINAL AND OTHER CORE-DOMINATED RADIO
SOURCES +

A.G. de Bruyn and R.T. Schilizzi
Radiosterrenwacht
Postbus 2, 7990 AA Dwingeloo, The Netherlands.

The maps that we will show you today probably have the lowest resolution
of any that you will see this week. They were made from data taken by
the Westerbork Synthesis Radio Telescope (WSRT), which has a maximum
baseline of 2.8 km, yielding a resolution of $3\overset{''}{.}5$ (λ 6 cm), $12\overset{''}{.}5$ (λ 21
cm) or 29" · (λ 49 cm). Because the WSRT is an EW-array the resolution in
declination is worse by a factor $1/\sin\delta$. This is unfortunate since
three interesting superluminal sources are close to the equator. For two
of these, namely 3C120 and 3C273 we nevertheless have obtained some interesting results.

Since 1981 the WSRT is increasingly being used in a new observing mode whereby all 14 telescopes are correlated with one another
(Noordam and de Bruyn, 1982). The large number of redundant spacings
thence provided are used to determine the phase and gain errors
contributed by each of the 14 telescopes as a function of time (hour-
angle). In addition, tropospheric wave front distortions on scales
smaller than the array are removed. The resulting data base is a time-
series of perfect 1-dimensional (complex) visibility curves with only
two unknowns: the absolute amplitude and the phase slope across the
array (= absolute position). Because there are far more spacings (66)
involved in the redundancy decomposition solution than there are
unknowns to solve for (14 telescopes + 13 visibilities), a useful
byproduct is the consistency of the solution, and hence the reliability
of the visibilities. With the digital line backend (Bos et al., 1981) we
now obtain an accuracy of 0.03-0.05% rms in amplitude and 0.02-0.03° in
phase depending, among other things, on frequency and bandwidth. The new
broadband continuum backend may even surpass this already excellent
performance. It is therefore possible to measure flux densities of weak
extended haloes around discrete sources, on scales of about 5 arcseconds
and larger, with an accuracy better than 0.1%. It is important to
realise that this can be done <u>without</u> knowledge of the brightness
distribution in the field of view.

In order to make a two-dimensional map from the sequence of
one-dimensional scans an 'aligning' routine is used, which is basically

+ <u>Discussion on page 438</u>

R. Fanti et al. (eds.), VLBI and Compact Radio Sources, 165–168.
© *1984 by the IAU.*

similar to SELFCAL (see Noordam and de Bruyn, 1982). However, an alignment scheme that uses the fact that the source centroid position and total flux are invariant under projection has been successfully used on several complex sources (e.g. 3C120). This scheme can do without a model for the source and is thus radically different from SELFCAL. Moreover, it is much less of a burden on the astronomer _and_ the computer!

In the remaining time we will present you some results of three ongoing programmes using the WSRT in its redundancy configuration. All sources in these programmes are core-dominated.

1)   Sample 1 contains 14 sources which belong to one or more of the following classes: BLLac type objects, low frequency variables and optically violently variables. Of the eight sources thus far reduced (BLLac, OJ287, 0735+17, 1308+32, 1358+62, 1219+28, CTA102 and 3C454.3) nearly all contain significant extended emission. One of the sources that comes closest to being a "perfect" point source is 1219+28 (W Coma) of which a map is shown in Fig. 1a. No extended emission with a peak intensity greater than 0.1% of the core is seen. However, on the short baselines (< 500 meter) an excess flux of about 5-10 mJy is recorded, which in a smoothed map shows up to the SW of the core. In BLLac an extended region, containing about 1% of the total flux (3.74 Jy at that time), is revealed after subtraction of an unresolved core (see Fig. 1b). The source boundaries are not well-defined since the emission fades into the noise level. However the size is at least 30" oriented roughly in position angle 20°, which is not significantly different from that of the VLBI structure (Phillips and Mutel, 1982). A third very active BLLac-type object is OJ287 which was found to contain a significant very diffuse halo. In this source the extended emission has a peak intensity of about 0.03% of the core flux (4.8 Jy at the time of the observation). If these sources are to be explained as the favourably oriented examples of normal lower luminosity elliptical radio galaxies whose core emission is relativistically boosted (Browne, 1983), then their beam Lorentz factors must lie between 4 and 8.

2)   Another continuing programme is the mapping of the 'classical' superluminal sources at all three wavelengths available. Results at 6 cm for 3C120, 179, 273, 279, 345 and NRAO140 have been published (Schilizzi and de Bruyn, 1983; henceforth SB). New 21 cm observations of 3C120 have since shown that there is faint emission in addition to the east and west lobes discovered at 6 cm. A smoothed map is presented in Fig. 2. It shows a diffuse structure several arcminutes in size centered 4!5 west and 2' south of the nucleus. There is little doubt that it is associated with 3C120. The EW-size of 3C120 is therefore some 7!5. In addition, significant emission exist to the north and south of the source. Its total dimension in those directions is not well determined but certainly measures 10'; it could be as large as 20'.. This revised estimate of the LAS of 3C120 puts it in the very top part of the LAS-z diagram shown by SB, even _without_ correcting for projection effects. It further exacerbates the problem of the intrinsic sizes of the superluminal sources, in the context of the directed relativistic beaming models. The large range in observed position angles in the outer structure of 3C120 suggests that the nuclear ejection axis undergoes a

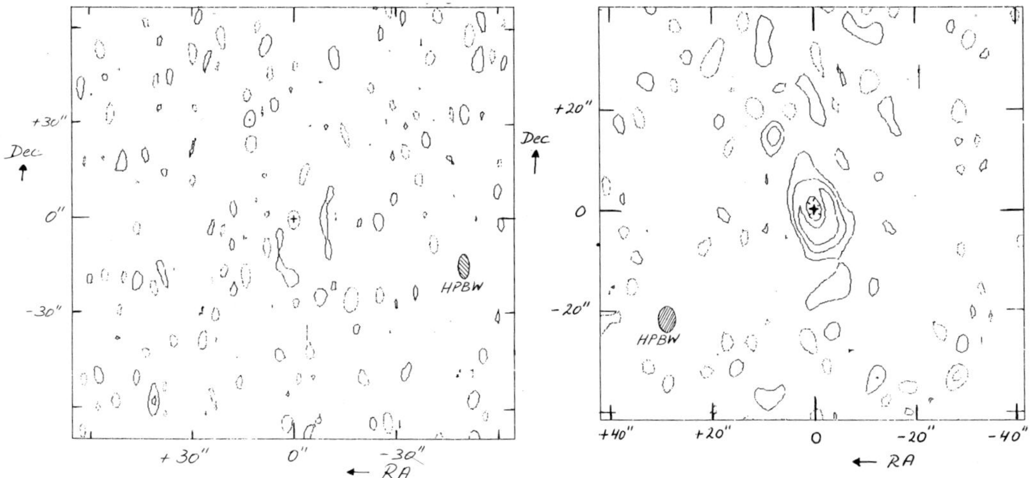

Fig. 1a) 6 cm map of W Coma (1219+28) after removal (at the cross) of a 1.91 Jy core. Contours are at ±0.04% (=0.8mJy) intervals.

1b) 6 cm map of BL Lac after removal (at the cross) of a 3.74 Jy core. Contours are at ± 0.02% (=0.75 mJy) intervals.

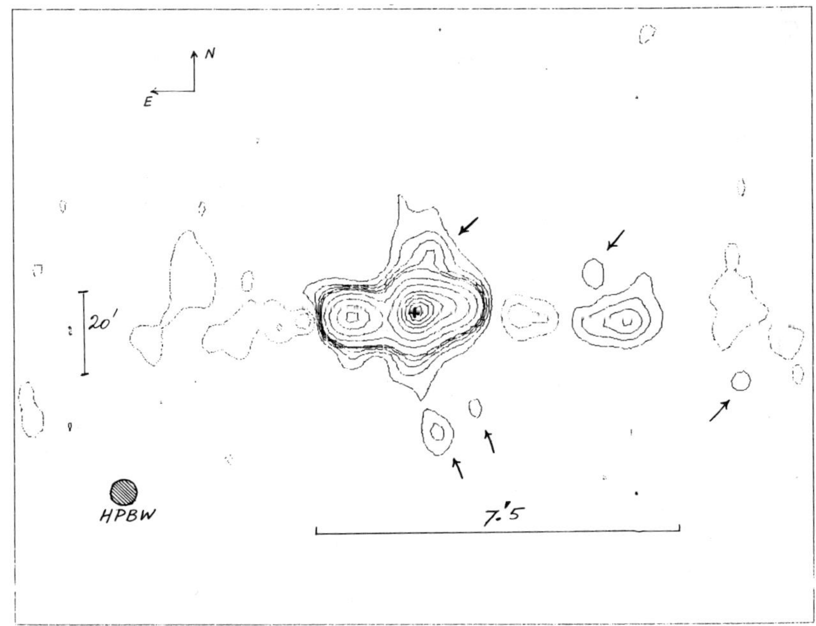

Fig. 2   Smoothed map of 3C120 at 21 cm after subtraction of a core (at the cross) of 3.25 Jy. The beam measures 34"x365" at half width; note that the declination scale has been compressed to yield a circular beam. Contours are at -5, -2.5, 2.5 (2.5) 12.5, 25, 50 (50) 300 mJy/beam. Arrows point to unrelated discrete background sources.

significant wobble or precession in the course of time, just as demanded
by the statistical results presented by SB. Our 21 cm and 49 cm results
on 3C345 do not show any emission features beyond the 20" region mapped
by the WSRT and VLA (SB, Perley et al., 1982; Rudnick and Jones, 1983).
The 6 cm, 21 cm and 49 cm results on 3C273 thusfar have not produced any
evidence for diffuse emission beyond 20" west or east of the core. Our
best limits at present are 10 mJy per beam (12"x300" stripscan) at 21 cm
but we expect to improve these by a considerable factor (de Bruyn and
Schilizzi, in preparation). But even these limits, compared to 16 Jy
total emission in the jet at 21 cm,, establish 3C273 as a unique one-
sided source (see also Perley et al., 1982; Browne et al., 1982).
3)   A third sample of sources being investigated by one of us (AGdB)
     are the brightest examples of the steep spectrum core sources,
namely 3C48, 3C147 and 3C286 (them being the primary WSRT calibrators).
Each of these sources contains, in addition to emission on the arcsecond
scale (Perley, 1982), extended emission on scales of 20"-60". This
transforms to linear dimensions of several hundred kpc. The total flux
density in these regions, however, is only 0.2-0.5% of the core flux
densities. Not much can be said as yet about the structure of these
components since they are not more than a few beams across. But they do
show that relativistic plasma (beams?) does manage to escape the inner
regions of these quasars (see also van Breugel, this symposium), and
that large amounts of energy resides in the outer regions.

     The Westerbork Synthesis Radio Telescope is operated by the
Netherlands Foundation for Radio Astronomy with the financial support of
the Organization for the Advancement of Pure Research (ZWO).

REFERENCES

Bos, A., Raimond, E. and van Someren Gréve, H.W., 1981, Astron.
     Astrophys. 98, 251.
Browne, I.W.A., 1983, M.N.R.A.S. 204, 23P.
Browne, I.W.A. et al., 1982, Nature, 299, 788.
Noordam, J.E. and de Bruyn, A.G., 1982, Nature 299, 597.
Perley, R.A., 1982, Astron. J. 87, 859.
Perley, R.A., Fomalont, E.B. and Johnston, K.J., 1982, Astrophys. J.
     255, L93.
Phillips, R.B. and Mutel, R.L., 1982, Astrophys. J. 257, L19.
Rudnick, L. and Jones, T.W., 1983, Astron. J. 88, 518.
Schilizzi, R.T. and de Bruyn, A.G., 1983, Nature, 303, 26.

THE LOW FREQUENCY VARIABILITY OF EXTRAGALACTIC RADIO SOURCES [+]

L. Padrielli
Istituto di Radioastronomia, CNR
Bologna

1. Introduction

The flux variability of extragalactic radio sources at decimetric wavelengths (L.F.V.) is now considered a classical astrophysical subject connected with compact radio sources.

The observational constraints, already shown in several reviews and universally accepted, can be summarized in a few points.

1) There is an association between sources showing L.F.V. and sources with other 'extreme' characteristics: flat spectrum radio sources (50% of them show dS/S>6%), high frequency variables, superluminal expanding sources, optically violent variables and highly polarized quasars.

However there are exceptions, L.F.V. sources not identified with quasars, others with steep radio spectrum and so on.

2) The 'light-curves' of the sources are very complex, as a sequence of maxima and minima where it is very difficult to determine a pre-event stable level. Therefore, it is almost impossible to state whether the variations we observe are emission or absorption events.

3) The average time scale of the L.F.V. is about 2 years (corrected to the proper frame of the source). That means that for the majority of variable quasars, with the usual 'causality arguments', we obtain brightness temperatures in the range 10^{14}-10^{16} K. These values exceed by 2-4 order of magnitude the 10^{12} limit for incoherent synchrotron radiation. The expected inverse Compton X-ray fluxes were not detected for several sources observed during their active phases.

Several groups of researchers are currently working to extend the amount of information available on this subject and to provide new input for theoretical models, focusing the attention on the spectral character of the L.F. variations. It is also very important to search for a possible relationship between high and low frequency enhancements and monitor, with high resolution (milliarcsecond scale) the structures of the sources simultaneously with the L.F.V.

+ Discussion on page 439

169

R. Fanti et al. (eds.), VLBI and Compact Radio Sources, 169–176.
© *1984 by the IAU.*

2. Multifrequency monitoring of L.F.V.

Several programs of multifrequency observations have started during the last few years: H.E. Payne et al.(1982) are monitoring 30 sources at 5 frequencies in the range 300–1400 MHz bandwidth; S.R. Spangler and W. Cotton are observing 10 sources in the frequency range from 1.4 to 89 GHz; H.D. Aller, M.F. Aller, P.E. Hodge and the Bologna group have been observing monthly, since 1980, about 60 sources at 0.4, 4.4, 8.0 and 14.4 GHz; in the last year the frequencies 1.4 and 2.7 were also added; W.B. McAdam is monitoring a big sample of $\sim$180 sources at 0.84 GHz.

We refer to the paper of B. Dennison (this volume) for detailed results of multifrequency observations. We wish only to point out that the L.F. variable sources can show two different spectral behaviours.
1) Some sources follow, at least qualitatively, the expectation of standard models of variability. The bursts are first seen at very high frequency and then they drift towards lower frequencies until they reach decimetric wavelenghts, with reduced amplitude. This is the case of BL Lac.
2) However, the majority of sources show a narrow L.F.V. bandwidth from 300 to 1000 MHz (for example DA 406). Other sources show, together with L.F.V., a high frequency activity that follows more or less the expectations of the standard models, leaving a mid-frequency gap around 1–3 GHz where the source is quite quiescent (3C 454.3).

Spangler and Cotton (1981) noted close coincidences between some decimetric and millimetric events in the last type of sources.

To evaluate the statistical significance of these coincidences we have used the sample of 50 sources simultaneously observed at 408 MHz at Bologna and at 8 GHz at Michigan University during the last 3 years.

For each source, the light-curves at the two frequencies were first interpolated to obtain regular monthly sampled functions and after they were correlated with each other. The distribution of correlation coefficients were compared with a sample obtained correlating the function at 8 GHz of a source with that at 0.4 GHz of an other source, chosen at random. The real distribution of the correlation coefficients shows no excess of high coefficients with respect to the random population. We can put a limit of the order of 5% to the occurrence of sources with simultaneous bursts at 8 GHz and 0.4 GHz. Also introducing a delay in time between the two frequencies we find a small (not significant) excess in the real distribution. This excess is mostly due to sources with a burst spectrum behaviour like BL Lac, and the occurrence of this kind of sources is not greater than 10%.

These data give no evidence for correlated 8 GHz and low frequency activity through the intermediate gap, but the same analysis should be made with data at higher frequency (>30 GHz).

3. The 18 cm VLBI monitoring program

We carried out 18 cm VLBI observations in two epochs of 23 L.F. variable sources to determine their structures, their physical sizes, and to investigate the structural changes connected with the flux variations (Romney et al. 1983).

The first observations were made in February 1980, using 7 telescopes, namely: Simeis USSR, Onsala Sweden, Effelsberg W.Germany, Artebeesthoek South Africa, Green Bank, Fort Davis and Owens Valley USA. The observations were repeated in October 1981 with the addition of the Jodrell Bank station.

The choice of the observing frequency of 18 cm represented a compromise between low frequency and high resolution. Unfortunately, this frequency is close to the 'intermediate frequency gap', but the resolution of few mas is good enough for detecting the structural changes expected by the current models based on relativistc motion along the line of sight.

The analysis of the first observations was done independently in different institutions and hybrid maps were obtained for all sources. All sources have been detected in the transatlantic baselines and all the compact features appear resolved in the longest baselines with extension of the order of 2 - 3 mas.

The most common morphology can be described as being a core-jet like structure, with linear sizes ranging from 5 to 50 pc. There are, however, also sources that present elongated emission regions symmetric about the core, that are not expected in these core dominated objects (fig.1). 60% of the sources have 'large' scale structures on a scale of few arcsecond. The comparison between arcsecond and milliarcsecond structures shows a great misalignment. This result is similar to that found by Browne et al. (1982), who studied a sample of core dominated sources.

The reduction of the second run of observations is still in progress and we show here only the preliminary results on the comparison between the two epochs.

The most spectacular variation of total power flux between the two epochs is shown by BL Lac, whose 18 cm total flux increased by a factor 2.3. In fig.2 we show the two epochs VLBI maps and the light-curves of the source at three different frequencies. It is possible to see a strong burst shifting with time toward lower frequencies. The map presents an extended small jet with the same flux density of about 600 mJy at both epochs. The increased central component is resolved in the longest baselines with an extension of 2.4 mas in p.a. 10°. The extension is almost the same in the two epochs.

An other example of structural change in the central component is shown by the source 0607-15 (fig.3). From the light-curves it is not completely clear whether the L.F.variation seen at 408 MHz corresponds to the event seen earlier at high frequency. Also in this case the jet is very similar in the two epochs and the decrease is due to the central component (4 mas extended in p.a. 60°).

Again a similar case: 1730-13 (fig.4), where the flux decreases in a broad frequency band from 0.4 to 8 GHz. The shape of the source is the same in the two epochs and the decrease is mostly due to the central component (2.3 mas). The minor differences of the structures cannot be considered significant, but it is worthwhile noting that this source shows a radio emission rather symmetric about the core.

Fig.5 shows the source DA 406. Its 18 cm total power flux showed a flux increase of ~15% between the two epochs, that corresponds to the strong flare seen at 408 MHz. If we interpret this variation in term of relativistic motion the 408 MHz time scale implies a brightness temperature of 2×10^{15} K with a kinematic factor $(\delta) \sim 12$ and an apparent expansion of about 0.7 mas between the two epochs[*]. The two maps are very similar: a single compact source increased ~10% in flux but with the same extension in the two observations (3 mas in p.a. 10-20°). An expansion of only 0.5 mas would cause a decrease of 40% in the fringe visibilities of the baselines with the S.African station.

3C 454.3 (fig.6) is an other interesting case. Unfortunately, 18 cm corresponds to the almost quiescent frequency gap for this source, but nevertheless, after the first VLBI measurement, the source had a strong burst at 408 MHz. The corresponding brightness temperature is of the order of 1.3×10^{15} with $\delta \sim 11$ and an expected apparent expansion of about 1 mas.

The two maps are very similar and the separation between the core (E component) and the jet (W component) is unchanged. From spectral information derived by 2.8 and 6 cm VLBI observations (Pauliny-Toth et al.1981) we expect to see mainly the jet at 408 MHz, and therefore we would expect an increase of 1 mas just in the separation of the two components. This expansion is excluded by the data at the level of 1 sigma. However the data can not exclude an expansion of 1 mas of the central component, that could contribute to the total L.F. flux by not more than a few (2- 3) Jy.

A similar case is NRAO 140 (fig.7). An interpretation of the L.F.V. in term of relativistic motion would predict an expansion of the 408 MHz emitting region of the order 0.7 mas.

The 18 cm map shows a core of about 3 mas that is resolved at 2.8 cm in two superluminal expanding components at a rate of about .10-.14 mas per year (Marscher and Broderick 1982). The second E component is a jet visible also at 2.8 cm. It has a steep spectral index, of the order of one, down to .4 GHz, while the other components are completely self-absorbed at this frequency (Marscher and Broderick,1981). Also in

---

[*] $H=100$, $q_o=1$

this case we have no direct evidence of an increase in the angular separation of the two components.

We do not wish to draw general conclusions from this comparison, because the data of only a few sources of the second run are completely reduced and the program could produce some surprises. However, the first impression is that the L.F.variations are not always accompanied by superluminal expansions. We would like merely to point out that the experimental landscape of the L.F.V. is quite complex. The models that invoke synchrotron electrons and magnetic fields relativistically ejected along the line of sight, fit many observational behaviours but surely they can not explain all the observed properties of these sources.

References

Browne I.W.,Orr J.L., Davis R.J.,Foley A, Muxlow T.W, Thomasson P, 1982 Mon.Not.R.astr.Soc, 198,673

Dent W.A., O'Dea C.P., Kinzel W.,1982 Workshop held at NRAO 21-22 April

Marscher A.P., Broderick J.J.,1981, Ap.J. 249,406

Marscher A.P., Broderick J.J., 1982 'Extragalactic Radio Sources' IAU Symp. 97

Payne H.E., Altshuler D.R., Broderick J.J., Condon J.J., Dennison B, O'Dell S.L., 1982, Workshop held at NRAO 21-22 April

Pauliny-Toth I.I.K., Preuss E., Witzel A., Graham D., Kellermann K.I., Ronnang B.,1981 Astron. J. 86,371

Romney J.D, Padrielli L., Bartel N., Weiler K.W., Ficarra A., Mantovani F., Kogan L., Matveyenko L., Moiseev I.G., Nicholson G, 1983, in press

Spangler S.R., Cotton W.D., 1981 A.J. 86,730

Spangler S.R., Cotton W.D.,1982, Workshop held at NRAO 21-22 April

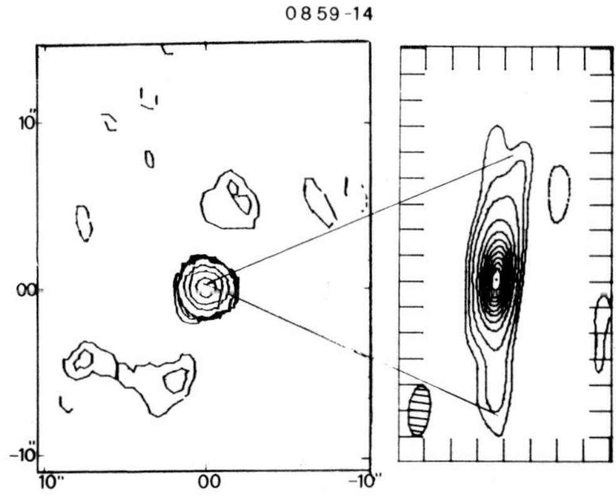

Fig. 1 20 cm VLA map (left) and 18 cm VLBI map (right) of the source 0859-14. In the VLBI map the first contour level corresponds to 5% of the peak flux, the distance between ticks corresponds to 5 mas.

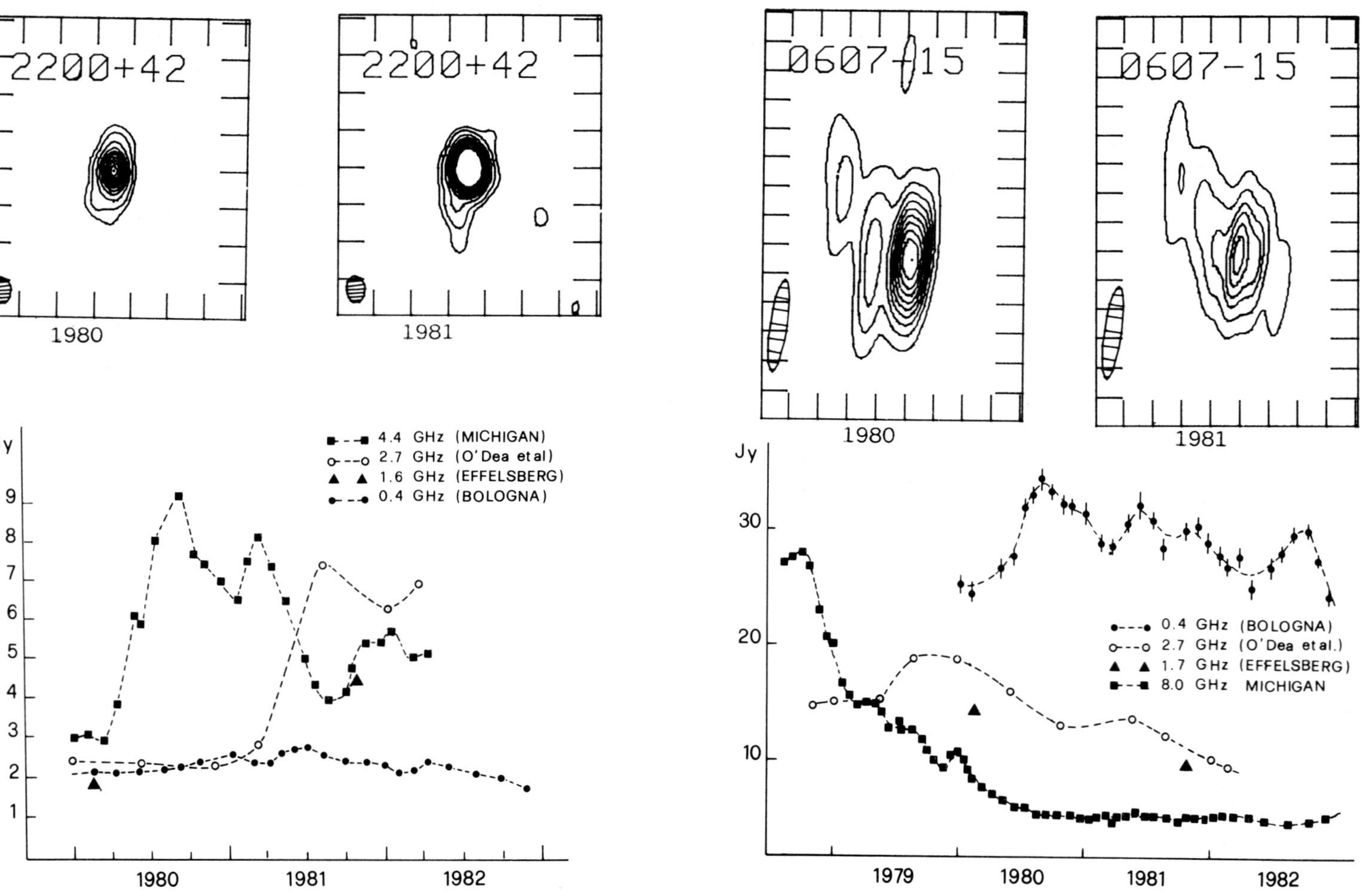

Fig.2-3 Two epochs VLBI maps of BL Lac (2200+42) and 0607-15. The distance between ticks corresponds to 5 mas. The contours are plotted at the same flux levels for both epochs. The light-curves at different frequencies are shown for each source.

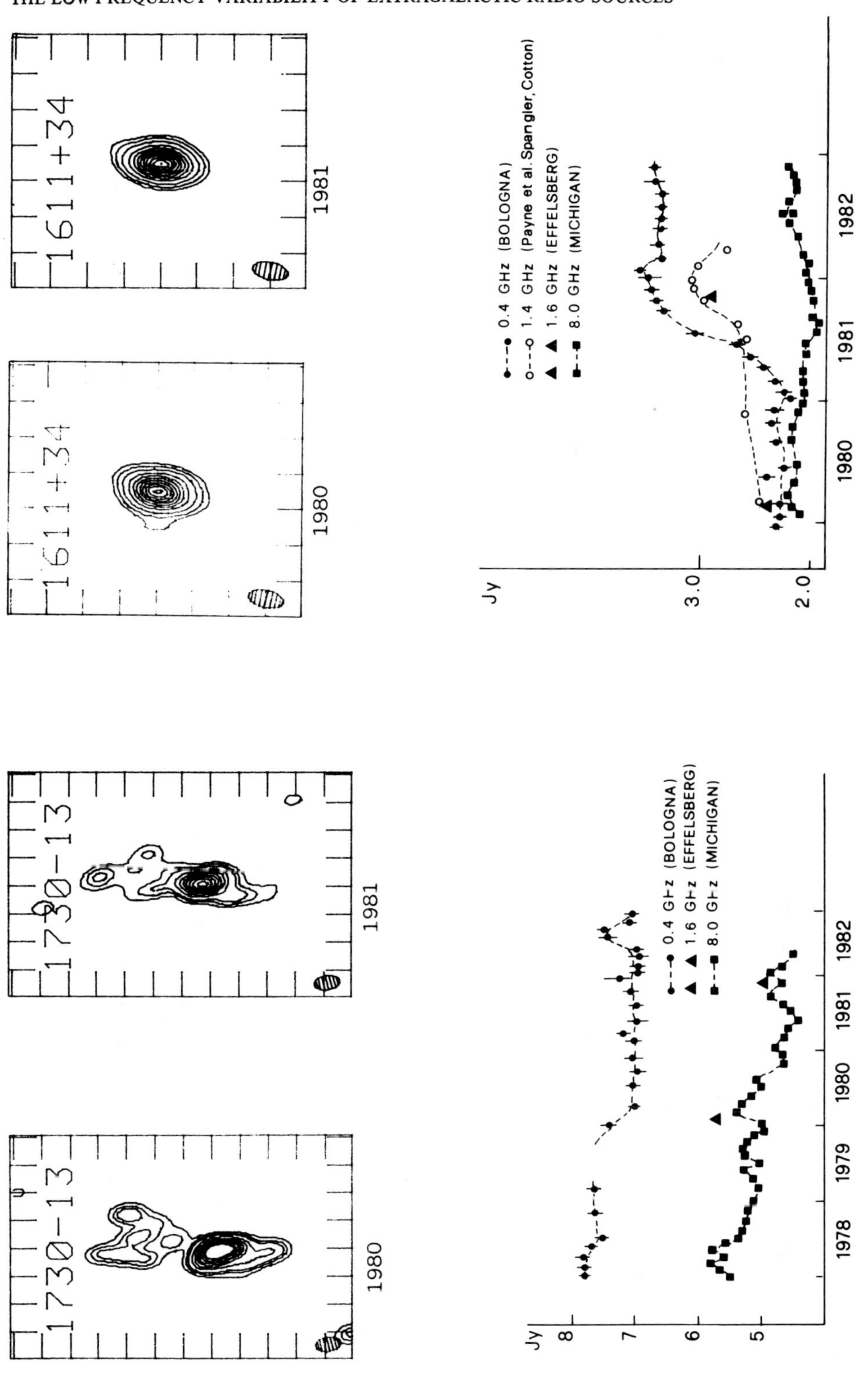

Fig. 4-5 Two epochs VLBI maps of 1730-13 and DA 406 (1611+34). The distance between ticks corresponds to 5 mas. The contours are plotted at the same flux levels for both epochs. The light-curves at different frequencies are shown for each source.

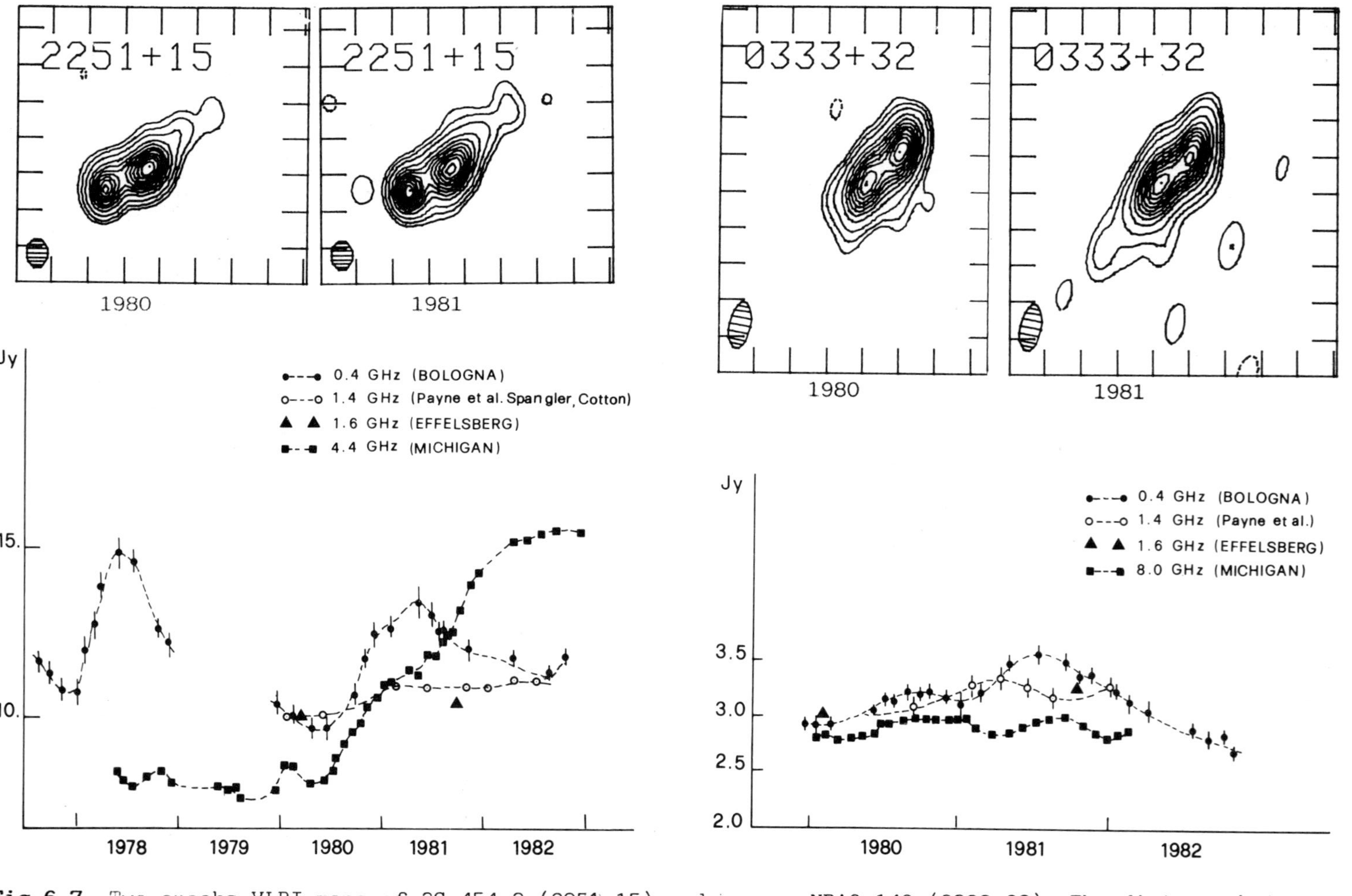

Fig.6-7 Two epochs VLBI maps of 3C 454.3 (2251+15) and NRAO 140 (0333+32). The distance between ticks corresponds to 5 mas. The contours are plotted at the same flux levels for both epochs. The light-curves at different frequencies are shown for each source.

DYNAMIC SPECTRA OF LOW-FREQUENCY VARIABLES [+]

Brian Dennison, J. J. Broderick, S. L. O'Dell, K. J. Mitchell
Physics Dept., Virginia Polytechnic Institute and State University
D. R. Altschuler
Physics Dept., University of Puerto Rico
H. E. Payne and J. J. Condon
National Radio Astronomy Observatory

A wide range of models have been proposed to account for low-frequency ($\lesssim$ 1 GHz) variability in cosmologically-distant radio sources. To address these models we (Payne et al. 1982) are monitoring the 0.3-1.4 GHz spectra of such sources. Results from three years of monitoring indicate that two distinct types of low-frequency variables may exist.

The first, and rarer, type is that exemplified by AO 0235+16. For both outbursts seen (Fig. 1), it appears that the amplitude was greatest at 1.4 GHz, with a decaying, and probably delayed, amplitude at lower frequencies. These characteristics and the time-dependent spectra (Fig. 2) suggest that the variations in AO 0235+16 result from relativistically moving clouds. The only other similar low-frequency variable is BL Lac (Aller and Aller 1982, Fanti et al. 1983). 'It is of considerable importance to determine whether variations of this type are characteristic of BL Lac objects in general.

Most low-frequency variables, however, exhibit time variable spectra of a decidedly different nature. We illustrate this with the case of 1611+34 (DA 406) which has undergone a fairly dramatic variation at low-frequencies (Fig. 3). From the lack of a large increase at 0.9 and 1.4 GHz, it is straightforward to show that if this event is a synchrotron outburst, then the causative particles must have an exceedingly steep spectrum above energies corresponding to 0.6 GHz (Cotton and Spangler 1979). Most interesting is the spectral evolution during the flux increase (Fig. 4). The dynamic spectra are not indicative of simple evolution of expanding synchrotron components, rather the spectra seem to suggest that major changes in opacity may be occurring. The spectral turn-up around ~0.3 GHz could be due to the larger size of the emitting region exceeding the opaque portion of the source. Qualitatively similar spectra are observed in other low-frequency variables. We are currently exploring theoretical models involving opacity changes.

This research is supported by NSF grants AST 79-25345 and AST 81-17864 to VPI&SU and NSF grant AST 80-18384 to UPR.

[+] Discussion on page 440

R. Fanti et al. (eds.), VLBI and Compact Radio Sources, 177–178.
© *1984 by the IAU.*

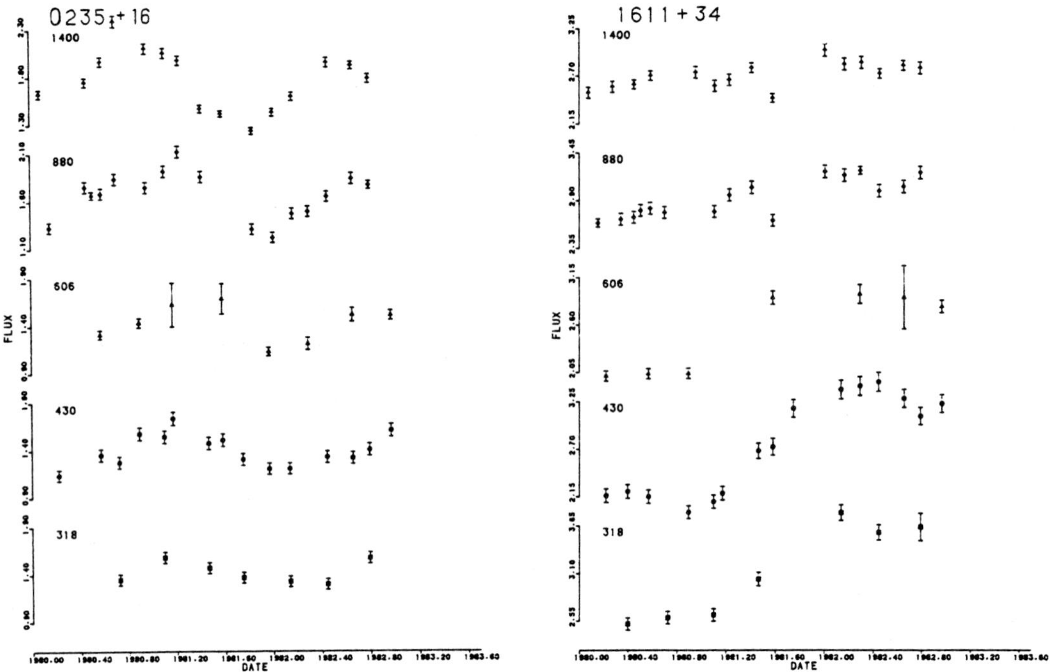

Fig. 1 - Light Curves of AO 0235+16.   Fig. 3 - Light Curves of DA 406.

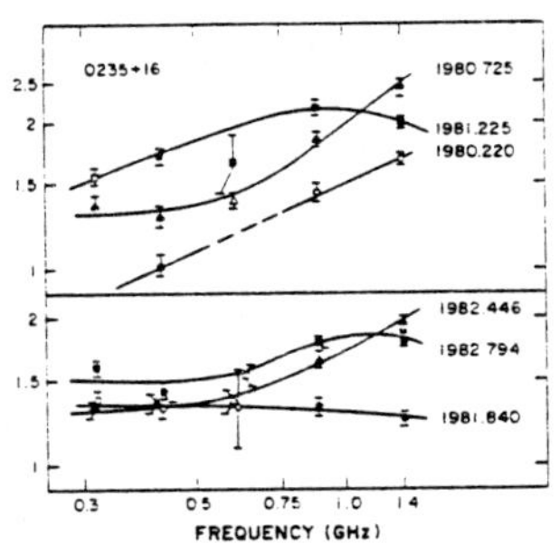

Fig. 3 - Dynamic Spectra
of AO 0235+16.

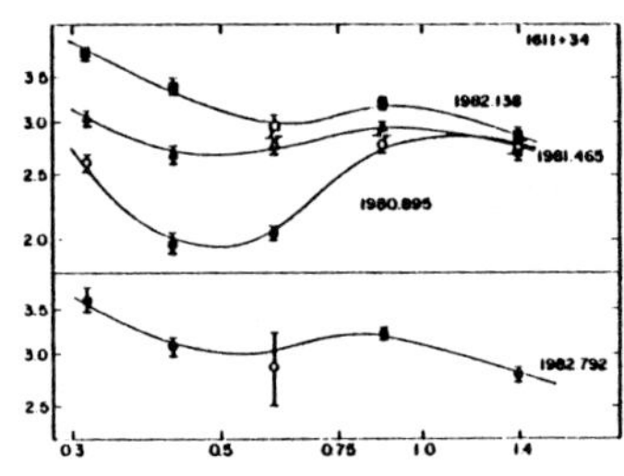

Fig. 4 - Dynamic Spectra
of DA 406.

REFERENCES

Aller, H. D., and Aller, M. F.: 1982, in <u>Green Bank Workshop on
    Low-Frequency Variability</u>, eds. W. D. Cotton and S. R. Spangler, 105.
Cotton, W. D., and Spangler, S. R.: 1979, Astrophys. J. Lett. 228, L63.
Fanti, C., Fanti, R., Ficarra, A., Gregorini, L., Montovani, F., and
    Padrielli, L.: 1983, Astron. Astrophys. 118, 171.
Payne, H. E. Altschuler, D. R., Broderick, J. J., Condon, J. J.,
    Dennison, B., and O'Dell, S. L.: 1982, in <u>Green Bank Workshop on Low-
    Frequency Variability</u>, eds. W. D. Cotton and S. R. Spangler, 9.

DAILY RADIO FREQUENCY OBSERVATIONS OF SELECTED OBJECTS

B.J. Geldzahler[1,3], K.J. Johnston[1], J.H. Spencer[1],
E.B. Waltman[1], W.J. Klepczynski[2], P.E. Angerhofer[2],
D.R. Florkowski[2], D.D. McCarthy[2], and D.N. Matsakis[2]
[1] E. O. Hulburt Center for Space Research, Naval Research
Laboratory, Washington, D. C. 20375
[2] U. S. Naval Observatory, Washington, D. C. 20390
[3] NRC/NRL Cooperative Research Associate

In Table I we present the list of 38 celestial objects that have
been observed since January 1978 at 2.7 and 8.1 GHz with the Green Bank
interferometer. The sources fall naturally into three categories: radio
stars, possibly Galactic sources, and extragalactic sources. SS433, Cyg
X-3, and each extrgalactic source is measured several times per day
while the other sources are measured once every three days. Reports on
the entire program will be found in Geldzahler et al. (1983a), and on
specific sources: SS433--Johnston et al. (1983a), BL Lac--Johnston et
al. (1983b), Cyg X-3--Geldzahler et al. (1983b) and elsewhere in this
volume), and CTA 26--Spencer et al. (1983).

We have defined for the variable sources a "rapidity index" which
gives the number of maxima/year. This index includes major outbursts as
well as "flickering". We also show in Table I the value of k ($-d$ (log
S_{max})/d(log λ)). The values of this index fall into three groups:
$k > 0$, $k \sim 0$, and $k \sim -0.4$. A uniform source that is initially optically thick and whose energy losses occur primarily through adiabatic
expansion should yield $k = -1$ (c.f. van der Laan 1966). We find that
$k > 0$ when we have an optically thin object such as SS433 or Cyg X-3
during outburst. To make the value of $k \sim -0.4$ more agreeable with the
standard model, we suggest the uniformity should be replaced by a variable opacity throughout the source. Finally $k \sim 0$ in those sources,
such as the "quiescent" Cyg X-3, where repeated, rapid flickering has
stretched and weakened the magnetic field in the immediate vicinity of
the source.

REFERENCES

Geldzahler, B.J. et al.: 1983a, to be submitted to A. J.
Geldzahler, B.J. et al.: 1983b, Ap. J. (Letters) in press.
Johnston, K.J. et al.: 1983a, A. J. in press.
Johnston, K.J. et al.: 1983b, Ap. J. (Letters) in press.
Spencer, J.H. et al.: to be submitted to Ap. J.
van der Laan, H.: 1966, Nature 211, 1131.

R. Fanti et al. (eds.), VLBI and Compact Radio Sources, 179–180.
© *1984 by the IAU.*

Table I. List of Program Sources

Source	ID	Observing Interval [a]	Redshift [b]	RI [c]	k [d]	Other Names
a. Radio Stars						
0236+610		5001-5007		I		LSI 61 +303
0323+285		4976-5006		--		UX Ari
0334+004		4971-5010		28. 1:		HR1099
1617-155		4977-5517		--		Sco X-1
1909+048		4067-5517		5. 7	+0. 53±0. 03	SS433
1956+350		4976-5125		--		Cyg X-1
2030+407		4971-5517		--	-0. 04[e] +0. 5	Cyg X-3
2259+586		4974-5129		--		GF2259+586
b. Possibly Galactic Objects						
0125+628		4976-5517		4. 0		G127. 11+0. 54
2013+370		4983-5517		3. 7		G74. 89+1. 22
c. Extragalactic Objects						
0224+671	Q	3942-5517		3. 6		
0235+164	BL	4971-5148		4. 1	-0. 42±0. 05	OD160
0237-234	Q	3941-5148	2. 223	1. 5		OD-263, PHL8462
0316+413	G	3942-4066		I		3C84
0336-019	Q	3942-5148	0. 852	6. 0	-0. 01±0. 03	CTA26
0355+508	EF	3941-5148		3. 1		
0402-362	Q	3941-4066	1. 417	8. 7:		
0727-115		5060-5148		24. 9:	-0. 09±0. 02	
0742+103	EF	3942-4060, 5010-5059		I		
0851+203	BL	3942-5148	0. 306::	3. 9	-0. 48±0. 01	OJ287
0923+392	Q	3942-4066	0. 699	I		C39. 25, DA267, OK340
0964+658	BL	3942-5517		2. 1	-0. 04±0. 02	
1226+023	Q	3943-5517	0. 158	I		3C273, 4C02. 32, NRAO400, ON044, DA324
1245-197	Q	3942-5517		0. 0		
1328+254	Q	3942-5517	1. 055	0. 0		3C287, 4C25. 43, NRAO424, OP247, DA345
1328+307	Q	3942-5517	0. 849	0. 0		3C286, 4C30. 26, NRAO425, OP348, DA346, CTA60
1502+106	Q	3942-5517	1. 833	2. 7	+0. 08±0. 03	OR103
1519-273	Q	3943-4067		11. 8:		
1641+399	Q	3941-5517	0. 595	0. 6		3C345, 4C39. 48, NRAO513, OS368, DA420
1749+701	BL	3943-5517		2. 4	-0. 03±0. 04	W1
1901+319	Q	3943-4066		I		3C395
2021+614	Q	5010-5059		I		
2037+511	Q	3943-4066	1. 686	I		3C418, 4C51. 12, NRAO636
2048+312	Q	4970-5517	3. 18	22. 3		CL4
2134+004	Q	4011-5517	1. 936	I		PHL61, DA553, OX057
2200+420	BL	3942-5517	0. 0688	5. 7	-0. 30±0. 09	BL Lac
2251+158	Q	3942-5517	0. 859	3. 9	-0. 12±0. 01	3C454. 3, 4C15. 76, OY185, NRAO701, DA506
2345-167	Q	3942-4066	0. 600	20. 6:		OZ-176

a As of 1 July 1983
b Emission line redshifts taken from Hewitt and Burbidge (1980)
c Rapidity Index: the number of maxima per year; I= variations exist by the value
  of RI cannot be determined with reliability, -- = no obvious variation, : = value
  is uncertain due to short time base
d k = (d log peak flux density)/(d log frequency)
e In Cyg X-3, $k \sim 0.5$ during optically thin outbursts and $k \sim -0.4$ during optically
  thick outbursts

AN OUTBURST OF OJ 287

Sen KIKUCHI and Makoto INOUE*
Tokyo Astronomical Observatory, Mitaka, Tokyo 181, Japan
*Nobeyama Radio Observatory, Nobeyama, Minamisaku, Nagano 384-13,
Japan

ABSTRACT: Optical and radio behaviours of OJ 287 in the recent outburst
are presented. The light curve is similar to that in the previous outburst in 1970-1971, although the maximum brightness is about 1 mag fainter. The intraday variations are found in both optical and radio regions.
We find no periodicity of time scales less than 0.2 days.

We have been made optical photometry and polarimetry of OJ 287 mainly
with the multichanel polarimeter attached to the 91cm reflector at the
Dodaira Station of the Tokyo Astronomical Observatory since 1973, and in
December 1982, we began to measure flux density at 10 GHz with the 45m
radio telescope of the Nobeyama Radio Observatory. On January 10, 1983,
the brightness in V was observed as 12.92 mag by Haarala et al. (1983).
Combining the results on December 27, we find that the abrupt increase of
the brightness during two weeks has amounted to 1.1 mag, which is comparable to that in the previous outburst in 1970-71. Then the brightness has
been fluctuating with a trend of becoming fainter.

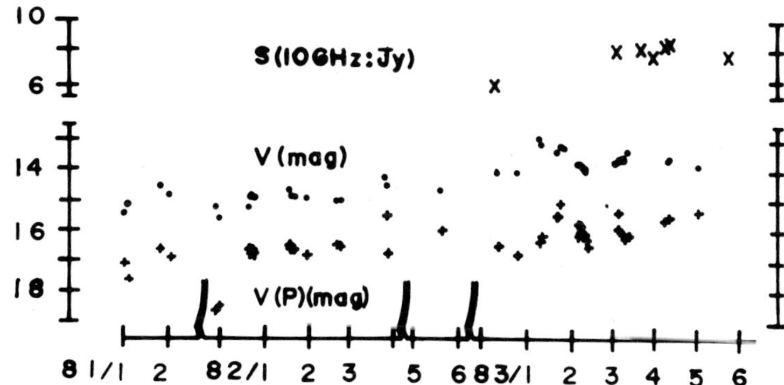

Figure 1.   Long-term variations of OJ 287. V(P) denotes the
polarized flux in V region in unit of magnitude.

181

R. Fanti et al. (eds.), VLBI and Compact Radio Sources, 181–182.
© *1984 by the IAU.*

The light curves in the both active phases are similar although the maximum brightness in the present outburst is about 1 mag fainter than that in the outburst in 1970-71. As shown in Figure 1, OJ 287 has been undergone an outburst also in the radio region. The optical polarization was not so weak as in the previous one except during the period of light maximum. On the contrary to the case in the previous outburst, the position angles varied around the direction preferrred in the basic and declining phases (Kikuchi et al. 1976, and Hagen-Thorn 1980).

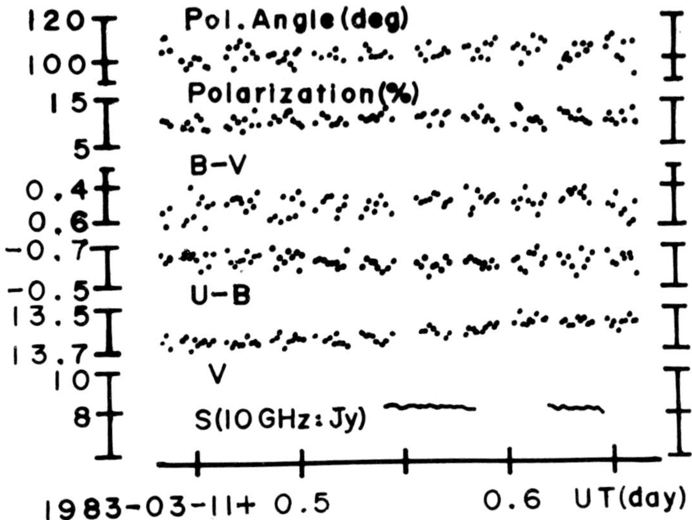

Figure 2. Intraday variations of OJ 287 on March 11, 1983.

The rapid changes of the brightness during a night were found several times. The largest variation was found on March 11, and the variations of observed V, colors and polarization properties are displayed in Figure 2 with those of flux density at 10 GHz. In the radio region, the rapid variation within a day is also confirmed and on March 9 the flux density at 10 GHz changed by 4% during 3 hours. Although Vaitaoja et al. (1983) suggested a 15.7 min periodicity in the optical and radio variations, we found no significant periodicity of time scales less than 0.2 days.

REFERENCES

Haarala, S., Korhonen, T., Sillanpaa, A., and Salonen, E. 1983, IAU
    Circ., NO. 3764.
Hagen-Thorn, V. A. 1980, Astrophys. Space Sci., **73**, 263.
Kikuchi, S., Mikami, Y., Konno, M., and Inoue, M. 1976, Publ. Astron.
    Soc. Japan, **28**, 117.
Vaitaoja, E., Lehto, H., Teericorpi, P., Haarala, T., Korhonen, T.,
    Valtonen, M., Terasranta, H., Salonen, E., Urpo, S., Tiuri, M.,
    and Piirola, V. 1983, TURKU-FTL-R26, Informo No. 62.

Evidence of Doppler Beaming in Variable Radio Sources [+]

T.H. Legg
National Research Council of Canada, Ottawa

Total flux observations of variable sources (Andrew et al. 1978) have been re-examined to test their consistency, or lack of it, with the Doppler beaming hypothesis of Scheuer and Readhead (1979).

Probably the most significant finding is that bursts from different variable sources, though differing in time scale, have profiles of the same shape. Fig. 1 shows burst profiles at λ=2.8 cm from six sources in which individual bursts are isolated from surrounding events. The data

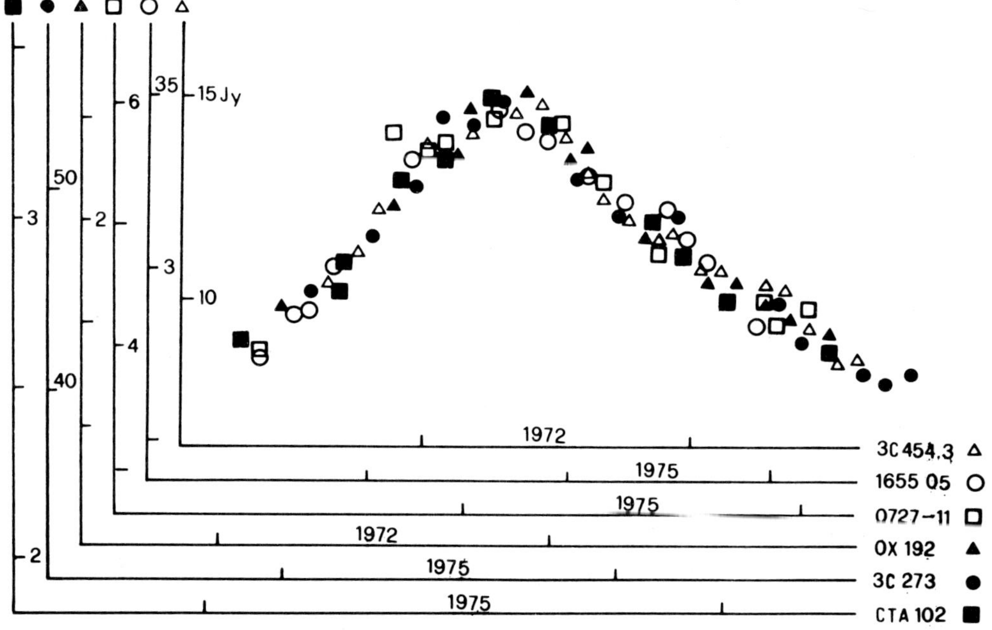

Fig. 1 - Burst profiles from six variable sources.

+ Discussion on page 440

R. Fanti et al. (eds.), VLBI and Compact Radio Sources, 183–185.
© 1984 by the IAU.

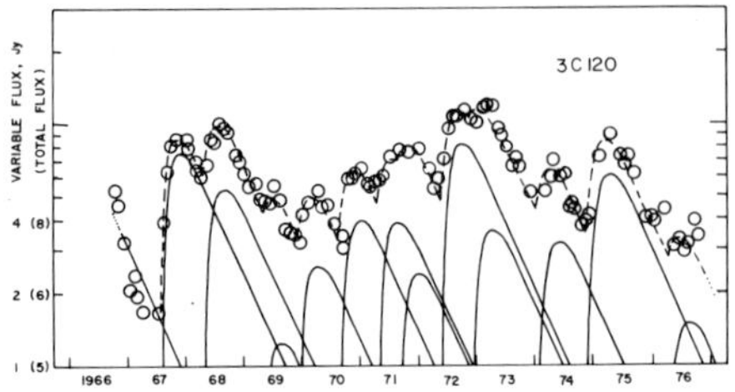

Fig. 2

are evidently consistent with various amounts of time-compression, such as would be imposed by Doppler beaming. The data also suggest a common physical mechanism in these sources.

The profiles of Fig. 1 are

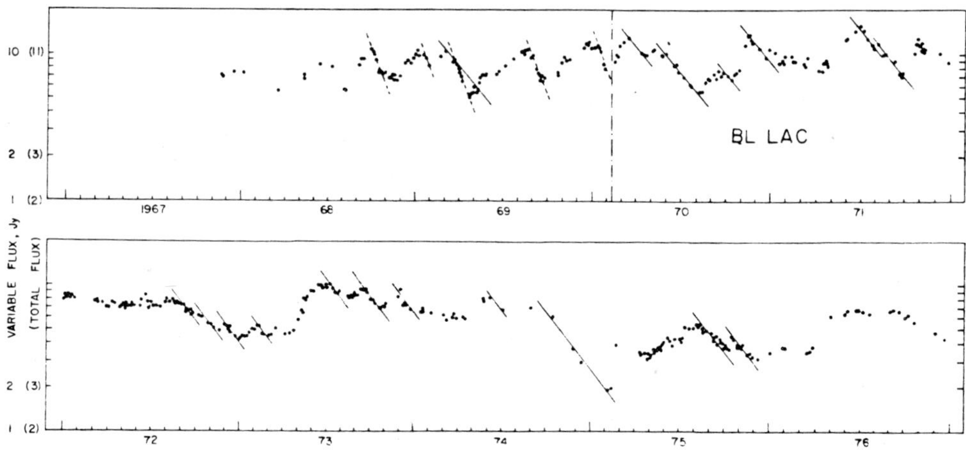

Fig. 3

fitted reasonably well with a function of the form: $t^n e^{-t/\tau}$, with $n \gtrsim 3$. The most distinctive feature of the profiles is an exponential decay. This is evident when the 'light-curve' is plotted on a logarithmic scale of flux, as for the 3C120 data shown in Fig. 2. Here, an attempt has been made to fit the observed data with burst profiles of various amplitude, but the same shape, superimposed upon a quiescent component of flux. For 3C120, burst of the same shape (and in particular the same exponential decay) are found to fit best for an assumed quiescent component of about 4 Jy.

An exponential decay is also evident in the light curve of Bl Lac (Fig. 3). Here, however, there is an interesting change in 1970, February, when the rate of decay of the bursts decreases suddenly and unmistakably, by a factor of about two. Though admittedly, the number of bursts observed in other sources is not large, Bl Lac is the only source to show this behavior. A plausible explanation might be that, because Bl Lac varies so rapidly, the angle of its axis to the line-of-sight would (on the Doppler beaming hypothesis) be small. Bl Lac would therefore be effected more noticeably by a change in this angle than would other sources.

It has been previously questioned (e.g. Andrew et al. 1978) whether rapid variability, such as in Bl Lac, is associated with bursts of large amplitude. Such a relationship would be expected with Doppler beaming and its existence was investigated by estimating the 'half-life' (i.e. the time for the flux of a burst to drop by a factor of two in the exponential part of the profile) of 24 variable sources. These constituted all of the sources with a measured redshift for which such an estimate could be made.

A weak anti-correlation was found between a corrected half-life, $t_{0.5}^{*}$ (half-life $\times (1+Z)^{-1}$), and the peak variable part of the luminosity of the 24 sources. The correlation is slightly better, though still not strong, between $t_{0.5}^{*}$ and the ratio of (peak) variable flux, F, to quiescent flux, Q. The data are shown in Fig. 4.

Despite the large scatter (part of which is due to poor statistical sampling of the variable flux), the slope of the least-squares regression line (solid) is close to the value 3.0 (indicated with a dashed line) that would be expected on the basis of Doppler beaming and the assumption that the sources have the same intrinsic ratio of variable to quiescent flux.

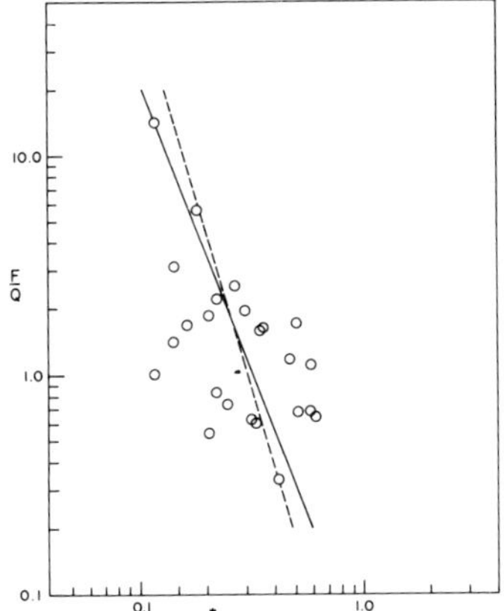

Fig. 4

REFERENCES

Andrew, B.H., MacLeod, J.M., Harvey, G.A., and Medd, W.J., 1978, A.J., 83, 863

Scheuer, P.A.G., and Readhead, A.C.S., 1979, Nature, 277, 182

EVIDENCE OF RELATIVISTIC BEAMING IN BL LAC-TYPE OBJECTS

Diana M. Worrall
Center for Astrophysics and Space Sciences
University of California, San Diego

Multifrequency observations of a variable extragalactic object, when all acquired within the inferred variability time scale of the source, can provide clues to the source's energy mechanisms. A guest observer program with the International Ultraviolet Explorer satellite has provided the focus for such measurements of a few BL Lac-type and related objects at frequencies in the radio, mm, IR, visual and UV. Earlier-epoch X-ray measurements are included in subsequent model fitting. This paper summarizes some of the observations in this program. Synchrotron self-Compton (SSC) models are then applied to the data, leading to the conclusions that the objects are relativistically beamed and that radio emission, at least below a frequency of $\sim$20 GHz, is from a separate source region.

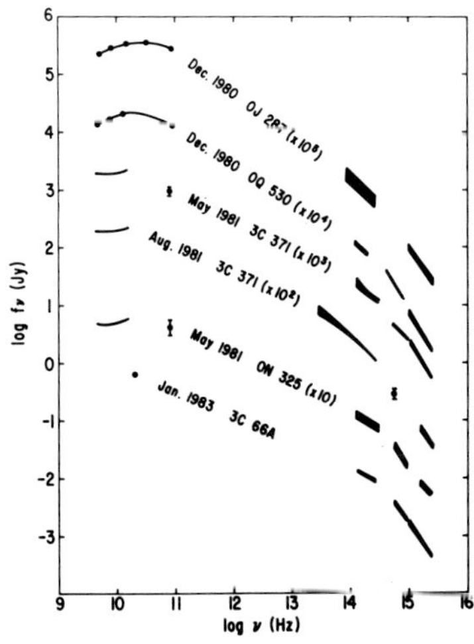

Figure 1. The "Simultaneous" multifrequency spectra display curvature between IR and UV frequencies.

The sources OJ 287, OQ 530, 3C 371, ON 325 and 3C 66A have been selected for their relatively low reddening. Only 3C 371 has a surrounding galaxy which is bright enough that it must be subtracted from the visual and IR data before the radiation of the central component can be studied (Sandage 1973). Contributors to the observational program include M. Aller, H. Aller, T. Balonek, F. Bruhweiler, F. Cordova, P. Hodge, B. Jones, W.H.-M. Ku, R. Leacock, K. Mason, K. Matthews, H.R. Miller, G. Neugebauer, A. Pica, J. Pollock, J. Puschell, J. Rodriguez-Espinosa, R. Rudy, M. Sitko, A. Smith, P. Smith, B.T. Soifer, W. Stein, J. Webb, W. Wisniewski and myself. The largest separation between same-epoch observations of a particular object has typically been that between the IR and the UV. For successive spectra of Figure 1, beginning with OJ 287, these separations were 1,1,8,12,8 and 9 days. Using published data, we infer that these should be within the time scales of any spectral-shape variability. For observa-

R. Fanti et al. (eds.), VLBI and Compact Radio Sources, 187–188.
© 1984 by the IAU.

tional details see Worrall et al. (1982;1983). Note that the objects all exhibit spectral curvature between IR and UV frequencies. Not shown in Figure 1 is the fact that previous-epoch Einstein Observatory X-ray measurements for OJ 287, 3C 371, On 325 and 3C 66A, and a HEAO-1 A-2 X-ray upper limit for OQ 530, lie above the visual to UV power-law extrapolations.

I have assumed that the core radiation is nonthermal and have applied SSC models to the data. The degree of relativistic beaming is determined by the value of $\delta = (1-\beta^2)^{\frac{1}{2}} /(1-\beta \cos \Theta)$, where β is the velocity in units of c and Θ is the angle of motion relative to the line of sight. The sources of no measured redshift, ON 325 and OQ 530, are conservatively assumed to have z=0.05. The observed IR-UV spectral curvature at ν_b is assumed to be due to electron energy losses. The first model assumes a homogeneous source in which the lifetime of electrons producing synchrotron radiation at frequencies $\lesssim \nu_b$ is at least as long as the light-crossing time of the source. Using the equations of Marscher et al. (1979) and Marscher (1983), the data of Figure 1, and previous-epoch X-ray measurements, it is found that all the sources require $\delta \gtrsim 10$, and no radiation at frequencies less than ~ 100 GHz is described by the model. A possibly more realistic model is presented by König1 (1981). The magnetic field and electron density decrease with increasing pathlength along the jet and electrons are continuously accelerated. Although precise values of δ are sensitive to spectra in the 10^{11}-10^{13} Hz band, where we have no observations, lower limits still suggest $\delta \gtrsim 2$ for ON 325 and 3C 66A and $\delta \gtrsim 10$ for the other sources. No radiation at frequencies less than ~ 20 GHz is described by the model (~ 100 GHz for OQ 530 and 3C 371). This is a consequence of the measurement of ν_b in the IR-UV frequency band. The model may describe flat-spectrum <20 GHz radio emission from sources for which $\nu_b > 2 \times 10^{15}$ Hz.

Assuming SSC emission, it is concluded that all of the five sources presented here are relativistically beamed and that radio emission, at least below a frequency of ~ 20 GHz, is from a separate source region. Both the homogeneous and König1 SSC models could be constrained by better knowledge of the relationship between variability time scales at different frequencies. Spectral coverage between 10^{11} and 10^{13} Hz, with temporal monitoring, could tie down parameter fits. Simultaneous mm and X-ray monitoring could also be an important diagnostic, as could X-ray spectral measurements.

This work was partially supported by NASA grant NAG 5-63.

REFERENCES
König1, A., 1981, Ap.J., 243, p 700.
Marscher, A.P., Marshall, F.E., Mushotzky, R.F., Dent, W.A., Balonek,T.J., and Hartman, M.F., 1979, Ap.J., 233, p 498.
Marscher, A.P., 1983, Ap.J., submitted.
Sandage, A., 1973, Ap.J., 180, p 687.
Worrall, D.M., et al., 1982, Ap.J., 261, p 403.
Worrall, D.M., et al., 1983, Ap.J., submitted.

A COMPARISON OF COMPACT AND EXTENDED RADIO SOURCES
IN THREE WAVEBANDS +

Lance Miller
Department of Astronomy, University of Edinburgh

I shall compare the luminosities in the X-ray, optical, and radio wavebands of two types of radio source - compact flat-spectrum sources and extended double sources - and attempt to place limits on models which unify these types.

Fig. 1 shows the correlation between Hβ emission-line luminosity and X-ray luminosity (taken from Zamorani et al. 1981, Owen et al. 1981, Tananbaum et al. 1983, and Fabbiano et al. 1983) obtained by combining various radio samples. The correlation is similar to that found for Seyfert galaxies (Elvis et al. 1978, Kriss et al. 1980) and for other quasars (Blumenthal et al. 1982, Reichert et al. 1982). Under the hypothesis that all the objects with broad emission-lines form a single class of object, it is reasonable to combine all these sources on one diagram, and the correlation then shows that the compact radio sources have the same range of the ratio $L_X/L_{H\beta}$ as the extended radio sources. Since the emission-lines are neither blue-shifted, nor shifted with respect to other emission or absorption lines in the objects, the emission-lines cannot be relativistically boosted. Thus, if the nuclei of the differing classes of radio source are intrinsically the same, the correlation shows that the X-rays are not relativistically boosted by a factor greater than 3 in any one class of object. Obscuration of the Hβ emission is unlikely to be important, since (a) extended double quasars show similar optical spectra and colours to compact quasars, and (b) there is no evidence for large obscuration in either radio galaxies (Saunders and Miller, 1983) or quasars (Puetter et al. 1981) from infra-red observations.

Fig. 2 shows the correlations between flat-spectrum nuclear radio emission and X-ray emission (Owen et al. 1981, Tananbaum et al. 1983, Fabbiano et al. 1983). The total radio luminosity has been used for the compact sources. For a given X-ray luminosity these have radio emission which is a factor about 30 stronger than the radio cores of the extended sources. Since the X-ray emission is not substantially beamed, the radio cores of compact radio sources are at most boosted by a factor 30 above the cores of extended double radio sources. This is

+ <u>Discussion on page 441</u>
189

R. Fanti et al. (eds.), VLBI and Compact Radio Sources, 189–192.
© *1984 by the IAU.*

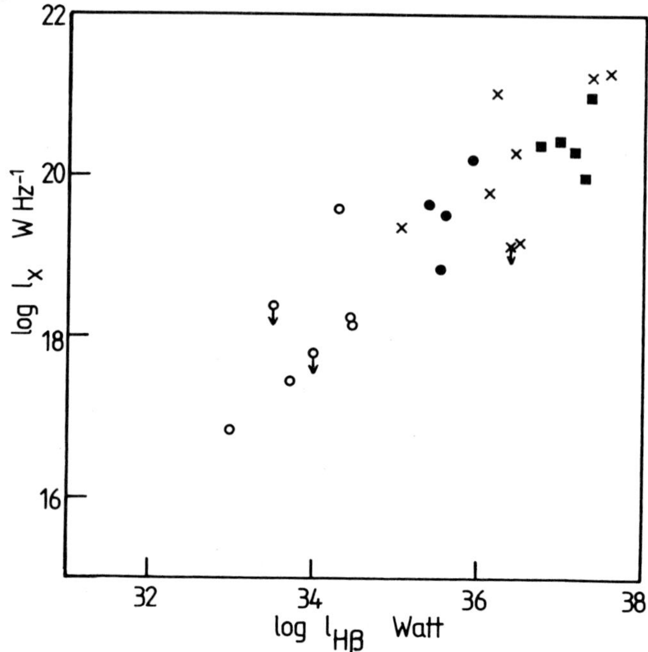

Fig. 1. The correlation between X-ray and Hβ luminosity for compact
quasars (crosses), extended quasars (filled squares), extended radio
galaxies with broad emission-lines (filled circles), and extended radio
galaxies with narrow emission-lines only (open circles).

insufficient to turn a double source into a compact source, since a
boosting factor > 300 would be required to swamp the extended emission
(Perley et al. 1982). The model of Orr and Browne (1982) would require
boosting factors > 1000. The radio properties are thus not consistent
with models where the only difference between extended double sources
and compact sources is that of relativistic boosting of the radio core
luminosity.

This is illustrated also in fig. 3, which shows X-ray luminosity
plotted against total radio luminosity at 6 cm, for the objects in fig.
2. All the classes of source lie on the same slope. Yet the fraction
of the total radio emission in the compact sources which can be
attributed to the lobes of an extended double source is typically less
than 3% (Perley et al. 1982). So, for a given X-ray luminosity, any
extended double-lobe structure in the compact sources is at least a
factor 30 weaker than in the double sources. In fig. 3, there are only
correlations for the radio galaxies and the compact sources. The
comparison is unlikely to be biased by source selection, however, since
the two samples of double sources used are complete to radio and
optical flux density limits, and there was no selection against X-ray
bright objects (Tananbaum et al. 1983, Fabbiano et al. 1983). Hence,

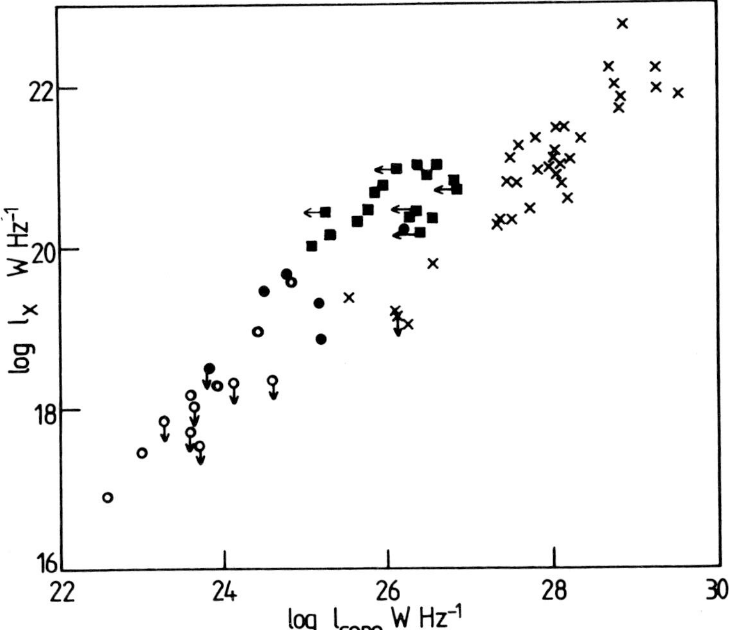

Fig. 2. The correlations between X-ray and flat-spectrum radio core luminosity.

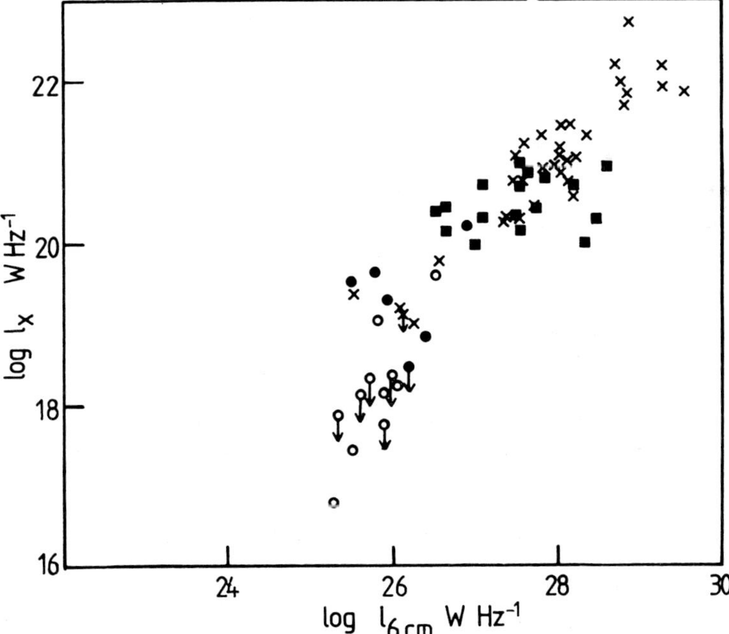

Fig. 3. The correlations between X-ray and total radio luminosity.

in the relativistic beaming models, the unfavourably oriented counterparts of the compact sources should be X-ray and optically bright, unless they are affected by obscuration, but any double-lobe structure should be relatively weak.

Finally, the compact sources might be intrinsically weaker sources which have been gravitationally lensed. Since most compact quasars show arcsec - scale radio emission (Perley et al. 1982, Browne et al. 1982) the most promising candidates for unlensed objects are the cores of extended double radio sources. Fig. 1 shows that the X-ray and emission-line regions would each have to be lensed by similar amounts, and fig. 2 shows that the radio emission would have to be lensed by a factor about 30 greater. The X-ray and emission-line regions would thus have to be the same size, and be either separated from or be larger than the radio region.

REFERENCES

Blumenthal, G.R., Keel, W.C. and Miller, J.S., 1982: Astrophys.
    J. 257, pp 499-508
Browne, I.W.A., et al., 1982: Mon. Not. R. astr. Soc. 198,
    pp 673-688
Elvis, M., et al., 1978: Mon. Not. R. astr. Soc. 183, pp 129-158
Fabbiano, G., Miller, L., Trinchieri, G., Longair, M.S. and Elvis, M.,
    1983: Astrophys. J., submitted
Kriss, G.A., Canizares, C.R. and Ricker, G.R., 1980: Astrophys. J. 242,
    pp 492-501
Orr, M.J.L., and Browne, I.W.A., 1982: Mon. Not. R. astr. Soc. 200,
    pp1067-1080
Owen, F.N., Helfand, D.J. and Spangler, S.R., 1981: Astrophys. J. 250,
    pp L55-L58
Perley, R.A., Fomalont, E.B. and Johnston, K.J., 1982: Astrophys. J.
    255, pp L93-L97
Puetter, R.C., Smith, H.E. and Willner, S.P., 1981: Astrophys. J.
    243, pp 345-368
Reichert, G.A., Mason, K.O., Thorstensen, J.R. and Bowyer, S.,
    1982: Astrophys. J. 260, pp 437-468
Saunders, R. and Miller, L., 1983: in preparation
Tananbaum, H., Wardle, J.F.C., Zamorani, G. and Avni, Y., 1983:
    Astrophys. J. 268, pp 60-67
Zamorani, G., et al., 1981: Astrophys. J. 245, pp 357-374

PASCHEN ALPHA AS A PROBE OF RELATIVISTIC BEAMING [+]

Richard Saunders
Cavendish Laboratory
Cambridge, U.K.

Is the theme of core-brightening by relativistic beaming, so useful for compact sources, helpful in explaining the nuclear properties of classical doubles? Certainly one can construct an attractive model – involving the usual beaming together with an anisotropic dust distribution at $r \lesssim 1$ pc – to explain the differences in nuclear radio and optical properties of radiogalaxies as due differences in orientation with respect to us. There are clear predictions about the strength of broad Pα. Preliminary results indicate that this kind of model is not correct.

The idea of relativistic beaming has proved extremely useful in accounting for core-jet morphology in compact sources. Can we fruitfully apply it to the nuclei of classical double radiogalaxies? There are sparse VLBI data on these sources, but one important piece of evidence suggests beaming may indeed be occuring : nuclear radio luminosity differs substantially between radiogalaxies with much the same extended, unbeamed properties like total luminosity and shape. Clearly, we could identify sources with bright nuclei as those with small values of angle θ between axis and line-of-sight. Then we must take account of the strong correlation existing between the radio luminosity ratio $P_{nucleus}/P_{total}$ and the strength of broad nuclear Balmer lines (see Hine and Longair 1979; extensive spectrophotometric work by Osterbrock's group and by the Cavendish group confirms this result). This correlation cannot be produced by the beaming of line emission because the peaks of all lines in a given radiogalaxy are at the same redshift. So we need a way of hiding broad lines except when θ is small : the obvious way is with dust – which reddens the visible – covering (or in) all the broad line region except for a pair of "channels" centred on the source axis. Note that there is no implicit requirement that the channel opening angle should be the same as the range of θ over which relativistic boosting occurs. The resulting model is shown in the Figure.

Is the model right? There are two immediate tests.

[+] Discussion on page 442   193

R. Fanti et al. (eds.), VLBI and Compact Radio Sources, 193–194.
© *1984 by the IAU.*

(a)   Compare the projected linear sizes of BLRGs (broad-line radio galaxies) and NLRGs (narrow-line RGs) – the BLRGs should be smaller on average. The observed difference is not statistically significant because of the combination of small numbers of BLRGs and of the great range in size of RGs in general.

(b)   Search for broad Pα in NLRGs. Pα is hardly reddened by dust and is often redshifted into the K band, and broad Pα has been looked for and found in a few quasars and BLRGs despite major technical difficulties. If the model is right, the new Cooled Grating Spectrometer on UKIRT should detect broad Pα in many radiogalaxies. Lance Miller and I have just attempted such a search. Despite serious weather and instrument problems, we detected broad Pα as expected in quasars but failed, with high significance, to find it in NLRG Cygnus A : our 3σ upper limit is 1.5×10^{-17} W/m^2/channel in each channel of width 1700 km/s. If Cygnus A were a sideways-on version of a BLRG of the same P_{total}, we would have found 3×10^{-16} W/m^2/channel. The conclusion from this preliminary result is that this kind of model is wrong.

REFERENCE

Hine, R.G. and Longair, M.S.: 1979, Monthly Notices Roy. Astron. Soc., 188, pp. 111-130.

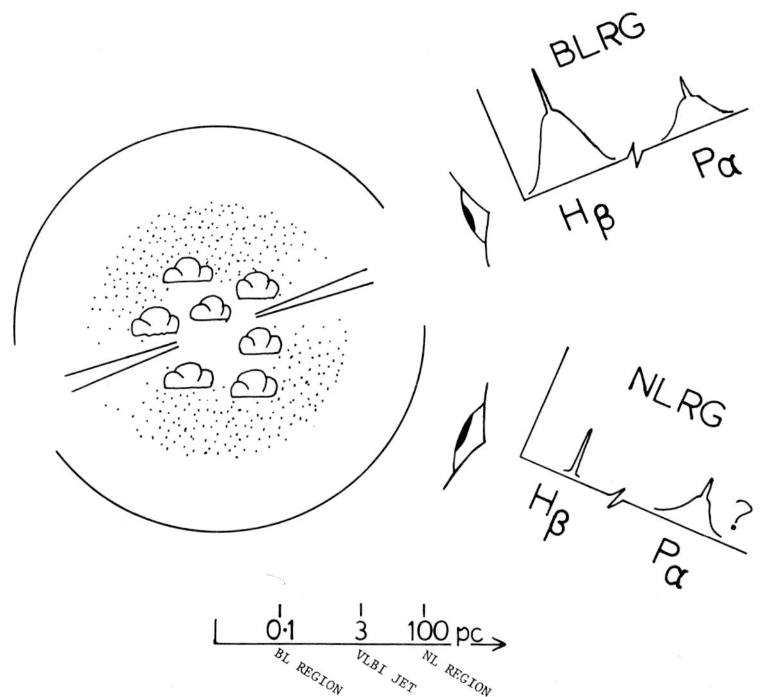

THE STATISTICS OF BENT JETS[+]

P. M. Allan
Kitt Peak National Observatory
950 North Cherry Avenue
Tucson, Arizona   85726

I present some results of a study of the statistics of bent jets. The general conclusion is that non relativistic jets appear less bent than they really are, while relativistic jets appear more bent.

The study of radio sources which do not have a perfectly linear structure is greatly complicated by projection effects which can make a relatively simple source appear quite complex. Attempts have been made to model particular sources with precessing jets (Gower et al. 1982) however, this procedure requires very good data to produce meaningful results. I have approached a somewhat simpler problem from the opposite viewpoint, i.e. given a random sample of bent jets, how will an average jet appear to an observer. The bent jet is represented by the idealization of two straight lines with an angle α between them. The bisector of the angle α defines the x axis of a coordinate system and the lines making the angle α lie in the x–y plane. If this angle is viewed from a position with angular coordinates (θ, ϕ) then the apparent angle of the bend α' is given by

$$\cos \alpha' = \frac{\cos \alpha - \sin^2 \theta \, \cos(\phi+\frac{\alpha}{2}) \, \cos(\phi-\frac{\alpha}{2})}{\sqrt{(1-\sin^2 \theta \, \cos^2(\phi+\frac{\alpha}{2})) \, (1-\sin^2 \theta \, \cos^2(\phi-\frac{\alpha}{2}))}}.$$

Using this expression one can calculate the probability of seeing an angle α' for a given intrinsic bend angle α. . The details of the calculation will be given elsewhere (Allan, in preparation) however a simple way of visualizing the result is to draw lines of constant α and α' on the surface of a sphere. The probability of seeing an angle in the range α' to $\alpha' + d\alpha'$ is proportional to the area between the lines (α, α') and $(\alpha, \alpha' + d\alpha')$. The main result of this is that if an obtuse angle α is observed from many random directions then it will appear to be bent less than α (i.e. $\alpha' > \alpha$) more often than it will appear to be bent more than α (see Figure 1). If α is an acute angle the reverse is true, but presumeably this is less significant for real bent jets.

+ Discussion on page 443

195

R. Fanti et al. (eds.), VLBI and Compact Radio Sources, 195–196.
© 1984 by the IAU.

Figure 1. Lines of constant (α, α') drawn on a projection onto a plane of the octant $0 < \theta < \pi/2$, $0 < \phi < \pi/2$ for the case $\alpha = 140°$. θ increases downwards and ϕ increases anticlockwise. The numbers indicate the value of α'.

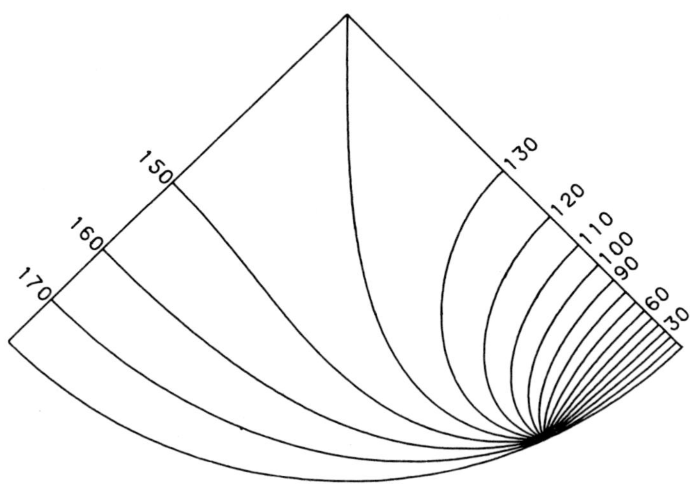

So far I have not mentioned the velocity of the jet and I have implicitly assumed that the jet is non relativistic. If a bent relativistic jet is observed with a limited dynamic range, then the bend will not be visible from certain directions as the Doppler beaming may make one side of the bend much brighter than the other side, thus when actually observed, only one side is seen and the source is not described as being bent at all. The directions where one side of the jet is lost are predominantly those directions from which the jet appears straighter than it really is (i.e. small ϕ), and so the effect is that on the average, bent relativistic jets will appear more bent than they really are. This effect has been appreciated for a while in that VLBI jets often appear very bent (Readhead et al. 1978). What has not been fully realized is that it can be impossible to 'straighten' the jet since when it is observed from a direction where it appears straight, half of the jet is lost in the noise.

When applied to a large complete sample of bent jets, these techniques should be capable of giving the distribution of intrinsic bend angles. They can also be applied to the large scale double sources where the central galaxy does not lie on the line joining the radio lobes (Ingham and Morrison 1975).

References

Ingham, W. and Morrison, P. 1975. Mon. Not. R. astr. Soc., 173, 569.
Gower, A. C., Gregory, P. C., Hutchings, J. B. and Unruh, W. G. 1982.
    Astrophys. J., 262, 478.
Readhead, A. C. S., Cohen, M. H., Pearson, T. J. and Wilkinson, P. N.
    1978. Nature, 276, 768.

EXPLANATIONS OF SUPERLUMINAL MOTION [+]

P.A.G. Scheuer
Mullard Radio Astronomy Observatory, Cavendish Laboratory,
Madingley Road, Cambridge CB3 OHE, U.K.

It is a sobering thought that Blandford, McKee and Rees (1977) wrote a review on this subject 6 years ago and that we are still considering much the same range of possibilities. Observationally, five things are now firmly established that were either unknown or less clear in 1977:

a)   Superluminal motion really does occur, and is very common (5 out of 7) among those sources that have been investigated. Of course the present sample is full of strong selection effects, which we shall never be able to define retrospectively.

b)   Most VLBI sources (including all or nearly all the superluminals) consist of an unresolved self-absorbed part (the "core") and a steep-spectrum "jet" on one side only.

c)   The motion relative to the "core" is reasonably uniform and <u>always outwards</u>. (Dr. Shaffer has just shown us an exception.)

d)   Disappearance of the middle ground. Until last year one could believe that VLBI jets are highly relativistic but that they slow down somehow and have $v \lesssim 0.1c$ by the time they are many kpc long. Now you must go the whole hog: either you are for fast jets, or you are for slow jets, all the way. The reason is this simple observation: wherever there is a large-scale jet, there is a VLBI nucleus, and wherever that has been mapped the VLBI jet points towards the large-scale jet. Now suppose that VLBI jets have $\gamma \gg 1$ but large-scale jets have $v \ll c$. Then .

(i) If VLBI jets are intrinsically one-sided, 50% of VLBI jets are relativistically beamed away from us, so the majority are invisible: we should observe many large scale jets with no VLBI cores.

(ii) If VLBI jets are intrinsically two-sided, the VLBI jet should point away from the large-scale jet in 50% of all cases.

[+] <u>Discussion on page 443</u>

R. Fanti et al. (eds.), VLBI and Compact Radio Sources, 197–205.
© 1984 by the IAU.

In either case there is a contradiction with observation.

e)   Low-frequency variability (at $\lambda \sim$ 1m) really does occur, and is also fairly common among VLBI sources.

I also note that recently VLBI observers have used $H_O = 55$ km s^{-1} Mpc^{-1}, thereby doubling all the apparent velocities!

The fact that the motion is always outwards excludes primitive versions of the "Christmas tree" model, in which superluminal motion was simulated by a row of lights flashing at random times. My list of types of model (not a comprehensive list!) therefore begins with

1.   <u>"Computer-controlled Christmas trees, with $\sim$ isotropic lights"</u>.

This includes "lighthouse" models in which a beam of radiation or particles sweeps over a screen (at arbitrarily large speed) and excites it temporarily. It also includes models in which, say, a shock front crosses some linear feature at a small angle, and the crossing point radiates.

It is not easy to construct physically plausible models of this kind in which the source position always moves away from the core, but it can be done. The real downfall of most such models is that, when the source motion is not perpendicular to the line of sight, the apparent source velocity is often inward because of light travel time effect - as shown by the standard formula (1) (below) with v > c.

A model that escapes this criticism is the filament of matter excited by something (say a spherical shock front) propagating out from the central engine. For any shock speed and filament orientation, the total of shock travel time + light travel time has a minimum for some position on the filament, and as minima are flat, the two radiating spots appear to separate initially at infinite speed, decelerating gradually. This model has always been rejected on the grounds that the observed speed is constant; at this meeting the Caltech group have presented evidence against constant velocity in 3C345, but they find acceleration not deceleration. This simple model fails to account for the general rule that there is one heavily self-absorbed component at the end of the line.

2.   <u>Relativistic beaming</u>

- in which the same radiating matter is seen at successive epochs and the apparent "superluminal" motion is all due to light travel time effects. For a jet of speed v, Lorentz factor γ, at angle θ to the line of sight, this yields an apparent speed given by the well-known formula

$$v_{app} = v \sin \theta / (1 - \frac{v}{c}\cos \theta). \tag{1}$$

This has been the favourite model ever since the one-sided VLBI structure (Item B) was discovered, and was worked out in considerable depth by Blandford and Königl (1979). Apparent speeds comparable with the maximum $\gamma v \cong \gamma c$ should occur only if $\theta \sim 1/\gamma$, i.e. for one source in several γ^2. Readhead and I (1979) pointed out that

(i) a very large fraction of radio quasars could be superluminal if the observed "core" were not really the stationary core at all, but the self-absorbed base of the relativistic jet itself. The strong selection effect in favour of small θ would then apply to the "core" emission too (slightly less strongly because of the flatter spectrum). While observational papers still refer to a "core" for lack of a better word, I think this hypothesis is now taken for granted in most discussions.

(ii) the model explains the strong apparent curvature in core-dominated sources as a projection effect. This had already been suggested earlier by the Caltech group, and has also become part of the conventional wisdom.

(iii) argument (i) cannot apply to optical and in particular to line emission. So, in a sample free from orientation bias, selected e.g. by optical line emission, only a fraction of the order of 1% should have strongly Doppler-boosted cores with $v_{app} \gg c$. That would be compatible with observation if a substantial fraction of optically selected quasars were really fairly weak radio sources, well below the limits of most catalogues, and appeared strong only when $\theta \ll 1$. Item (iii) has been attacked vigorously from two directions.

On the one hand, there are statistically significant spectroscopic differences between radio-loud and radio-quiet quasars (see e.g. Osterbrock 1982 and Wills 1982). I think this is sufficient to establish that a large proportion of optically selected quasars are a physically distinct group of radio-quiet quasars; the spectroscopic differences would be hard to explain away as orientation effects.

On the other hand, Browne and others state that the superluminal sources are just the cores of ordinary double radio sources, on the grounds that all the core-dominated sources have some diffuse structure on both sides of the core if one looks hard enough. (The unique exception is 3C273.) In some sense they must be right. Even theoretically it would be astonishing if the radio jets vanished into the blue yonder without ever interacting with their surroundings. The questions at issue are not qualitative but quantitative: how powerful are the radio lobes associated with typical superluminals, and what distributions of Lorentz factor γ can occur in a sample of radio sources free from orientation bias, e.g. a sample selected by the flux in radio lobes alone. Orr and Browne (1982) find that the statistics of core fluxes are compatible with jets of $\gamma = 3.7$ in the 3C sample, while samples selected at higher frequencies (where the cores contribute more to the total flux) are compatible with $\gamma = 5$. I think there would be statistical trouble if the Lorentz factor were typically 7, and very

severe trouble if it has to be 10 or 15, corresponding to putting
H_o = 55 km s^{-1} Mpc^{-1} in the VLBI data.

Orr and Browne's samples are radio selected, and corrections
have to be made for the selection effect due to the core contribution
to the flux. Optically selected samples were considered by Strittmater
et al. (1980) and Condon et al. (1981). They agree that the slow but
inexorable increase in numbers with decrease in radio core flux,
predicted by all relativistic beaming models, does not occur. Some
quasars - particularly the optically very luminous ones - are radio
loud, and the rest are very radio-quiet indeed. Lacking maps, we do
not know how much of this flux comes from VLBI cores, but some of the
brighter sources have flat radio spectra, and on the basis of these
alone we should expect to see more of the quasars at the 10 mJy level.
One escape route from this statistical threat is to note that the samples
were selected largely by optical continuum flux, which should be
contaminated with enough relativistically beamed optical flux from the
jet to invalidate the statistics. The anticorrelation between the
equivalent widths of broad emission lines and optical luminosity
(Baldwin et al. 1978; Wampler et al. 1983) could be attributed to
beamed optical continuum. If that is so, we must expect to see an
excess of bright radio cores in a sample selected by optical magnitude.
I have used the spectra published by Osmer and Smith (1980) to make
rough estimates of C IV λ1549 equivalent widths of radio sources in
their optically selected sample, in the range 1.8 < z < 2.4, and
magnitude brighter than 18.5, and compared them with flat-spectrum
Parkes quasars at high flux densities, selected from a large area of
sky but subject to the same magnitude and redshift limits: for a
majority of these, equivalent widths and continuum fluxes have been
measured by Baldwin et al. and Wampler et al. The statistics are
very poor, and all I can say is that the data do not encourage the
hypothesis that some of the optical flux is beamed along with the
radio flux. However, we also have to remember that there is a
positive correlation between intrinsic radio power and line strength
(e.g. Hine and Longair 1979).

The statistical evidence mostly revolves around the flux densities
of VLBI cores, whose own apparent jet speeds have not been measured.
Crucial experiment No. 1, of monitoring the VLBI motion of a complete
sample selected without orientation bias, has not been done. It was
not possible before the routine operation of the Mk 3 VLBI because
most of the cores are faint, and it may not be done convincingly till
the VLBA is in action. The sample would have to be selected either
by the radio flux of large-scale double structure or by optical,
preferably line emission. Crucial experiment No. 2 is to look for
superluminal motion in the VLBI jet of a source such as Cyg A, whose
axis we have good reason to believe lies near the plane of the sky.
In principle that can be done now, but in practice it seems that the
core of Cyg A is peculiarly hard to map (Linfield 1981; Kellermann
et al. 1981).

My reading of the statistical evidence is that we should look for
mechanisms that can operate over a larger range of inclinations θ than
the simplest beaming model. One hypothesis (mentioned by Rees 1981) is
that the jet has angular width $> 1/\gamma$ but power that fades outwards on
angular scales $\sim 1/\gamma$, so that we still see a one-sided structure. I
have computed the appearance of such models, and find that they produce
extremely fat jets (Fig. 1), so fat that I think this model is already
excluded by existing VLBI maps. A variation is the "shotgun" model,
in which the central engine ejects things along several distinct
lines within a fairly narrow cone, and we see only that nearest the line
of sight. However, as a majority of quasars have VLBI cores, we should
have a good chance of seeing a few cases with two or more jets pointing
in quite different (projected) directions.

3. Gravitational lenses

Splitting of a single source by a gravitational lens is excluded
by the spectral difference between VLBI components (Fact B). However,
a distributed mass, such as a galaxy, can magnify real motions and
at the same time increases the observed flux ($\propto$ magnification^2) thus
creating a strong selection effect in favour of magnified velocities.
As Dr. Subramanian will speak about this theory, I will only remind
you of the traditional difficulties which I hope he will address. The
difficulties are greatest for the nearest sources. 3C120 ($z = 0.032$),
BL Lac ($z = 0.07$) and 3C273 ($z = 0.158$). The nearer the source, the
stronger the gravitational lens must be (in the sense of mass per unit
area) to achieve substantial magnification. But the probability of
finding a galaxy along the line of sight to an object with $z \lesssim 0.15$
is exceedingly small, and, furthermore, no sign of an intervening galaxy
has been found in these cases.

At this point I respectfully mention

4. Non-cosmological redshifts

- but say no more as I am not aware of any striking new developments
in this field.

I do want to say something about

5. Computer-controlled Christmas trees with beamed lights

I know of only one published model in this class: the magnetic
dipole model, proposed in original form by Sanders (1974) and
elaborated by Bahcall and Milgrom (1980). It does not fit easily into
current pictures of what goes on in quasar nuclei, and indeed, at a
recent workshop at Jodrell Bank, Sanders himself demolished the model.
Yet in one respect it is conspicuously successful, in that it predicts
superluminal speeds over a large range of orientations θ, and furthermore the predicted speeds are never 1.1c or 1.5c but always $> 4c$, more
or less as observed. These desirable features are not peculiar to the

Fig. 1.   How a non-uniform conical jet would look. Assume the axis
of the cone makes angle θ with the line of sight, that the emitting
material is ejected with uniform speed v, and that the emissivity varies
as $r^{-n} \exp(-\sin \psi/\sin \psi_0)$ where ψ is the angle between the cone axis
and the velocity of the emitting material, and r its distance from the
core; n and ψ_0 are parameters of the model. The r^{-n} dependence
ensures that all isophotes have the same shape, given, in polar
coordinates R,A on the sky by

$$R^{n-1} \propto \int_0^\pi \sin^{n-2}\chi(1 - \frac{v}{c}\cos \chi)^{-(2+\alpha)} \exp(-\sin \psi/\sin \psi_0)d\chi$$

where $\cos \chi \equiv \cos A \sin \theta \sin \chi + \cos \chi$; χ represents the angle
between the velocity of an element of source material and the line
of sight. The diagrams show isophote shapes for various θ, n and
ψ_0, and $\gamma = (1 - v^2/c^2)^{-\frac{1}{2}}$.

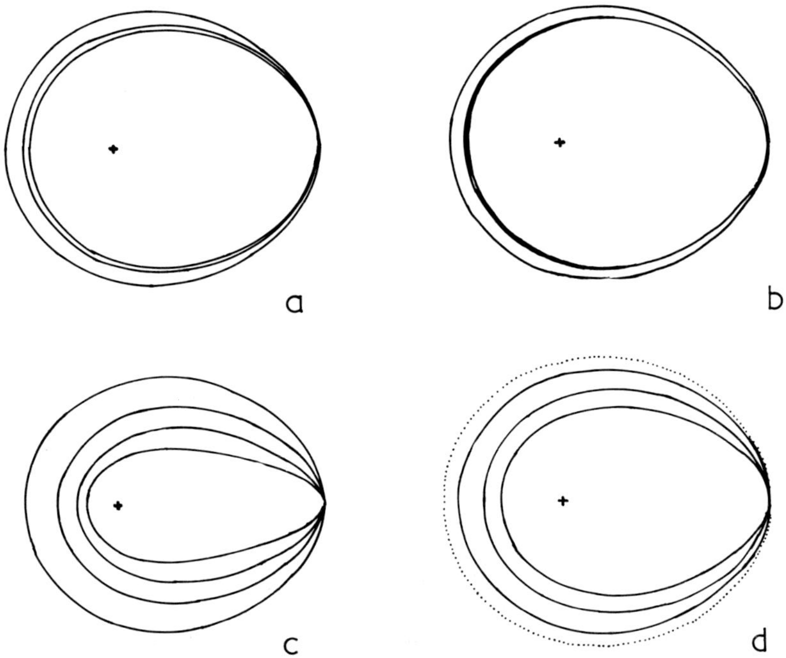

In Figures (a),(b) and (c), n=3 and spectral index α=1; θ is least for
the outermost isophote.
      (a) has γ=10, ψ_0=5°; θ= 5°,10°,20°.
      (b) has γ= 5, ψ_0=10°; θ=10°,20°,40°.
      (c) has γ= 5, ψ_0= 5°; θ= 5°,10°,20°,40°.
      (d) has γ=10, ψ_0= 5°; α=0.3; the full lines show n=3, θ= 5°,10°,20°;
        the dotted line , n=4, θ=5°.

geometry of a dipole magnetic field, and I think we can incorporate
them in kinematic schemes that may have plausible underlying physics.
What we need is bursts of high-γ ejecta travelling along a <u>diverging</u>
bundle of trajectories. In the picture I want to consider (Fig. 2),
the observer's line of sight makes angle θ with the x-axis, and sources
of radiation move with speed $v \simeq c$ along curves that start off like
$(y^2 + z^2)^{1/2} = Ax^n$. (For a dipole field line, n = 1.5). As in the
dipole model, after a burst of high activity in the central engine,
the observer sees ejecta travelling tangentially to his line of sight
on different trajectories in quick succession. As in Scheuer and
Readhead, the core flux is limited by self-absorption, so the base
of the jet bundle is always visible at about the same place, but
maxima of energy output appear to propagate along the optically thin
jet at a speed which is straightforward to compute, and for which
we can find explicit formulae in particular cases, e.g.

$$\frac{v_{app}}{c} = \tan\theta \; \{\frac{2c}{v}(\sec^3\theta - 1)\cos\theta\,\mathrm{cosec}\,\theta - 3 - 2\tan^2\theta\}^{-1} \quad \text{for } n = {}^3/2 \qquad (2)$$

$$\frac{v_{app}}{c} = \tan\theta \; \{\frac{c}{v}[\sec^2\theta + \mathrm{cosec}\,\theta \; \ln(\tan\theta + \sec\theta)] - 1 - \sec^2\theta\}^{-1}$$

$$\text{for } n = 2 \qquad (3)$$

$$\frac{v_{app}}{c} = (1 - \cos\theta)(\frac{c}{v}\theta - \sin\theta)^{-1} \quad \text{for circular arcs} \qquad (4)$$

and, for $\gamma \gg 1$, $\theta \ll 1$

$$\frac{v_{app}}{c} \sim \{\frac{n}{(n-1)2\gamma^2\theta} + \frac{n-1}{2n-1}\theta\}^{-1}, \quad \text{with a maximum value}$$

$$\sim \gamma(1 - 1/2n)^{\frac{1}{2}}. \qquad (5)$$

In my picture (unlike the dipole model) the trajectories do not close
up; in fact, their precise form at large θ is unimportant, since
the expansion of the source material makes its luminosity unobservably
small. For this reason we do not expect to see the far side of the jet,
even if the jets are intrinsically two-sided, and we expect to see the
jets feebly if at all when θ is really large (where 'really large' might
mean 15° or 45°). Note that the apparent total luminosity of such a
diverging jet does NOT vary in proportion to $(1+(v/c)\cos\theta/1-(v/c)\cos\theta)^{2+\alpha}$,
like simple undirectional jets. It is expected to look much fainter
at large θ, but in a manner which depends on the decrease in power of
the source material as it spreads out at large θ, a manner which cannot
be computed in any straightforward way because it depends on additional
physical assumptions.

Figure 2. Thick lines represent the trajectories of emitting matter, which is observed when its velocity is directed within $1/\gamma$ of the line of sight.

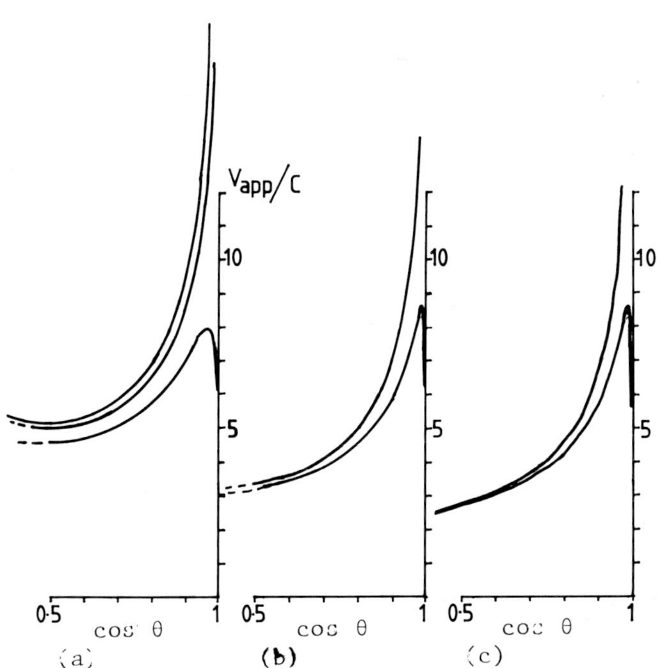

Figure 3. Apparent velocity versus cos θ, for models of the type illustrated in Figure 2, with trajectories of the forms: (a) $y = Ax^{3/2}$ (b) $y = Ax^2$ and (c) circular arcs. The top curve in each diagram represents emitting matter with v=c, the bottom curve represents emitting matter with $\gamma=10$, and the middle curve in Figure 3(a) represents $\gamma=20$.

The apparent velocities shown in Fig. 3 illustrate that the
results do not depend critically on the form of the trajectories.
They are plotted against cos θ, to illustrate the distribution of
velocities expected for a sample free from bias in orientation; a
sample selected for strong core fluxes would of course crowd up
towards small θ.

Such a model is quite flexible, literally as well as
metaphorically. The trajectories could be twisted, resulting in a
displaced and/or bent apparent jet, and the trajectories could
themselves be moved by, say, pressures from a surrounding fluid.
However, if the trajectories move too freely and too fast, there is
a risk of generating apparent motion towards the core.

References

Bahcall, J.N. & Milgrom, M., 1980. Astrophys. J., 235, 24.
Baldwin, J.A., Burke, W.L., Gaskell, C.M. & Wampler, E.J., 1978.
     Nature, 273, 431.
Blandford, R.D. & König1, A., 1979.   Astrophys. J., 232, 34.
Blandford, R.D., McKee & Rees, M.J., 1977.   Nature, 267, 211.
Condon, J.J., O'Dell, S.L., Puschell, J.J. & Stein, W.A., 1981.
     Astrophys. J., 246, 624.
Hine, G. & Longair, M.S., 1979.   Mon. Not. R. astr. Soc., 188, 111.
Kellermann, K.I., Downes, A.J.B., Pauliny-Toth, I.I.K., Preuss, E.,
     Shaffer, D.B. & Witzel, A., 1981. Astron. Astrophys., 97, L1.
Linfield, R., 1981.   Astrophys. J., 244, 436.
Orr, M.J.L. & Browne, I.W.A., 1982.   Mon. Not. R. astr. Soc., 200, 1067.
Osmer, P.S. & Smith, M.G., 1980.   Astrophys. J. Suppl. Ser., 42, 333.
Osterbrock, D., 1982. In "Extragalactic Radio Sources", IAU Symposium
     No. 97, ed. Heeschen & Wade, p. 369.
Rees, M.J., 1981.   In "Origin of Cosmic Rays", ed. Setti, Spada &
     Wolfendale (see p. 149).
Sanders, R.H., 1974.   Nature, 248, 390.
Scheuer, P.A.G. & Readhead, A.C.S., 1979.   Nature, 277, 182.
Strittmatter, P.A., Hill, P., Pauliny-Toth, I.I.K., Steppe, H. &
     Witzel, A., 1980.   Astron. Astrophys., 88, L12.
Wampler, E.J., Gaskell, C.M., Burke, W.L. & Baldwin, J.A., 1983.
     Astrophys. J., (in press).
Wills, B., 1982.   In "Extragalactic Radio Sources", IAU Symposium
     No. 97, ed. Heeschen & Wade, p. 373.

PHYSICS OF RELATIVISTIC JETS ON SUB-MILLIARCSECOND SCALES [+]

Martin J. Rees
Institute of Astronomy
Madingley Road, Cambridge CB3 OHA
England

1. INTRODUCTION

The observed superluminal components have (deprojected) lengths
of $\sim 10^{20}$ cm, and imply relativistic bulk motions on these scales.
There are, however, persuasive reasons for attributing the primary
energy production to scales $10^{14} - 10^{15}$ cm. Moreover, the initial
bifurcation and collimation must also be imposed on these small scales
if the long-term stability of the jet axis in extended sources is due
to the gyroscopic effect of a spinning black hole (Rees 1978). The
issues I shall address in this talk are: how the jet gets from $\sim 10^{15}$cm
to $\sim 10^{19}$ cm; and what VLBI data can tell us about the properties of
galactic nuclei on scales below $\sim 10^{19}$ cm — scales where optical and
X-ray studies provide some evidence, but where there is no short-term
hope of achieving spatial resolution.

2. INTERNAL PHYSICS OF JETS ON SCALES $10^{15} - 10^{19}$ cm

The stuff ejected from $\sim 10^{15}$ cm may be electron-positron plasma
or "ordinary" electron-ion plasma; reasons for seriously considering
the former option are discussed elsewhere (e.g. Rees 1981, Guilbert,
Fabian and Rees 1983). Energy can also be transported via Poynting
flux — either as a large-scale field carried out with the particles
(a directed MHD wind) or as low-frequency wave modes (cf. Rees 1971) —
and this can, in principle, swamp the flux associated with the kinetic
energy of the charged particles themselves. I shall show that there
are severe constraints on e^+-e^- jets, particularly in quasars.

Suppose that a relativistic jet carries an energy flux $10^{45}L_{b45}$
erg s^{-1}, and has a cone angle $\theta(r)$. The region $r \lesssim 10^{15}$ cm may
correspond to the "strong-field" domain around a black hole (the
Schwarzschild radius r_s being related to mass by $r_s = 3 \times 10^{13}(M/10^8 M_\odot)$cm),
where the jet is energised by relativistic electrodynamic processes,
and shaped by a disc or torus aligned with the hole's spin axis. I
shall not discuss this region, but shall consider $r = 10^{15}r_{15}$ cm $\gtrsim 1$,

R. Fanti et al. (eds.), VLBI and Compact Radio Sources, 207–214.
© 1984 by the IAU.

where the jet can be assumed to be "coasting" - carrying a fixed total
(internal plus bulk) energy - except insofar as internal radiative
losses occur, or it interacts with the local environment (§ 3).

Annihilation constraints

Suppose that a fraction F_+ of the jet's luminosity is carried by
e^+-e^- plasma, and that this (essentially charge-neutral) plasma moves
outward with a bulk Lorentz factor γ_b; suppose further that the mean
random motion has a Lorentz factor γ_r. The total particle flux along
the jet is then proportional to $F_+ L_{b45} \gamma_b^{-1} \gamma_r^{-1}$ and the particle density
in the comoving frame is proportional to $F_+ L_{b45} \gamma_b^{-2} \gamma_r^{-1} \theta^{-2}$. To check
whether the particles will interact with each other (and cool and/or
annihilate) we can calculate a "fiducial" cross-section σ_* such that a
typical particle would undergo any process with at least this cross-
section on the outflow timescale. We find

$$\sigma_* / \sigma_T = 0.02 \; F_+^{-1} \; L_{b45}^{-1} \; \theta^2 \; \gamma_b^3 \; \gamma_r \; r_{15} \tag{1}$$

The mean number of times N_{Scat} that a photon in the jet would be
scattered (with $\sigma \cong \sigma_T$) before escaping is

$$N_{Scat} = \sigma_T / \sigma_* \; \max \; [50 \; F_+ \; L_{b45} \; \gamma_b^{-1} \; \gamma_r^{-1} \; r_{15}^{-1} \; ; 1] \tag{2}$$

The two options in the square brackets correspond to the cases when
the photon can or cannot diffuse through an angle $\sim \theta$, and thereby
escape from the jet in the outflow timescale.

These numbers imply that, if $F_+ L_{b45} \cong 1$ (corresponding to a jet
powerful enough to energise a strong double source or a superluminal
component), then an e^+-e^- jet would be very optically thick at $r_{15} \cong 1$,
unless γ_b and γ_r (in some combination) are sufficiently large. However,
γ_r cannot itself be large, because the avoidance of runaway compton
losses implies

$$N_{Scat} \; \gamma_r^2 \lesssim 1 \tag{3}$$

This holds provided that there is even a trivial supply of soft photons
within the jet. If N_{Scat} is itself $\gtrsim 1$, this implies that the jet must
be "internally cold", in the sense that the particles are sub-
relativistic in the frame moving with Lorentz factor γ_b. But the
annihilation rate ($\sigma_{ann} V$) for sub-relativistic pairs is then $\sim \sigma_T c$.
This means that, on the outflow timescale, annihilation would reduce
$F_+ L_{b45}$ until σ_T / σ_* (equation (1)) were ~ 1.

This argument tells us that the energy cannot emerge as e^+-e^-
plasma unless the bulk Lorentz factor is high, the requirement being

$$\gamma_b \gtrsim 3.5 \, F_+^{1/3} \, L_{b45}^{1/3} \, \theta^{-2/3} \, r_{15}^{-1/3} \tag{4}$$

If the jet material carried a magnetic field whose strength, in the comoving frame, was at least equivalent to equipartition with the random kinetic energy, then a simple argument shows that for $L_{b45} \cong 1$ and $r_{15} \cong 1$, synchrotron cooling would be rapid enough to reduce γ_r to a value ~ 1 on the outflow timescale. However, synchrotron cooling would be inhibited by reabsorption, and this would permit electrons and positrons with $\gamma_r > 10$ to persist if there were no other losses. But condition (3) implies that Compton processes would in any case ensure that γ_r was reduced to ~ 1 if N_{Scat} were large. Synchrotron reabsorption therefore does not offer an "escape clause" to (4).

We can therefore conclude that the energy flowing along a jet at $r_{15} \cong 1$ must be in <u>ordered</u> motion or in Poynting flux: if the electrons were injected with high γ_r, (each therefore with total energy $\sim \gamma_r \gamma_b m_e c^2$) then the bulk of their energy would quickly be converted into a photon beam, with cone angle max $[\theta; \gamma_b^{-1}]$, only $\sim \gamma_b m_e c^2$ (i.e. $\sim \gamma_r^{-1}$ of the initial particle energy) surviving as directed kinetic energy.

Compton drag constraints

The same central engine which initiates the jets is also, presumably, the source of the primary continuum luminosity in the optical and X-ray bands; the particles in the jet will then exchange momentum with the photon flux via Compton scattering. The characteristic Compton timescale for a relativistic electron or positron in an isotropic photon field with energy density $L_{ph}/4\pi r^2 c$ can be expressed as

$$t_{comp}/t_{dyn} \cong 0.02 \, L_{ph45} \, r_{15} \, \gamma^{-1}, \tag{5}$$

where $t_{dyn} = r/c$ is the dynamical timescale for relativistic outflow. This simple result tells us that, if $L_{ph45} \gtrsim 1$ (as it is in quasars) the interaction with the ambient radiation is very important; however, the nature of this interaction depends on the radiation's angular distribution. Some radiation escapes directly from the central source (which may have $r_{15} < 1$); part, however is scattered by gas at larger radii, or absorbed, and re-emitted isotropically mainly as emission lines. If the radiation was all directed outward from a small central source, or were itself collimated, the jet would tend to be <u>accelerated</u>; on the other hand, if the radiation were more nearly isotropic it would slow down a relativistic outflow on a timescale $\lesssim \gamma_b^{-1} \, t_{comp}$, where t_{comp} is given by (5). (I have written an <u>inequality</u> here to allow for the dependence on γ_r discussed by O'Dell (1981).)

A test particle released from rest into a flux of radiation would initially be accelerated at a rate $\sim c \, t_{comp}^{-1}$. However, when the particles are in a relativistic beam, what is relevant is the angular distribution and intensity of the radiation in the comoving frame.

There are two effects:

(i) The intensity in the beam frame of radiation directed along the beam is down by γ_b^{-2}; this fact, together with the time dilation factor, means that the terminal γ_b achievable by the pressure of radiation whose intensity falls off as r^{-2} scales only as $L_{ph}^{1/3}$ (see, for instance, Phinney (1982)).

(ii) When the radiation comes from a range of angles, then aberration will shift the apparent directions of some photons into the forward hemisphere, if γ_b is sufficiently high. For an angular distribution $I(\theta)$, axisymmetric around the beam direction, there will be a critical value of γ for which

$$\int_{o}^{\pi} I'(\theta')\,\cos\theta'\,\sin 2\theta'\,d\theta' = 0, \tag{6}$$

where dashes denote quantities transformed into a frame with Lorentz factor γ.

In a realistic quasar environment, radiation drag sets a significant limit on γ_b, even if the initial acceleration owes nothing to radiation pressure. This is because, when $\gamma_b \gg 1$, the acceleration due to radiation from behind can be outweighed by the drag due to a much smaller isotropic component. Let us define $L(r)$ as the luminosity which, in effect, emerges isotropically from shells with radii in the range $r - 2r$. Models for quasar emission lines imply $L(10^{18}\ \mathrm{cm}) \gtrsim 0.1 L_{ph}$ (where L_{ph} here denotes the total photon luminosity); we have no direct handle on $L(r)$ at smaller radii, but models of the intercloud medium suggest that its Thomson depth could be $\gtrsim 0.1$ (cf. § 4 below). This radiation provides a drag which limits γ_b to

$$(\gamma_b(r) - 1) < 0.02\ F_+^{-1}\ L_{ph45}^{-1}\ (r_{15})\ r_{15}. \tag{7}$$

We cannot quantify this further until models of the galactic nucleus allow $L(r)$ to be estimated for the whole range 10^{15} cm $- 10^{19}$ cm. However, it is clear that (7) sets a significant upper limit on γ_b, especially for a pure e^+-e^- beam (for which $F_+ \cong 1$). This result, in conjunction with the earlier argument (relation (4)) that γ_b <u>must</u> be large if an e^+-e^- beam is to carry a high energy flux without annihilating, sets severe constraints on this type of beam. (The Compton drag on an electron-ion beam, given by (7) with $F_+ = 2(m_e/m_p) \cong 10^{-3}$, is much less severe.)

I will just mention here a different context in which radiation pressure could be important in driving jets: the "cauldron" model of Begelman and Rees (1983a,b). Here the jet contains ordinary thermal plasma and radiation; it (and its surroundings) are so optically thick that the radiation can be treated as a fluid. If the radiation pressure is sufficiently high — so that (p_{rad}/c^2) exceeds the rest mass density of the matter — then relativistic outflows can be achieved.

3. INTERACTIONS WITH CLOUDS AND THE INTERCLOUD MEDIUM

Evidence on the medium through which the jet propagates comes primarily from the broad optical emission lines. These imply clouds with $n_e \cong 10^{10}$ °K, $T \cong 10^4$ °K, and a covering factor $\gtrsim 0.1$ (though a very small volumetric filling factor). The clouds must be confined by a rarified intercloud medium: they could be in pressure balance with it, but are more probably moving supersonically through it. If the intercloud medium is at X-ray emitting temperatures, then for each r we know that

$$(n_e(r))^2 \; r^3 \leqslant 10^{69} \; cm^{-3} \tag{8}$$

(the volume depending on the X-ray luminosity or limit). Its temperature is likely to be determined by a balance between Compton cooling and heating (Krolik, McKee and Tarter 1981).

For the jet to be in pressure balance with the intercloud medium,

$$P_{ext} \cong 3 \times 10^4 \; \theta^{-2} \; r_{15}^{-2} \; L_{45} \mathcal{M}^{-2} \; dyne \; cm^{-2} \; , \tag{9}$$

$\mathcal{M}$ being the (relativistically generalised) Mach number. Confinement is required if $\theta < \mathcal{M}^{-1}$, and this can be supplied by external pressure only if the intercloud medium has

$$nT > 3 \times 10^{16} \; r_{15}^{-2} \; L_{45} \tag{10}$$

This is possible at $\sim 10^{18}$ cm, but is only marginally compatible with (8) at $\sim 10^{15}$ cm.

Another characteristic temperature of importance is the <u>virial temperature</u>. For a central object of 10^8 M_8 solar masses,

$$T_{virial} \cong 3 \times 10^{10} \; r_{15}^{-1} \; M_8 \; °K \; . \tag{11}$$

Relation (11) applies until the radius r gets so large that the dominant-mass encompassed within it is no longer just the central hole but the stellar content of the galactic core: T_{virial} then (for larger r) levels of at $10^6 - 10^7$ °K, depending on the stellar velocity dispersion. The intercloud medium cannot be in quasi-static equilibrium if its temperature is below T_{virial}. If Comptonisation prevents the temperature from rising above $\sim 10^9$ °K, the intercloud medium must be dynamic rather than quasi-static for $r_{15} < 30 \; M_8$. We do not know whether the intercloud medium is undergoing chaotic infall, or is an outflowing wind, but in neither case would the argument leading to (10) be applicable.

The problem of confinement — and the constraints discussed in § 2 — would all be evaded if the power were mainly in the form of Poynting flux (only a small fraction of L_b being contributed by particle kinetic energy). This could be low frequency waves, or an axisymmetric toroidal flux spun off a central compact object. This energy could be converted into relativistic particles at the location of the VLBI components.

In this connection, it is interesting that there is a characteristic scale in galatic nuclei of $\sim 10^{19}$ M_8 cm, this being the radius at which the central mass ceases to dominate the dynamics in accordance with equation (11). Outside this radius, quasi-static hot gas can exist; at smaller radii this is problematical. It is therefore a natural scale at which a free jet encounters quasi-static gas in the galactic potential, and may be relevant to the scale of the blobs revealed by VLBI data, and to the sharp bending of jets sometimes seen on this scale.

[It has been argued from Faraday rotation considerations that the radio-emitting relativistic particles in the VLBI components are themselves a mixture of electrons and positrons. If so, the positrons must presumably have been created not "in situ" but at smaller r. Enough positrons to produce the radio emission would have been transported out along the beam even if $F_+ \cong 0.01$, the Poynting flux being used "in situ" to accelerate them from $\gamma \cong 10$ to the values $\gamma \cong 1000$ required for synchrotron radiation in the GHz band.]

In any case, the jet's ram pressure would not be sufficient to influence the trajectories of dense fast-moving clouds, still less to destroy them. Thus, the covering factor due to such clouds (along the jet axis) must definitely be < 1; otherwise the jet momentum would be isotropised before getting out to the scale relevant to VLBI observations. It would seem inevitable that some of the jet's energy is tapped (and converted into relativistic particles) on passage through the line-emitting region. If optical observations could be made with micro-arc-second resolution, they would reveal not only a central optical con-tinuum, but also streaks of optical synchrotron radiation delineating the jet's path.

The structure of jets on scales $10^{15} - 10^{19}$ cm, if we could probe it in the same detail that the VLA provides for scales a million times larger, would no doubt prove just as complex — there would be entrain-ment of surrounding gas, bending by transverse pressure gradients, and shocks where the jet impinges on "broad line" clouds. We do not know how well collimated the jets are — there is really no evidence that θ is small on scales < 10^{19} cm — but one general statement can be made. The flow patterns would not simply be a scaled down version of those seen on larger scales, because one key number — the ratio of radiative cooling times ($\propto r^2$ for a simple diverging jet) to dynamical times ($\propto r$) — is proportional to r rather than being scale-dependent. Con-sequently, the flows on small scales would tend to be less elastic and more dissipative: they are less likely to maintain a high internal pressure, and would dissipate more energy if bent through large angles.

4. INDUCED SCATTERING

The VLBI maps reveal brightness temperatures up to $\gtrsim 10^{11}$ °K, corresponding to $\gtrsim 20\ m_e c^2/k$. For radiation of such high intensities, the ordinary Thomson scattering cross-section is enhanced by a factor $\sim kT/m_e c^2\ \phi^2$ owing to induced effects (ϕ being the scattering angle). It is unlikely that the gas surrounding the jet has a Thomson depth $\tau_T > 1$; on the other hand, a value in the range 0.01 - 0.1 is entirely compatible with two-phase models for the medium in the broad-line-emitting region (and the observed optical polarisation would require $\tau_T \cong 0.1$ if it were due to electron scattering). Thus, the effects of induced scattering are likely to be significant.

Wilson (1982) discussed induced scattering in detail; he even proposed a model whereby intensity peaks would exhibit apparent super-luminal motions along a jet-like feature in a source with a small central component varying in intensity, but with "no moving parts". This model is hard put to account for _all_ the superluminal data we now have, but the possibility of extra effects due to induced scattering should be borne in mind in interpreting radio maps. To distinguish reliably between genuine motions and features which are merely "echoes" due to induced scattering may have to await detailed maps with similar resolution at different frequencies. The main reason why induced effects cannot offer the entire explanation of superluminal motions is that they are strongly frequency dependent — if the only data we had was at 5 GHz they would offer a plausible escape clause to those unwilling to accept bulk flows at $> 0.98c$.

5. CONCLUSIONS

My specific conclusions are:

(i) Relativistic beams are probably initiated on scales $\lesssim 10^{15}$ cm; but there are significant constraints on the form in which the energy is transported out to the much larger ($> 10^{19}$ cm) scales where super-luminal effects are observed. If the energy were carried by electrons and positrons the beam would need to start off with a high γ_b — other-wise the required pair density is so high that most would annihilate.

(ii) The quasi-isotropic radiation field in quasars at $r \lesssim 10^{18}$ cm (due to the line-emitting clouds, etc.) exerts a Compton drag which precludes high γ_b for an e^+-e^- beam (though is less serious if there is more inertia per pair).

(iii) The external medium through which these small-scale jets propagate is likely to be inhomogeneous and in a chaotic rather than stationary state: it is therefore unlikely that the beams on these scales are confined by external gas pressure.

(iv) It is a plausible inference from (i) - (iii) above, and from general theoretical considerations, that the power emerges mainly as directed Poynting flux, rather than primarily as particle kinetic energy.

(v) The jets cannot avoid some interactions with the "broad-line" clouds — some of the observed non-thermal optical continuum may result from dissipation behind shock waves in this region, rather than from the ($\lesssim 10^{15}$ cm) nucleus itself.

(vi) The scale probed by VLBI, and where the jets are often observed to bend sharply, may be the domain where T_{virial} falls to $\sim 10^6 - 10^7$ $^\circ$K (eqn (11)), and the external medium is no longer dominated by the central massive object but by the ordinary potential well of the surrounding galaxy.

(vii) The intercloud medium may provide a sufficient Thomson optical depth for induced scattering to be important in modifying the appearance and variability of high-surface brightness radio components.

More generally, one must realise that the outflowing material and energy must undergo complex interactions with its surroundings before attaining the distance from the central power source at which VLBI techniques can probe it. Evidence from other wavebands, and improved theoretical understanding, are necessary before we can understand the range of scales $10^{15} - 10^{19}$ cm. Cynics may chide theorists for retreating to still smaller scales, now that the VLBI data have progressed beyond the stage of merely confirming superluminal motions, and are starting to reveal complex "weather" on milli-arc-second scales. But, on the contrary, it is surely gratifying that VLBI data can tell us about the primary energy source, and be related to observations in other wavebands.

Acknowledgements

I am grateful to Mitch Begelman, Roger Blandford, Colin Norman and Sterl Phinney for many discussions about jets.

REFERENCES

Begelman, M.C. and Rees, M.J. 1983a in "Astrophysical Jets" ed. A. Ferrari
    and A.G. Pacholczyk, p. 215 (Reidel, Dordrecht).
Begelman, M.C. and Rees, M.J. 1983b, MNRAS (in press)
Guilbert, P.W., Fabian, A.C. and Rees, M.J. 1983, MNRAS (in press).
Krolik, J.H., McKee, C.F. and Tarter, C.B. 1981, Astrophys.J. 249, p.422.
O'Dell, S.L. 1981, Astrophys.J.(Lett.) 243, L.147.
Phinney, E.S. 1982, MNRAS, 198, p. 1109.
Rees, M.J. 1971, Nature, 229, p. 312 (errata, p. 510).
Rees,, M.J. 1978, Nature 275, p. 516.
Rees, M.J. 1981 in "Origin of Cosmic Rays" ed. G. Setti et al. p. 139
    (Reidel, Dordrecht)
Wilson, D.B. 1982, MNRAS, 200, p. 881.

Quasars and Seyfert 1 galaxies appear to have UV excesses which have been modelled as a black body with temperature $\sim$ 25,000 K which Malkan (1983) has interpreted as radiation from the surface of a thin disk. This requires that the black holes be extremely large (M $\gtrsim 10^9 M_\odot$) and unfortunately in the case of the radio quasars predicts linear polarisation orthogonal to what is observed. A large fraction of the power liberated by the gas as it spirals inwards may be dissipated in a tenuous corona above the disk. This is of interest because Seyfert 1 galaxies and quasars seem to radiate from $\sim$ 0.1 to $\sim$ 0.5 of their bolometric power as hard ($\gtrsim$ 50 keV) X-rays or γ-rays. If the source spectrum, perhaps created non-thermally, extends beyond 0.5 MeV then electron-positron pairs can be created and these will annihilate so as to maintain a Thomson scattering optical depth of order unity. This is an alternative route to a Comptonised power law spectrum (Guilbert, Fabian and Rees 1983, preprint). Electron-positron pairs, unencumbered by proton inertia can be blown off by radiation pressure and may ultimately be collimated to form a jet. Jets may also be launched and collimated by magnetic torques acting upon the disk. If the poloidal field emerging from the surface of a Keplerian disk makes an angle of less than 60° with the radial direction then gas will be flung out along the field lines. As it moves radially outwards, its inertia will cause the field to become increasingly toroidal and thereby create a magnetic pinch (Blandford and Payne 1982).

When the accretion rate is increased ($\dot{M} \gtrsim 10\ \dot{M}_E$) radiation pressure can inflate the inner disk and produce a radiation and rotation supported torus (e.g. Jaroszyński, Abramowicz and Paczyński, 1980). Within this torus, the gas pressure is negligible and electron scattering opacity predominates. The diffusive heat flux must therefore be $-\underline{g}c/\kappa_T$ where $\underline{g}$ is the local effective gravity and κ_T the Thomson opacity. A radiation torus may be analogous to a giant early-type star, possessing a core around the pressure maximum where most of the binding energy resides, and an extensive envelope bounded by a photosphere. A radiation torus need not accrete gas steadily and if it is established by a very rapid episode of fuelling, it may settle down to a quasi-static configuration slowly deflating on a thermal timescale as photons diffuse outwards at roughly the Eddington limit. If the torus is large enough, the effective temperature can fall to a limiting value $\sim$ 25,000 K where Helium recombines and the opacity changes rapidly. It is tempting to associate this with the UV excess. Radiation tori are similar in many regards to the massive objects originally postulated by Hoyle and Fowler and may suffer the same fate - i.e. be shown to be dynamically unstable. There are possible axisymmetric local instabilities caused by unfavourable entropy and angular momentum gradients (e.g. Seguin 1975, Kandrup 1982). These presumably evolve to create marginally stable convection zones just as in a star. More threatening are non-axisymmetric instabilities. Papaloizou and Pringle (1983, preprint) have recently demonstrated that a toroidal configuration known to be marginally stable to axisymmetric disturbances possesses global, non-axisymmetric dynamical instabilities. It will apparently destroy itself in a few orbital periods unless non-linear terms can saturate the instability at a low amplitude. It is not

yet known if this is a general property of these tori. Furthermore, it
is not at all clear that tori can evolve towards stable or marginally
stable states even if they exist and that the rate of internal energy
generation through viscous dissipation can always be balanced by heat
transport.

The most relevant property of radiation tori to extragalactic radio
sources is that they possess funnels. It has been widely speculated
that this is the site of jet collimation. This idea seems difficult to
sustain because most of the observed radio jets are associated with
comparatively faint galactic nuclei. They are certainly not radiating
at anything like the Eddington limit of a $10^8 M_\odot$ black hole (10^{46} erg s^{-1}).
Furthermore it seems that radiation drag prevents the outflow from ever
being relativistic and well collimated (e.g. Sikora and Wilson 1981).
This does remain a possible mechanism for jet production in the case of
radio quasars that are not known to be superluminal (and also in SS 433)
but there seems to be no need to invoke a separate mechanism in these
cases. When the energy production becomes very large the radiation may
be trapped in an outflowing optically thick jet perhaps collimated through
a pair of nozzles (Begelman and Rees 1983).

When the accretion rate falls below the Eddington value then most
of the power released may be extracted electromagnetically from the
spinning black hole by an axisymmetric magnetic field held in place by a
disk or ion-supported torus (e.g. Thorne and Blandford, 1982). The
details of this mechanism have been investigated further by MacDonald
and Thorne (1982) and Phinney (1983). In particular, it has been shown
that if the magnetosphere contains plasma that is free to move along the
field, then inertial effects reduce the maximum extractable energy from
roughly fifteen per cent of the rest mass of the hole to roughly three
per cent. Nevertheless, this is still ample to supply the minimum
energy requirements of all known radio sources with black holes of
mass $M \lesssim 10^9 M_\odot$. One major problem with this mechanism is to understand
the interaction of the orbiting gas with the field that it encircles.
This is crucially different from the analogous problem with an accreting
magnetised neutron (e.g. Ghosh and Lamb 1978) because the magnetic
geometry will not cause accreting gas to flow to high latitude around
a black hole. Most of the magnetosphere should therefore be quite low
density which is a necessary condition for this mechanism to operate.

For further details on these and related models see Rees, Begelman
and Blandford 1981, Rees 1982, Blandford 1983.

3. BEAMING

As is well known, the discovery of superluminal expansion within
compact radio sources suggests that they involve relativistic outflow and
this in turn implies that the radio emission is highly anisotropic (e.g.
Scheur, these proceedings). Our classification of a particular radio
source is then strongly influenced by our orientation with respect to it.

Figure 1. Simple models of emission features of compact radio sources
a) Individual plasmoid moving with speed β along a direction making an
angle $\theta = \cos^{-1}$ to the line of sight. b) Several identical plasmoids
or equivalently an optically thin jet. c) Conical shock wave (with cone
angle 51°) moving with speed β 0.99 through a non-relativistic jet seen
at a given coordinate time in the jet frame. d) The same, but seen in
the frame in which the shock structure is at rest. Rays destined for
the observer are emitted along a direction making an angle $\theta' = \cos^{-1}\mu'$
to the axis. The observer sees the projection of the shape of the shock
structure on a plane normal to this direction. e) Numerical simulation
of an axisymmetric jet kindly supplied by Drs. Norman, Smarr and Winkler.
The Mach number is 3 and the jet density is one tenth that of the surroundings. The dark crosses are strong shock waves. f) Schematic representation of features likely to be present in a real jet.

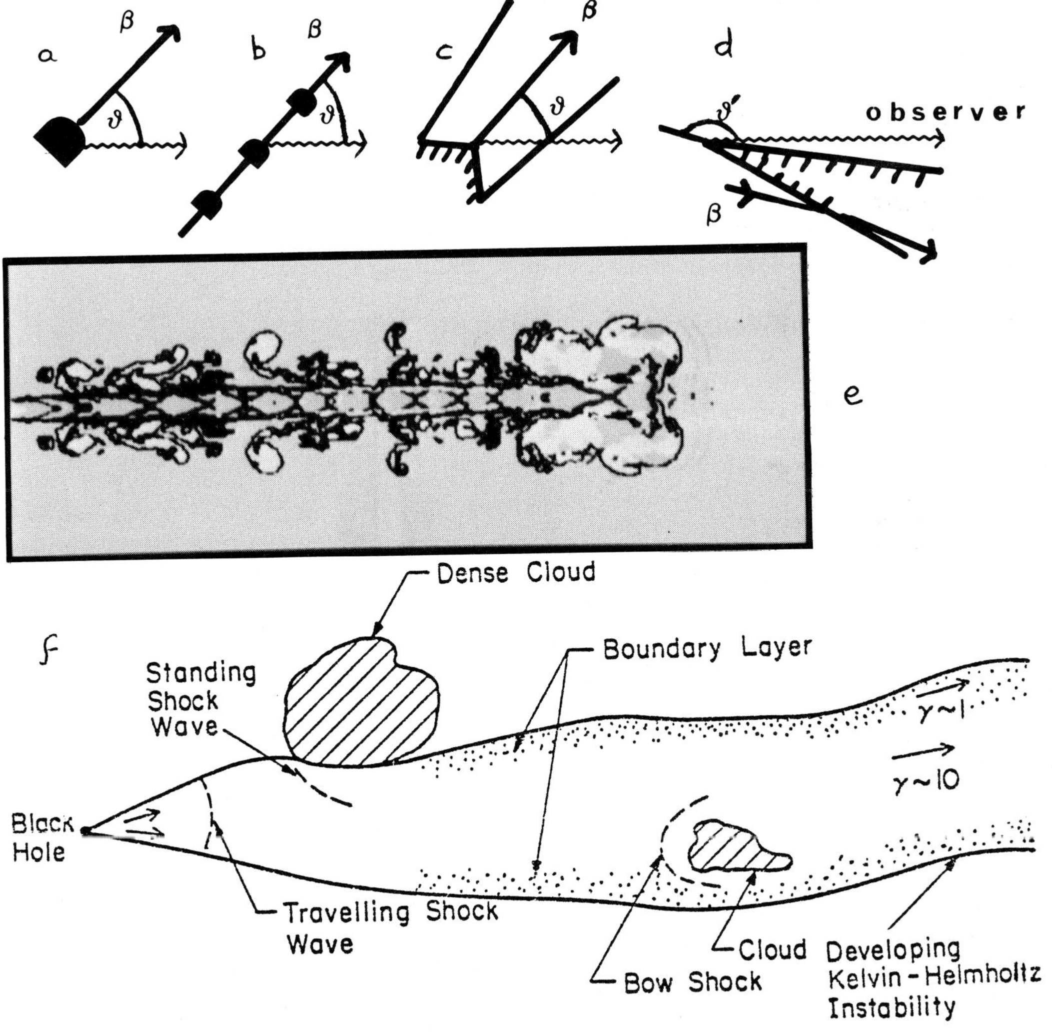

In particular, the powerful compact radio sources are postulated to be beamed in our direction. In order to test this idea quantitatively and to determine the nature of the unbeamed objects it is necessary to have a model of the emitting region. To date, most models have been far simpler kinematically than we can reasonably expect the sources to be (see Fig. 1). They can also be misleading.

The prototype is a plasmoid moving uniformly with speed $\beta c = (1 - \gamma^{-2})^{\frac{1}{2}}c$ at an angle $\theta = \cos^{-1}\mu$ to the observer direction. The observed velocity on the sky in units of c is $\beta_0 = \beta(1 - \mu^2)^{\frac{1}{2}}(1 - \beta\mu)^{-1}$ which can be as large as $\gamma\beta$. Now assume that the plasmoid is unchanged as it moves. Radiation destined for the observer is emitted in the plasmoid frame at an angle $\theta' = \cos^{-1}\mu'$ to the velocity where $\mu' = (\mu - \beta)(1 - \mu\beta)^{-1}$. The shape seen by the observer in his frame is simply the projection of the plasmoid seen from the angle θ' in its frame as may readily be verified by Lorentz transformation (e.g. Terrell 1966). So, in order to evaluate the observed flux and brightness, we must evaluate the intensity $I'_{\nu'}(\nu', \mu')$ in the source frame and Lorentz transform it into the observer frame using

$$I_\nu(\nu, \mu) = \mathcal{D}^3 \, I'_{\nu'}(\nu' = \nu/\mathcal{D}, \mu') = \mathcal{D}^{3+\alpha} \, I'_{\nu'}(\nu, \mu')$$

where $\mathcal{D} = \gamma^{-1}(1 - \beta\mu)^{-1}$ is the Doppler factor and α is the spectral index (Cosmological corrections can be inserted at this stage by replacing $\mathcal{D}$ with $\mathcal{D}(1 + z)^{-1}$ etc.). The observed flux density $S(\mu)$ is then obtained by integrating the observed intensity $I_\nu(\nu, \mu)$ over the solid angle subtended by the observed source.

The intensity in the source frame is computed by integrating the emissivity $j'_{\nu'}$ over the distance x' measured from the surface of the source.

$$I'_{\nu'}(\nu', \mu') = \int dx' \, j'_{\nu'}(\nu', \mu', x') e^{-\int^{x'} \kappa'(\nu', \mu', \xi')d\xi'}$$

where

$\kappa'(\nu', \mu', x')$ is the absorption coefficient.

For an optically thin source at distance D, $S = D^{-2} \mathcal{D}^{3+\alpha} \int j'_{\nu'}(\nu, \mu', x) dV' \propto (1 - \beta\mu)^{-(3+\alpha)}$ where V' is the proper volume of the source (This expression is incorrect if the source varies or there is absorption.)

A useful way to quantify the degree of beaming inherent in a particular model is to compute the probability P(S) that we are oriented so as to observe a flux density in excess of S. For individual plasmoids whose emissivity changes slowly on their light crossing times, it takes an interval $dt_0 = \mathcal{D}^{-1}dt'$ of observer time to receive radiation emitted in an interval dt' of source proper time. As S increases monotonically with μ,

Figure 2. Probability that the flux density from a given source will exceed a value S for five different models considered in the text. In all five models the emitting features move with space velocity $\beta = 0.99$ and $\alpha = 0$. The flux is normalised to the value S that would be observed from an angle $\cos^{-1}\beta$ where the features could be observed to move with their maximal superluminal velocity $\beta_0 = 7.0$ in all except case (b) where the normalisation is arbitary.
a) Individual plasmoids of constant power. b) Individual plasmoids in which the spectral power varies as the inverse cube of radius. c) Superposition of many plasmoids (or steady jet). d) Superposition of plane shocks moving along a non-relativistic jet. e) Superposition of conical shocks moving along a non-relativistic jet. The cone angle in the frame of the jet is 51°.

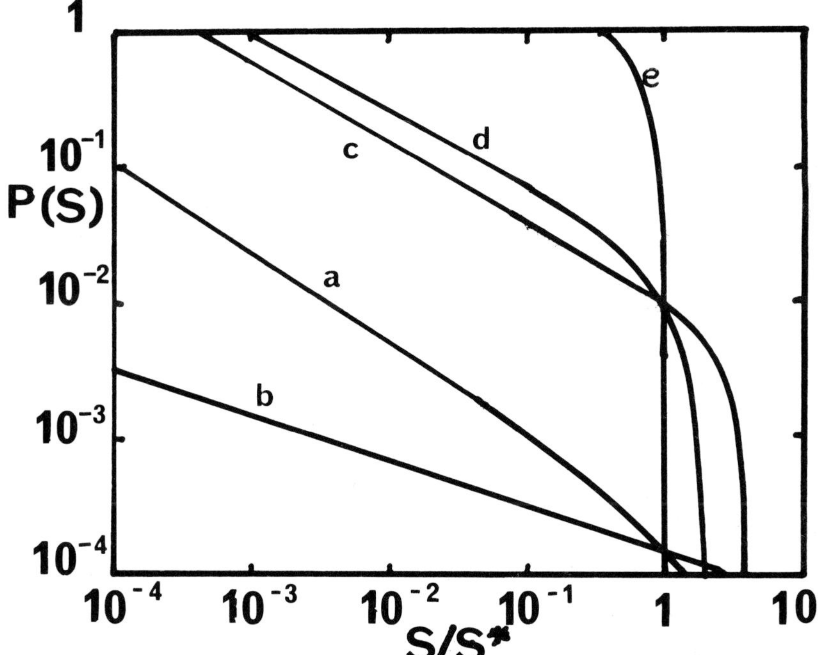

$$dP(S) \propto (1 - \beta\mu)d(1 - \mu), \quad 0 \lesssim \mu \lesssim 1$$

$$P(S) = \frac{(1 - \beta)^2}{2\beta} \left[\frac{S_{max}}{S}\right]^{\frac{2}{3+\alpha} - 1} ; \quad (1 - \beta)^{3+} S_{max} < S < S_{max}$$

See Figure 2. (We ignore plasmoids moving away from us).

In computing this distribution, we have assumed that each plasmoid has constant emissivity in its frame for a fixed proper time i.e. in moving over a finite range of radius. However, suppose for example, that the source weakens as a power of radius.

i.e. $S \propto r^{-n} (1 - \beta\mu)^{-(3+\alpha)}$ as the radius increases by a factor $\gg \gamma^{\frac{2(3+\alpha)}{n}}$.

The source can then be seen out to a radius $r \propto S^{1/n}(1 - \beta\mu)^{\frac{-3+\alpha}{n}}$. The probability of observing a source with flux greater than S satisfies

$$P(S) \propto \int r(1 - \beta\mu)d(1 - \mu) \, S^{-1/n} \propto \int (1 - \beta\mu)^{1 - \frac{3+\alpha}{n}} d\mu$$

In this case, the slope of the $P(S)$ variation is dictated by the radius exponent n. Furthermore, if $n > 3+\alpha$, then a flux-limited sample will be dominated by jets moving away from us!

In these two examples, we have only included those sources that we can see and we preferentially omit those beamed towards us because we observe them for a relatively shorter time than those moving with larger angle θ. Now, the cores of core-jet sources appear to be permanently bright and might therefore comprise a sum of several plasmoids which can merge together to form a continuous jet. In this case we have a stationary pattern through which the synchrotron-emitting plasma is flowing. If we do not observe individual features then there will be no superluminal expansion. The emissivity and absorption coefficient must be transformed from their values in the frame co-moving with the plasma and designated with a bar

$$j_\nu(\nu, \mu) = \bar{j}_{\bar{\nu}}(\bar{\nu} = \nu/\mathcal{D}) \mathcal{D}^2$$

$$\kappa(\nu, \mu) = \bar{\kappa}(\bar{\nu} = \nu/\mathcal{D}) \mathcal{D}^{-1}$$

The emergent intensity can then be computer from the equation of transfer. For an optically thin source, the flux is $S = \mathcal{D}^{-2} \int dV \, \dot{j}_{\bar{\nu}}(\nu) \propto (1 - \beta\mu)^{-(2+\alpha)}$ and

$$P(S) = (1 - \mu) = \left(\frac{1 - \beta}{\beta}\right) \left| \left(\frac{S_{max}}{S}\right)^{\frac{1}{2+\alpha}} - 1 \right| \, ; \, (1 - \beta)^{2+\alpha} S_{max} \lesssim S \lesssim S_{max}$$

The flat spectra of compact sources are probably caused by the superposition of self-absorbed synchrotron spectra. Simple jet models (e.g. Blandford and Königl 1979, Reynolds 1983) are very similar to optically thin sources with $\alpha = 0$.

Already, these three models make quite different predictions about the nature and number of the unbeamed sources. A somewhat more realistic model introduces further differences. The radio power that we observe from jets is probably derived from the kinetic energy of the flow and the most natural way to do this is with a shock wave. A shock wave necessarily introduces a difference between the velocity β of the stationary pattern which dictates the speed of observed superluminal motion and the velocity of the emitting plasma which is responsible for the Doppler boosting.

For simplicity, set $\alpha = 0$ and assume an ultrarelativistic equation of state ($p = 1/3\rho$). If the front is perpendicular and moves with speed β in excess of the sound speed $3^{-\frac{1}{2}}$ through a slowly moving jet, the post-shock velocity (in the shock frame) is $1/(3\beta)$ (e.g. Landau and Lifshitz 1969). The emissivity in the pattern frame satisfies $j'_{\nu'} \propto (1 + \mu'/3\beta)^{-2}$. As before, if there are several identical shocks, $S \propto (1 + \mu'/3\beta)^{-2} (1 - \beta\mu)^{-2}$. The emission has an angular distribution similar to that from plasmoids moving with a speed $\beta - (1 - \beta^2)/2\beta$. However, real jets are more likely to produce oblique shock waves which deflect the flow without changing its speed so much. Oblique shocks are naturally in numerical simulations of non-relativistic jets (Norman, Smarr and Winkler 1983). In the frame of the shock, the emission is beamed along the backward direction thereby making the observed emission far more isotropic. This effect can be seen quite clearly in Fig. 2 where we plot P(S) for a conical shock with cone angle 51° in the jet frame (but only 10° in the frame of the shock). Although there are $\sim 2\gamma^2 \sim 100$ sources for every one observed to have an observed speed $\beta_0 \sim 7$, these sources are no more than four times fainter instead of being up to 1000 times fainter in the case of a plane shock (Blandford and Lind, 1983, in preparation).

Realistic source models should contain features moving with a range of speeds, directions and intrinsic powers, and what we observe depends upon an integration over these quantities. It is clear from the above that the observed power from an individual source can just as easily be dominated by features moving at angle $\sim 60°$ to the line of sight as by those moving at 5° and so the apparent failure to find further examples of superluminal expansion reported by Readhead in these proceedings need not be disturbing. In addition one should be cautious about making purely statistical arguments about the necessity or irrelevance of beaming in compact radio sources until we understand the source structure better (Scheuer and Readhead 1979, Browne, these proceedings).

4. GRAND UNIFIED THEORY OF ACTIVE GALACTIC NUCLEI

Unified theory interprets compact radio sources as being a small fraction of active galactic nuclei with jets pointed towards us. The association of unbeamed BL Lac objects with intermediate power radio galaxies seems increasingly attractive (Browne, 1983, Moore, Angel and Wardle 1983 preprint). However, the identification of unbeamed compact quasars with either optical or extended radio quasars is proving harder to sustain. A resolution of these difficulties is possible if we regard superluminal expansion more as a symptom of relativistic outflow and less as a measurement of the angle between the jet and the line of sight. The very attractive beaming hypothesis can then shelter behind a barricade of increasingly sophisticated source models, that are in any case necessary to interpret the maps presented here by Pearson and others (cf. § 3.)

When we try to go further and unify the powerful radio sources with
the totality of active galactic nuclei, then the important unifying idea
is a massive black hole. Now the light travel time across a $\sim 10^8 M_\odot$
black hole is ~ 1 hour and it is perhaps surprising that most active
nuclei are steady on timescales far longer than this and we might
conclude that the black holes themselves need some secluding. It is
then fortunate that the three modes of accretion distinguished in § 2
probably do not allow us to view the black hole directly, either because
the power is released as an essentially invisible electromagnetic
energy flux or because there is an extensive electron scattering
atmosphere.

The two most important parameters controlling accretion onto a
massive black hole are its accretion rate in units of the Eddington
value $\dot{m}$ and its mass M. Radiation tori with $\dot{m} \gtrsim 10$ and large photospheres
can be associated with the optical quasars and in the case of smaller
holes (M $\lesssim 10^7 M_\odot$), with Seyfert I galaxies. Their power law continua
can be produced by the Comptonisation of infrared synchrotron and thermal
photons by mildly relativistic electrons in a hot corona perhaps heated
by radiation dominated gas flowing outwards along the funnels. Radiation tori probably accelerate a poorly collimated wind. The comparatively weak radio sources may be produced when this is decelerated
within the galactic nucleus.

As the accretion rate is lowered, the disk will thin and the
relative importance of non-thermal process will increase. These are the
radio quasars and again for lower hole masses, the Seyfert II galaxies.
At quite small values of $\dot{m}$, non-thermal mechanisms dominate and indeed
most of the power may be derived directly from the hole. These are the
radio galaxies. All of this can be set in an evolutionary context by
noting that high accretion rates were far more common in the past.

Seyfert galaxies are associated with spirals and radio galaxies
with ellipticals. A corrollary of the above identification is that
spirals should generally have lower hole masses than ellipticals. In
particular the brightest radio quasars are most likely to be associated
with ellipticals although less powerful optical quasars may be predominantly spiral. Fur further references see Blandford (1983).

ACKNOWLEDGEMENTS

I thank the Director of the Institute of Astronomy, Cambridge for
hospitality during the writing of this paper. I am indebted to
Tony Readhead, Martin Rees, Peter Scheuer and especially Sterl Phinney
for invaluable discussion. Financial support by the National Science
Foundation under grant AST 82, 13001 and the Alfred P. Sloan Foundation is
gratefully acknowledged.

REFERENCES

Begelman, M.C.,and Rees, M.J.: 1983, Mon.Not.R.astr.Soc (in press).
Blandford, R.D.: 1983, Proceedings of 9th Texas Symposium 1983
    (in press).
Blandford, R.D., and Königl, A: 1979, Astrophys.J., 232, p. 34.
Blandford, R.D., and Payne, D.G., 1982, Mon.Not.R.astr.Soc.,
    199, p. 883.
Browne, I.W.A.: 1983, Mon.Not.R.astr.Soc. (in press).
Ghosh, P., and Lamb, F.K.: 1978, Astrophys.J.Lett., 223, L83.
Ipser, J., and Price, R.M.: 1982, Astrophys.J., 255, p. 654.
Jaroszynski, M., Abramowicz, M., and Paczyński, B.: 1980, Acta
    Astronom., 30, p. 1.
Kandrup, H.: 1982, Astrophys.J., 255, p. 691.
Landau, L.D., and Lifshitz, E.M.: 1969, Fluid Mechanics, Pergamon
    Press.
MacDonald, D., and Thorne, K.S.: 1982, Mon.Not.R.astr.Soc., 198,
    p. 345.
Malkan, M.: 1983: Astrophys.J. (in press).
Maraschi, L., and Treves, A.W.: 1977, Astrophys.J.Lett., 21B,
    p. 413.
Norman, M.L., Smarr, L.L., and Winkler, K.M.: 1983, Proceedings
    of Turin Conference on Astrophysical Jets, ed. Ferrari and
    Pacholczyk.
Pacholzyk, A.G., and Stoeger, W.: 1983, Preprint Steward
    Observatory.
Phinney, E.S.: 1983, Proceedings of Turin Conference on Astro-
    physical Jets, ed. Ferrari and Pacholczyk.
Pringle, J.E.: 1981, Ann.Rev.Astron.Astrophys., 19, p. 137.
Rees, M.J.: 1966, Nature, 211, p. 468.
Rees, M.J.: 1982, "Extragalactic Radio Sources", ed. Heeschen and
    Wade (Reidel, Dordrecht, Holland).
Rees, M.J., Begelman, M.C., and Blandford, R.D.: 1981, Ann.N.Y.
    Acad.Sci., 357, p. 254.
Reynolds, S.P.: 1982, Astrophys.J., 256, p. 13.
Schmid-Burgk, J.: 1978, Astrophys.Sp.Science, 56, p. 191.
Sequin, F.: 1975, Astrophys.J., 197, p. 745.
Shklovsky, I.S.: 1968, Astr.Zh.,45, P. 919.
Sikora, M., and Wilson, D.B.: 1981, Mon.Not.R.astr.Soc., 196,
    p. 257.
Stockman, H.S., Angel, J.R.P., and Miley, G.: 1979, Astrophys.J.
    Lett., 277, L55.
Takahara, F., Tsuruta, S., and Ichimaru, S.: 1981, Astrophys.J.
    251, p. 26.
Terrell, J.: 1966, Science, 154, p. 1281.
Thorne, K.S., and Blandford, R.D.: 1982, in "Extragalactic Radio
    Sources", ed. Heeschen and Wade, (Reidel Dordrecht,Holland).

A NUCLEAR REGULA-TORI MODEL FOR QUASARS AND SEYFERTS[+]

Mitchell C. Begelman
Joint Institute for Laboratory Astrophysics
University of Colorado and National Bureau of Standards
Boulder, Colorado 80309 USA

Although VLBI reveals structure in active galaxies on scales as small as a few parsecs, this is still likely to be several orders of magnitude larger than the scales on which jets are accelerated. We must therefore rely on theory, bolstered by indirect evidence, to deduce the behavior of plasma immediately surrounding the "central engine". One hypothesis which has received much attention recently postulates that jets emerge from thick tori of gas which surround massive black holes. The low optical and X-ray luminosities of radio galaxies imply that tori in these objects would have to be supported by ion pressure[1], while tori in quasars and Seyferts may be supported by radiation pressure. Rees et al[2] have proposed that two parameters, α(the ratio of viscous stress to pressure) and $\dot{M}/\dot{M}_E$ (the ratio of the accretion rate to the value associated with the Eddington limit), are sufficient to determine the general characteristics of a torus, independent of the mass of the black hole. While existing observations of radio jets do not contradict the existence of such a homology, systematics involving the broad emission lines in quasars and Seyferts suggest that the homology may fail in these objects, at least insofar as conditions at $\lesssim 1$ parsec reflect conditions in the torus. I report here on how the onset of nuclear reactions in a radiation-supported torus may break the homology, and regulate the properties of the broad line regions in quasars and Seyferts.

The maximum temperature, pressure and density in tori around black holes less massive than a few times $10^8 M_\odot$ is determined by the onset of nuclear burning; around more massive holes, self-gravitation is the limiting factor[3]. The limiting density $\rho_{nuc} \sim 3\times10^{-5}$ gm cm^{-3} implies a maximum mass which can settle into the torus at any one time. If matter is fed into the torus at too high a rate, then nuclear reactions will drive strong convection, which will drain away the excess mass by enhancing the turbulent viscosity and perhaps by driving a wind as well. These "nuclear regula-tori" will resemble the supercritical accretion tori discussed by Paczynski and Wiita[4] and others. For an accretion rate much greater than the Eddington value, the torus will need to have a low efficiency[5], which implies that it inflates to a radius exceeding $\dot{M}/\dot{M}_E$ times the Schwarzschild radius. However, for $\dot{M}/\dot{M}_E \gtrsim 5000 \ (M_{hole}/10^8 M_\odot)^{-\frac{1}{2}}$

+ Discussion on page 447

R. Fanti et al. (eds.), VLBI and Compact Radio Sources, 227–228.
© *1984 by the IAU.*

the photospheric temperature drops below the threshold for helium
recombination, and the required degree of inflation cannot be maintained[6].
When this happens, the torus produces more energy than it can radiate
away, and a supercritical wind must develop.

The velocity of this wind, and the density and radius at which it
becomes optically thin, are reminiscent of the parameters deduced
observationally for the broad line regions of quasars and Seyferts.
Filaments of line-emitting gas may condense out of the general flow
through thermal or radiation driven instabilities, as has been discussed
by other authors. Because the crossing time for material in the broad
line region is $\sim 10^2$ years, and the efficiency of accretion during the
wind phase is $\lesssim 10^{-3}$, the wind need not operate for more than one per cent
of the time. Indeed, if the corresponding accretion occurred continuously
the black hole would double its mass in less than 10^5 years. More
plausibly, the eruption of the wind shuts off the supply of mass to the
torus, which slowly drains until another burst of rapid accretion and
wind generation occurs. A quasar undergoing a windy episode may be
marked by an anomalously soft spectrum, though not by a significantly
different luminosity. Statistically, one might expect several of the
known quasars to be in a windy phase, although this would not be the
case if the broad emission lines tended to be weak during these periods.
In between windy episodes, the spectrum of the torus may be considerably
harder, and could be partially responsible for exciting the lines. However, it is likely that much of the photoionizing radiation is supplied
by the jets, which would persist during the windy phases.

In summary, the properties of the broad line regions in quasars and
Seyferts may provide a sensitive probe of gasdynamical conditions on
much smaller scales, where tori regulated by nuclear reactions may be
responsible for producing the observed VLBI jets. A more detailed report
on this research will appear elsewhere.

REFERENCES

1. Rees, M.J., Begelman, M.C., Blandford, R.D. and Phinney, E.S., 1982,
     Nature 295, pp. 17-21.
2. Rees, M.J., Begelman, M.C. and Blandford, R.D., 1981, *Proceedings of
     the Tenth Texas Symposium on Relativistic Astrophysics* (NY Acad.
     Sci.: New York), pp. 254-286.
3. Wiita, P.J., 1982, Astrophys. J. 256, pp. 666-680.
4. Paczynski, B. and Wiita, P.J., 1980, Astron. Astrophys. 88, pp. 23-31.
5. Abramowicz, M.A., Calvani, M. and Nobili, L., 1980, Astrophys. J.
     242, pp. 772-788.
6. Blandford, R.D., 1983, *Proceedings of the J. Wilson Festschrift*,
     Univ. of Illinois, in press.

STABILITY PROPERTIES OF VLBI JETS[+]

M. Birkinshaw
Mullard Radio Astronomy Observatory, Cavendish Laboratory,
Madingley Road, Cambridge CB3 OHE, U.K.

ABSTRACT

The growth lengths of Kelvin-Helmholtz instabilities are
calculated for three vortex-sheet bounded beams with v = 0.9c. These
beams survive the action of the instability for distances of several
hundred beam radii, but no prediction of a single dominant mode of
breakup can be made.

MODEL AND METHOD

It is of interest to estimate the Kelvin-Helmholtz instability
growth lengths for fluid-dynamical beams of the type that may be
responsible for observed VLBI structures. Some investigation of this
problem has already appeared (Ferrari, Trussoni & Zaninetti 1981), but
this work was done in the temporal domain, so that some assumption about
the convection velocities of the modes of instability is required in
order to estimate their growth lengths. In highly unstable situations,
such as jets, it is better to work in the spatial domain, when estimates
of the survival length for a beam appear directly, and this method is
used here.

The calculation described here considered the first-order stability
of a vortex-sheet bounded cylindrical beam with bulk velocity v = 0.9c
(γ = 2.3). An adiabatic equation of state was assumed for the beam and
the ambient medium, and the effects of gravity and magnetic fields were
ignored. The speed of sound in the ambient medium was assumed to be
750 km/s, and stability was considered for beams with densities 0.1,
1.0 and 10 times that of the external medium (cases L, E and H
respectively). The dispersion relation obtained in the linear stability
analysis was solved for various perturbation frequencies, and the
results were interpreted as curves of the rest-frame e-folding
instability length, λ_e, as a function of the wavelength, λ_o, for each
assumed azimuthal mode number, n. n describes the variation of

+ Discussion on page 448

229

R. Fanti et al. (eds.), VLBI and Compact Radio Sources, 229–230.
© *1984 by the IAU.*

solutions around the cylindrical beam: $n = 0$ are pinching modes, $n = 1$ are helical modes, and $n > 1$ are fluting modes (Hardee 1979). For any n, a wide range of modes are obtained; these may be labelled by the radial mode number, N, which describes the radial variation of solutions ($N = 0$ have least variation and are the ordinary modes of Ferrari et al. 1981; $N > 0$ are their reflection modes). In general the $N = 0$ mode is the least unstable and appears at the greatest λ_o; higher N modes are more unstable and appear at smaller λ_o.

RESULTS

The results of this analysis may be summarized as follows. All the (n,N) modes investigated in all three beams are unstable. The minimum growth lengths for any observed wavelength are given by the lower envelope of the stability curves as

$$\lambda_e^{min}/R \sim \begin{Bmatrix} 550 \\ 190 \\ 70 \end{Bmatrix} (\lambda_o/R)^{0.2} \qquad \begin{Bmatrix} H \\ E \\ L \end{Bmatrix}$$

where R is the beam-radius. If the periods of variations of the fluid parameters imposed by the (growing) perturbations are assumed to be of the order of the sound crossing time for the beam, then the low-order (small n,N) modes occur with characteristic wavelengths and e-folding lengths

$$\begin{aligned}
\lambda_o/R &\sim 1100 & \lambda_e/R &\sim 2000 & &H \\
\lambda_o/R &\sim 350 & \lambda_e/R &\sim 600 & &E \\
\lambda_o/R &\sim 100 & \lambda_e/R &\sim 200 & &L
\end{aligned}$$

Thus if low-order modes are excited, the beams will only be affected by Kelvin-Helmholtz instabilities on scales greater than a few hundred beam radii, but the beam stability is always dominated by the highest-frequency modes that are assumed to be excited. Since the variation of λ_e^{min} with λ_o is fairly slow, however, the survival distances are not strongly dependent on the maximum input frequency. It is clear that higher-density (higher inertia) beams are more stable to Kelvin-Helmholtz instabilities than lower-density beams, and hence are to be preferred in explaining the survival of VLBI beams to kpc scales; for any one (n,N) mode the minimum e-folding distance increases as the square root of the beam overdensity.

REFERENCES

Ferrari, A., Trussoni, E. & Zaninetti, L., 1981.   Mon. Not. R. astr. Soc., 196, 1051.
Hardee, P.E., 1979.   Astrophys. J., 234, 47.

TURBULENT CASCADE IN VLBI JETS

L. Zaninetti (1), G. Pelletier (2)

(1) Istituto di Fisica Generale, Torino, Italy
(2) Groupe d'Astrophysique, Grenoble, France

ABSTRACT

In the framework of continuous beam from small (VLBI region) to large scales (VLA region), we deduce some typical quantities of VLBI-jets like maximum luminosity and time-scales of formation of turbulence in the hypothesis of turbulent cascade triggered by relativistic Kelvin-Helmholtz instabilities.

I. Relativistic Kelvin-Helmholtz instabilities

The MHD instabilities of a cylindrical jet have been studied in detail, see Ferrari et al. (1981); the envelope of reflecting modes can drive a strong hydrodynamic turbulence on wavelengths smaller than the radius, R_J, of the jet. Thus we will adopt a relativistic growth rate γ_p, and sound velocity C_s, $= C / \sqrt{3}$ (C = light velocity):

$$1) \quad \gamma_p \simeq \frac{C_s}{R_J \gamma_L} \qquad \text{with scale of production } \lambda_p \simeq 0.5 \, R_J$$

where γ_L is the Lorentz factor of the relativistic bulk motion (this approximation is good only for $\gamma_L < 10$).

II. Determination of luminosity

We assume that a fraction close to unity of the sonic perturbations, produced on large scales by Kelvin-Helmholtz instabilities supplies smaller scale turbulence. We model the transfer of turbulent energy spectrum in terms of the wavenumber k (Pelletier and Zaninetti (1983)) with the hypothesis of Kraichnan cascade in which the turbulent energy spectrum $W(k) \propto k^{-3/2}$; this happens if $C_s < 5 \, V_A$ (V_A = Alfven velocity).

R. Fanti et al. (eds.), VLBI and Compact Radio Sources, 231–232.
© *1984 by the IAU.*

The power density supplied to the cascade is:

$$2)\quad Q \simeq \frac{C_s^2}{V_A R_J \gamma_L}\; W_M$$

where W_M is the density of magnetic energy. We thus obtain the maximum contribution to the total luminosity, in terms of the main parameters of the VLBI jet: D_J = length (D_1 = 1pc units), B = magnetic field of equipartition (B_{-2} = 10^{-2} gauss units) and n = density (n_{+2} = 10^{+2} part/cm^3):

$$3)\quad L \simeq 5.6\;10^{39}\;\left(\frac{R_i}{D_J}\right)\;(B_{-2})^3\;(n_{+2})^{-\frac{1}{2}}\;(D_{+1})^2\;\left(\frac{\beta}{\gamma_L}\right)\;\text{ergs/sec}$$

where β is the ratio between total kinetic and magnetic pressure. This maximum luminosity is in agreement with the total luminosities of VLBI jets as given by Preuss (1983).

III. Time scale of formation of the turbulence

The time, τ , required for a perturbation of size $< R_J$ to reach the dissipation range is of the order of $\approx \gamma_L R_J/C_s$. This time could be related to the time scale of variability of VLBI jets that are greater than a month. Setting $C_s = C/\sqrt{3}$ we obtain:

$$4)\quad \tau \simeq 70\,\gamma_L\left(\frac{R_i}{D_J}\right)\;(D_{+1})\;\text{months}$$

References

Ferrari, A., Trussoni, E., Zaninetti, L.: 1981, M.N.R.A.S. 196, 1051
Pelletier, G., Zaninetti, L.: 1983, submitted to Astron. Astrophys.
Preuss, E.: 1983, Workshop on Astrophysical Jets, Reidel, Dordrecht

A RELATIVISTIC WIND-TYPE MODEL FOR THE GENERATION OF VLBI JETS [+]

A. Ferrari [1], R. Rosner [2], E. Trussoni [3], and K. Tsinganos [2]

[1] Istituto di Fisica Generale, Universita' di Torino, Italy
[2] Harvard-Smithsonian Center for Astrophysics, Cambridge, USA
[3] Istituto di Cosmo-geofisica del C.N.R., Torino, Italy

ABSTRACT

We discuss wind-type solutions for flows from accretion funnels, and show under what physical conditions such flows can become supersonic and relativistic already very close to the stagnation point within the funnel. The acceleration is due to radiation emitted by the funnel walls, while the location of the transonic points is also affected by the geometrical shape of the funnel's cross-section.

I. INTRODUCTION

VLBI observations and variability time scales suggest that the central engine powering active galactic nuclei and the associated radio sources has dimensions of the order of our solar system ($\lesssim 10^{14}$ cm). In addition, supersonic jets ejected from active nuclei appear to be accelerated and collimated inside the deep cores: in some cases, the process can be extremely efficient, as suggested by superluminal expansion. These facts suggest the existence of accretion disks orbiting around massive black holes (Rees 1982). Several models have been proposed along these lines, wherein acceleration is produced either by radiation pressure forces or by electromagnetic processes in the vicinity of such disks.

Here we discuss some preliminary results of a hydrodynamical study of steady flows emanating from accretion funnels. We demonstrate the importance of the geometrical shape of the funnel in determining the final flow pattern, and discuss how, for identical boundary conditions, more than one physical solution may be allowed, some of which involve shocks. The general framework of this model has been presented elsewhere (Ferrari et al., 1983). Here we address the acceleration of an optically-thin, isothermal wind in the relativistic regime, assuming a simplified disk structure and radiation field within the funnel, as appropriate to VLBI jets and, perhaps, superluminal sources.

II. PHYSICAL PARAMETERS AND FLUID EQUATIONS

The conservation equations for particle number and energy-momentum give the following single equation for the dimensionles flow speed $\beta=v/c$ along the axis z of the jet:

$$\frac{d\beta^2}{d\xi} = \frac{1}{\gamma^4\left(1 - \frac{\beta_{so}^2}{\beta^2}\right)}\left[2\beta_{so}^2\gamma^2\frac{1}{A(\xi)}\frac{dA(\xi)}{d\xi} - \frac{b}{\xi^2} + \frac{D}{\rho c^2}\right] , \tag{1}$$

+ Discussion on page 449

R. Fanti et al. (eds.), VLBI and Compact Radio Sources, 233–236.
© 1984 by the IAU.

In this notation, $\xi = z/z_0$ is the dimensionless coordinate along the stream-lines, $A(\xi)$ is the cross-sectional area of the funnel, β_s is the dimensionless sound speed v_s/c, $b = GM/2z_0^2c^2$, and the subscript "o" indicates quantities calculated at the base of the funnel z_0, which we fix as the flow's stagnation point (for example, $z_0 = 10^5$). Note that an isothermal equation of state has been assumed.

In Eq. (1) the first term on the right represents the effect of transverse pressure from the boundaries on the flow in terms of an integral average over the cross-sectional area (this requires a prompt response of the plasma to transverse perturbations). In this sense, the mathematical model is called "quasi-two-dimensional". The second term represents the gravitational attraction of the central mass, while the third represents the effects of nonthermal momentum deposition, which we assume to be due to radiation. A complete treatment of the problem would involve solving a radiative transfer equation; however, restricting ourselves to the case of a radiation field in an optically-thin plasma, we shall use the results of Schmidt-Burgk (1978):

$$\frac{D}{\rho c^2} = \frac{gb\gamma^3}{A(\xi)}\left[H(\xi)(1-\beta)^2 - \beta[J(\xi)-H(\xi)](1+f)\right], \qquad (2)$$

where J, H, K, are the zeroth, first and second moments of the radiation field (averaged over frequencies), as measured in its rest frame. We recall that J is related to the total flux and H to the collimated part, and that K is the angular distribution of the collimated radiation:

$$f = \frac{K-H}{J-H}, \quad g = \frac{J_0}{J_{Edd}}. \qquad (3)$$

The presence of an anisotropic radiation field is due to the effect of the funnel walls (Sikora 1981).

We assume, with Piran (1982), a non-spherically symmetric shape for the funnel. Immediately above the stagnation point the funnel has a constant opening angle, i.e., $A(\xi) \propto \xi^2$; beyond this region, the opening angle increases parabolically, and $A \propto \xi^4$. Finally, at the exit of the funnel, we consider a sudden expansion, with $A \propto \xi^{2n}$, and n possibly very large. Further out, we allow for the possibility that n decreases again, perhaps to n = 1, if external conditions lead to recollimation (via magnetic or thermal pressures).

In this geometry, we can write the radiation field moments in the following form:

$$H(\xi) = \frac{1}{A(\xi)}\left(1 - \alpha(\xi,\epsilon)\right), \quad f(\xi) = \frac{1}{A(\xi)}\left(1 - \frac{2}{3}\alpha(\xi,\epsilon)\right), \quad \alpha(\xi,\epsilon) = \frac{1-\epsilon}{\cosh\left(\frac{\xi-1}{\xi_c-1}\right)^m} \qquad (4)$$

for $m \gtrsim 1$, where ϵ is a coefficient between 0 and 1 defining the level of collimation, and ξ_c is a typical (dimensionless) distance of the order of the disk geometrical thickness above which the radiation field becomes essentially isotropic. These expressions are valid within the accretion funnel; above the disk, we simply assume that the radiation field decays spherically. Finally, the remaining moment $J(\xi) \propto 1/A(\xi)$.

III. RESULTS

We shall not present all details of our calculations here; instead we shall simply focus on the qualitative trends of the solutions of the hydrodynamic equation (1) as applied to our astrophysical problem. The steady-state solutions are found by determining the positions of the zeroes of the term enclosed by square brackets on the right-hand-side of Eq. (1). The relevant solutions form a subset of those which make a transition from subsonic to supersonic flow at these points, because some of these critical points correspond to flow topologies which do not correspond to physical solutions. As discussed by Habbal and Tsinganos (1983) and by Ferrari *et al.* (1983), it is straightforward to find the physically relevant critical points, and then to integrate the corresponding solutions. A sample topology is given in Fig. 1 for a mildly relativistic case, with $T = 10^8$ K and $M = 10^8$ $M_\odot$.

For values of the physical parameters typical of active galactic nuclei, the Parker-type wind would become supersonic at a distance $z_p = GM/2\beta_{so}^2 c^2$, i.e., very far from the core, contrary to the observations quoted above. The presence of a sudden expansion of the funnel substantially modifies this classical solution, introducing new critical points near the location at which the maximum expansion occurs, so that the first two terms on the right side of Eq. (1) are in balance. The flow becomes supersonic, and collimated, at the exit of the throat of the disk. In addition, if the radiation field within the funnel is sufficiently intense (corresponding to $L \gtrsim 0.5$ L_{Edd}) and well-collimated ($\epsilon \gtrsim 0.7$), an additional transonic point appears before the exit of the funnel.

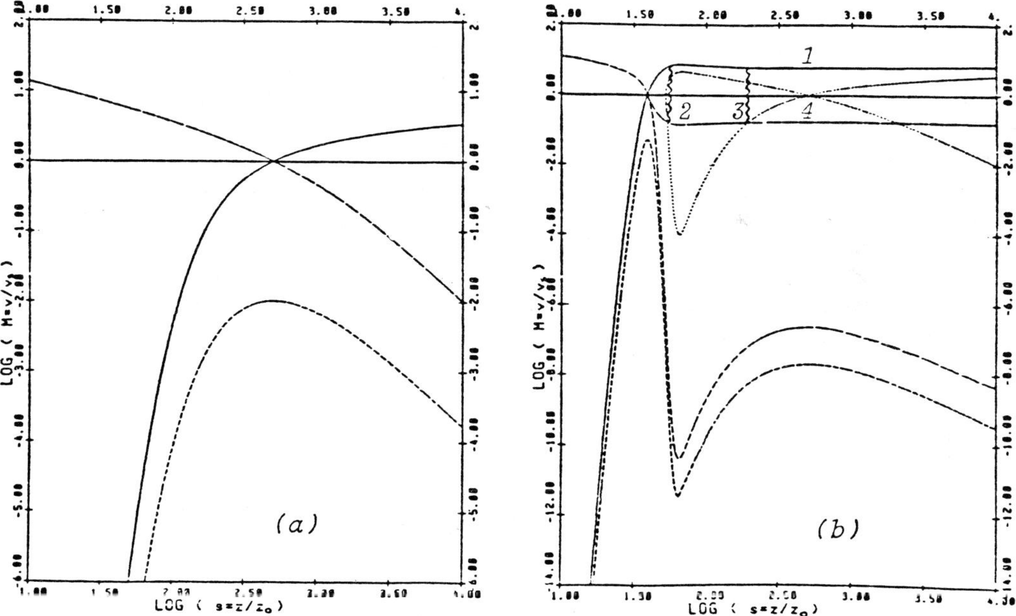

Fig. 1 - Topologies for mildly relativistic winds.

(a) Wind without nonthermal momentum addition ($D = 0$ and $A(\xi) \propto \xi^2$); the Parker-type critical point occurs at $\xi = 500$.   (b) Wind with nonthermal momentum addition peaked at $\xi = 50$; one branch [1] becomes supersonic at $\xi = 39$; two additional solutions exist, which revert to subsonic flow at shocks [2,3], and again become supersonic at the Parker-type critical point [4].

The existence of additional transonic points also offers the possibility of distinct degenerate solutions corresponding to the same boundary and initial conditions. As shown in Fig. 1 (see also Ferrari *et al.*, 1983), steady solutions exist in which the flow becomes supersonic initially at the inner critical point, and then reverts to subsonic flow by an isothermal shock discontinuity, reaching the branch crossing the second transonic point. These configurations can be obtained in our problem for ϵ in the range 0.7 to 0.8 and for $L \sim L_{Edd}$. The presence of solutions with shocks is very interesting because they may be related to enhanced particle acceleration and variability effects when boundary conditions allow the flow to jump from one branch to another.

Finally, we have studied the conditions under which the flow can become relativistic after crossing the critical point. The geometrical effect does not seem to be able to provide enough momentum addition, unless one considers implausibly large expansion factors for the cross-sectional area A, and the quasi-two-dimensional theory becomes questionable. However, the radiation energy density inside the funnel can be much larger than that corresponding to the Eddington luminosity because of the geometrical concentration of the radiation field emitted from the walls in a narrow funnel (Piran 1982). In that case, radiation pressure suffices and, using $L \cong 50 \, L_{Edd}$, allows to reach $\beta \gtrsim 0.9$.

IV. CONCLUSIONS

We have shown how straightforward wind theory allows a rather complete analysis of the hydrodynamics of flows from active galactic nuclei. A number of important effects have yet to be included in our analysis, such as the effect of rotation of the flow about its axis (which is likely to enhance collimation); the role played by magnetic fields (especially in determining the boundary conditions at the exit of the disk throat); the possibility of mass addition to the flow (because mass is likely to be added to the jet flow by "evaporation" and entrainment of matter from the funnel walls); and, finally, departures from isothermality. For example, the use of a more general polytropic equation increases the complexity of the topologies, and, following some preliminary results, tends to increase the momentum addition requirements to reach large supersonic velocities. However, even within the limits of the present work, our results appear to be a good guide for interpreting experimental data and for suggesting directions in the numerical simulations.

Acknowledgements: The authors were partially supported by C.N.R grant PSN 82-015 and NSF grant AST 83-03522.

REFERENCES

Ferrari, A., S.R. Habbal, R. Rosner and K. Tsinganos, 1983, *Ap. J. Lett.* (in press).
Habbal S.R., and K. Tsinganos 1983, *J. Geophys. Res.* **88(A3)**, 1965.
Piran, T. 1982, *Ap. J. Lett.*, **257**, L23.
Rees, M.J. 1982, in **Extragalactic Radio Sources** (Dordrecht, D. Reidel), p. 211.
Schmidt-Burgk H. 1978, *Astrophys. Space Sci.* **56**, 191.
Sikora, M. 1981, *Mon. Not. R. Astr. Soc.*, **196**, 257.

JETS AND BROAD EMISSION LINE REGIONS

Peter Barthel, Colin Norman and George Miley
Leiden Observatory, Leiden, The Netherlands

In recent years an elegant hypothesis for a unified scheme for compact and extended radio sources, and also radio quiet QSOs and radio loud Quasars has been developed (Scheuer & Readhead 1979, Orr & Brown 1982). In this scheme the relative orientation of (relativistic) beam and observer is thought to explain the observed source morphology. We feel that observations suggest examining compact and extended radio sources from still another "angle".

The most significant parameters here to explain differences between types of active galaxies, are the nuclear environment and the duty cycle of the jet.

There are several compelling reasons for considering possible interactions between radio jets and the Broad Line Region (BLR) of active galaxies:
1. VLBI jets and BLRs have similar sizes ($1-10$ pc) and energetics ($10^{43}-10^{44}$ erg/sec).
2. Radio jets interact with galactic material on larger scale (kpc):
   *the narrow line region is morphologically related to the radio structure in NGC4151 (Heckman & Balick, 1983) and other Seyferts (Wilson 1983);
   *strong, sharp bending of jets has been observed in high-z quasars with steep-spectrum cores (Barthel & Lonsdale 1983a, 1983b);
   *in ComaA a radio jet interacts with clouds, producing local particle acceleration, photo ionisation as well as jet bending and decollimation (Miley et al., 1981);
   *NeII lines are well correlated with the spiral jet structure in the Galactic Center (Ekers et al., 1983).

With these facts in mind we noted the following properties of various types of active galaxies:
1. Concerning space densities: BLRGs/Sy1 $\sim$ Quasars/QSOs $\sim 10^{-1}-10^{-2}$
2. BLRGs and Sy1 are similar in having bright nuclei and strong emission lines; however, FeII/Hβ is weak in BLRGs and strong in Sy1s (Osterbrock 1982);
3. Radioloud quasars and radio quiet QSOs are similar in having very

R. Fanti et al. (eds.), VLBI and Compact Radio Sources, 237–238.
© *1984 by the IAU.*

bright nuclei and strong emission lines; however, FeII/Hβ weak in most quasars and strong in QSOs (Osterbrock 1982);

4. The radioloud quasars with strong optical FeII have *compact* radio structure, the extended radio quasars have weak FeII (Miley & Miller 1979).

In other words: in active galaxies possessing extended radio sources the optical FeII is not excited.

From studies of BLR excitation mechanisms (Collin-Souffrin et al., 1980) we know that the BLR is characterized by two distinct ionization systems: a standard photo ionisation region (I), containing Lyα, CIII, HeI, HeII, and a collisionally excited, partially ionized region (II), containing most of the Balmer lines, the optical FeII, MgII and OI. Region II is hot (10^4 K) and dense (10^{11}cm^{-3}).

A way to look at the properties just mentioned is to reason that if a radio jet is stopped its energy may dissipate and excite the low ionisation FeII in dense parts of the BLR by collisional heating by turbulent jet velocities. Commonly observed line shifts of $\pm$ 10^4 kms^{-1} of these low-ionisation species with respect to the mean redshift (Gaskell 1983) provides evidence for this jet hypothesis.

In summary: although Doppler boosting and aspect angles may play a role in explaining differences between types of active galaxies, the nuclear environment is likely to be an important factor.

Acknowledgements.

PDB was supported by the Netherlands Foundation for Astronomical Research (ASTRON) with financial aid from the Netherlands Organization for the Advancement of Pure Research (ZWO). He also acknowledges travel support from the Leidsch Kerkhoven-Bosscha Fonds.

References.

Barthel, P.D. & Lonsdale, C.J. 1983a, M.N.R.A.S., in press.
Barthel, P.D. & Lonsdale, C.J. 1983b, in preparation.
Collin-Souffrin, S. et al., 1980, Astron. Astrophys. 83, 190.
Ekers, R.D. et al., 1983, Astron. Astrophys., 122,143.
Gaskell, C.M., 1983, preprint.
Heckman, T.M. & Balick, B., 1983, preprint.
Miley, G.K. & Miller, J.S. 1979, Ap.J. (Lett.) 228, L55.
Miley, G.K. et al., 1981, Ap.J. (Lett.) 247, L5.
Orr, M.J.L. & Browne, I.W.A. 1982, M.N.R.A.S. 200, 1067.
Osterbrock, D.E., 1982, in Extragalactic Radio Sources, IAU Symp. 97, p. 369.
Scheuer, P.A.G. & Readhead, A.C.S., 1979, Nature 277, 182.
Wilson, A.S., 1983, in Highlights of Astronomy, IAU General Assembly 1982, p. 467.

A NEW GEOMETRY FOR SUPERLUMINAL MOTIONS[+]

Marco Salvati
European Southern Observatory, Garching, FRG *), and
Istituto di Astrofisica Spaziale, Frascati, Italy
*) present address

ABSTRACT

In the context of superluminal motions, the statistical difficulties intrinsic to the canonical jet geometry are alleviated if the relativistic flows occur on a plane screen, orthogonal to the source axis. The strongest superluminal effects should be detected when the source lies in the plane of the sky, rather than pointing at the observer. The polarization properties, the simplest deviations from the basic geometry, and the connection with other non-thermal phenomena are also indicated.

Jets have proved to be a powerful, unifying concept in the interpretation of extragalactic radio sources. The global schemes proposed by Scheuer and Readhead (1979) and Orr and Browne (1982) suggest that randomly oriented jets and their anisotropic emission can explain major morphological differences as well as short-time-scale phenomena. In particular, the superluminal motions detected with VLBI (Cohen and Unwin, 1982; Phillips and Mutel, 1982) offer the most direct evidence of relativistic flows, and the information conveyed by their kinematics is more straightforward to interpret than dynamics or radiation transfer. A proper modelling of this specific phenomenon would then help in assessing the soundness of the general picture.

The most obvious difficulty which jets have to face - and which is relevant to a broader context than just superluminal motions - is the very tight alignment between line of sight and jet axis, required for strong light-travel-time effects and large differences with respect to the classical case. The alignment would entail peculiar statistical properties, easy to recognize even in the limited sample of sources which can be used to this purpose: the expectations, however, are not met in practice. The most evident inconsistency is provided by 3C 273 (Cohen and Unwin, 1982), at whose redshift one should observe several tens of radio quiet optical analogues. Further arguments are derived from the extended radio structure associated with superluminal sources. Deprojection with the angle implied by the apparent velocity would systematically put these

[+] Discussion on page 449

239

R. Fanti et al. (eds.), VLBI and Compact Radio Sources, 239–242.
© 1984 by the IAU.

sources at the upper end of the known size distribution (Schilizzi and de Bruyn, 1983). Reversing the argument, one would not expect superluminal motions in classical extended doubles, while a neat counterexample has been discovered by Porcas (1981).

The suggestion which can be deduced from these discrepancies, together with the simplicity and attractiveness inherent in the idea of relativistic flows, is one of changing the geometry and retaining the basic mechanism. The purpose of this paper is to describe and discuss - and to some extent to provide with a physical background - a non-canonical configuration where the relativistic motion occurs on a two-dimensional screen instead of a one-dimensional jet; thus the solid angle of the "useful" lines of sight, down which the desired effects are observable, is increased substantially.

In the configuration which we envisage, a standard, relativistic jet impinges on a screen which is approximately orthogonal to the flow direction. The distance from the impact point to the jet's origin is of the order of several m.a.s.; hence a possible outer jet and - a fortiori - a possible extended lobe are far downstream, and their existence depends on the jet's ability to drill its way out. At any rate, even if a tunnel is established through the screen, a certain level of "steady" interaction and emission is expected at the intersection point. We identify this emission as the steady radio core, with respect to which outbursts and motions are measured. The screen itself is an important ingredient of our model; for the time being, we can picture it as a cloud, or a collection of clouds, whose upstream boundary is somehow sharply defined. A basic requirement for superluminal effects is that the generalized sound speed for waves along this boundary be very close to the speed of light; hence the volume inside the screen must be filled with an anisotropic relativistic fluid, which dominates dynamically the first and thinnest layers of screen material. A possible scenario has a toroidal magnetic field associated with the jet (Rees, 1982).

Now suppose that a perturbation occurs, for instance, an extra lump of matter is expelled together with the jet, or the flow properties are subject to some modification which the screen can "sense"; if extra energy is deposited at the jet-screen intersection, a circular wave is excited which sweeps the screen's boundary at a speed βc, $\beta \simeq 1$. We propose that radio emission from the tip of such waves is responsible for the superluminal phenomena. Let ψ be the angle between line of sight (observer to source) and jet axis (quasar to lobe); then, as long as $1-\beta \ll 1-\sin \psi$, the light-travel-time effects and the projection effects conspire to produce on the sky the image of a superluminally expanding ring. At the oberver's time t, the ring diameter is $2ct/\cos \psi$, and the steady core is projected excentrically, at a distance $ct(1-\sin \psi)/\cos \psi$ from the receding edge. The brightness distribution along the ring can be evaluated under simple assumptions about the perturbation decay. If the particles are isotropic, notwithstanding the strong anisotropy of the field an almost flat distribution is obtained, which contrasts with the very collimated structures observed in real life. Material outflow at a

speed comparable to the perturbation speed would only produce radian-wide
collimation angles, as it happens in the context of canonical, freely
expanding jets; in both configurations the most straightforward remedy is
to assume a strong internal anisotropy; narrow images are obtained of
jets if their opening angle $<< 1/\Gamma$, and of circular waves if the particle
pitch angle distribution $<< \cos \psi$. In effect, in the proposed scenario
one cannot invoke turbulence and field reconnection as the source of
particle energy, since the field is dominant and very regular. One can
expect, however, a radial compression of the field, which implies induced
electric fields and accelerated particles only in the meridian planes. If
kinetic instabilities are not fast enough to isotropize the particle
velocities during a wave transit time, from a given line of sight one
should observe only the intersections of the full ring with the plane
containing that line of sight and the jet axis. A final remark is in
order about the possible values of $\cos \psi$: we have tacitly assumed that it
be small and positive, i.e., that we are looking at the screen from the
inside. In this case the putative VLBI jet would extend from the core in
the same direction as the outer jet, in accordance with the observations.
But a priori one would expect an equal number of cases where $\cos \psi$ is
small and negative, and the VLBI jet points in the opposite direction. We
must therefore assume that the screen is opaque, at least in the region
around the jet-screen intersection where the superluminal phenomena
occur.

If our picture has any bearing to reality, the fastest motions
should be observed when $\cos \psi \simeq 0$, i.e., in those sources which lie in
the plane of the sky. It is impossible to distinguish superluminal
sources by means of their extended emission, and classical doubles have -
if anything - a larger than average chance of showing the effect. The
role of the Doppler enhancement is drastically reduced: in particular,
the one-sidedness of jets must be taken as a real one-sidedness, which,
at a distance of several kpc from the center, is perhaps more acceptable
than the "Doppler favouritism" hypothesis (Rees, 1982). Note also that
the a priori probability of a one-jet source showing superluminal
velocities in excess of nc is $1/n$, to be compared with $1/2n^2$ of the
canonical configuration; thus, with a conservative choice of H_o, 3C 273
becomes unlikely to just the 20% level. The fate of our model will of
course depend on the agreement with the very detailed measurements which
are now becoming feasible. Roberts (1984) reported the first polarization
studies at m.a.s. resolution, and modern VLBI techniques allow a model-
free monitoring of position angle and separation of the individual knots
as a function of time (Moore, 1984). We can anticipate that, apart from
small scale irregularities due to the unevenness of the screen surface,
accelerated motions are expected because of the concave shape of the
screen looked at from the inside. Also, changes in position angle with
separation are easily explained with a small helicoidal, distance-
dependent twist of the field lines; the intrinsic bending would then be
amplified by projection effects as in the canonical case. The crucial
tests will be provided by polarimetry; we definitely expect a strong
linear polarization. On a m.a.s. scale the magnetic field direction
should be independent of the bending, and orthogonal to the large-scale

jet axis. More generally, our model relies heavily on the properties of
the magnetic field in accounting for the superluminal motions; it is
tempting to identify this feature - rather than the Doppler enhancement -
as the link between superluminal effects and other extreme non-thermal
manifestations, which has been pointed out by Dennison et al. (1981) on
empirical grounds.

Much physics, and many constraints and/or information, underlie the
properties of the screen which we have just postulated; a model of these
properties consistent with current ideas on the environment of quasars is
beyond the scope of this paper, and will be tackled elsewhere.

REFERENCES

Cohen, M.H., and Unwin, S.C.: 1982, IAU Symposium 97, Extragalactic Radio
      Sources, ed. D.S. Heeschen and C.M. Wade (Dordrecht, Reidel), p.
      345.
Dennison, B., Broderick, J.J., Ledden, J.E., O'Dell, S.L., and Condon,
      J.J.: 1981, Astron. J. 86, p. 1604.
Moore, R.: 1984, this Symposium.
Orr, M.J.L., and Browne, I.W.A.: 1982, M.N.R.A.S. 200, p. 1067.
Phillips, R.B., and Mutel, R.L.: 1982, Astrophys. J. (Letters) 257, p.
      L19.
Porcas, R.W.: 1981, Nature 294, p. 47.
Rees, M.J.: 1982, IAU Symposium 97, Extragalactic Radio Sources, ed. D.S.
      Heeschen and C.M. Wade (Dordrecht, Reidel), p. 211.
Roberts, D.: 1984, this Symposium.
Scheuer, P.A.G., and Readhead, A.C.S.: 1979, Nature 277, p. 182.
Schilizzi, R.T., and de Bruyn, A.G.: 1983, Nature 303, p. 26.

VLBI OBSERVATIONS OF THE GRAVITATIONAL-LENS IMAGES OF Q0957+561[+]

M. V. Gorenstein, I. I. Shapiro
Harvard-Smithsonian Center for Astrophysics

N. L. Cohen
Cornell University

R. J. Bonometti, E. E. Falco
Massachusetts Institute of Technology

A. E. E. Rogers
Haystack Observatory

J. M. Marcaide
Max-Planck-Institut fur Radioastronomie

We have conducted a series of VLBI observations of the gravitational-lens images of the quasar Q0957+561 (Walsh et al., 1979), utilizing the Mark III VLBI data acquisition system (Rogers et al., 1983). The goals of our observations are to (1) map the milliarcsecond structure of the A and B images, (2) detect the predicted third image of the quasar, and (3) determine the time delay between the images. We will use these results to constrain the mass distribution of the lens and, possibly, cosmological constants.

1. THE VLBI STRUCTURE OF THE IMAGES

The quasar, Q0957+561, is a radio source, albeit a weak one (Greenfield, Burke, and Roberts, 1980; Pooley et al., 1979). Our VLBI observations in February 1980 at 13 cm wavelength, with three antennas (Gorenstein et al., 1983a), revealed two images, each with a partially resolved core with a full-width at half-maximum (FWHM) of about 1 milliarcsec (mas), and each elongated toward position angle (p.a.) 20 deg. The flux densities of these cores were 22 ± 2 and 18 ± 2 mJy for the A and B images, respectively.

+ Discussion on page 450

R. Fanti et al. (eds.), VLBI and Compact Radio Sources, 243–246.
© 1984 by the IAU.

More extensive observations in March 1981, again at 13 cm wavelength, but with six antennas, detected in each core a weak "inner" jet whose center was displaced from the core by about 3 mas, also along p.a. about 20 deg. These observations also confirmed the existence in each image of a larger "outer" jet whose center was displaced about 50 mas from the core along p.a. about 15 deg, as first reported by Porcas et al. (1981).

Are these resolved brightness distributions consistent with their being images of a single quasar? Given that the 50 mas extent of each image is very small compared to the size of the "lens", the images must be related by a linear transformation of coordinates. The transformation is specified by four parameters: two linear magnifications (one with negative sign to account for the parity reversal of one of the images with respect to that of the source) and two position angles along which the respective magnifications apply. From the data collected in February 1980, which confirmed this hypothesis quantitatively (Gorenstein et al., 1983a), we determined preliminary values for these magnification parameters. We are now carrying out a similar analyis for the data collected in March 1981.

2. ADDITIONAL IMAGES OF Q0957+561

In general, a transparent, non-singular, bounded lens should produce an odd number of images of a point source (e.g. Burke, 1981). Detailed models of the gravitational lens for Q0957+561 (Young et al., 1981) show that a third image should appear between the B image and the galaxy, G1, primarily responsible for the multiple imaging and located about 1 arcsec north of the B image (Stockton, 1980).

The data collected in March 1981 allowed us to perform a sensitive search for additional compact sources of radiation in a 1 arcsec region that included the B image and the G1 galaxy. We used the bright B image as a phase reference to allow us to coherently add visibilities obtained on the most sensitive baselines. The search revealed a third compact component, designated G', that appears near the center of the G1 galaxy. The component's flux density of 0.6 ± 0.1 mJy, extent of less than 2 mas, and position all seem consistent with it being either radio radiation from the center of the galaxy or the third image of the quasar (Gorenstein et al., 1983b).

In May and June of 1983 we performed a new series of VLBI observations of Q0957+561 at 13, 6, and 3.6 cm wavelengths. The results of these observations will help clarify the nature of G' by allowing us to compare the spectral indices of A, B and G'. Moreover the additional data taken at 13 cm wavelength can be combined coherently with those from March 1981 to increase the detection sensitivity by about 40%.

3. THE TIME DELAY

Fluctuations in the brightness distribution of the quasar must produce corresponding, "time-delayed", fluctuations in the images. But the time delays are different for the different images due to the combination of differences in the geometric paths of the rays for the different images and of differences in the speeds of traversal of the rays along these paths. The detection of correlated, time-delayed, variations between images may provide a useful new means of determining the distance to an object (Refsdal, 1964).

A prime goal of our reobservation of Q0957+561 at 13 cm wavelength in May 1983, was the detection of any superluminal motion of the inner jets away from their respective cores. If we assume that we are observing images of the same jet, and, further, that the jet moves on a simple ballistic trajectory, then the measurement of the core-jet separations at the March 1981 and May 1983 epochs is sufficient in principle to determine the apparent epoch of ejection of the inner jet in each image. The difference in these epochs of ejection is the time delay. This idea was independently suggested by Vanderriest (1982).

In the case of Q0957+561 a single epoch of observation may be sufficient to determine the _sign_ of the time delay. The presence of the outer jet in each image determines the relative one-dimensional magnification along the jet axis. If the ratio of the separations between the cores and the inner jets is not consistent with the ratio of the separations of the cores and the outer jets, than we can reasonably attribute the discrepancy to motion of the inner jet. Preliminary examination of the March 1981 data suggests that the inner jet appeared first in the A image; however, we cannot as yet attach any significance to this result as we have yet to consider the errors in our estimates of the core-jet separations. First-epoch observations at the 6 and 3.6 cm wavelengths, if they yield detections of the inner jet, may provide increased accuracy in determining its position relative to the core.

4. MODELING THE LENS

Young _et al._ (1981) presented their models of the mass distribution of the lens that accounted for their optical data. Our analysis of one of their detailed models shows that the estimates of some of their parameters are so highly correlated (Falco _et al._, 1983) that these estimates are not "robust", but are very sensitive to systematic error. We intend to use our VLBI data in conjunction with all of the other available data to constrain more reliably the parameters of simple models of the lens system. For example, a three-parameter quadrupole shear (Falco _et al._, 1983) seems adequate to describe the contribution of the cluster of galaxies to the lens.

ACKNOWLEDGEMENTS

We thank the staffs of the participating observatories for their indispensable aid. The Harvard and MIT experimenters were supported in part by NSF grants PHY-82-43330 and AST-83-00796. This work also represents one phase of research at Jet Propulsion Laboratory performed under NASA contract NAS7-100. This paper is also scheduled to appear in the Proceedings of the 24th Liege International Astrophysical Symposium.

REFERENCES

Burke, W. L.: 1981, Ap.J. (Letters), 244, L1.

Falco et al.: 1983, in preparation.

Gorenstein et al.: 1983a, in preparation.

Gorenstein, M. V., Shapiro, I. I., Cohen, N. L., Corey, B. E., Falco, E. E.,Marcaide,J. M., Rogers, A. E. E., Whitney, A. R., Porcas, R. W., Preston, R. A., and Rius, A.: 1983b, Science, 219, 54.

Greenfield, P. E., Burke, B. F., and Roberts, D. H.: 1980, Nature, 286, 865.

Pooley, G. G., Browne, I. W. A., Daintree, E. J., Moore, P. K., Noble, R. G., and Walsh, D.: 1979, Nature, 280, 461.

Porcas, R. W., Booth, R. S., Browne, I. W. A., Walsh, D., and Wilkinson, P. N.: 1981, Nature, 289, 758.

Refsdal, S.: 1964, MNRAS, 128, 307.

Rogers, A. E. E., Hinteregger, H. F., Levine, J. I., Nesman, E. F., Whitney, A. R., Clark, T. A., Ma, C., Ryan, J. W., Corey, B. E., Counselman, C. C., Herring, T. A., Shapiro, I. I., Knight, C. A., Shaffer, D. B., Vandenberg, N. R., Lacasse, R., Mauzy, R., Rayhrer, B., Schupler, B. R., and Pigg, J. C. 1983, Science, 219, 51.

Stockton, A.: 1980, Ap.J. (Letters), 242, L141.

Vanderriest, C.: 1982, Astron. Astrophys., 106, L1.

Walsh, D., Carswell, R. F., and Weymann, R. J.: 1979, Nature, 279, 381.

Young, P., Gunn, J. E., Kristian, J., Oke, J. B., and Westphal, J. A. 1981, Ap.J., 244, 736.

1038+528 A, B: PHASE-REFERENCE AND SPECTRAL-INDEX MAPS [+]

J. Marcaide[1], I. Shapiro[2], N. Cohen[2,3], B. Corey[4], W. Cotton[5], M. Gorenstein[2], A. Rogers[4], J. Romney[1] and B. Rönnäng[6]

[1]Max-Planck-Institut für Radioastronomie
[2]Harvard-Smithsonian Center for Astrophysics
[3]Cornell University
[4]Haystack Observatory
[5]National Radio Astronomy Observatory
[6]Onsala Space Observatory

On 1981 March 17-18 we undertook MkIII VLBI observations of the quasars 1038+528 A, B (Owen et al. 1979; Owen et al. 1980) with an array of 7 telescopes operating simultaneously at $\lambda 3.6$ and $\lambda 13$ cm with right circular polarization reception at each wavelength. Because the sources are $\sim 33''$ apart they could be observed simultaneously at every telescope. Thus the corrupting contributions of the propagation medium and the instrumentation were approximately the same for each of the quasars, hence allowing us to calibrate the structure phase of B with respect to a reference point chosen in the map of A using the expression

$$\phi_B^S (t) \simeq \phi_B (t) - [\phi_A (t) - \phi_A^S (t)] - [\phi_B^G (t) - \phi_A^G (t)]$$

where ϕ_B and ϕ_A are the observed fringe phases, ϕ_B^G and ϕ_A^G are the geometric contribution with respect to the reference points chosen in each map and ϕ_A^S is the structure phase contribution with respect to the reference point chosen in the A map.

Use of this calibrated phase in the construction of the map of B preserved the information of the position of this map with respect to the reference point chosen in the map of A. Unfortunately the reference points in the $\lambda 3.6$ and $\lambda 13$ cm maps of A cannot be related with a similar technique because of the corrupting contribution of the propagation medium, mainly the ionosphere. Hence to register the maps at the two wavelengths we sought plausible simultaneous registrations of the two quasar maps using the positional constraints obtained at each wavelength from phase-reference mapping. We found that to within about 0.1 mas the registration shown in Fig. 1 was the only plausible one. Any shifts larger than 0.1 mas of the $\lambda 3.6$ cm maps with respect to the $\lambda 13$ cm maps along directions parallel to the structures of each of the A and B quasars produced 'unreasonable' spectral-index morphologies for one or the other quasar:

<u>+ Discussion on page 450</u>

R. Fanti et al. (eds.), VLBI and Compact Radio Sources, 247–248.
© *1984 by the IAU.*

$\alpha \gg 2.5$ ($S \propto \nu^{\alpha}$), gradients in the spectral index perpendicular to the axis of the structure, spectral indices larger at the edges of the map than in its center, etc.

A result shown in Fig. 1 which is immediately striking is that the distance between the peak brightness points in the two maps at λ13 cm is about 0.7 mas ($\sim$ 6 pc at the distance of the A quasar for H_0 = 60 km s^{-1} Mpc^{-1}, q_0 = 0) shorter than the corresponding distance at λ3.6 cm. Using λ18 cm data we ruled out plasma refraction as the cause of this difference. We hypothesize that for the A quasar, which is a 'core-jet' source (Marcaide, 1982), the location of peak brightness may depend on wavelength as $k\lambda^{\beta}$ (k, normalization constant) with respect to the point defined by λ = 0 (apex of the jet – see Blandford and Königl (1979) for a related discussion). Using several other experiments we have determined that, for β, the range is $0.8 < \beta < 2$.

With the registration shown in Fig. 1 we obtain the spectral-index maps shown in Fig. 1. (The four times larger beam of λ13 cm was used to convolve the λ3.6 cm map prior to making the spectral-index map).

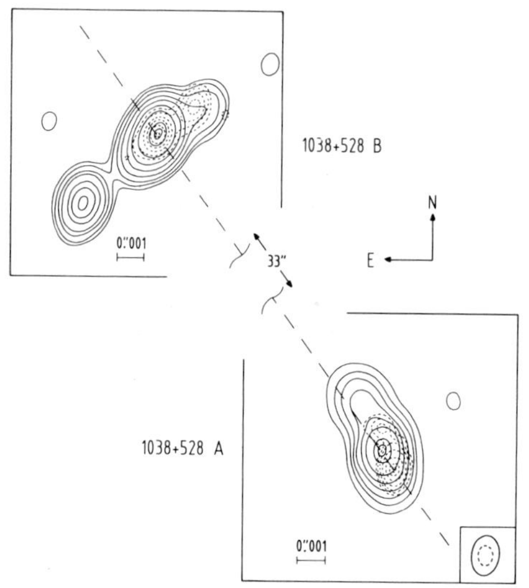

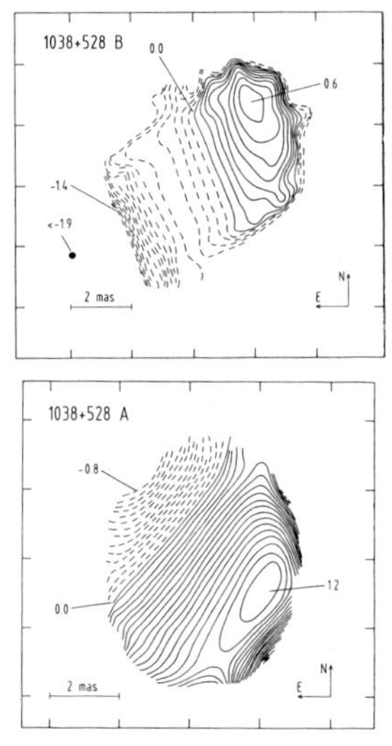

Fig. 1: Brightness distribution of 1038+528 A, B at λ3.6 cm (dashed line) and λ13 cm (continuous line). This particular registration is believed to be correct to within 0.1 mas.

Fig. 2: Spectral-index maps of 1038+528 A, B constructed using the registration of the brightness distributions in Fig. 1.

REFERENCES

Blandford, R.D. and Königl, A.: 1979, Astrophys. J. <u>232</u>, 34.
Marcaide, J.M.: 1982, Ph.D. thesis, Massachusetts Institute of Technology
Owen, F.N., Porcas, R.W. and Neff, S.G.: 1978, Astron. J. <u>83</u>, 1009
Owen, F.N., Wills, B.J. and Wills, D.: 1980, Astrophys. J. Lett. <u>235</u>, L57

GRAVITATIONAL LENSES AND VLBI STRUCTURES [+]

K. Subramanian, D. Narasimha and S.M. Chitre
Tata Institute of Fundamental Research
Homi Bhabha Road
Bombay 400 005, India

The double quasar Q0957 + 561 A,B along with bright radio arches
and VLBI structures is modelled using a gravitational lens consisting of
an elliptical galaxy and a cluster. The effective time-delay between
components A and B comes out to be about a year and this enables one to
distinguish between intensity fluctuations resulting from intrinsic quasar variations and minilensing by low mass stars.

The double quasar Q0957 + 561 is composed of two components with
identical redshifts of 1.41 and a separation of 6.15" (Walsh, Carswell
and Weymann, 1979). Young et al (1981) have modelled the lens action
produced by the combined gravitational effects of the galaxy and the
background cluster at $Z = 0.36$, to produce the observed separation between the components A,B and their intensity ratio of 1.3. The VLBI
observations of the double quasar by Porcas et al (1981) has revealed
core-jet type structures with extensions of 46 $\pm$ 1 mas and 56 $\pm$ 2 mas
respectively for the components A, B and the respective position angles
of 21° $\pm$ 1° and 17° $\pm$ 1°. Any lens model must also explain these observations.

We consider the combined gravitational effect of the elliptical
galaxy and the background cluster by adopting a density profile
$(r^2_c + r^2)^{-3/2}$ (r_c , core-radius) for both the galaxy and cluster mass
distributions. The lens parameters can be so arranged that the second
image B1 is about 75% as bright as image A, while the third image B2,
which is typically 0.5" from B1 is about 2 magnitudes fainter. For a
spherically symmetric distribution of matter the positon angles of the
core-jets are expected to be oppositely directed in relation to the line
through the images. When we attempt to model the VLBI features, it turns
out that the image configuration in which the axes of the core-jets are
aligned almost parallel is obtained only for a narrow range of source
positions close to the critical curves across which transition from one
to three images occurs. The lens parameters are: Core-radius, r_g= 3kpc,
$r_{c\ell}$ = 170 kpc; Velocity dispersion, σ_g = 287 km/sec, $\sigma_{c\ell}$ = 1010 km/sec;
Ellipticity, ε_g = 0.6, $\varepsilon_{c\ell}$ = 0.8.

[+] Discussion on page 451   249

R. Fanti et al. (eds.), VLBI and Compact Radio Sources, 249–250.
© *1984 by the IAU.*

With respect to the centre of the galaxy chosen as the origin, the right ascension and declination of the cluster-centre are respectively -31" and -8.65". This choice of the lens parameters yields average linear amplification $\sim$ 2 and time delay: $t_{B1} - t_A = 1.198$ yrs, $t_{B2} - t_{B1} = 0.0117$ yrs. The lens model implies a linear magnification by a factor of the order 2-3. It is tempting to speculate that future VLBI observations of this object may reveal existence of apparent superluminal motion of components of order $\leq$ 3c. The calculated time delay also indicates that image A should show intensity variation ahead of B by 1.198 years, while Young et al. (1981) report the corresponding value of 5.88 years from their calculations. We believe this variance of our calculated time-delays from those of Young et al. is because of an error of sign in their potential time-delay calculation.

During the monitoring period since its discovery, component A has not varied, while component B which was about 70% as bright as component A in early January 1980 has become comparable and in fact, slightly brighter than A since November 1982 (Gott, 1983). Such a behaviour is difficult to explain on the basis of intrinsic variation in the quasar, if the time delay between A and B is indeed of the order of a year, for in that case component A would have shown intensity variation during the past four years. This points to the possibility that the light from the split images of the quasar is influenced by stars passing in front of the beam causing the images to fluctuate on the time-scale of few years (Chang and Refsdal 1979, Gott 1981). The pattern of intensity variation seems to indicate the minilensing events being operative in the case of Q0957 + 561 B. We find that stars in the mass range $10^{-3}M_\odot < M < 10^{-1}M_\odot$ in the lensing galaxy cause fluctuations in the quasar intensity over a time-scale of 1.5 - 15 years assuming a transverse velocity of $\sim$ 300 km/sec for the low-mass star. Observations over a long time-scale will be able to distinguish between the time-variation arising from mini-lensing and intrinsic quasar variation.

It is thus possible to model the double quasar along with the arched lobe extending from A with complete absence of the structure around B and also the VLBI features associated with A and B.

References
Chang, K., and Refsdal, S.: 1979, Nature, 282, pp. 561.
Gott, R.J.: 1981, Ap.J., 243, 140.
Gott, R.J.: 1983, American Scientist, 71, pp. 150.
Porcas, R.W., Booth, R.S., Browne, I.W.A, Walsh, D., and Wilkinson, P.N.:
     1981, Nature, 289, pp. 758.
Walsh, D., Carswell, R.F., and Weymann, R.J.: 1979, Nature, 279, pp. 381.
Young, P., Gunn, J.E., Kristian, J., Oke, J.B., and Westphal, J.A.:
     1981, Ap. J., 244, pp.736.

RADIO NUCLEI IN NEARBY GALAXIES

E. Preuss
Max-Planck-Institut für Radioastronomie, Bonn, Germany Fed. Rep.

This review is an attempt to summarize VLBI continuum observations at cm- and dm-wavelengths of active galactic nuclei at distances $\lesssim$ 100 Mpc (H_O = 50 km s^{-1} Mpc^{-1}). 'Nearby galaxies', thus defined, are close enough for achieving the highest possible spatial resolution. Galaxies at these distances, however, typically do not show extreme and rare forms of nuclear activity such as powerful radio sources, the cores of which are relatively easy to map with VLBI, and which are therefore the subject of most of the VLBI work done so far (see e.g. Preuss, 1983). Nearby active galaxies show rather more 'ordinary' forms of nuclear activity; they include a few of the weaker classical 'radio galaxies', but most of them are Seyfert galaxies and mildly active 'normal galaxies'. Their total radio emission is typically weak (P(5 GHz) $\lesssim$ 10^{31} erg s^{-1} Hz^{-1}) and so are their compact radio nuclei (if any). The highest available sensitivity is therefore required for their study and the current instrumental performance is just becoming sufficient to tackle the strongest of them in the hope of obtaining maps.

The linear resolution for a fringe spacing of 0.''001 at a distance of 20 Mpc is about 0.03 pc $\sim$ 9 x 10^{16} cm, i.e. a few 100 Schwarzschild radii of a 10^9 M$_{\odot}$ object. This means that VLBI of nearby galaxies is sensitive to size scales which are certainly comparable to those of broad line emission regions and which may even be characteristic of accretion discs or other fundamental scales near the origin of radio sources. VLBI observations of radio nuclei in nearby galaxies have a bearing on a number of fundamental topics such as the central energy source in active nuclei, the process providing the energy supply to the extended emission regions well outside the nuclear environment, the radiation mechanism and the relation between manifestations of nuclear activity in different spectral regimes. In fact the nature of the basic physical constituents just mentioned is not clear in many cases. Given the braod range in power $\gtrsim$ 10^8 :1 and in other characteristic properties of otherwise similar nonthermal radio sources, the important question arises: Can most of the apparent variety be understood in terms of scaling the characteristic properties of a few universally operating mechanisms such as the 'central engine' and the energy supply process to outer emission regions; or more specifi-

R. Fanti et al. (eds.), VLBI and Compact Radio Sources, 251–256.
© *1984 by the IAU.*

cally: Is accretion of material onto massive compact objects (see e.g.
Blandford, this volume) the primary energy releasing process operating
in many, if not all, active nuclei? Or are, e.g. bursts of star formation (Biermann and Fricke, 1977) more important for weaker forms of nuclear activity? To what extent is continuous collimated outflow in the
form of 'jets' a universal phenomenon governing the evolution and structure of nonthermal radio sources? Universality of the principal mechanisms is often suspected or hypothesized but it is only now that we can
go about checking this assumption observationally for the low power radio
nuclei.

Straightforward questions to be answered by observations and relevant to the more fundamental topics mentioned, are the following: What
are the size and power scales in nearby radio nuclei? Are they smaller
than the cores of strong sources? Are they elongated and is their direction related to those of larger scale features or any characteristic
directions revealed in other spectral ranges? How does their detailed
structure compare with more powerful radio cores? Are they simpler? How
does their flux density and structure depend on frequency and time?

In what follows I will try to summarize results of pertinent VLBI
observations which have been made so far. Table 1 lists 40 'nearby galaxies' and it is intended to include all objects from VLBI papers which
have come to my attention and which meet the following 2 criteria:
a) they are detected at least at one wavelength on an angular scale
$\lesssim$ 0.''05 and b) the object is closer than 100 Mpc ($50/H_0$) or the total
radio emission is weak (P(5 GHz) $\lesssim 10^{31}$ erg s^{-1} Hz^{-1}).

Table 1 is based on the results of VLBI observations made at wavelengths between 2.8 and 21 cm and reported in 23 papers, 16 of which have
appeared during the past 2 years. 14 papers deal with individual objects
(M81, M82, CENA, M87, NGC315, NGC1052, M104, NGC3894, see refs in table 1),
3 report results of pilot observations of mixed selections (van Breugel
et al., 1981; Graham et al., 1981; Preuss et al., 1977), the remaining
6 papers report results for samples selected by the following criteria:
Seyfert galaxies (Neff & de Bruyn, 1983), broad line emission regions
(Preuss & Fosbury, 1983), pairs of galaxies (Biermann et al., 1981), optically bright galaxies with strong core sources (Crane, 1979, Jones et
al., 1981) and spiral galaxies (Hummel et al., 1982). The only relatively
nearby object (distance 110 Mpc) mapped at 1.3 cm (Readhead et al., 1983)
is the strong radio core in NGC1275.

At least 28 (70%) of the objects in Tab. 1 are at distances $\lesssim$ 100 Mpc.
The list includes 3 (M81, M82, CENA) of the 179 galaxies known to be
within 10 Mpc (Kraan-Korteweg & Tammann, 1979).

A breakdown of all galaxies listed according to optical morphological
and spectroscopic type reads as follows:

	no emission lines	emission lines (no Seyfert spectrum)	Seyfert galaxies
E/SO	17	6	1
S	4	1	6
Other	2	1	2

Considering the normal (non-Seyfert) galaxies: there are 23 (74%) early type (E/SO) and 5 Spiral galaxies. This is certainly statistically significant in the sense that relatively strong radio nuclei ($\lesssim$ 0$\overset{\prime\prime}{.}$05) are predominantly associated with early type galaxies. A certain statistical significance of table 1 results from the fact that a major fraction of it is effectively the combined detection yield of a number of well defined samples of bright galaxies associated with strong radio cores on a $\sim$ 1" scale. This can be shown by tracing back the routes of source selection to the radio surveys (see e.g. Ekers, 1981) of galaxies by means of single dishes and local interferometers. See also the contribution by Bagri & Ananthakrishnan (this volume) who find most of their scintillating components (IPS at 327 MHz) in nearby galaxies of type E/SO.

What are the characteristic sizes of the smallest radio components detected in the objects listed? At least 34 (85%) show structure on a scale $\lesssim$ 1 pc, 16 (40%) on a scale $\lesssim$ 0.1 pc, and 3 (M81, M82, CENA) on a scale $\sim$ 0.01 pc. The median correlated flux density of the 20 sources detected at 6 cm on a scale $\lesssim$ 1 pc is about 90 mJy. The 10 brightest radio nuclei with scale sizes $\lesssim$ 1 pc in normal galaxies (excluding classical radio galaxies and Seyfert galaxies) are N1052, N4278, N2911, M104, N3894, N3998, M89, M81, M82 and N6500. M81 is the best studied object with very small scale structure and the paper by Bartel et al. (1982) is the first one based on observations with the broad band (56 MHz) Mark III VLBI system. Almost 100% of the nuclear emission in M81 originates in an elongated region of extent $\sim$ 1000 x 4000 A.U. (scale $\sim$ 10^{16} cm). Even smaller is the radio nucleus (SAG A) in our own Galaxy with a size < 100 A.U. $\sim$ 10^{15} cm (see Lo this volume).

The radio luminosity of the core sources in objects of table 1 (plus SAG A) covers a range of 10^{7}:1 typified by the following objects: SAG A: 10^{34}, M81:10^{38}, M104:10^{39}, M87:10^{40}, and NGC315:10^{41} erg/s. Peak brightness temperatures are typically a few 10^{10} K.

The conclusion to be drawn from flux and size measurements is: Very compact nuclei with sizes $\lesssim$ 1 pc occur quite commonly in the centres of low luminosity active galaxies over a wide range of power scales. There is a high VLBI detection probability for flat spectrum radio sources in any kind of galaxy. This supports the idea that the prevailing radiation mechanism is synchrotron emission of relativistic electrons. It looks as if intrinsically weak objects have smaller radio cores than strong radio sources. To rule out the possibility that this is a selection effect, a cross-check on the powerful (distant) sources with baselines larger than the earth-diameter (Satellite VLBI) is needed.

Table 1: Nearby Galaxies detected with VLBI (as of June 1983)

Name	Name	Distance (Mpc)	Optical type		L.A.S.	S.A.S.	References VLBI	other	Remarks
(1)	(2)	(3)	(4)	(5)	(6)	(7)	(8)	(9)	(10)
0017+296	N0076	153	Compact			5	1		
0046+316	N0262	90	S/S0	SEY2	<0."5	1 *	2,3,4	33	Mk348; variable, HI detected
0055+300	N0315	100	E		30' D	1 *	5,6		Giant radio galaxy
0238-084	N1052	28	E	EM	20" D	5 *	7,8,31,36	33	HI detected,variable
0240-002	N1068	22	S	SEY2	14" D	50*	1,4,9	13 33	M77, 3C71
0305+039	N1218	172	S0	EM	2."5	5 *	1	10	3C78
0359+229	N1497		S0			5	1		
0609+710	Mk3	83	S	SEY2	1."5 D	30	2,4	11	4C70.05
0645+744	Mk6	106	S0	SEY1	1" D	50	2,4	12	IC450
0840+504	N2639	67	S		0."7	20	14		
0931+103	N2911	60	S0	EM	<1"	5	1,3,7		
0951+693	N3031	3.25	S	SEY1	>1"	1 *	1,9,15,20	35	M81
0951+699	N3034	3.25	Irr	EM	15"	1	1,2,8,16	17	M82, 3C231
1122+39	N3665	40	S0		30" D	20	7	18	dust lane
1139+267	N3826	181	E			5	1,3		
1146+596	N3894	66	E		2'	1 *	19,31,32		in galaxy pair
1155+557	N3998	24	S0		4' D	20	14		variable
1208+396	N4151	19	S	SEY1	10" D	20	2,7	13 21 22	in galaxy pair
1216+061	N4261	44	E		9' D	1	1,9		3C270
1217+29	N4278	21	E	EM	≲1"	5 *	1,7,8,31	12	HI detected
1222+131	N4374	22	E		2.4 D	1 *	1,9	10	M84, 3C272.1,dust lane
1228+126	N4486	22	E	EM	50" D	1 *	23,24,25, 37	26 27	M87, 3C274, VIR A 10' radio halo
1233+128	N4552	22	E		<1"	5 *	1,7,9	10	M89, variable
1237-113	N4594	18.6	S	EM		5	8,28		M104, "Sombrero"
1254+571	Mk231	248		SEY1	10"	1	2,4	12	variable
1317-12	N5077	50	E	EM		20	7		
1322-427	N5128	5	E		10° D	1 *	29,36		CENA, dust lane
1348+339	N5318	85	S0?			5	1,3		brightest in a group
1351+405	N5353	46	S0		<0."2	10	14		
1353+054	N5363	22	Irr		5"	5	1,3,14		
1353+186	Mk463	304		SEY2	1."3	50	4	12	double nucleus
1426+276	N5635	78	S		<1"	5 *	1	10	variable source
1430+365	N5675	86	S		2."5	5 *	1,3	10	
1553+246		260	E			20	3,19		in galaxy pair
1717+490	ARP102B	150	E	SEY1		1	19		in galaxy pair
1753+183	N6500	64	S		5."3	1 *	1,3,7, 19	10 30	in galaxy pair
2116+262	N7052	99	E			5	1,3		
2303+338	N7485		S0			5	3		
2322+282		138	S0			5	1,3		
2337+268	N7728	189	E		4"	5 *	1,3	10	NRAO 716

Meaning of columns (where not self-explanatory):
column (3): Distances taken from the references given or from Huchra et
            al. (1983); c/H_0 = 6000 Mpc, q_0 = 0.5
column (5): SEY1 and SEY2: Seyfert type 1 and 2, EM = Emission lines
            present
column (6): L.A.S. = largest radio angular size, D = double source
column (7): S.A.S. = smallest angular scale on which radio structure has
            been detected by VLBI; a star "*" means: VLBI structure is
            elongated with position angle given in one of the references

References to table 1

1 Jones et al. (1981a)
2 Preuss & Fosbury (1983)
3 Crane (1979)
4 Neff & de Bruyn (1983) and
  this volume
5 Linfield (1981)
6 Preuss et al., in prep.
7 van Breugel et al. (1981)
8 Shaffer & Marscher (1979)
9 Preuss et al. (1977)
10 Jones et al. (1981b)
11 Wilson et al. (1980)
12 Ulvestad et al. (1981)
13 Wilson & Ulvestad (1982)
14 Hummel et al. (1982)
15 Kellermann et al. (1976)
16 Geldzahler et al. (1977)
17 Kronberg et al. (1981)
18 Kotanyi & Ekers (1979)
19 Biermann et al. (1981)
20 Bartel et al. (1982)
21 Johnston et al. (1982)
22 Booler et al. (1982)
23 Pauliny-Toth et al. (1981)
24 Cotton et al. (1981)
25 Reid et al. (1982) and this volume
26 Owen et al. (1980)
27 Charlesworth & Spencer (1982)
28 Graham et al. (1981)
29 Preston et al. (1983)
30 Hummel et al. (1983)
31 Jones et al. (1983)
32 Wrobel, this volume
33 Bagri & Ananthakrishnan, this volume
34 Pedlar et al. (1983)
35 Crane, this volume
36 Kellermann et al. (1975)
37 Kellermann et al. (1977)

What is known about the morphology of the objects in table 1? For
about 6 of them (M87, N315, Mk348, N3894, N4278, N1052) there are VLBI maps
available and models of the brightness distribution for another 10 objects.
There is enough information to determine the position angle of any elongated or jet-like structure in 15 cases (indicated by '*' in column 7 of
table 1). At least 24 (60%) of all objects listed show extended ($\gtrsim$ 1")
radio structure (column 6), in 12 cases in the form of double structure
associated with galaxies of E/S0 or Seyfert type. The list includes 5 of
about 15 Seyfert galaxies which are known to have double structure on the
$\sim$ 1 kpc scale (Ulvestad, 1983). A comparison of small and large scale
structure in these objects leads to the conclusion: Elongated, linear
structure of radio nuclei and their alignment with larger scale features
seems to be the rule, rather than the exception, also for mildly active
galaxies. Jones et al. (1981) report alignment between pc- and kpc scale
features for N1218, N4374, N5675, N6500, and N7728. See also Jones et al.
(1983) and Neff & de Bruyn (this volume). The maps or models of the radio
nuclei in nearby galaxies are as yet not sufficiently detailed to allow a
proper comparison with the core structures of powerful radio sources. There
is also as yet no information available on the structural variability of

these objects. In any case there is a growing body of evidence that the
extended radio sources associated with mildly active nuclei or Seyfert
nuclei are powered by the same mechanism as radio galaxies or quasars, by
collimated outflow ('jets') of nonthermal material originating on scales
< 1 pc. Both pieces of evidence, the small sizes and the linear structure
of the radio nuclei in nearby galaxies speak against the star burst hypothesis but, rather, favour accretion models with quasi-continuous outflow. But at this stage of the observations it is to early to tell whether this will turn out to be the case universally.

REFERENCES

Bartel, N. et al. 1982. Ap.J. 262:556
Biermann, P. & Fricke, K.J. 1977. Astron. Astrophys. 54:461
Biermann, P. et al. 1981. Ap.J.Lett. 250:L49
Booler, R.V., Pedlar, A., Davies, R.D. 1982. MNRAS 199:229
Breugel van, W. et al. 1981. Astron. Astrophys. 96:310
Charlesworth, M., Spencer, R.E. 1982. MNRAS 200:933
Cotton, W.D., Shapiro, I.I., Wittels, J.J. 1981. Ap.J.Lett. 244:L57
Crane, P.C. 1979. Astron. J. 84:281
Ekers, R.D. 1981. 'The Structure and Evolution of Normal Galaxies', p.
    169, eds. Fall & Lynden-Bell, Cambr. Univ. Press
Geldzahler, B. et al. 1977. Ap.J.Lett. 215:L5
Graham, D.A., Weiler, K.W., Wielebinski, R. 1981. Astron.Astrophys.97:388
Hummel, E. et al. 1982. Astron. Astrophys. 114:400
Hummel, E., van Gorkom, J.H., Kotanyi, C.G. 1983. Ap.J.Lett. 267:L5
Johnston, K.J., Elvis, M., Kjer, D., Shen, B.S.P. 1982. Ap.J. 262:61
Jones, D.L., Sramek, R.A., Terzian, Y. 1981a. Ap.J. 246:28
Jones, D.L., Sramek, R.A., Terzian, Y. 1981b. Ap.J.Lett. 247:L57
Jones, D.L., Wrobel, J.M., Shaffer, D.B. 1983. Ap.J. in press
Kellermann, K. et al. 1975. Ap.J.Lett. 197:L113
Kellermann, K. et al. 1976. Ap.J. Lett. 210:L121
Kellermann, K. et al. 1977. Ap.J. 211:658
Kotanyi, C.G., Ekers, R.D. 1979. Astron. Astrophys. 74:156
Kraan-Korteweg, R.C., Tammann, G.A. 1979. Astron. Nachr. 300:181
Kronberg, P.P., Biermann, P., Schwab, F.R. 1981. Ap.J. 246:751
Linfield, R.P. 1981. Ap.J. 244:436
Neff, S.G., de Bruyn, A.G. 1983. subm. to Astron. Astrophys.
Owen, F.N., Hardee, P.E., Bignell, R.C. 1980. Ap.J.Lett. 239:L11
Pauliny-Toth, I. et al. 1981. Astron. J. 86:371
Pedlar, A. et al. 1983. MNRAS 202:647
Preston, R. et al. 1983. Ap.J. 226:L93
Preuss, E. et al. 1977. Astron. Astrophys. 54:297
Preuss, E. 1983. 'Astrophysical Jets', p. 1, eds. Ferrari & Pacholczyk,
    D. Reidel Publ. Company: Dordrecht
Preuss, E., Fosbury, R.A.E. 1983. MNRAS in press
Readhead, A. et al. 1983. Ap.J. 265:107
Reid, M. et al. 1982. Ap.J. 263:615
Shaffer, D.B., Marscher, A.P. 1979. Ap.J.Lett. 233:L105
Ulvestad, J.S., Wilson, A.S., Sramek, R.A. 1981. Ap.J. 247:419
Ulvestad, J.S. 1983. Subm. to Ap.J.
Wilson, A. et al. 1980. Ap.J.Lett. 237:L61
Wilson, A.S., Ulvestad, J.S. 1982. Ap.J. 263:576

RADIO ACTIVITY IN BRIGHT E/SO GALAXIES: NGC 3894

J.M. Wrobel
Owens Valley Radio Observatory
California Institute of Technology

NGC 3894 is a 13th magnitude E/SO galaxy (de Vaucouleurs, de Vaucouleurs & Corwin 1976; Nilson 1973) whose recession velocity (Kelton 1980) implies a distance of 44 Mpc if $H_o=75$ km s^{-1} Mpc^{-1}. The radio emission from NGC 3894 is compact-core-dominated and, as shown in Fig. 1, this compact core is surrounded at 1.5 GHz by weak diffuse emission with a characteristic size of $\sim$2 arcmin (26 kpc) (Wrobel & Heeschen 1983). If the compact core radiates isotropically, then its radio luminosity is $\sim$10^{40} erg s^{-1}.

This galaxy was observed at 5 GHz with six stations of the US VLBI Network. The data were recorded in Mk II format, processed on the five-station CIT/JPL correlator and calibrated in the standard fashion. For details see Wrobel, Jones & Shaffer (1983). Fig. 2 shows a map of the galaxy, made using the MEM routine of S.F. Gull. The emission is basically double and is elongated at a position angle of $\sim$-47 degrees, which differs from the orientation of the galaxy's isophotal minor axis by $\sim$23 degrees. Two gaussian components were fit to the visibility data; these components are separated by $\sim$5 mas (1 pc), have strengths of $\sim$0.4 and $\sim$0.2 Jy, and have sizes of $\lesssim$2 mas ($\lesssim$0.4 pc) and $\sim$4 mas ($\sim$0.8 pc) x $\lesssim$2 mas ($\lesssim$0.4 pc). Spectral decomposition of the pc-scale emission from NGC 3894, in combination with the X-ray detection reported by Biermann et al. (1981), would permit an investigation of the case for superluminal motion in this bright E/SO galaxy.

The radio emission from NGC 3894 is clearly well collimated on a pc scale (Fig. 2), but is amorphous in appearance on a 30-kpc scale (Fig. 1). The large scale radio emission may be inherently uncollimated, perhaps because of disruption of an intrinsically weak beam by its interaction with the host galaxy's tenuous interstellar medium (Jenkins 1982); and/or because of the interaction of NGC 3894 with its binary companion galaxy, NGC 3895 (Holmberg 1937). The recession velocity difference between these two galaxies is only 11 km s^{-1} (Kelton 1980) and their projected separation is about a galactic diameter (Dressel & Condon 1976; see Fig. 1).

R. Fanti et al. (eds.), VLBI and Compact Radio Sources, 257–258.
© *1984 by the IAU.*

REFERENCES

Biermann, P., et al., 1981, Astrophys.J.(Letters), 250, L49
de Vaucouleurs, G., de Vaucouleurs, A., & Corwin, H.G., 1976 2RCBG
Dressel, L.L., & Condon, J.J., 1976, Astrophys.J.Suppl., 36, 53
Holmberg, E., 1937, Annals of the Observatory of Lund, No. 6
Jenkins, C.R., 1982, M.N.R.A.S., 200, 705
Kelton, P.W., 1980, A.J., 85, 89
Nilson, P., 1973, Uppsala General Catalog of Galaxies
Wrobel, J.M., & Heeschen, D.S., 1983, in preparation
Wrobel, J.M., Jones, D.L., & Shaffer, D.B., 1983, in preparation

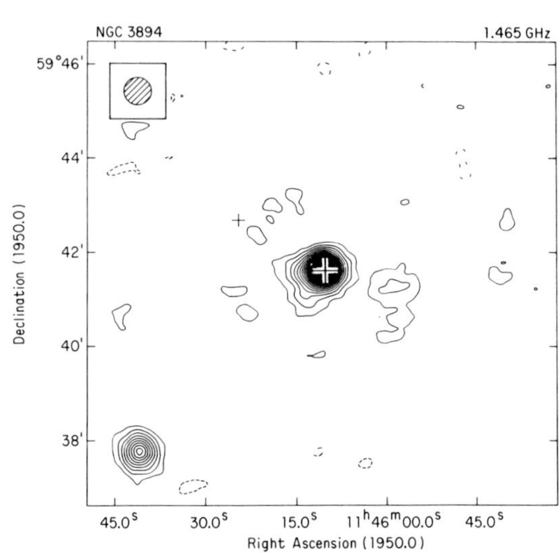

Fig. 1: A 1.5-GHz CLEANed map of NGC 3894. The contour interval is 2 mJy per beam area. Equally spaced contours are plotted; negative ones are dashed. The large cross marks the location of the compact core of NGC 3894, which was removed from the data prior to mapping. The small cross marks the optical position of NGC 3895. The source 6 arcmin SE of NGC 3894 has no optical counterpart visible on the PSS.

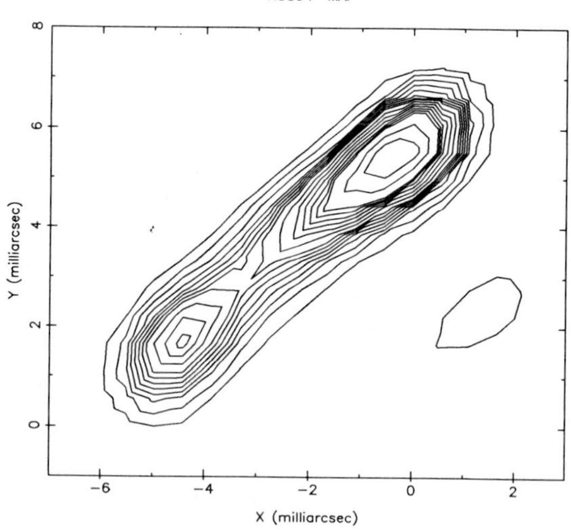

Fig. 2: A 5-GHz MEM map of NGC 3894. Contours at 2, 4, 6, 8, 10, 12, 14, 16, 20, 24, 28, 32, 36, 40, 60 and 80% of the peak intensity are plotted. The tick marks are separated by 1 mas (0.2 pc).

THE STRUCTURE OF THE RADIO NUCLEUS OF M81 [+]

P. C. Crane
National Radio Astronomy Observatory

R. M. Price
Department of Physics and Astronomy
University of New Mexico

New observations of the radio nucleus of the nearby bright spiral
galaxy M81 (NGC 3031) show that the structure of the nucleus is
considerably more complex than previously thought. The radio nucleus
has a slightly inverted, variable spectrum (de Bruyn et al. 1976;
Crane, Guffrida, and Carlson 1976). The variability and VLBI
observations (Kellermann et al. 1976; Jones, Sramek, and Terzian 1981)
both indicate a linear dimension of ~1500 AU. The recent VLBI
observations of Bartel et al. (1982) determined that nearly 100% of the
emission originates in an elongated region with linear dimensions of
1000-4000 AU. Peimbert and Torres-Peimbert (1981) have classified the
optical nucleus of M81 as Seyfert type 1.5 (the weakest known) with the
narrow-emission-line region extending over ~5" (Münch 1959).

The new radio observations reported here were obtained at 4885 MHz with
the Very Large Array in the A configuration. The position and flux-
density calibrators were 0836+710 and 3C286, respectively. The data
were self-calibrated, and four components were identified from the
final maps. The parameters obtained from Gaussian fits are summarized
in the Table. The Figure shows a cleaned map of components 3 and 4
(contour interval of 80 µJy); components 1 and 2 were subtracted prior
to mapping. The dynamic range obtained is 1600:1.

Earlier observations lacked the resolution and dynamic range to detect
components 2-4. No optical counterparts of components 2 and 3 are
known; component 2 is at least 300 times as large as the broad-
emission-line region. Component 4 coincides with the narrow-emission-
line region and a significant fraction of the emission ($\geq$ 10%) may be
thermal (Osmer, Smith, and Weedman 1974; Peimbert and Torres-Peimbert
1981). Component 3 is offset from component 1 by 1."45 (22.9 pc) at a
position angle of 209°, 33° less than that of the rotation axis (Rots
and Shane 1975) and 20°-50° less than the position angles of the major
axes of the VLBI sources measured by Bartel et al. (1982). The
properties of components 2-4 will be clarified as additional
observations become available.

+ Discussion on page 452   259

R. Fanti et al. (eds.), VLBI and Compact Radio Sources, 259–260.
© 1984 by the IAU.

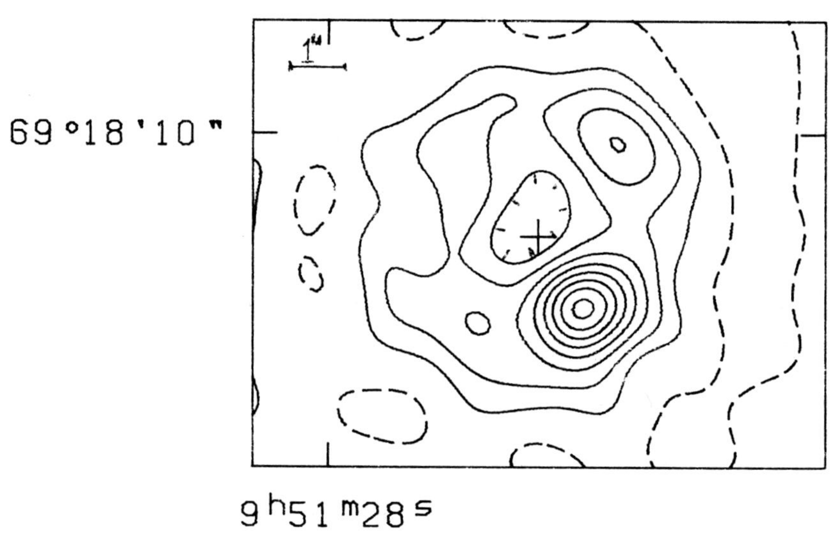

STRUCTURE OF THE RADIO NUCLEUS OF M81

Component	S_{4885}	α_{1950}	δ_{1950}	$\theta(L)$
1	69.6±0.9 mJy	09^{h}51^{m}27^s.315	69°18'08".17	0".00 (0pc)
2	17.9±1.4	27.313	08.16	0.24±0.01 (3.8)
3	0.8±0.1	27.183	06.90	0.68±0.07 (10.7)
4	2.2±0.8	27.404	08.62	4.4±1.2 (69)

REFERENCES

Bartel, N., Shapiro, I.I., Corey, B.E., Marcaide, J.M., Rogers, A.E.E., Whitney, A.R., Capallo, R.J., Graham, D.A., Romney, J.D., and Preston, R.A. 1982, Ap.J., 262, 556.

Crane, P.C., Giuffrida, T.S., and Carlson, J.B. 1976, Ap.J.(Letters), 203, L113.

de Bruyn, A.G., Crane, P.C., Price, R.M., and Carlson, J.B. 1976, Astr.Ap., 46, 243.

Jones, D.L., Sramek, R.A., and Terzian, Y. 1981, Ap.J., 246, 28.

Kellermann, K.I., Shaffer, D.B., Pauliny-Toth, I.I.K., Preuss, E., and Witzel, A. 1976, Ap.J.(Letters), 210, L121.

Münch, G. 1959, Pub.A.S.P., 71, 101.

Osmer, P.S., Smith, M.G., and Weedman, D.W. 1974, Ap.J., 192, 279.

Peimbert, M., and Torres-Peimbert, S. 1981, Ap.J., 245, 845.

Rots, A.H., and Shane, W.W. 1975, Astr.Ap., 45, 25.

SUB ARC SECOND COMPONENTS IN NEARBY BRIGHT GALAXIES AT METRE-WAVELENGTHS [+]

D.S. Bagri and S. Ananthakrishnan
Radio Astronomy Centre, Tata Institute of
Fundamental Research, P.O. Box 8,
Ootacamund 643 001, India.

150 nearby bright galaxies have been observed for compact components at 327 MHz by the method of Interplanetary Scintillation (IPS) using the Ooty radio telescope; scintillation was detected in 28 of these galaxies indicating compact components with size < 30 mas to 300 mas having flux density > 100 mJy. Comparison of our data with other results at high frequencies indicates that these compact components are associated with the central regions of the galaxies.

In Table 1 we present results on five Seyfert galaxies. In NGC 262 only the extended component of size 38 mas (50 mJy) at 18 cm seen by Jones et al. (81 Ap.J.,246,28) appears to be responsible for the scintillating flux; but, in NGC 1068 it is the smallest of the components of size $\sim$ 70 mas (85 mJy) at 2 cm seen by van der Hulst et al. (82 Ap.J., 261, L59) which seems to contribute to the scintillating flux at 92 cm. In NGC 1052, the compact component is very weak and shows an inverted spectrum.

Source	S_c (Jy)	Θ_c (mas)	Remarks
N 262	0.23 $\pm$ 0.06	< 100	Obs. at large elongation only.
N 1052	0.3 $\pm$ 0.15	$\sim$ 100	Inverted spectrum below $\sim$ 3 GHz
N 1068	1.0 to 1.5	100	Elongated in PA 200°
N 1275	4.6 to 5.6	$\lesssim$ 30	Multiple scintillating components;
3C 120	1.5 to 2.5	$\sim$ 50	may be variable

Table 1 IPS data on five Seyfert galaxies

In NGC 1275 and 3C 120, from the break in the IPS power spectra shown in Fig. 1 (a,b) it is clear that the structure of the scintillating components in these sources is complex. The total flux spectra of both the sources along with the spectra of compact components using VLBI data show the presence of two compact components; one of size $\sim$ a few mas, and the other $\sim$ 15 mas. Since the overall size of the source derived from the IPS spectra is < 30 mas it appears that the weak broad IPS

[+] Discussion on page 452

R. Fanti et al. (eds.), VLBI and Compact Radio Sources, 261–262.
© *1984 by the IAU.*

spectrum (Fig.1a) arises due to the few mas component and the strong
narrow spectrum, due to the 15 mas component. From this we have derived
fluxes in both the components A and B (Fig.1c) at 327 MHz using the area
under each part of the spectrum. Similarly, using available VLBI data
we have derived fluxes for the two compact components in 3C 120 (Fig.1d).

From the above, it appears that it is possible to separate very
compact components in radio sources at metre wavelength using the IPS
method. More observations of similar sources as well as careful modelling of the source and solar wind parameters are necessary to establish
the technique.

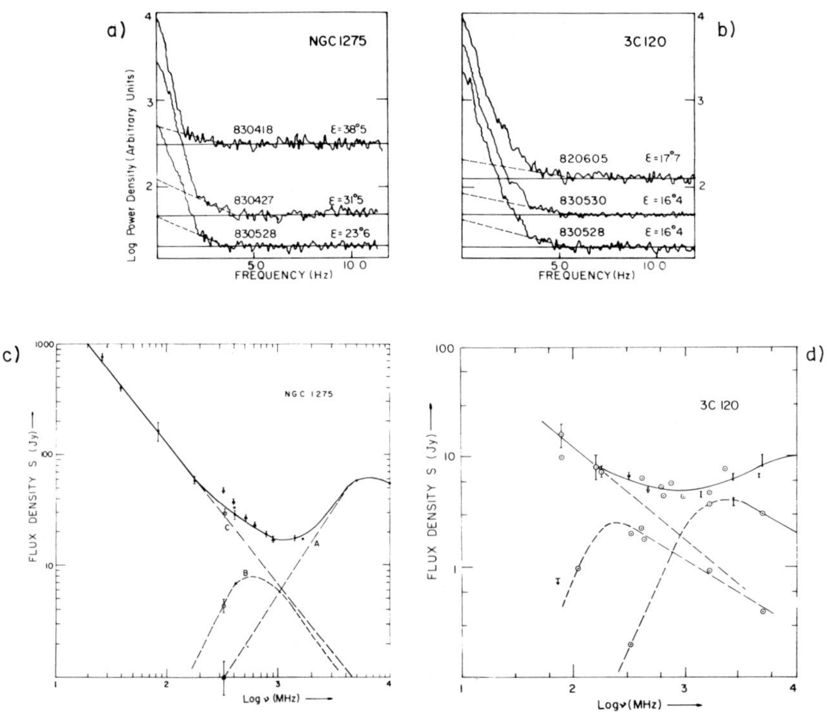

Fig.1.(a) and (b) : IPS power spectra
      (c) and (d) : Flux density spectra

THE RADIO CORE OF MARKARIAN 348 +

S. G. Neff and A. G. de Bruyn
Netherlands Foundation for Radio Astronomy
Dwingeloo, The Netherlands

The type 2 Seyfert galaxy Markarian 348 (NGC 262) is a normal early-type disc galaxy (S0/a or Sa) surrounded by a very large HI envelope. The continuum radio source in the galaxy is unresolved by conventional interferometers on scales as small as an arcsecond, is variable on time scales of a few months, and is quite strong by Seyfert standards (300 mJy). Mkn 348 is relatively nearby, allowing one to study very small-scale structure in the galaxy (at the assumed distance of 60 Mpc (H_o = 75 km/sec/Mpc), one light year subtends an angle of about one mas). Both the relative proximity and the radio source strength make it a good candidate for VLBI studies.

European VLBI observations of Mkn 348, presented here, show that the nuclear radio source is triple and linear in nature, is oriented along position angle 168 degrees, and has a total size of about 50 parsecs (see figure 1). Figure 2 shows the fit of the map CLEAN components to the visibility data. The map displayed accounts for all of the flux density from Mkn 348 at the time of the experiment; we therefore conclude that there is no significant emission outside some 0.2 arcseconds (60 pc).

Total flux density monitoring with the WSRT over a four-year interval show that the source is highly variable at 6cm, but only moderately so at 21cm, on time scales of a few months. A new outburst began in early 1982 as can be seen in figure 3. The variability provides indirect evidence for the existence of sub-mas core in Mkn348. Structural and spectral data combined indicate the the existence of an inverted-spectrum core and an optically thin steep-spectrum extended region (a jet?). Figure 4 shows our suggested spectral decomposition for the radio source in mid-1982.

These results support the hypothesis that radio sources in both type 1 and type 2 Seyfert galaxies are powered in the same manner as radio galaxies and quasars, by directed outflow of non-thermal material from a compact energetic nucleus.

A fuller account of this work will appear in Astronomy and Astrophysics.

+ Discussion on page 452

263

R. Fanti et al. (eds.), VLBI and Compact Radio Sources, 263–264.
© *1984 by the IAU.*

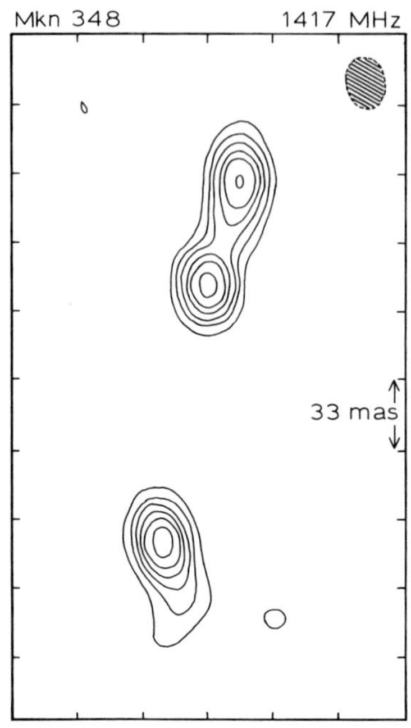

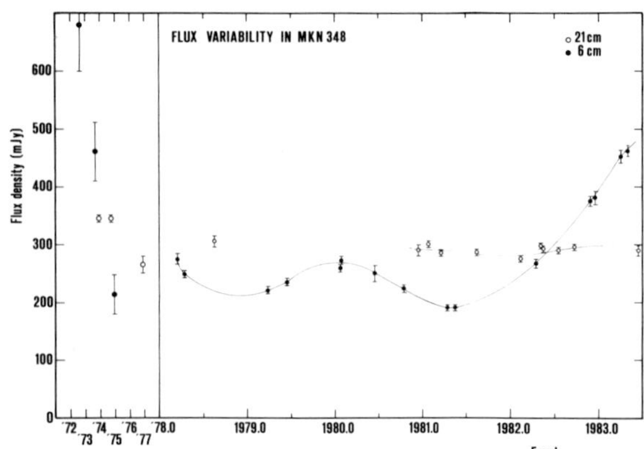

Figure 3. Radio variability in Mkn 348 at 6 and 21 cm, over a period of several years.

Figure 1. Map of Mkn 348, made using the European VLBI network at 21 cm.

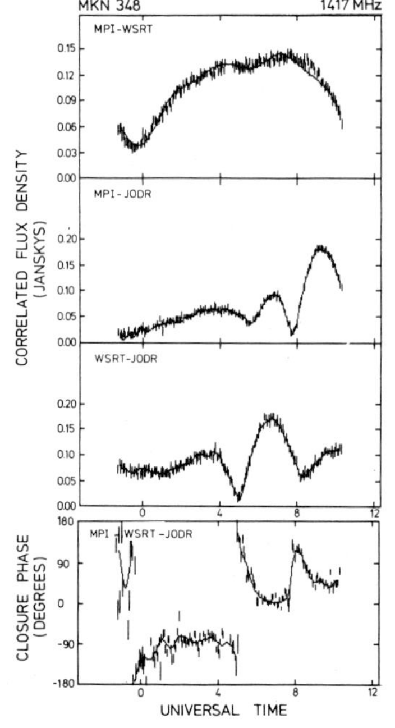

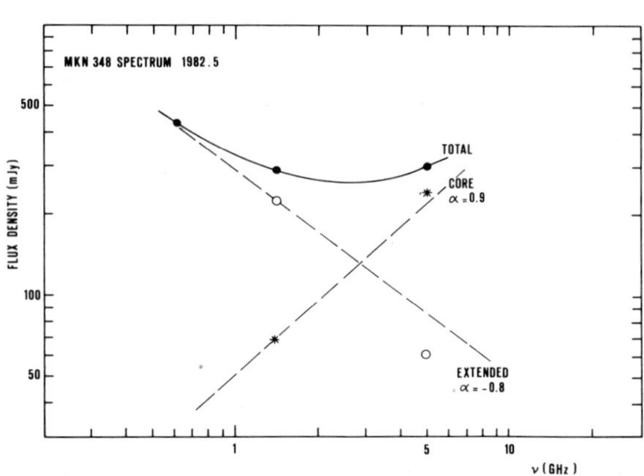

Figure 4. Suggested spectral decomposition of the radio source in Mkn348, mid-1982.

Figure 2. Visibility curves for Mkn 348. The line through the data shows the fit of the map in figure 1.

THE GALACTIC CENTER [+]

K.Y. Lo
Astronomy Department
California Institute of Technology

The center of our Galaxy contains an extremely compact nonthermal radio
source. For the first time, elongation in the source structure has been
detected. The long axis is nearly aligned with the minor axis of the
Galaxy. Recent high resolution observations of the ionized gas within
the central 3 parsecs suggest that matter may be falling in towards the
center. This has interesting implications on the processes within our
Galactic nucleus.

1. INTRODUCTION

At a mere distance of 10 Kpc, the center of our Galaxy can be studied
at unsurpassed linear resolution and detail. An example of this is
shown in Figure 1: a 1"-resolution λ=6 cm VLA map of Sgr A West, the
ionized gas region within 1.5 pc of the center (Lo and Claussen 1983).
All the details in the intricate distribution of the ionized gas would
subtend only 1" at the distance of M31, the nearest external galaxy with
a galactic nucleus.

Besides its proximity, the Galactic Center is interesting because of a
variety of phenomena which suggest past and present activities:

(a) Large scale expanding motions in the HI and molecular gas that may
suggest past explosions of extraordinary energy (cf Oort 1977);
(b) The extremely compact nonthermal radio source, (e.g. Lo 1982), a
possible signature of a massive collapsed object (Rees 1982);
(c) Large velocity dispersion of the ionized gas within the central
parsec, indicating the possible existence of massive point mass (Lacy
et al 1980);
(d) Time variable e^+e^- annihilation γ-ray line observed towards the
center, suggesting a compact source of e^+ (e.g. Lingenfelter and Ramaty
1982);
(e) Very broad ($\Delta V \sim 1500$ Km/s) 2.06μ HeI line, that might indicate mass
outflow (Hall et al 1982);
(f) The "spiral-like" pattern of ionized gas within the central 3 pc,

[+] Discussion on page 453   265

R. Fanti et al. (eds.), VLBI and Compact Radio Sources, 265–273.
© *1984 by the IAU.*

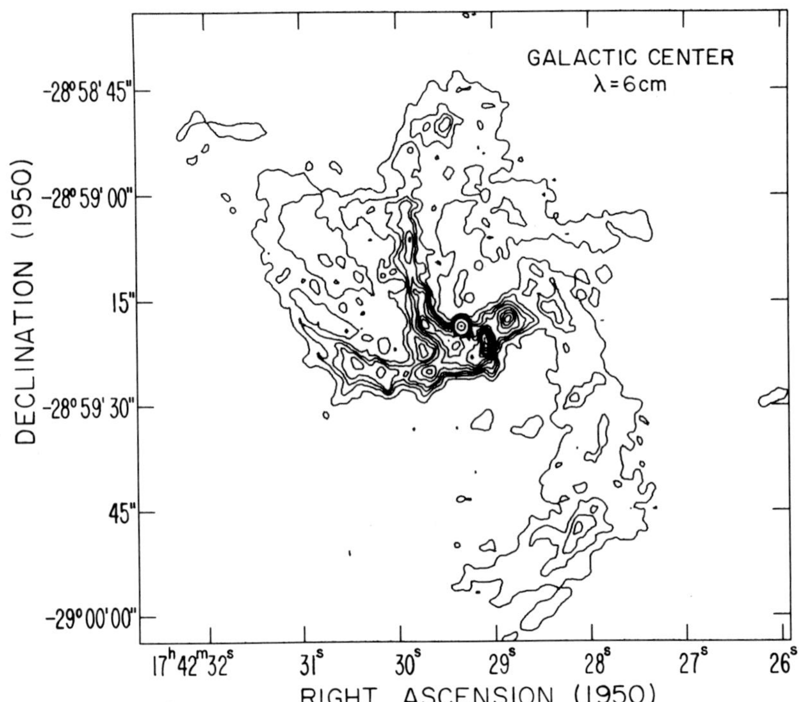

Figure 1
The compact radio source is at the map center.

first noted by Ekers et al (1983).

Radio continuum observations of the Galactic Center have previously been
reviewed by Lo (1982). Here, we discuss the recent VLBI results on the
compact source and high resolution VLA observations of Sgr A West. We
present a new interpretation of the observations of the ionized gas and
a model to account for its origin. Finally, we discuss the implications
of the observations on the understanding of the Galactic Center.

2. VLBI OBSERVATIONS OF THE COMPACT RADIO SOURCE

Observations prior to 1978 were made at $\lambda \geq 3.6$ cm using the Mark II
recording system with sensitivity ~ 0.1 Jy (5σ). The results have been
given by Lo et al. (1981). The principal result was that the measured
size scale of the source varies as λ^2 where λ is the observing wavelength
This could be explained by either broadening of the intrinsic source size
by interstellar scattering (Davies et al. 1976) or the optical depth of
thermal electrons within the source (Brown et al. 1978; Lo et al. 1981).
In any case, the intrinsic source structure was not known, and the
possibility of interstellar scattering implies that the observed source

size may only be an upper limit to the intrinsic size.

As both the interstellar scattering and the optical depth of thermal
electrons decrease with shorter wavelength, short wavelength observations
were emphasized subsequently. This became more practicable with the
availability of the $\sim$5 times more sensitive Mark III recording system
(Rogers et al. 1983), and improvements in the receivers. Table I summarizes a series of 2.8 cm (Lo, Backer and Cohen, unpublished) and
1.35 cm (Lo, Backer, Moran, and Cohen, unpublished) observations carried
out prior to 1983.

Table I

Wavelength	Date of Observations	Telescopes*
2.8 cm	1980, April	K,G,O
	1981, June	K,G,O
1.35 cm	1981, December	K,G
	1982, June	K,G,O
	1982, December	K,G,O

* K = Haystack 36.5-m telescope, G = NRAO 42-m, O = OVRO 40-m

Success of this series of measurements was severely limited by bad
weather at one or more telescopes during the observations. Nevertheless,
we obtained the following: At λ= 2.8 cm, the scale size was $\sim$0.01" which
is consistent with the λ^2-extrapolation from longer wavelength. On the
other hand, at λ= 1.35 cm, the scale size was >0.003" (the inequality was
due to calibration uncertainty), much larger than the extrapolated
scattering size. This has important implications because it suggests
that at 1.35 cm, we are observing the intrinsic source structure. The
1.35 cm result has been confirmed by measurements of Kellermann and
Ekers (private communication).

On May 23, 1983, we carried out a 6-station λ = 3.6 cm observation (Lo,
Backer, Cohen, in preparation) using the Mark III system in mode B. While
the calibrator source, NRAO 530, was detected on all baselines, the compact radio source in Sgr A was detected only convincingly on the relatively short baselines formed by the NASA 64-m telescope at Goldstone
(DSS14), the OVRO 40-m and the Hat Creek 25-m telescopes (HCRK).

The amplitude data (Figure 2) were fitted to two simple models: a symmetric Gaussian brightness distribution and an elliptical Gaussian distribution. The baseline most sensitive to the difference of the two
models is DSS14-HCRK (cf Figure 3), because of the relatively large
rotation of the baseline.

The better fitting elliptical Gaussian model has the following parameters:
the major axis (half-power) diameter is $\sim$0.015" and an axial ratio of 0.25
with the long axis oriented at a position angle (PA) of 107°. The PA
is in agreement with the results of Jauncey et al. (private communica-

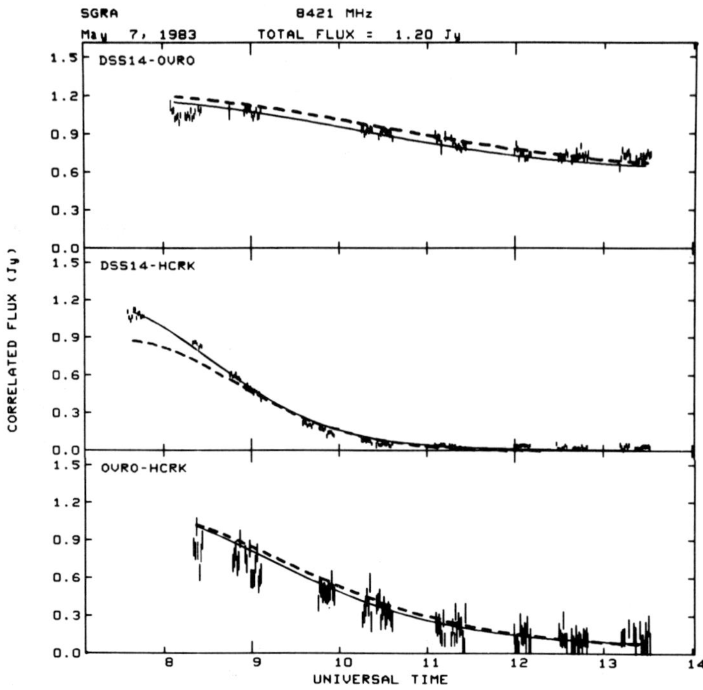

Figure 2

The lines are the model visibility amplitudes.   The dashed line is the
symmetric Gaussian model.   The models are indistinguishable on the
other 2 baselines.

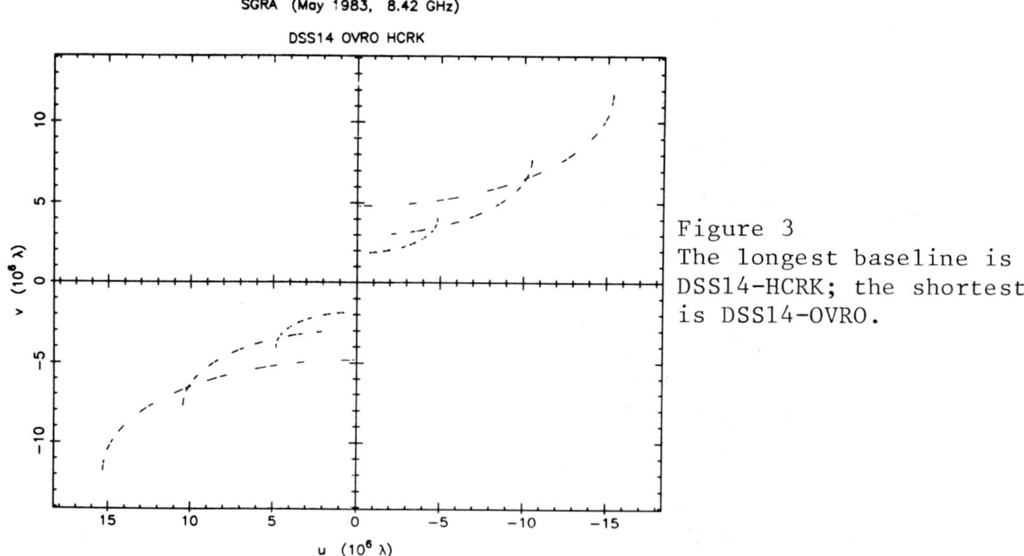

Figure 3
The longest baseline is
DSS14-HCRK; the shortest
is DSS14-OVRO.

tion) who used a single N-S baseline in Australia.

The 0.001" core of the compact radio source reported by Kellerman et al. (1977) and Geldzahler et al. (1979) was not detected at an upper limit of 0.01 Jy (6σ).

3. NATURE OF THE COMPACT RADIO SOURCE

The known properties of the compact source is summarized in Table II.

Table II

Scale Size	$\sim 5 \times 10^{14}$ cm
Position Angle of Elongation	$\sim 107°$
Spectral Index	~ 0.2, up to 115 GHz
Radio Luminosity	$\sim 2 \times 10^{34}$ erg/s
Turnover Frequency	$\geqq 115$ GHz
Brightness Temperature	$\sim 4 \times 10^{8}$ K
Flux Variability $\Delta S/S_{min}$	< 1 (1975-1982)
Upper Limit to Mass	$< 5 \times 10^{6}$ $M_{\odot}$

They are different from those of any known galactic radio sources associated with stellar objects: pulsars, binary stars, or young supernova remnants (Lo et al. 1981). Rees (1982) suggested that the source could be due to low-level accretion onto a $\sim 10^{6}$ $M_{\odot}$ black hole. While the compact radio source has a dimension $\sim 10^{3}$ times the Schwarzschild radius of such a black hole, it has a very modest radio luminosity of ~ 10 $L_{\odot}$. Based on energetics alone, it is therefore not possible to exclude a stellar object as the underlying energy source.

Hence, the mass of the radio source is a critical parameter to determine.

4. MASS DETERMINATION

The ideal method of mass determination is to observe the distribution of the velocity dispersion of the stars around the radio source. This has to be done at the infrared wavelength because of the large obscuration towards the center. However, attempts at measuring the stellar velocity dispersion have been unsuccessful so far.

An independent constraint on the mass of the radio source is its proper motion. If the radio source is massive, no motion is expected, whereas a stellar mass object would have high space motion and thus measurable proper motion. First results of this difficult measurement (Backer and Sramek 1982) are consistent with the secular parallax expected for an object at rest at the Galactic Center.

A third approach is via near infrared spectroscopic observations of the ionized gas within the central parsec. High spatial resolution 12.8μ

[NeII] observations (Lacy et al. 1980) have revealed much about the
motion of the ionized gas there. If the motion is due to gravity alone,
the velocity dispersion could be used to infer the total mass. Lacy et
al. (1980) inferred that there may be $\sim 10^7$ $M_\odot$ of mass within the central
parsec, either all in stars, or half in stars and the rest in a central
point mass. However, the origin, the motion, the ionization state and
the disposal of the gas require explanations (Lacy et al. 1982).

5. RADIO OBSERVATIONS OF THE IONIZED GAS

Ekers et al. (1983) first revealed the "spiral" distribution of the
ionized gas in Sgr A West in their large-scale study of Sgr A. Subsequently, Lo and Claussen (1983) obtained a 1" resolution map that adds
much detail to the previous maps.

The high resolution map of Sgr A West (Figure 1) shows the following
features:

(a) The overall distribution of the ionized gas is dominated by 3
bright "arms" extending north, east and west from the center. By "center"
we mean the region immediately surrounding and including the compact
radio source.
(b) Features of lower surface brightness are found on the periphery of
the distribution. They are the northern continuation of the east arm, an
incomplete loop to the north-east of the center, and most prominent of
all, the south arm which is apparently joined to the west arm.
(c) The south arm is in fact not connected to the west arm. Where the
two arms meet there is a break as well as a large difference in the
surface brightness of the arms. Also, there is a large velocity difference between the two arms (cf Lo and Claussen 1983). The south arm
continues north, past the west arm, and is most likely connected to the
tip of the north arm. Previous interpretation (e.g. Ekers et al. 1983)
of the north, west and south arms as forming a "spiral" pattern is no
longer compelling.
(d) The compact radio source while not at the peak of the thermal
emission, is situated at the centroid of the overall distribution.

6. VELOCITY FIELD OF THE IONIZED GAS

The velocity field of the ionized gas (Figure 4) is given by the [NeII]
observations (Lacy et al. 1980). The systematic variation along the
north, south and east arms is obvious. Decomposition of the complicated
profiles within the central 30" into 3 velocity components makes it possible to identify 3 streams of gas, each corresponding to one of the 3
bright arms (Lo and Claussen 1983). Lines are drawn in Figure 4 to outline the loci of similar velocity. They also suggest the current positions of the gas trajectories which appear to converge to or diverge
from the center. Note that the streams do not meet at a single point and
the compact radio source is offset $\sim$2" (0.1 pc) from the streams.

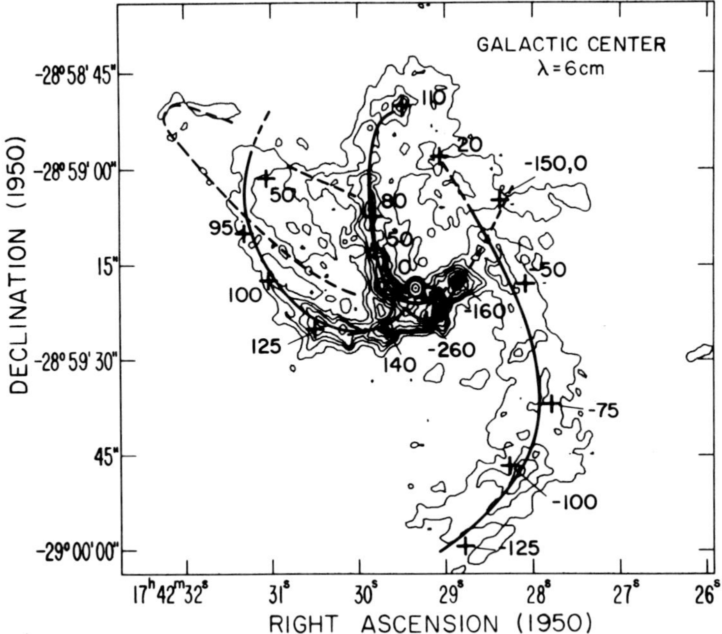

Figure 4

The numbers next to the crosses are the radial velocities
(Km/s) of the [NeII] emission measured at those points.

There exists some symmetry in the velocity field: the north and south
arms have velocities of opposite sign, which is also true for the velocities of the east and west arms. However, for the north and south arms,
the magnitude of the velocity decreases towards the center, while for the
east and west arms, the maximum velocity is observed near the center.
This suggests that the north and south arms could be part of a rotating
"ring" about the center whereas the east and west arms could not be part
of the same structure.

7. PROPOSED MODEL OF THE IONIZED GAS

The morphology of the ionized gas is a transient phenomenon since the
dynamic time scale is $\sim 10^4$ years. It must be the result of systematic
motion- a combination of rotation, ejection and infall (cf Ekers et al.
1983).

Precessing twin-beam ejection such as that proposed by Brown (1982) cannot explain the complicated morphology. To account for the 3 bright arms,
3 precessing nozzles operating simultaneously in 3 independent directions

are required. Blandford (private communication) suggested that the
streams result from ionized gas driven outwards along the field lines of
a spinning magnetized object. In general, however, ejection by a single
object cannot explain why the 3 streams of gas do not meet at a single
point near the center.

Ekers et al. (1983) suggested tidal distortion of molecular clouds as one
of several possible explanations. Since we can now identify the 3 streams
of gas near the center, the tidal distortion picture has become more
attractive. The Roche density of the central parsec is $\sim 10^8$ cm^{-3}, much
higher than the ionized gas density, making the tidal effect very strong.

The gas moves from north to south along the south arm, which orbits about
the center, and is probably connected to the north arm which spirals in
towards the center. The east and west arms are independent streams falling towards the center along paths roughly normal to the plane of orbit
of the south and north arms (cf Lo and Claussen 1983).

The gas must originate as low angular momentum molecular clouds in the
vicinity of Sgr A West. As it falls within 1.5 pc of the center, it is
ionized by the central ionizing source, which is most likely a collection
of O,B stars (Lacy et al. 1982). A single central ionizing source is
not excluded, however.

8. IMPLICATIONS

The central issue here is whether the compact radio source is the signature of a massive collapsed object. The supporting evidence is compelling, but circumstantial. Based on energetics alone, we cannot
exclude a stellar object. The latest VLBI results indicate that the compact radio source is elongated in a direction nearly parallel to the
rotation axis of the Galaxy (PA = 122°). If the alignment has a physical
origin, then it would be hard to argue that the compact source is simply
a peculiar stellar radio source within the central star cluster. Thus,
it is even more urgent now to obtain a detailed VLBI map of the compact
radio source.

The compact radio source may be dependent on its environments. Recent
high resolution observations have led to a clearer picture of the spatial
distribution and kinematics of the ionized gas within the central few
parsecs. This is important since it may be possible to use the morphology
and the velocity field of the ionized gas to probe the central mass
disbribution and check the existence of a central point mass.

The most natural explanation of the ionized gas in Sgr A West is that it
is tidally distorted molecular gas falling in towards the center. The
transient nature of the ionized gas, however, implies that the processes
within the Galactic Center may not occur under steady-state conditions.
Thus, variability of the activities with time is indicated, especially
if the central luminosity is dependent on the accretion of interstellar
matter.

If the proposed infall is verified, then we have for the first time
direct observational evidence of infall of matter towards the center of
a galaxy. Accretion of interstellar matter by a central black hole has
been proposed as the energy source for much of the activities seen in
quasars, radio galaxies and Seyfert nuclei. In the center of our Galaxy,
we may be witnessing directly the path of accretion.

This research is supported by NSF grants AST82-10259 and AST82-10828.

REFERENCES

Backer, D.C., Sramek, R.A.: 1982, Astrophys. J., 260, pp. 512-519.
Becklin, E.E., Gatley I., Werner, M.W.: 1982, Astrophys. J., 258, pp.
    135-142.
Brown, R.L., Lo. K.Y., Johnston, K.J.: 1978, Astron. J., 83, pp. 1594-1597.
Brown, R.L.: 1982, Astrophys. J., 262, pp. 110-119.
Davies, R.D., Walsh, D., Booth, R.: 1976, Monthly Notices Roy. Astron.
    Soc., 177, pp. 319-333.
Ekers, R.D., van Gorkom, J., Schwarz, U.J., Goss, W.M.: 1983, Astron.
    Astrophys., 122, pp. 143-150.
Geldzahler, B.J., Kellermann, K.I., Shaffer, D.B.: 1979, Astron. J.,
    84, pp. 186-188.
Hall, D.N.B., Kleinman, S.G., Scoville, N.J.: 1982, Astrophys. J. (Lett.),
    260, L53-L58.
Kellermann, K.I., Shaffer, D.B., Clark, B.G., Geldzahler, B.J.: 1977,
    Astrophys. J. (Lett.), 214, L61-L62.
Lacy, J.H., Baas, F., Townes, C.H., Geballe, T.R.: 1979, Astrophys. J.
    (Lett.), 227, L17-L20.
Lacy, J.H., Townes, C.H., Geballe, T.R., Hollenbach, D.J.: 1980, Astrophys.
    J., 241, pp. 132-146.
Lacy, J.H., Townes, C.H., Hollenbach, D.J.: 1982, Astrophys. J., 262,
    pp. 120-134.
Lingenfelter, R.E., Ramaty, R.: 1982, The Galactic Center, AIP Conf. Proc.
    No. 83 (ed. Reigler, G.R. and Blandford, R.D.), pp. 148-159.
Lo, K.Y., Cohen, M.H., Readhead, A.C.S., Backer, D.C.: 1981, Astrophys. J.
    249, pp. 504-512.
Lo, K.Y.: 1982, The Galactic Center, AIP Conf. Proc. No. 83, (eds.
    Reigler, G.R., and Blandford, R.D.), pp. 1-8.
Lo, K.Y., Claussen, M.J.: 1983, submitted to Nature.
Oort, J.H.: 1977, Ann. Rev. Astron. Astrophys., 15, pp. 293-361.
Rees, M.J.: 1982, The Galactic Center, AIP Conf. Proc. No. 83, (eds.
    Reigler, G.R. and Blandford, R.D.), pp. 166-176.
Rogers, A.E.E. et al: 1983, Science, 219, pp. 51-54.

MARK III VLBI OBSERVATION OF PULSARS [+]

N. Bartel[1], R.J. Cappallo[2], M.I. Ratner[1],
A.E.E. Rogers[2], I.I. Shapiro[1], A.R. Whitney

[1]Harvard-Smithsonian Center for Astrophysics
[2]NEROC Haystock Observatory

1. INTRODUCTION

Mark III VLBI observations of the pulsars PSR 0329+54 and PSR 1133+16
were made at 2.3 GHz using antennas with diameters and locations as
follows: 100m, Effelsberg, West Germany (but only for SPR 0329+54); 43m
Green Bank, WV, USA; and 40m, Big Pine, CA, USA. The Mark III processor
at the Haystack Observatory was "gated" to compute visibility amplitudes
and phases as a function of pulsar longitude. This method allowed a)
an improvement of the signal to noise ration, by as much as a factor of
ten in the case of PSR 1133+16, and b) an interferometric investigation
of the pulse structure.

2. PULSAR ASTROMETRY

Observations of the pulsars and of compact extragalactic sources 0355+508
(NRAO 150) and 1119+183, each nearby in the sky to one of the pulsars,
yielded the following positions (epoch 1981.21) and uncertainties given
as the sum of our estimates of sistematic errors:
PSR 0329+54: $\Delta\alpha$(VLBI-PTA) = 0^S03 $\pm$
PSR 1133+16: α(1950) = $11^h33^m27\overset{\cdot}{.}^S409$ $\pm0^S03$, δ(1950) = $16°07'38".9$ $\pm0".3$.
These positions differ from those determined from pulse-time-of arrival
measurements (PTA) by Manchester and Taylor (<u>A.J.86</u>, 1953, 1981) as
follows:
PSR 0329+54: α(1950) = $03^h29^m11\overset{\cdot}{.}^S0133\pm0\overset{\cdot}{.}^S0005$, δ(1950) = $54°27'37".300\pm0".005$
PSR 1133+16: $\Delta\alpha$(VLBI-PTA) = $0^S056\pm0^S008$, $\Delta\delta$(VLBI-PTA) = $-0".9\pm0".3$.
The errors shown are the root-sum-squares of the errors given above and
the statistical standard errors given by Manchester and Taylor. The differences between the VLBI and PTA values, if significant, may be due mostly to the difference in orientation of the VLBI reference frame from that
in which the timing positions are expressed.

3. UPPER LIMIT OF APPARENT PULSAR SIZE

In order to account for the total flux density variations of PSR+54, the
correlated flux densities of all three interferometers were divided by
the corresponding correlated flux densities for the interferometer with
the shortest baseline. The width of a circular Gaussian determined by a
least-squares fit to these normalized amplitudes is $0.23^{\pm0.13}_{0.23}$ mas, where

+ Discussion on page 455

275

R. Fanti et al. (eds.), VLBI and Compact Radio Sources, 275–276.
© *1984 by the IAU.*

the error was estimated by varying the gain of each station independently
by 15%.

UPPER LIMIT OF OFF-PULSE CONTINUUM EMISSION

The data for PSR 0329+54 were recorrelated using a gating window corresponding to 0.8 of the pulsar period, centered between the pulses. All
of these "off-pulse" data were coherently integrated ($\sim$2.9 hours total)
by using the technique of phase referencing. No continuum emission was
detected at the position of the pulsar above a correlated flux density
level of 2.5 mJy (3σ). A separate limit was determined for the approximately two minute period during which the received pulses were strongest.
This limit is less than 0.1 percent (4σ) of the flux density of the peak
of the average pulse during that period.

5. INTERFEROMETER PHASE WITHIN THE PULSE STRUCTURE

In order to search for position offsets between the emission regions of
the three components in the average pulse profile of PSR 0329+54, the
Green Bank-Effelsberg interferometer phase-delays for these components
were compared with each other (Figure 1). Upper limits (3σ) on any
differences between the positions of the three components are :
$\Delta\alpha$ < 20 μas, $\Delta\delta$ < 50 μas. An offset of 20 μas corresponds to an offset
of $\sim$ 7x10^6 km (less than five solar diameters)at an assumed distance of
2.3 kpc for PSR 0329+54. Also no significant phase differences were detected within individual pulses, which were partitioned into 75 bins each
of 715 μs extent.

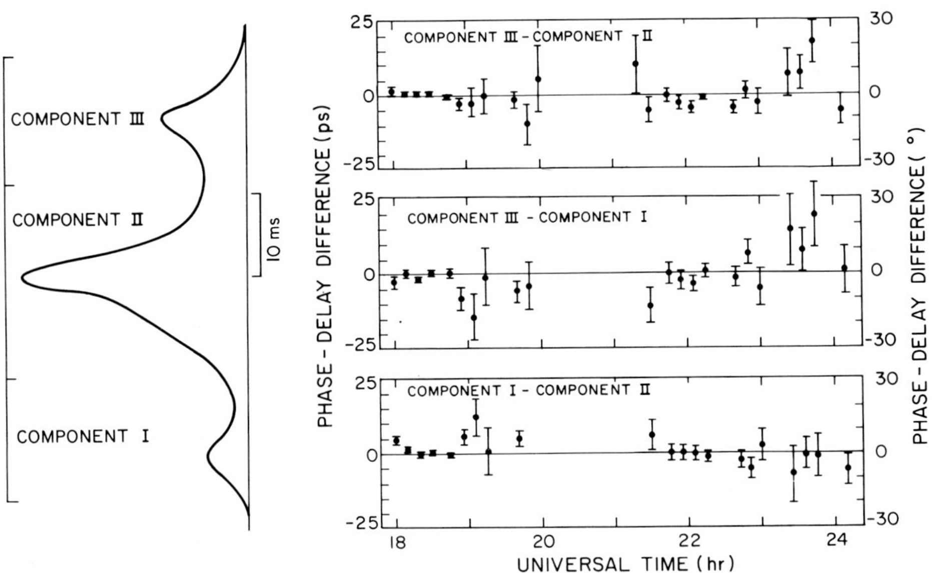

Figure 1: Differences of interferometer phases for each pair of components in the average pulse profile of PSR 0329+54 from observations with
the Green Bank and Effelsberg antennas. Error bars shown signify standard deviations (statistical).

VLBI OBSERVATIONS OF STELLAR BINARY SYSTEMS +

R. L. Mutel
Dept. Physics & Astron., Univ. of Iowa, Iowa City, IA

J. F. Lestrade*
Bureau de Longitudes, Paris, France

R. A. Preston and A. E. Niell
Jet Propulsion Laboratory, Pasadena, CA

R. B. Phillips and J. Webber
Haystack Observatory, Westford, MA

*On leave at JPL.

Since 1982 July we have been engaged in a systematic investigation of the milliarcsecond structure of the radio emission from several binary star systems, mainly in the RS CVn class (Hall, 1976). The first few observations utilized the MkII VLBI recording scheme, but no useful data were obtainable because of the relatively low flux levels involved ($\overline{S} \leqslant 30$ mJy). Beginning in 1982 December, we have been using the MkIII system with large telescopes and have successfully detected seven binary systems (UX Arietis, HR 1099, Algol, II Peg, σ Crb, and the distant system LSI 61°303). A summary of the three experiments already completed is given in Table 1.

The scientific goals are twofold: (1) to measure the source sizes and brightness temperatures which, when combined with polarization and spectral information, should allow a detailed model of the radio emission mechanism and (2) to make precise position measurements relative to nearby extragalactic objects so that future optical positions from the Hipparcos program can be used to tie the optical and radio reference frames together at the $\sim$ 2 mas level (cf. Preston, Lestrade, and Mutel, 1983). In this paper we report preliminary progress which has been made in the astrophysics area, since we have not yet attempted any astrometric measurements.

During the first two VLBI observations, the sources were barely resolved and only upper limits to the sizes could be made. During the first experiment (1982 December), a moderate outburst of HR 5110 was detected, yielding a brightness temperature limit $T_B \geqslant 4 \times 10^8$ K. The second experiment (1983 February) provided interesting lower limits to the brightness temperatures ($T_B \geqslant 1.4 \times 10^{10}$ K for UX Arietis and $T_B \geqslant 2.9 \times 10^{10}$ K for HR 1099). The 1983 March observation showed clear

R. Fanti et al. (eds.), VLBI and Compact Radio Sources, 277–279.
© 1984 by the IAU.

resolution of a very large (S ~ 400 mJy) outburst of the source HR 1099 on continental U.S. baselines, so that an accurate source size of ℓ ~ 4 x 10^{11} cm could be measured. This is about equal to the size of the primary star (KO IV, diameter = 6 $R_\odot$). The data rule out a coherent process, such as cyclotron maser mechanism (Melrose and Dulk, 1982) at least for the outbursts which we observed.

If we assume that incoherent emission from a gyrosynchrotron process is responsible for the emission (Owen et al., 1976), then the brightness temperature limits constrain the magnetic fields to $\overline{B}$ ≤ 6 gauss and the energetic electron densities to 10^6 ≤ n_e 10^8 cm^{-3} and energies $\overline{E}$ ~ 1 MeV. The measured values of circular (~ 5 - 10%) and linear (< 2%) polarization during the 1983 February events on HR 1099 and UX Arietis are also consistent with this picture.

This research is supported by NASA grants NAGW-386 and NAS7-918.

Table 1

Summary of Results

Date	Telescopes	Freq. (GHz)	Sources	Flux (mJy)	Size (mas)	T_B(K)	Ref.
19 Dec 82	DSS13, DSS14, GRAS, OVRO	8.4	HR 5110	32	≤ 1.4	⩾ 4 E 8	a
13 Feb 83	VLA, OVRO, NRAO	1.65	UX Arietis	95	≤ 2.1	⩾ 1.4 E 10	b
			HR 1099	~ 200	≤ 2.1	⩾ 2.9 E 10	b
20 Mar 83	DSS14, DSS13, OVRO, WEST, GRAS	8.4	UX Arietis	~ 12	*	*	c
			HR 1099	~ 400	0.7	2 E 10	c
			Algol	30-48	≤ 1.4	⩾ 4 E 8	c
			σ Crb	10	*	*	c
			LSI 61°303	18	*	*	c
			II Peg	7	*	*	c

Telescope identification (diameter): DSS13 (26 m), DSS14 (64 m), Goldstone, CA; OVRO (40 m), Owens Valley, CA; GRAS (26 m), Ft. Davis, TX; WEST (18 m) Westford, MA; NRAO (43 m), Green Bank, WVA.

References: (a) Lestrade et al. (1983a), (b) Mutel et al. (1983), and (c) Lestrade et al. (1983b). * = too weak for detection on long baselines (unresolved on DSS14-OVRO).

REFERENCES

Hall, D.S.: 1976, in IAU Colloquium 29, "Multiply Periodic phenomena
    in Variable Stars," W.S. Fitch, Ed. (Reidel: Dordrecht).
Lestrade, J.F., Mutel, R.L., Preston, R.A., Scheid, J.A., and
    Phillips, R.B.: 1983a, submitted to Astrophys. J. (Lett.)
Lestrade, J.F., Mutel, R.L., Preston, R.A., and Phillips, R.B.:
    1983b, in preparation.
Melrose, D.B. and Dulk, G.A.: 1982, Astrophys. J. 259, p. 844.
Mutel, R.L., Doiron, D.J., Lestrade, J.F., and Phillips, R.B.:
    1983, submitted to Astrophys. J.
Owen, F.N., Jones, T.W., and Gibson, D.M.: 1976, Astrophys. J.
    (Lett.) 210, p. L27.
Preston, R.A., Lestrade, J.F., and Mutel, R.L.: 1983, Proc. of
    Hipparcos Thinkshop, Asiago, Italy, May 1983.

THE SEPTEMBER 1982 RADIO OUTBURST OF CYG X-3

B.J. Geldzahler[1,4], K.J. Johnston[1], J.H. Spencer[1],
W.J. Klepczynski[2], F.J. Josties[2], P.E. Angerhofer[2],
D.R. Florkowski[2], D.D. McCarthy[2], D.N. Matsakis[2], and
R.M. Hjellming[3]
[1]E. O. Hulburt Center for Space Research, Naval Research
Laboratory, Washington, D. C. 20375
[2]U. S. Naval Observatory, Washington, D. C. 20390
[3]National Radio Astronomy Observatory, Socorro, N.M. 87801
[4]NRC/NRL Cooperative Research Associate

Cyg X-3 underwent a series of giant radio outbursts beginning on
September 28, 1982 (Geldzahler _et al._ 1983). The flux densities at 2.7
and 8.1 GHz (11.1, 3.71 cm respectively, see Figure 1) were measured
with the 2.4 km baseline of the Green Bank interferometer once every
three days before October 5, 1982 (= JD 244 5248) and three times daily
thereafter.

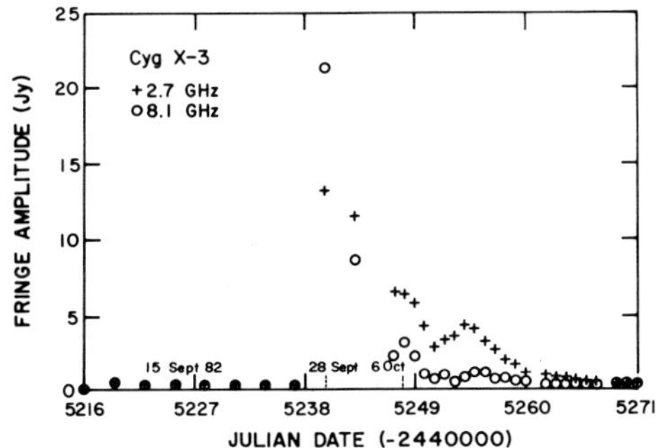

Date	Measured Size arc seconds		Frequency GHz	Comments
Oct 10.04	0.32 x 0.22 pa 9°		1.660	VLBI
	±0.10 0.07			
Oct 9.20	0.10 x 0.03 4°		average of	VLA
	±0.02 ± 0.02 ±4°		22 and 15	
	0.33 x 0.24 0°		1.465	VLA
	±0.10 0.20 ±20°			
Oct 20.02	0.25 x 0.06 178°		average of	VLA
	±0.02 0.04 2°		22 and 15	
	0.40 x 0.23 23°		1.465	VLA
	±0.10 0.14 ±34°			

The maximum flux density of Cyg X-3 during the outburst was at least
22 Jy. We can only set a lower limit on the maximum because of our
three day sampling interval before October 5. The September 1982 outburst was not a single, unique event, but on ~ Oct. 5 and again on Oct.
9 there is clear evidence other outbursts began. This is similar to the
mid-September 1972 outburst (c.f. Hjellming 1973). The 1982 event
sequence ended on about Nov. 4, again comparable to the mid-September
1972 events as to duration.

Cyg X-3 was observed at 20, 6, 2, and 1.3 cm with the B configuration of the VLA on 7, 8, 9, and 20 Oct. 1982. Spectra were obtained on
all four days. For 9 and 20 Oct. 1982, we had a sufficient number of
VLA antennas to make maps. In Table 1, we present the results of source

281

R. Fanti et al. (eds.), VLBI and Compact Radio Sources, 281–282.
© _1984 by the IAU._

size measurements derived from VLA data. The results from the 1.3 and 2 cm data are presented because they were obtained with the highest angular resolution ($\sim$ 0.2 arcsec). Assuming that the outburst began on Sept. 27.5, which is accurate to a day, we infer from the Oct. 9 and 20 VLA result that the linear expansion proceeded at a rate of $0\rlap{.}''010 \pm 0\rlap{.}''002$/day. This is a minimum expansion rate derived from Gaussian source sizes (Case 1) and the knowledge of the date of the outburst. Had we modeled the data as a uniform sphere (Case 2), for example, the expansion rate would be almost twice as great. Dickey's (1983) lower limit of 11.6 kpc to Cyg X-3 coupled with the angular expansion rate yields an apparent linear expansion speed of $\gtrsim 0.7$ c (Case 1) or $\gtrsim 1.3$ c (Case 2). The radio emission seems to be beamed since at 1.3 and 2 cm the axial ratio of the source is 4:1.

If the energy transport mechanisms in Cyg X-3 and SS433 (Hjellming and Johnston 1981) are similar and if both beams in Cyg X-3 are observable, then the beam velocity is $\gtrsim 0.35$ c (0.63 c) relative to the central source. We favor a model for Cyg X-3 describing a two-sided emission because of analogies with Sco X-1 and SS433 and because of the lack of motion of the brightness centroid ($\Delta\theta < 0\rlap{.}''05$).

At a distance of 12 kpc, the radio luminosity, on September 27.5-the peak of the outburst, was $\gtrsim 10^{34}$ erg sec^{-1}. The total radio energy involved in the event sequence was then $\gtrsim 10^{40}$ erg. These values are quite similar to those derived for the 1972 outbursts. As Gregory et al. (1972) point out for the 1972 events, the total energy is substantially less than that involved in a supernova.

Cyg X-3 joins the ranks of SS433 (c.f. Hjellming and Johnston 1981) and Sco X-1 (Fomalont et al. 1983) as a galactic source whose expansion has been measured. Cyg X-3 is clearly more similar to SS433 where v_{exp} = 0.26 c rather than Sco X-1 where $v_{exp} < 0.0001$ c. However, in SS433, the expansion velocity and jet collimation arises from radiation pressure and line locking due to absorption of photons below the Lyman edge by L α transitions (Milgrom 1979) whereas in Cyg X-3 the expansion results from an explosive event. The same is true when comparing the ratio of luminosity to the Eddington luminosity in that for Cyg X-3 and SS433 $L/L_E \simeq 1$ whereas for Sco X-1, $L/L_E \ll 1$. As Rees et al. (1982) point out, the stellar scale objects may be scaled down versions of active galactic nuclei and quasars.

REFERENCES

Dickey, J.M.: 1983, Ap. J. in press.
Fomalont, E.B., Geldzahler, B.J., Hjellming, R.M., and Wade, C.M.: 1983, Ap. J. in press.
Geldzahler, B.J. et al.: 1983, Ap. J. in press.
Gregory, P.C. et al.: 1972, Nature 239, 114.
Hjellming, R.M.: 1973, Science 182, 1089.
Hjellming, R.M. and Johnston, K.J.: 1981, Ap. J. (Letters) 246, L41.
Milgrom, J.: 1979, Astr. Ap. 78, L9.
Rees, M.J., 1982, "Extragalactic Radio Sources", Reidel

THE RATE OF COMPONENT SEPARATION IN SCO X-1 [+]

B.J. Geldzahler[1,3], E.B. Fomalont[2], R.M. Hjellming[2], and C.M. Wade[2]

[1]E. O. Hulburt Center for Space Research, Naval Research Laboratory, Washington, D. C.   20375
[2]National Radio Astronomy Observatory, Socorro, N.M. 87801
[3]NRC/NRL Cooperative Research Associate

Sco X-1 was observed in the A configuration of the VLA on 4 Feb. 1981 and 18 Mar. 1982 (Fomalont et al. 1983). The radio map at 1.5 GHz, made with the 1982 data, is shown in Figure 1. The central component is nearly coincident with the binary and has an angular size of less than 0″1. A 0″5-resolution map of the SW component of Sco X-1, shown in Figure 2, displays only the brightest peaks in the two emission regions. The brightest part of the emission is located at the extreme edge of the component, and its location and shape are reminiscent of hot spots in radio lobes associated with luminous extragalactic radio sources. Although the NE component is only slightly resolved at this resolution, it is extended in approximately the same position angle as that of the entire source and that of the SW component.

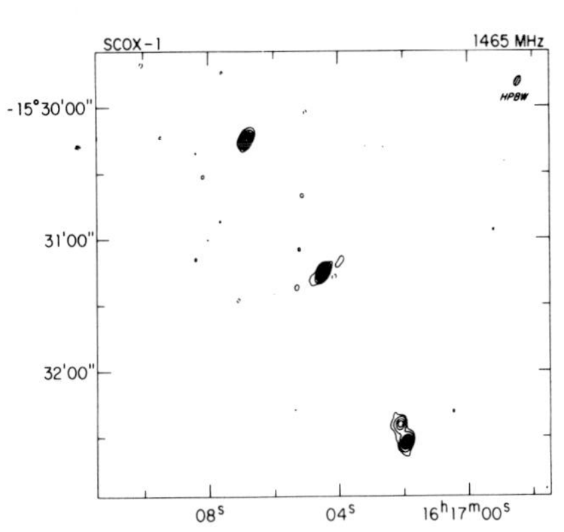

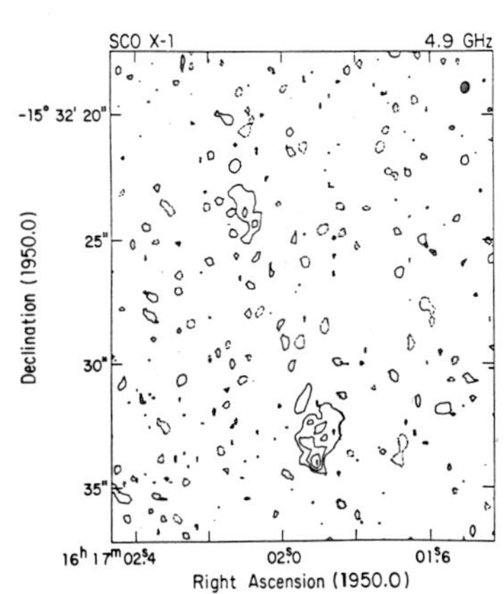

The results of the component position analysis from the Gaussian fits can be summarized as:

1. No change in the separation of the central component from the NE component to a limit of 0″015 (< 32 km s^{-1} at d = 500 pc).
2. No change in the separation of the central component from the bright peak in the SW component to a limit of 0″13 (< 277 km s^{-1}).

[+] Discussion on page 456

283

R. Fanti et al. (eds.), VLBI and Compact Radio Sources, 283–284.
© 1984 by the IAU.

3. No variability of intensity in the NE and SW components between 1981 and 1982. There may be some indication of variability on longer time scales.

The dual-opposing beam model has been used extensively for explaining jets and hot spots in extragalactic sources and this model has already been suggested as a good one for Sco X-1 (Blandford and Rees 1974). The major problem with the application of the beam model to Sco X-1 is the difficulty in transporting sufficient energy to supply the components with the observed radiative energy while not transporting excessive momentum consistent with the observed upper limit of 32 km s^{-1}. The lower limit of the external density, ρ_e, needed to keep the lobe from moving faster than 32 km s^{-1} is: $\rho_e > 3.2 \ [c(\gamma^3+1) \ (\gamma+1)/(v_b \ \varepsilon \ \gamma^4)]$ cm^{-3} where v_b = particle velocity in beam at working surface, ε = conversion efficiency of kinetic energy in beam to radiative energy at the working surface. Even with a relativistic, efficient beam the lower limit to the external density needed to restrain the NE component is uncomfortably large.

The beam model can work if the flow velocity is relativistic, if the conversion of bulk kinetic energy into radiative energy at the working surface is efficient (> 10%) and if the density of material into which the beam impinges has a density of > 10 cm^{-3} which is much larger than that of the typical interstellar region (e.g. Turner 1979). Other models in which most of the energy of the component is contained internally are even less attractive.

The small linear size of the NE component cannot be maintained without confinement of the relativistic electrons generated at the working surface. The confinement could be inertial if the component contains a high density gas which may be needed to stop the beam. This region may also contain a magnetic field strength as large as 0.1 G; hence the radiative age of the electrons may be as short as several years and the confinement problem significantly reduced.

Generalization of these results to extragalactic, luminous radio sources is speculative. The possible high efficiency of the Sco X-1 beam may suggest that the typical conversion efficiency of bulk kinetic energy into relativistic electrons and magnetic field approaches 100%. The possible high density region at the termination of the beam in Sco X-1 may also suggest that entrainment of the thermal matter by the beam and the depositing of the material near the lobes can occur. Finally, the magnetic field near the working surface may be much larger than that calculated from the usual equipartition considerations.

REFERENCES

Blandford, R.D. and Rees, M. J.: 1974, M.N.R.A.S. 169, 395.
Fomalont, E.B., Geldzahler, B.J., Hjellming, R.M., and Wade, C.M.: 1983, Ap. J. in press.
Turner, B.E.: 1979, in The Large-Scale Characteristics of the Galaxy, IAU Symp. 84, ed. W. B. Burton (Reidel, Dordrecht), 257.

RADIO FLARES FROM CIRCINUS X-1 [+]

G. D. Nicolson
National Institute for Telecommunications Research
C S I R
Johannesburg

The galactic x-ray source Circinus X-1 is notable for its periodic
flares at x-ray, optical, infrared and radio wavelengths. The flares
recur at 16.6 day intervals, but there is considerable variability from
one cycle to another. (See Dower et al, 1982, and references therein
for a detailed review of the properties of Circinus X-1).

The radio flares at cm wavelengths have been shown to be consistent
with a synchrotron source undergoing adiabatic expansion (Haynes et al,
1978). Circinus X-1 is therefore a member of a small group of galactic
x-ray sources such as Sco X-1, Cyg X-3 and SS433 which all have marked
similarities to extragalactic radio sources. Radio flares in these
objects probably originate from mass accretion onto neutron stars and
black holes and provide an important starting point for testing models
of variable synchrotron radio sources. In particular the periodic
nature of the flares from Circinus X-1 makes this object especially
amenable to coordinated multiwavelength studies and to VLBI campaigns.

Flux measurements of Circinus X-1 have been made since February 1978
with the 26m telescope at Hartebeesthoek. Observations were made at
the expected flare times for 97 of the 120 cycles of 16.6 days that
have elapsed since then. Twenty-seven flares ranging from 0.3 to 1.65
Jy were detected. The flares generally had several peaks and lasted
between one and three days. There was a distinct tendency for flares to
occur in groups, interspersed with low states when the flaring level
dropped below the typical detection level of 100 mJy. Two such low
states of four and ten months duration were observed. A third low
state began in September 1981 and was still in progress in mid 1983.

For fourteen flares the onset time could be determined with an accuracy
of less than one hour, and the results were used to determine an
ephemeris for the radio flares. This revealed that the period has been
decreasing at a rate of 1×10^{-5}. The rms deviation of flare times from
the computed ephemeris was less than 2 hours and sets a tight constraint
on the timing mechanism for the flares. This result is consistent with
the suggestion that Circinus X-1 is a binary system with a large

+ Discussion on page 457   285

R. Fanti et al. (eds.), VLBI and Compact Radio Sources, 285–286.
© 1984 by the IAU.

eccentricity (Murdin et al, 1980).

The results also suggest the existence of a longer term periodicity. Strong flares of long duration all occurred at intervals of 117 days plus or minus one 16.6 day cycle and at the complementary phase the flares were weaker and of shorter duration.

The current low state has frustrated recent attempts to make detailed VLBI observations. Monitoring is continuing so that further work can be initiated when flaring resumes.

References.

Dower, R.G. et al: 1982, Astrophys. Jnl. 261, pp. 228-250.
Haynes, R.F. et al: 1978, Monthly Notices Roy. Astron. Soc. 185, pp. 661-671.
Murdin, P. et al: 1980, Astron. Astrophys. 87, pp. 292-298.

THERMAL RADIO STARS AND HIGH-RESOLUTION RADIO OBSERVATIONS [+]

Alan E. Wright
Division of Radiophysics, CSIRO, Sydney, Australia

1. INTRODUCTION

The main purpose of this talk is to outline a few of the problems
in our understanding of thermal radio stars that may be resolved by the
large arrays. Several radio telescopes in operation (such as the VLA
and MERLIN arrays) or under construction (such as the Australia Telescope) have both the required resolution and sensitivity.

The basic model explaining the thermal radio stars was first
described by Seaquist and Gregory (1973) and developed by Olnon (1975),
Panagia and Felli (1975) and Wright and Barlow (1975). More recently
(Wright 1981), I have reviewed the model and discussed the various
classes into which the stars can be grouped.

2. THE CLASSICAL MASS-LOSS STARS

The classical mass-loss stars, such as γ Vel, P Cyg and ζ Pup, are
well described by the simple model. Mass loss rates determined from
the total radio fluxes also agree well with those determined by other
methods (see e.g. Morton and Wright, 1978; Abbott et al., 1980;
Bieging et al., 1982). The suggested flux variability of P Cyg and
9 Sgr (Abbott et al., 1981) has been shown to be caused by incorrect VLA
reduction procedures (see White and Becker (1982) for a discussion).
Measurement of the angular diameter and total flux density gives a value
for the wind temperature directly. Furthermore, measurements of the
angular diameter at different frequencies give information about the
temperature structure throughout the wind.

3. THE SYMBIOTIC-SLOW NOVA STARS

Observations of objects in the other main group of thermal radio
star, the symbiotic-slow nova class, do not in general agree as well
with the predictions of the simple model. They have radio spectra which
are moderately - but significantly - steeper than the expected $\alpha = 0.60$.
And many objects show a flattening of the spectrum at frequencies
$\gtrsim 10$ GHz, indicative of an optically thin region. The star H1-36 (see
Fig. 1) shows both effects.

A steeper spectrum indicates either (i) that the wind density decreases faster than r^{-2} or (ii) that the wind temperature decreases outwards. A cooling wind could be produced by expansion cooling becoming
important in its outer regions. But this cannot produce spectral indices steeper than $\alpha \approx 0.65$ (White and Becker, 1982). On the other hand,

+ Discussion on page 457   287

R. Fanti et al. (eds.), VLBI and Compact Radio Sources, 287–288.

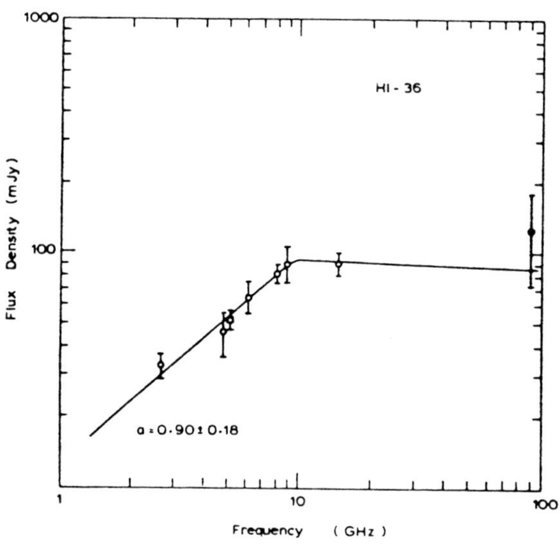

Figure 1.   The radio spectrum of
H1–36.   The line represents a
best fit to the data at the
lower frequencies but has been
forced to have a slope of
$\alpha = -0.1$ at high frequencies.

a steep density profile may be caused by (i) variable stellar mass loss
rates, (ii) acceleration of the wind or (iii) non-spherical symmetry of
the outflowing gas.   There is not space here to discuss these possibilities in detail.

Turning to the question of the optically thin spectrum at high
frequencies: the straightforward interpretation would be that mass loss
has recently stopped in these stars.   However, observations have found
no evidence that either the turnover frequency or the flux is decreasing
with time, as predicted.   On the contrary, the fluxes of H1–36 and V1016
Cyg appear to be *increasing*.   A more plausible explanation for the flat
spectrum is that the stellar wind is neutral (and thus transparent)
close to the star.   Obviously then this same star cannot cause ionization of the outer regions of the wind.   There must be another hot star
in the stellar system.   Just such a model has been proposed by Allen
(1983) for H1–36.   Briefly, a compact star accretes mass from a wind
flowing outwards from a companion M giant.   The resulting ultra-violet
radiation ionizes most of the wind, except for the dense regions close
to the M giant and a "shadow zone".   An angular size estimate would
severely test this model.

References

Abbott, D.C., Bieging, J.H., Churchwell, E.: 1981, *Astrophys. J.* 250, 645.
Abbott, D.C., Bieging, J.H., Churchwell, E., Cassinelli, J.P.: 1980,
     Astrophys. J. 238, 196.
Allen, D.A.: 1983, *Proc. Astron. Soc. Aust.* (in press).
Bieging, J.H., Abbott, D.C., Churchwell, E.: 1982, *Astrophys. J.* 263, 207.
Morton, D.C., Wright, A.E.: 1978, *Mon. Not. R. Astron. Soc.* 182, 47P.
Olnon, F.M.: 1975, *Astron. Astrophys.* 39, 217.
Panagia, N., Felli, M.: 1975, *Astron. Astrophys.* 39, 1.
Seaquist, E.R., Gregory, P.C.: 1973, *Nature Phys. Sci.* 245, 85.
White, R.L., Becker, R.H.: 1982, *Astrophys. J.* 262, 657.
Wright, A.E.: 1981, "Thermal Radio Emission from Stars". Presented at IAU
     Second Asian-Pacific Regional Meeting, Bandung.
Wright, A.E., Barlow, M.J.: 1975, *Mon. Not. R. Astron. Soc.* 170, 41.

THE EVOLUTION OF THE COMPACT RADIO STRUCTURE IN SS433 OVER A 16-DAY
PERIOD [+]

R.T. Schilizzi[1], J.D. Romney[2], and R.E. Spencer[3]
[1]Netherlands Foundation for Radio Astronomy, Dwingeloo, NL
[2]Max-Planck-Institut für Radioastronomie, Bonn, FRG
[3]Nuffield Radio Astronomy Laboratories, Jodrell Bank, UK

INTRODUCTION

     SS433 has been under intensive study for the past five years in
almost all wavelength bands of the electromagnetic spectrum. This
peculiar object is generally regarded (Beer 1981) as being a binary
system composed of a main sequence star losing mass via Roche lobe
overflow to a massive accretion disk associated with a compact object,
probably a neutron star. The binary period is 13.1 days. Supercritical
accretion onto the disk causes about 10^{-6} $M_\odot$/year of ionised matter to
be ejected in the form of jets with a relatively constant velocity of
0.26 c along the disk axis. The disk (or the inner part of it) precesses
with a period of about 164 days, although there is evidence that this
may not be constant. The half angle of the precession cone is ~20° and
its axis lies at an angle of ~80° to the line of sight. The main
sequence star loses mass at a rate of 10^{-4} to 10^{-6} $M_\odot$/yr into a stellar
wind with the result that a relatively dense environment surrounds the
binary system.
     The jets have been traced by radio means from distances from the
centre of the binary system of 1.5×10^{14} cm (5 milliarcsec) out to
~1.5×10^{17} (3 arcsec) and by X-ray imaging out to ~10^{20} cm (30 arcmin).
There is a suggestion that the jets reach the shell structure of W50
which lies 1 degree away from SS433 (Geldzahler et al 1980; Downes et al
1981). (A distance of 5 kpc to SS433 has been assumed). Analysis of data
for simultaneous radio and X-ray flares in SS433 by Seaquist et al
(1981), as well as long term monitoring of the radio flux density by
Johnston et al (1981) and Geldzahler et al (these proceedings) suggests
that radio flare generation occurs at distances of 10^{14-15} cm from the
central binary, perhaps where the beams or jets encounter inhomogeneities in the stellar wind. This evidence thus suggests that
radio emission is detectable only at distances two orders of magnitude
larger than the dimensions of the binary system and implies that, in the
radio, we are looking at the smoke from SS433 rather than the fire
itself.
     VLA observations by Hjellming and Johnston (1981a, 1981b) have been
crucial in resolving some of the angle ambiguities in the kinematic
model of Abell and Margon (1979) which was introduced to explain the
movement in the optical emission lines as function of epoch. These radio

+ Discussion on page 458   289

R. Fanti et al. (eds.), VLBI and Compact Radio Sources, 289–296.
© *1984 by the IAU.*

observations also demonstrated that the moving material was, in fact, being expelled with a proper motion of ~9 milli-arcsec/day. More detailed evidence that the radio emitting blobs lie along trajectories predicted by the kinematic model has come from VLBI and MERLIN work (Schilizzi et al 1981, 1982, 1983; Romney et al 1983, Niell et al 1981; Spencer 1979), but a case of non agreement has also come to light (Spencer and Waggett, these proceedings).

SS433 has been of considerable interest to astronomers working in active galactic nuclei because of the similarity of its gross properties to those of extragalactic radio sources (Hjellming and Johnston 1982) - it has an active "nucleus", variable flux density, knotty jets, relativistic motion and polarised extended emission. The fact that at VLBI resolutions we can observe individual blobs of emission affords us the possibility of studying the detailed evolution of these blobs as a function of epoch. This is the subject of this contribution.

OBSERVATIONS

Six epochs of observations at three day intervals were scheduled at 4990 MHz in December 1982 using networks of up to 5 telescopes. Due to combinations of inclement weather, equipment malfunction, short coherence times arising from the use of rubidium oscillators at Westerbork and Jodrell Bank, and a not overly strong source for the Mk II recording system, fringes were not obtained on all baselines. Table 1 summarizes the observations.

Table 1

Date	JD- 2445000	Telescopes*	Baselines with fringes detected
6 Dec 82	5310.1	O,E,W,J,G	OE,OW,EW,EJ,WJ
9	5313.1	O,E,W,J,G	OE,OW,EW,EJ
12	5316.1	O,E,W,J,G	OE,OW,EW,EJ
15	5319.1	E,W,N	EW
18	5322.1	E,W,J	EW
21	5325.1	E,W,J	EW,EJ,WJ

* O: Onsala, E: Effelsberg, W: Westerbork, J: Jodrell Bank, G: Green Bank, N: NRL, Maryland Point.

RESULTS AND DISCUSSION

Figure 1 displays the visibility curves for the most sensitive, and shortest, baseline E-W (fringe spacing ~50 mas) for the six epochs. (The full data will be published elsewhere.) It is clear that the first three epochs (6, 9, and 12 December) form a group, and the last three (15, 18 and 21 December) another group, there being a significant injection of energy into the source between 12 and 15 December.

The first group of observations is characterised by a decreasing fringe contrast around 1800 GST indicative of expanding structure. The

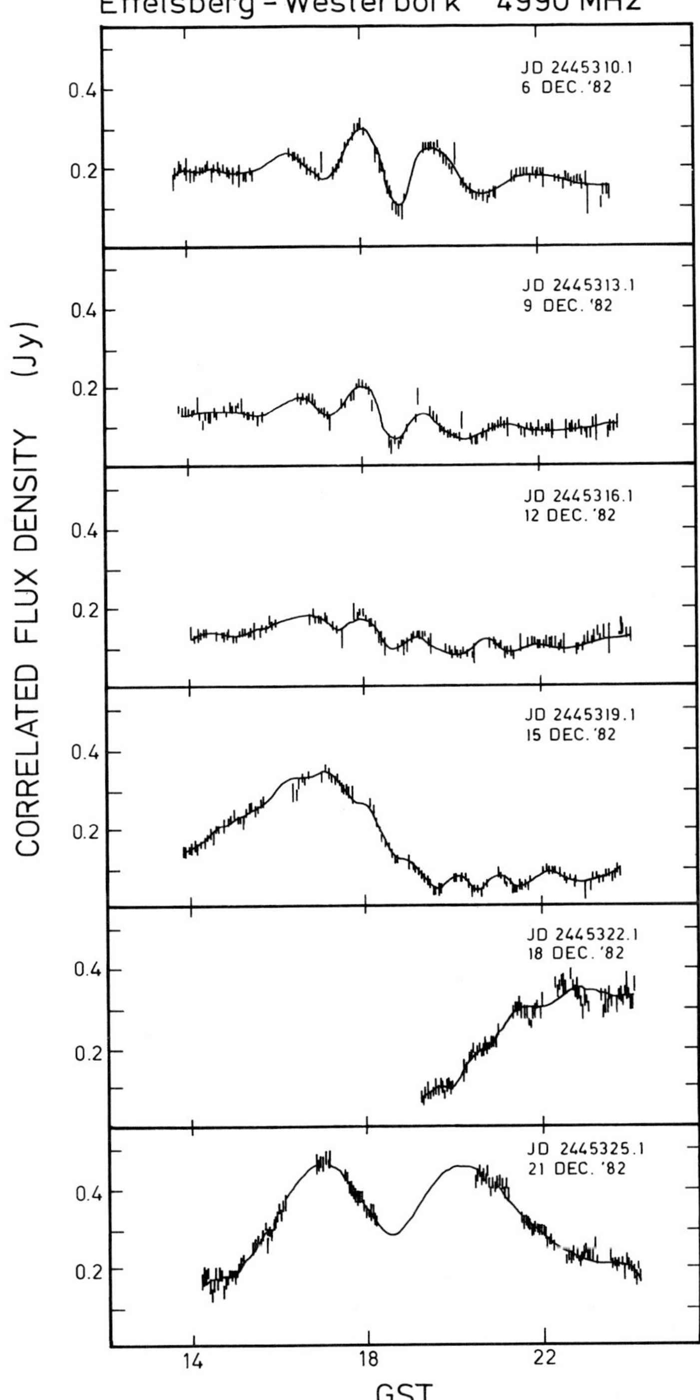

Figure 1 : Correlated flux densities for SS433 at 4990 MHz on the Effelsberg–Westerbork baseline for the 6 epochs of measurement. The curves drawn through the data represent models for the radio structures which have been derived from the whole dataset at each epoch. (see Figure 3 and text).

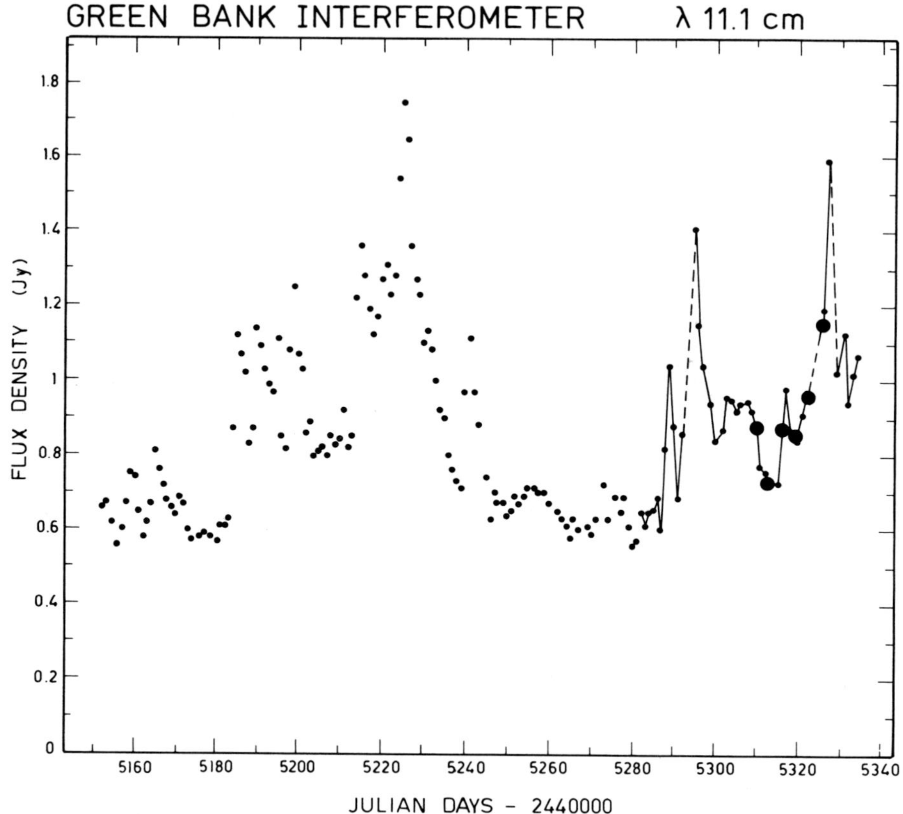

Figure 2: Total flux density for SS433 at λ11.1 cm (Johnston et al, private communication). The large dots indicate when VLBI measurements were made (6,9,12,15,18,21 December 1982). Dashed lines join points spaced at more than 1 day.

correlated amplitudes on the longer baselines E,W to O,J (fringe spacings ~10 milli-arcsec) are consistent with a "core" flux density of ~150 mJy on 6 December decreasing to ~100 mJy on 9 and 12 December. This can also be seen in a similar decrease in the peak correlated flux density on the E-W baseline. The total flux density at 11.1 cm (Johnston et al, private communication, figure 2) decreased from 6 to 9 December and then increased on 12 December to a peak on 13 December. In the period $1400 \leq GST \leq 1600$ on both 6 and 9 December there is some evidence for small amplitude oscillations in the data which are not reproduced by the model. If real, these may be related to the oscillations seen in the data for $1900 \leq GST \leq 2400$ on 15 and 18 December which can be modelled as an outlying component.

The models derived for the first three epochs (Figure 3) show one-sided emission typical of nuclear structure in extragalactic sources - a compact, relatively bright component (core) with an elongated, lower brightness component (jet) pointing towards an even lower brightness diffuse component. The maps in Figure 3 have been scaled so that the top

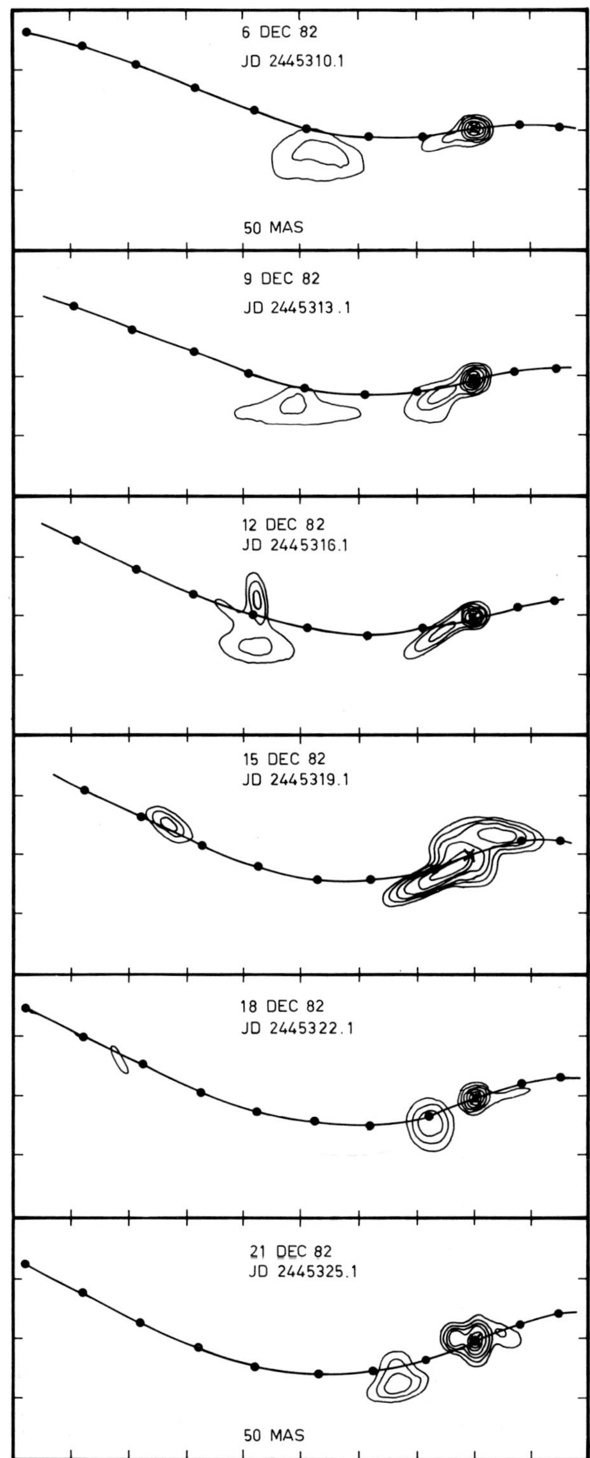

Figure 3: Contour maps of models for SS433 at the 6 epochs of measurement. Contour levels are 2, 5, 10, 20, 50 and 80% of the peak intensity. North is up, east is the left; the beam is 10x10 milli-arcsec. The curves drawn through the structures represent the expected trajectories of blobs on the kinematic model of Abell and Margon (1979) with parameters from Margon et al (1980); the dots are spaced by 6 days. A cross marks the assumed centre of the structure.

contour is 80% of the peak flux density in each map; this masks the change in flux density in individual components from epoch to epoch. However figure 4 displays the evolution of the flux densities of the model components. The jet appears to remain constant in length from December 6 to 9, but the diffuse component appears to move outwards. The proper motion derived is of order 10 milli-arcsec/day, in agreement with the results of Hjellming and Johnston (1981b). Both the diffuse component (blob 2) and the jet appear to lie somewhat ahead of the trajectory predicted from the kinematic model with parameters according to Margon et al (1980). However, when one considers the interferometer resolution the agreement is quite satisfactory. Moreover, absolute registration of the model trajectories on the maps is not possible, so that plausible assumptions, such as placing the centre of the trajectory on the most compact peak, have had to be made. From the proper motion, the most likely date of origin of blob 2 is JD2445295, at a time when a relatively steep spectrum (soft) flare was in evidence (see Figure 2).

Remarkable changes occur between the third and fourth epochs of observation. Figure 1 shows that the correlated flux density increased between December 12 and 15 over a wide range of hour angles (1400 $\leq$ GST $\leq$ 1900), peaking at GST ~ 1700 at a value ~150 mJy higher than the peak value three days earlier. Moreover, the total flux density at 11.1 cm increased by ~250 mJy between 9 and 13 December before falling back by ~100 mJy by 15 December. This minor flare is likely to be the source of the flux increase seen in the VLBI data at the fourth epoch; however it is not understood why the flare had no effect on the VLBI data at the third epoch. A peak at a particular hour angle suggests flux enhancement in an elongated structure rather than in a compact core, and the broad nature of the peak suggests a range of position angles in the elongated structure. A brightening in the core alone would have manifested itself in a general increase in flux density at all hour angles. Flux density measurements at Westerbork rule out the possibility of a short period flare between 1400 and 1900 GST.

The model in Figure 3 depicts a jet dominating the structure, with a broad component in the reverse direction. The length of the jet is -30 milli-arcsec, so that it appears that in the 3 days between epochs 3 and 4, the radiating electrons in the flare have filled the pre-existing channel, like air passing down a blow-pipe, at -9 milli-arcsec/day.

The short period oscillations in the amplitudes are modelled by an outlying component (blob 1) which was probably generated during a flare on day JD2445289 (Figure 2). This was a relatively hard flare; it may be that hard radio flares produce components which are more compact and long lived than soft flares. As noted earlier, there is some evidence for an outlying component in the data for the first two epochs.

There is disappointingly little data for the fifth epoch, but what there is, suggests that the radio structure bifurcated between epochs 4 and 5. The model (Figure 3) shows that a bubble (blob 3) has formed at the end of where the jet was three days earlier; the jet itself has apparently faded, but the data are too sparse to be certain. The oscillations in the amplitudes can again be fitted by including the outlying component, blob 1, in the model.

By the last epoch the correlated amplitudes in the range 2000 $\leq$ GST

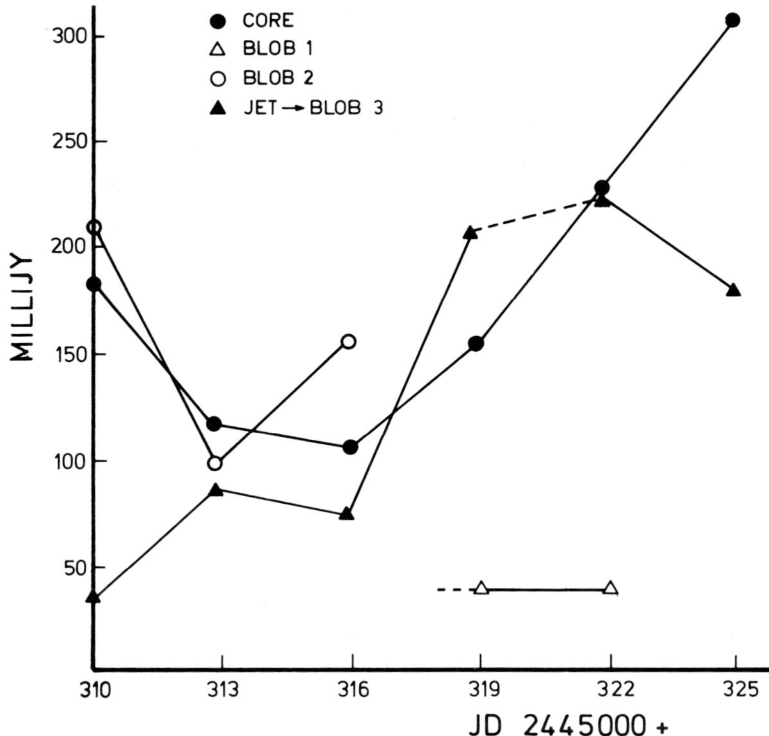

Figure 4: Model component flux densities as a function of epoch.

$\leq$ 2400 have again changed dramatically, and in addition the overall flux
density has increased by ~100 mJy consistent with an increase in the
core flux as the peak of the JD2445327 flare approaches (Figures 2 and
4). Blob 3 has moved outwards along the trajectory by ~30 milli-arcsec,
again consistent with the generally accepted proper motion. A jet to the
east has appeared in p.a. ~85° which does not lie on the trajectory.
However the position angle is influenced by noisy data on the long
baselines E-J, W-J and may not be correct. What is clear is that the
strong elongated component to the west at epoch 4 has faded to a mere
shadow of its former self by epochs 5 and 6. Note that this one-sided
emission is in the same sense as was found at the first three epochs.
There is some evidence for the outlying component, blob 1, in short
period oscillations in the visibilities, but these are not of sufficient
amplitude to require blob 1 in the model.

Figure 4 depicts the variation in model flux density for the major
compact radio components in SS433. The core decreases in a regular
manner for the first three epochs and then increases in accord with the
increasing total flux density going into the flare of JD2445327 (Figure
2). The eastward "jet" and "blob 2" vary up and down in flux density at
the first 3 epochs, the magnitude of the variation being a measure of
the uncertainty in the value for these rather low brightness extended
components. The jet brightens by ~150 mJy between epochs 3 and 4 and
remains at the same level after the bubble (blob 3) has been detached

from the blow-pipe (i.e. jet) at epoch 5, then decreases by ~50 mJy by
the final epoch.

CONCLUSIONS

These can be summarized as follows:
1) the proper motion of individual blobs is clearly seen,
2) the production of blobs is related to flare activity,
3) the evolution of blobs can be very different from one side of the
   core to the other,
4) the observations suggest that 'hard' radio flares produce blobs which
   are more compact and long-lived than 'soft'flares,
5) there is evidence of a "blow-pipe effect" in which the energy of a
   flare is pumped into an elongated cavity (the pipe) via particles
   moving at 0.26 c until a self contained radio emitting blob (the
   bubble) is formed and detaches itself to move ballistically away at
   0.26 c.

ACKNOWLEDGEMENTS

The authors wish to thank the staffs of the EVN, Green Bank and
Maryland Point telescopes for observing assistance; the staffs of the
MPIfR, NRAO and CIT VLBI processors for carrying out the processing;
Drs. T.J. Pearson, S.C. Unwin and D.L. Jones for assistance at CIT; Dr.
J.M. Benson for data reduction at NRAO; Dr. K.J. Johnston for
communicating total flux density data ahead of publication; and Dr. D.L.
Jauncey for stimulating discussions.

REFERENCES

Abell, G.O., Margon, B. (1979) Nature 279, 701.
Beer, P. (1981) (ed.) Vistas Astron. 25.
Downes, A.J.B., Pauls, T., Salter, C.J., Astr.Astrophys. 103, 227.
Geldzahler, B.J., Pauls, T., Salter, C.J. (1980) Astr.Astrophys. 84,
     237.
Hjellming, R.M., Johnston, K.J. (1981a) Nature 290, 100.
Hjellming, R.M., Johnston, K.J. (1981b) Ap.J. 246, L141.
Hjellming, R.M., Johnston, K.J. (1982) Proc.IAU Symp. 97, p.197.
Johnston, K.J. et al (1981) Astron.J. 86, 1377.
Margon, B., Grandi, S.A., Downes, R.A. (1980) Ap.J. 241, 306.
Niell, A.E., Lockhart, T.G., Preston, R.A. (1981) Ap.J. 250, 248.
Romney, J.D., Schilizzi, R.T., Fejes, I., Spencer, R.E. (1983) in prep.
Schilizzi, R.T., Miley, G.K., Romney, J.D., Spencer, R.E. (1981) Nature
     290, 318.
Schilizzi, R.T., Fejes, I., Romney, J.D., Miley, G.K., Spencer, R.E.,
     Johnston, K.J. (1982) Proc.IAU Symp.97, p.205.
Schilizzi, R.T., Romney, J.D., Spencer, R.E., Fejes, I. (1983) Proc.
Workshop on Astrophys. Jets (Torino) p. 157.
Seaquist, E.R., Gilmore, W.S., Johnston, K.J., Grindlay, J.E. (1982)
     Ap.J. 260, 220.
Spencer, R.E. (1979) Nature 282, 483.

ANOMALOUS EJECTION IN SS433 +

R.E. Spencer and P. Waggett
Nuffield Radio Astronomy Laboratories, Jodrell Bank
Macclesfield, Cheshire

A series of MERLIN observations at $\lambda 6$ cm of the peculiar object SS433 were made during the spring/summer of 1982. The maps obtained by hybrid mapping show that an elongated structure was formed, extending to ∿1 arcsec in length before evolving into two symmetrically placed knots of radio emission straddling a central unresolved core. The structures can be compared with the loci of radio emission expected on the twin jet model for SS433. The knots C and D clearly visible on the map of 820514 (= JD2445104, Figure 1) fit the locus well and were ejected on JD2445085 ±2 at the same time as a strong outburst in the X band total power (Johnston et al. 1983). Knots A and B were ejected earlier but do not lie on the expected locus, even if allowance for the nodding motions proposed by Katz et al. 1982 is made.

We can try and explain this anomalous ejection as follows:-

(i)   Ejection in an unusual direction,
(ii) Ejection with a high velocity - a locus with velocity 0.4c instead of the normal 0.26c will fit the observed structure in figure 1.
(iii) Bending subsequent to initial ejection. This requires a pressure gradient $\dfrac{dP}{dz} = \dfrac{u(\gamma\beta)^2}{R}$ where u is the energy density of the knot, γ the Lorentz factor, $\beta = v/c$ and R the radius of curvature. If this is caused by static gas pressure then a pressure gradient (assuming equipartition in the knot) such that $\Delta(nT) > 10^8$ K cm^{-3} over the distance from the observed to the expected position is needed. This can easily be supplied by the X-ray emitting gas close to SS433.
(iv) Acceleration of the knots after initial ejection.

Margon (private communication) finds that there was no unusual optical behaviour during our observations. If, as seems likely, that the early stage of ejection/radio knots is closely associated with the moving optical clouds then (ii) can be ruled out, and (i) can only occur if the anomalous ejection is in the plane of the sky.

Explanation (iii) requires similar conditions on both sides of the

+ Discussion on page 458   297

R. Fanti et al. (eds.), VLBI and Compact Radio Sources, 297–298.
© *1984 by the IAU.*

object (knots A and B) and also any pressure gradient would have to be quickly reduced so that the following knots C and D remain unaffected, infact in a time less than the minimum sound crossing time of the gas.

The standard ephemeris gives the observed position angle some 120 days before elongation was first observed on 820420 and so would require ejection with a very low initial velocity and long radio lifetime if acceleration is to explain the results. In addition, observations with the EVN network on 9 Dec 1981 at λ6 cm also show an anomalous position angle, (Romney et al. 1983) outside the maximum range allowed by the precessing jet model.

It seems to us therefore that the most likely explanation is that the knots were ejected with the wrong position angle close to the plane of the sky, though a more detailed investigation of the relationship between the optical and radio emission is required.

REFERENCES

Johnston, K.J., Geldzahler, B.J., et al. 1983. Ap.J. Submitted.
Katz, J.I., Anderson, S.F., Margon, B. and Grandi, S.A., 1982.
     Ap.J. 260, 780.
Romney, J.D., Schilizzi, R.T., Fejes, I. and Spencer, R.E., 1983.
     in preparation.

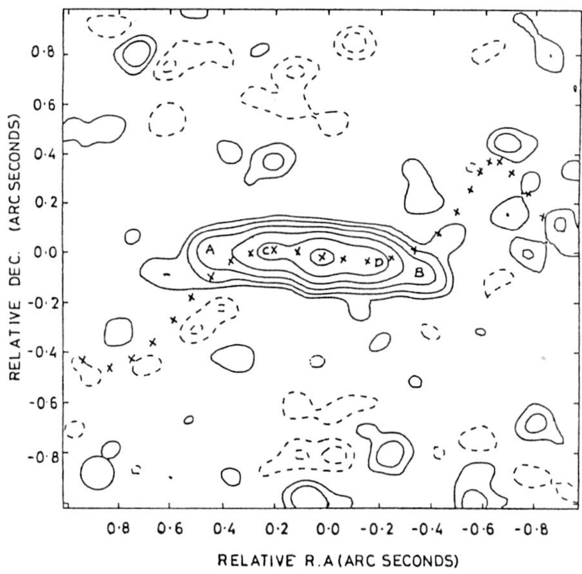

Figure 1 SS433 at λ6 cm on 820514, the beam is shown in the LH corner. Contours are at 4, 8, 16 etc. mJy/beam.

JETS IN MOLECULAR CLOUDS [+]

Arieh Königl
Institute for Advanced Study
Princeton, New Jersey 08540, U.S.A.

There is now growing evidence that the cosmic jet phenomenon manifests itself in a remarkable way in regions of active star formation embedded in dense molecular clouds. The first indications for oppositely directed, supersonic outflows from young stars were provided by molecular line observations (most notably of CO) which detected spatially separated regions of redshifted and blueshifted emission in association with embedded infrared sources. About twenty sources of this kind have been identified so far, and more are continuously being discovered; they typically have radii $\sim 10^{18}$ cm, velocities $\sim 10\text{-}50$ km s^{-1}, dynamical ages $\sim 10^4$ yr, and energies $\sim 10^{46}\text{-}10^{47}$ erg s^{-1} (see Bally and Lada 1983 for a review). Statistical arguments indicate that energetic outflows of this type are probably a common feature in stellar evolution, and that they occur in both massive and low-mass stars. Direct evidence that the outflows in many cases are highly collimated was subsequently provided by the detection of high-velocity Herbig-Haro objects (optical emission clumps with typical masses $\sim 10^{-5} M_\odot$) along the axes of the bipolar CO lobes. Proper-motion measurements are now available for a number of these objects (e.g., Herbig and Jones 1981), and they invariably reveal that the velocity vectors (of typical magnitudes 200-400 km s^{-1}) point away from the central star. The clumps are often found to consist of many sub-condensations which move independently with disparate speeds, but which nevertheless travel in the same general direction with an angular spread $\lesssim 10°$. Finally, radio continuum observations (e.g., Cohen et al. 1982) and deep CCD images (e.g., Mundt and Fried 1983) have shown that the collimation of the outflows is already well established on scales of $\lesssim 10^{15}$ cm.

The morphology of the bipolar emission sources detected in molecular clouds bears a striking resemblance to that of double radio lobes and jets observed on vastly larger scales in extragalactic sources. In both cases, collimated outflows are detected over several decades of scale on opposite sides of a compact source. In each case, the outflows terminate in extended emission regions which are formed when the jets ram against the ambient medium. And, in both cases, distinct emission knots are found along the jets which are most likely associated with shock waves.

[+] <u>Discussion on page 459</u>

299

R. Fanti et al. (eds.), VLBI and Compact Radio Sources, 299–301.
© *1984 by the IAU.*

The analogy with extragalactic sources has proven useful in formulating
a unified interpretation of molecular-cloud outflows (Königl 1982). In
turn, the study of these relatively nearby outflows may provide valuable
information on the jet mechanism which is not available from observations
of extragalactic objects. For example, the molecular sources are the
first instance of jets where the motion of the head of the jet and the
transverse expansion of the shocked jet material can actually be measured. The kinematic and spectroscopic data on Herbig-Haro objects likewise provide a unique input for general models of emission knots in jets
(cf. Blandford and Königl 1979). In fact, recent detections of shocked
optical line emission from nonthermal continuum knots in a number of radio jets (e.g., Brodie et al. 1983) support the hypothesis that the knots
in stellar and in galactic jets have similar interpretations. Finally,
the ability to measure the density and temperature structure of interstellar clouds by means of various atomic and molecular tracers again
could prove useful for a general study of the interaction between jets
and their environment.

   Why do jets form in protostellar environments? One likely explanation is that they are produced by a combination of the same two ingredients that were originally invoked by Blandford and Rees (1974) to account
for the production of extragalactic jets - namely, an isotropic stellar
wind and a flattened mass distribution which confines it. Mass outflows
are common in pre-main-sequence stars, and direct spectroscopic evidence
for stellar winds has been reported already for several of the central
stars in bipolar sources. Evidence is also rapidly accumulating from
continuum infrared and from molecular line measurements for the presence
of clumpy, high-density "tori" which could constrain the transverse expansion of the jets (e.g., Schwartz et al. 1983). One can argue on theoretical grounds (Königl 1982) that a stellar wind which propagates into
an anisotropic density distribution could become unstable to the formation of transonic nozzles if the ambient density scaled with radius roughly
as $\rho \propto r^{-2}$. Density distributions of this form are expected in isothermal
clouds, and have in fact been inferred to exist in the vicinity of bipolar
sources by molecular line observations as well as on statistical grounds
(see Bally and Lada 1983). The anisotropy in the density distribution
could have been induced by rotation, but an alternative interpretation
is suggested by the correlation found in some of the sources between the
bipolar axis and the direction of the ambient magnetic field as deduced
from polarization measurements. In this picture, which could perhaps
be relevant also in extragalacitc contexts, the density gradient is induced by magnetic stresses which inhibit cloud contraction normal to the
field lines.

   Radio continuum observations with $\lesssim 1''$ resolution are particularly
valuable for the study of molecular-cloud jets. Such observations have
already yielded detailed maps of jets on scales $\lesssim 10^{15}$ cm that are surprisingly similar (e.g., in revealing curvature effects) to VLBI maps
of extragalacitc jets. A number of sources are unresolved on these
scales, but their optically thick thermal spectra indicate that they may
be produced in massive outflows which become neutral already on scales

$\sim10^{14}$ cm (e.g., Simon et al. 1983). These observations may also help clarify a number of still outstanding problems, such as the discrepancy that often appears between the momentum discharge rate inferred from direct measurements of the central outflow and the substantially higher rate required to drive the expansion of the outer emission lobes.

REFERENCES

Bally, J., and Lada, C.J.: 1983, Astrophys. J. 265, pp. 824-847.
Blandford, R.D., and Königl, A.: 1979, Astrophys. Letters 20, pp. 15-21.
Blandford, R.D., and Rees, M.J.: 1974, Monthly Notices 169, pp. 395-415.
Brodie, J., Königl, A., and Bowyer, S.: 1983, Astrophys.J. 273 (October 1 issue).
Cohen, M., Bieging, J.H., and Schwartz, P.R.: 1982, Astrophys.J. 253, pp. 707-715.
Herbig, G.H., and Jones, B.F.: 1981, Astron. J. 86, pp. 1232-1244.
Königl, A.: 1982, Astrophys. J. 261, pp. 115-134.
Mundt, R., and Fried, J.W.: 1983, Astrophys. J., submitted
Shwartz, P.R., Waak, J.A., and Smith, H.A.: 1983, Astrophys. J. 267 pp. L109-L114.
Simon, M., et al.: 1983, Astrophys. J. 266, pp. 623-645.

INTERSTELLAR SCATTERING [+]

James M. Cordes
Cornell University

INTRODUCTION

Fine scale electron density fluctuations in the interstellar medium
(ISM) are manifest as scintillations and temporal broadening of pulsar
signals and as angular broadening of galactic and extragalactic sources.
Although scattering off the fluctuations is often a nuisance for
conventional studies of radio sources, analysis or searches for
interstellar scintillations (ISS) can lead to information about the ISM
or radio sources that otherwise would not be obtainable. For example,
the length scale probed by ISS in the ISM is typically 10^{11} cm and the
characteristic angular size for quenching ISS is less than 1 μarc sec.

In this paper we discuss the distribution of scattering material in
the Galaxy as probed by ISS observations of a large sample of pulsars.
The parameters of a two-component model for scattering material are
given and the typical scattering angle is predicted as a function of
galactic latitude, path length through the Galaxy, and frequency. The
predictions indicate that pulsars sample most, if not all, scattering
material in or near the Galaxy because the observed scattering of
extragalactic objects conforms to the predictions. The results imply the
absence of significant scattering material in intergalactic space. We
also discuss how ISS can be used to determine space velocities of a large
sample of pulsars, corresponding to proper motions of $\sim$mas/yr.

PULSAR SCATTERING AND SCINTILLATIONS

Temporal broadening of pulsar signals appears as an asymmetric tail
on pulse shapes whose time constant τ is related to the scattering
angle θ (FWHM) and the pulsar distance L by

$$\tau \sim L\theta^2 / 8c. \tag{1}$$

Pulsar scintillations appear as intensity variations with characteristic
frequency and time scales, Δf and Δt. The so-called decorrelation

+ Discussion on page 460   303

R. Fanti et al. (eds.), VLBI and Compact Radio Sources, 303–307.
© 1984 by the IAU.

bandwidth Δf and the scattering time τ are related by

$$2\pi\Delta f\tau = 1 \tag{2}$$

because both depend on path-length differences. Pulsar ISS is consistent with δn_e having a power-law wavenumber spectrum of the form

$$P_{\delta n_e}(q) = Qq^{-\alpha}, \quad \alpha \sim 3.7 \pm 0.3 \tag{3}$$

(Rickett 1977; Armstrong and Rickett 1981; Armstrong, Cordes, and Rickett 1981; Cordes, Weisberg, and Boriakoff 1983). The exponent α is best constrained by the scaling of scintillation bandwidth with observation frequency. If α is assumed to be equal to 11/3, then the coefficient Q can be estimated as

$$Q = L^{-11/6}f^{11/3}\Delta f^{-5/6} \tag{4}$$

with units of $(\text{meters})^{-20/3}$; Q is the same quantity as C_n^2 referred to by Rickett. Rather than showing a constant value, within errors, measurements show large systematic variations of Q with galactic latitude, longitude, and distance as well as random variations from object to object. Errors in the pulsar distance scale can account for only a factor of five range in Q whereas the actual range is four orders of magnitude.

GALACTIC DISTRIBUTION OF SCATTERING MATERIAL

Pulsars with $|b| > 10^{\circ}$ show values of Q in the range $10^{-3.5\pm0.5}$ while objects with $|b| < 5^{\circ}$ cover the range $10^{-3.5}$ to 1. As a function of path length, Q begins to systematically increase for $L > 2$ kpc for pulsars with $|b| < 5^{\circ}$ in the first quadrant of the Galaxy. Cordes, Weisberg, and Boriakoff (1983) have used these results to deduce that scattering material exists in two components: Component A with scale height $H_A > 0.5$ kpc, $Q \sim 10^{-3.5}$, and volume filling factor ~ 1; Component B wherein Q is large (1-100) only in clumplike regions whose filling factor is very small, $\sim 10^{-4}$, and which have small scale height, $H_B \sim 0.1$ kpc. The pulsar measurements suggest that the component B clumps are more common inside the solar circle, as is consistent with the angular broadening measurements of extragalactic sources by Dennison(1983)

SCATTERING ANGLES

The scattering angle for an extragalactic source is

$$\theta \text{ (FWHM)} = 0.133 \, \{L_{kpc}Q\}_{eff}^{3/5} \, f_{GHz}^{-11/5} \quad \text{arc sec} \tag{5}$$

where the bracketed quantity is a path-length integrated value. Because of the evident inhomogeneity of scattering material, θ varies at least as fast as $\theta \propto L^2$ (for $b \sim 0^{\circ}$ and $-90^{\circ} < \ell < 90^{\circ}$) rather than varying as $L^{\frac{1}{2}}$.

We estimate roughly

$$\theta_{FWHM} \; f^{11/5} \sim \begin{cases} 0.53 \;\; \text{arc sec} & |b| < 0\overset{o}{.}6 \\ 0.035|\sin b|^{-3/5} \;\; \text{arc sec} & 0\overset{o}{.}6 < |b| < 3^o - 5^o \\ (0.66 - 1.1)|\sin b|^{-3/5} \;\; \text{mas} & |b| > 3^o - 5^o \end{cases} \qquad (6)$$

for paths in the first quadrant around $\ell \sim 45^o$ where most of the pulsar measurements have been made; in this expression we have assumed a maximum path length through scattering material of 10 kpc. The scattering angle is deterministic at high latitudes but is statistical at low latitudes, depending on the probability of encountering a region of large Q. The dropoff of probability with b associated with the scale height of the "B" component causes the apparent discontinuity in θ at $|b| \sim 3^o$. Figure 1 shows the scattering angle versus b along with angular diameters of extragalactic sources that show λ^2 dependence. Figure 2 shows the frequency dependence of predicted and measured angles. The hatched region is bounded by inferred scattering angles of two pulsars, PSR1541+09 with b = 46° and 1 kpc distance above the galactic plane and PSR1641-45 with b = -0$\overset{o}{.}$1 and 4.9 kpc total distance. Also shown are angular sizes for synchrotron-Compton sources with designated maximum brightness temperatures.

The conclusions to be made are that 1) although the galactic center source Sgr A and various maser sources (Sgr B2, W49) show much larger sizes than the most scattered pulsar, the greater distances of these sources suggest that their scattering diameters are in accord with an extrapolation from the pulsar results; 2) The minimum observable angular size for b = 90° is $\sim$1 mas at 1 GHz; and 3) The 0.07 and 0.1 GHz VLBI observations of Resch (1974) are compatible with there being little or no scattering outside the Galaxy along lines of sight to distant QSO's.

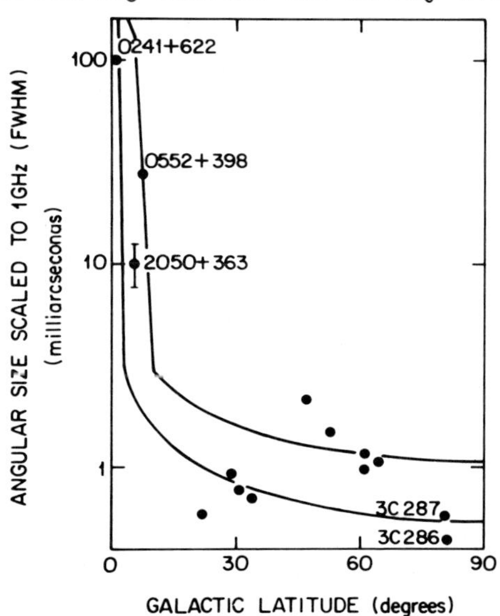

Figure 1   Scattering angles versus galactic latitude. The lines bracket the predicted range of angles. Note that the sources 0241, 0552, and 2050 are at galactic longitudes of 138°, 172°, and 79°. References are given in Cordes and Simonetti (in preparation).

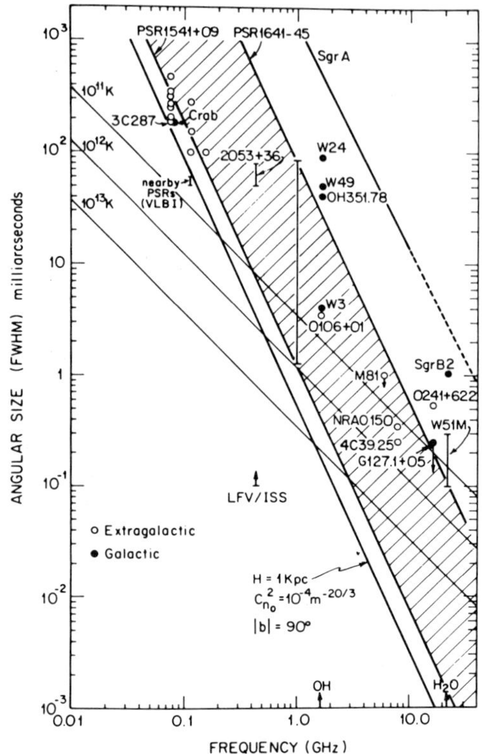

Figure 2. Scattering angles versus frequency. The diagonal line labeled with a scale height H is the predicted minimum angular size of the scattering disk when looking toward b = 90°. This line is uncertain because the scale height is not well constrained from pulsar observations, but the uncertainty is probably no more than a factor of two.

ISS AND PULSAR PROPER MOTIONS

Scintillations have a characteristic time scale Δt related to the speed and spatial scale $(\propto (L\Delta f)^{1/2})$ of a diffraction pattern across a telescope. Lyne and Smith (1982) have shown that the velocity estimate

$$V_{ISS} \propto (L\Delta f)^{1/2} / \Delta t \tag{7}$$

is 66% correlated with the velocity derived from interferometric proper motion measurements for 20 objects. Turning this around, one can make single epoch ISS observations of a large number of pulsars and determine their transverse speeds to an accuracy limited by the projected Earth velocity, differential galactic rotation, the unknown distribution of scattering material along the line of sight, and by the uncertainty in the pulsar distance scale. Improvements can be made by 1) observing binary pulsars whose known orbital motion will calibrate the proportionality in (7); 2) making multiple epoch estimates of Δt to remove the effects of the Earth's motion and thereby determine the vector pulsar velocity; 3) comparing $V_{ISS} \propto L^{1/2}$ with $V_{PM} \propto L$ from interferometer measurements to check the pulsar distance scale. The achievable accuracy is probably ±10 km/sec.

SUGGESTIONS FOR FURTHER OBSERVATIONS

The previous discussion leads to the following suggestions for new observations: 1) VLBI on distant pulsars to compare measured angular sizes with those predicted from scintillation measurements; disagreement is expected if dominant scattering material is near the pulsar or near the observer; 2) VLBI mapping of a highly scattered maser or pulsar to see if the visibility function is cylindrically symmetric for comparison with Sgr A; asymmetry can occur in media with a preferred direction defined by (e.g.) a velocity flow; 3) interferometric proper motions on more pulsars to calibrate the ISS velocity technique; 4) an undirected search for ISS; any sources found would necessarily have angular structure on the micro-arcsecond level and brightness temperatures well in excess of 10^{15} °K.

This research was supported by the Cal Tech President's Fund and by the National Astronomy and Ionosphere Center which runs Arecibo Observatory under contract with the National Science Foundation.

REFERENCES

Armstrong, J.W. and Rickett: 1981, M.N.R.A.S., 194, pp. 623-638.
Armstrong, J.W., Cordes, J.M., and Rickett, B.J.: 1981, Nature, 291, pp. 561-564.
Cordes, J.M., Weisberg, J.M., and Boriakoff, V.: 1983, Ap.J., submitted.
Dennison, B.: these proceedings.
Lyne, A.G. and Smith, F.G.: 1982, Nature, 298, pp. 825-827.
Resch, G.: 1974, Ph.D. Thesis, Florida State University.
Rickett, B.J.: 1977, Ann. Rev. Ast. Ap., 15, pp. 479-504.

INTERSTELLAR BROADENING OF COMPACT LOW GALACTIC LATITUDE RADIO SOURCES

Brian Dennison, M. Thomas, J. J. Broderick
Physics Dept., Virginia Polytechnic Institute and State University
R. S. Booth
Onsala Space Observatory, Chalmers University of Technology
Robert L. Brown, and J. J. Condon
National Radio Astronomy Observatory

1. INTRODUCTION

Scattering of radio waves off inhomogeneities in electron density
in the interstellar medium can produce an apparent broadening in the
angular diameter of an intrinsically compact background radio source.
The magnitude and distribution of this effect at low galactic latitudes
($|b|<5^\circ$) is not well known, although several cases suggest substantial
broadening in certain directions, such as the Cygnus X region (Anderson
et al. 1972), and the galactic center (Davies, Walsh, and Booth 1976).
Large scattering in the plane is consistent with the scintillation
properties of pulsars seen through substantial thicknesses ($\gtrsim 1$ kpc) of
the galactic disk.

2. SURVEY RESULTS

To further study interstellar broadening and its galactic longitude
dependence, we recently surveyed 30 low-latitude compact extragalactic
sources using the Bonn 100-m radiotelescope and the Jodrell Bank Mk IA
radiotelescope as a VLB interferometer at 408 MHz. Most of the sources
were taken from the Clark and Crawford (1974) survey of small-diameter,
low-latitude sources. In preliminary NRAO Green Bank interferometer
observations, we identified for special emphasis those Clark and Crawford
sources which are unresolved on baselines of $\sim 3 \times 10^5$ λ and $\sim 10^6$ λ at 2695
and 8085 MHz. (The Bonn-Jodrell Bank baseline is $\sim 10^6$ λ at 408 MHz). The
VLBI results for these sources are shown in Fig. 1.

The sources are grouped in two ranges of galactic longitude. In
the longitude range, $195^\circ < \ell < 235^\circ$ we found two measurably broadened
sources (as evidenced by their gaussian visibility curves), eight cases
in which the effects of broadening were not apparent in the visibility
data (either because the sources were unresolved, or complex in structure),
and one completely resolved source (in which fringes were not detected
with any projected spacing). The completely resolved source is seen
through a major HII region (IC 2177), in which heavy scattering is
probably occurring.

309

R. Fanti et al. (eds.), VLBI and Compact Radio Sources, 309–312.

Fig. 1 - Measured Angular
Broadening Diameters and
Limits, plotted in galactic
coordinates. Circles denote
broadening measurements.
Circles with exterior rays
denote lower limits, and
stars significy upper limits.
Diameters shown are proportional to measured or limited
angular broadening diameters
(FWHM).

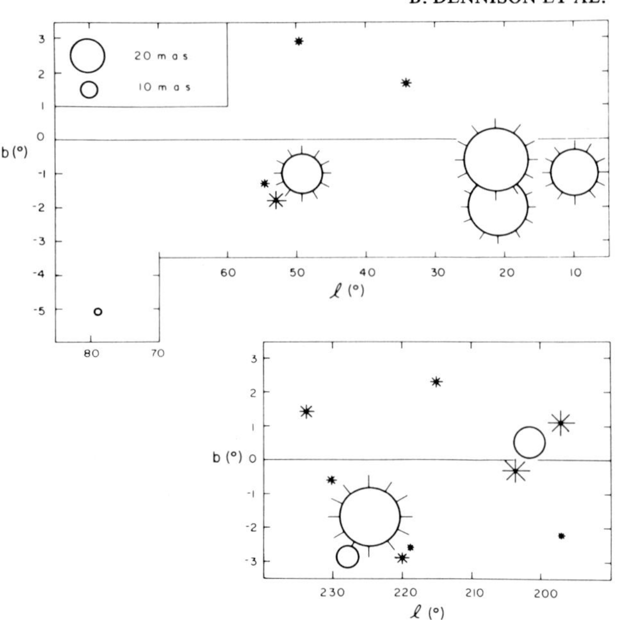

Significantly, four of eight sources observed in the range $10^\circ < \ell$
$< 60^\circ$ were completely resolved. Binomial statistics confirm an excess of
resolved sources in this longitude range, with 93% statistical confidence.
This is interpreted as due to angular broadening, since these same sources
were unresolved on baselines having comparable angular resolution at 2695
and 8085 MHz. (The results obtained using the larger sample of sources,
including those partially resolved at 2695 and 8085 MHz, further support
this conclusion. However, since these sources have not been shown to be
intrinsically compact for a 10^6 λ interferometer, they are not included
in this discussion.)

The magnitude of the scattering occurring in the 10°-60° longitude
range is appreciable, well in excess of an extrapolation of the formula
shown to be valid at latitudes $> 10^\circ$ by Duffett-Smith and Readhead (1976).
We therefore postulate that a separate, low-scale height, distribution of
scattering material is responsible for the observed broadening at low
latitudes. Major path-to-path variations in the observed broadening
suggest that the scattering material occurs in clouds. We have therefore
investigated models in which such clouds are distributed with low scale
height a) in a screen in the inner Galaxy and b) throughout the galactic
disk. Interception by one or more clouds would produce heavy scattering
such that an intrinsically compact source would appear totally resolved in
our observations.

In Fig. 2, we show the area covering factor as a function of galactic
latitude for the models best fitting the data. Although the free parameters (scale height of scattering clouds, areal cloud density at $b=0^\circ$)
are not highly constrained (due to the limited amount of data) it is
likely that the area covering factor at very low latitudes ($|b| \lesssim 1^\circ$) is

appreciable. This is particularly true if the scale height of scattering
clouds is comparable to that of population I phenomena ($\sim$80 pc), as might
be expected.

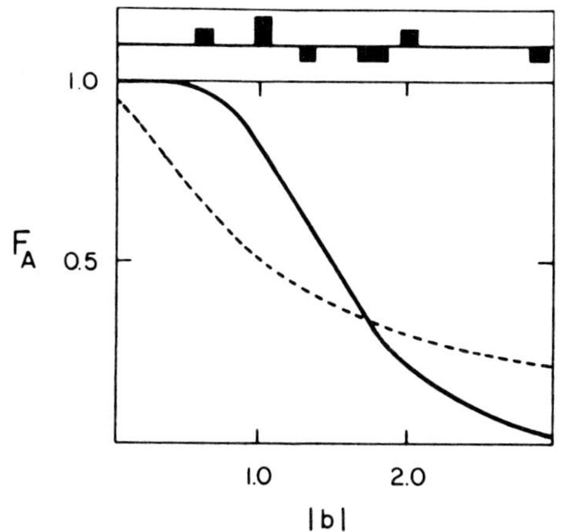

Fig. 2 - Covering factor versus
galactic latitude for models in
which the scattering occurs in
clouds in the inner Galaxy (solid
line), and in clouds distributed
throughout the galactic disk (dashed
line). The results from the sources
in the range $10^{o}<\ell<60^{o}$ are shown at
the top, above the line if resolved,
and below the line if not.

III.   IMPLICATIONS FOR COMPACT LOW-LATITUDE RADIO SOURCES

     This result has important consequences for low-latitude sources (e.g.
masers) seen through the inner Galaxy or large extents of the galactic
disk. Our data strongly suggest (but do not prove conclusively) that the
spot sizes observed in distant OH maser sources are significantly broadened
by interstellar scattering (Burke <u>et al</u>. 1968). Bowers <u>et al</u>. (1980) noted
that five of six OH/IR stars seen through the inner galaxy are lacking
small scale structure often found in these objects. This result, if it
is due to scattering, is completely consistent with our data and the models
depicted in Fig. 2.

     This survey provides an indication as to how much scattering is to be
expected along various lines of sight in the galaxy, and thus has applicability to studies of compact sources seen at low latitudes. The broadening
of CL 4 (Geldzahler and Shaffer 1981) is surprisingly large for its latitude
of 8°, and is probably due to the Cygnus Loop.

     Geldzahler and Shaffer (1981, 1982) have studied the compact radio
source G127.11+052 which appears very close to the center of the SNR
G127.1+0.5. The lack of scattering in G127.11+052 ($\lesssim$ 0.15 mas at 10.65
GHz - Geldzahler and Shaffer 1982) is not too surprising in view of our
results in the 195° - 235° longitude range. It is somewhat surprising
that heavy broadening does not occur in the SNR. A possibility that
should be investigated is that the scattering is suppressed at 10 GHz
because the rms phase fluctuation is $\lesssim$ 1 radian. At lower frequencies
the scattering could be quite heavy, however, if the effective turbulent
scale is quite small.

More thorough low-latitude broadening surveys, involving better matched resolution, and much greater dynamic range are in progress.

We thank the staffs of the Max-Planck-Institut für Radioastronomie and the Nuffield Radio Astronomy Laboratories for telescope time and technical assistance. This research was supported by a grant from the Research Corporation, with partial support by NSF grants AST 79-25345 and AST 81-17864. The NRAO is operated by Associated Universities under contract with the NSF.

REFERENCES

Anderson, B., Conway, R. G., Davis, R. J., Peckham, R. J., Richards, R. J., Spencer, R. J., and Wilkinson, P. N.: 1972, Nature, Phys. Sci. 239, 117.

Bowers, P. F., Reid, M. J., Johnston, K. J., Spencer, J. H., and Moran, J. M.: 1980, Astrophys. J. 242, 1088.

Burke, B. F., Moran, J. M., Barrett, A. H., Rydbeck, O., Hansson, B., Rogers, A.E.E., Ball, J. A., and Cudaback, D.: 1968, Astron. J. 73, S168.

Clark, D. H., and Crawford, D. F.: 1974, Australian J. Phys. 27, 713.

Cordes, J.: 1982, in Green Bank Workshop on Low-Frequency Variability, eds. W. D. Cotton, and S. R. Spangler, NRAO: Green Bank, 63.

Davies, R. D., Walsh, D., and Booth, R. S.: 1976, Mon. Not. Roy. Astron. Soc. 177, 319.

Dennison, B.: 1982, in Green Bank Workshop on Low-Frequency Variability, eds. W. D. Cotton and S. R. Spangler, NRAO: Green Bank, 71.

Duffett-Smith, P. J., and Readhead, A.C.S.: 1976, Mon. Not. Roy. Astron. Soc. 174, 7.

Geldzahler, B. J., and Shaffer, D. B.: 1981, Astrophys. J. 248, 132.

Geldzahler, B. J., and Shaffer, D. B.: 1982, Astrophys. J. Lett. 260, L9.

CIRCUMSTELLAR OH MASERS

R.S. Booth, P.J. Diamond
Onsala Space Observatory
S-439 00 Onsala, Sweden
R.P. Norris
Nuffield Radio Astronomy Laboratories
Jodrell Bank, Macclesfield, Cheshire SK11 9DL, United Kingdom

ABSTRACT

Synthesis maps of stellar OH maser emission have revealed that the OH lies in expanding spherical shells typically about 10^{16} cm in diameter. From the maps and the expansion velocity, derived from the OH spectrum, stellar mass loss rates may be determined. Typical values are 10^{-5} $M_{\odot}$/ yr. An important application of the stellar OH masers is in the estimation of stellar distances.

1. INTRODUCTION

Stimulated emission from the molecules OH, H_2O and SiO is found to be associated with late type, usually M, stars and supergiants, most of which are long period variables. Such stars are cool (photospheric temperatures $\sim$ 2500 K) and have a large IR excess, indicating a dust envelope, which is attributed to mass loss. They are often detected only at IR wavelengths.

The OH masers associated with late type stars (or OH/IR stars) radiate most strongly in the 1612 MHz line of the ground state Λ-doublet, although weaker emission in the so called main lines at 1665 and 1667 MHz is not uncommon. The 1612 MHz spectrum consists of emission occurring in 2 sharply peaked velocity intervals spaced by between 10 and 30 km s^{-1} (see Fig. 1) and many objects of this type have been discovered in 1612 MHz OH line surveys through their characteristic signature (e.g. Johansson et al., 1977).

A further important property of the OH emission is its variability, the line intensity being strongly correlated with the stellar light curve (in the case of identified objects) but having a phase lag of order 50 days (e.g. Harvey et al., 1974). This correlation indicates that the maser pump mechanism is intimately dependent on the stellar flux. In the case of the 1612 MHz sources there is a phase lag between the red and blue shifted OH peaks, an observation first reported by Shultz, Sherwood and Winnberg (1978).

R. Fanti et al. (eds.), VLBI and Compact Radio Sources, 313–318.
© *1984 by the IAU.*

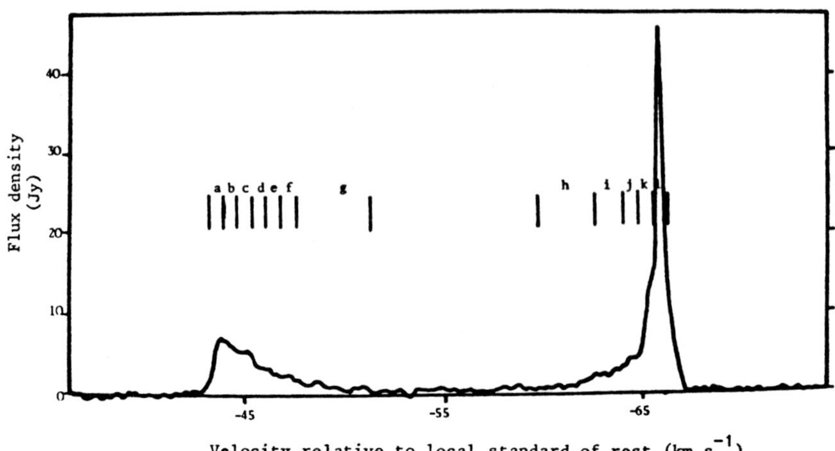

Fig. 1: 1612 MHz spectrum of OH 127.8-0.0.

The nature of the 1612 MHz spectral profile and the phase lag measurement strongly support the suggestion (e.g. Reid et al., 1977) that the
OH emission comes from a spherical, uniformly expanding circumstellar
shell. The strong outer spectral peaks are then the result of the high
maser gain produced in the long constant velocity paths along the line
of sight on the approaching and receding sides of the shell. If we
assume that the OH emission is pumped by the stellar flux, then at all
points in the shell it should emit and vary approximately in phase.
However, at the earth we will observe a phase lag between the emission
from different parts of the shell because of the difference in distance
travelled by the emission. In particular the phase lag between the blue
and red shifted peaks will give a direct measurement of the shell diameter.

The intensity of the maser emission observed from the earth will be a
function of the coherent path length along the line of sight and will
decrease with increasing shell radius. The observed emission should
obey the expression

$$a(v) = r(1 - (v/v_e)^2)^{\frac{1}{2}}$$

where   a(v)   is the shell radius at velocity v
        r   is the overall shell radius
and v_e is the expansion velocity of the shell.

At Jodrell Bank, combining data from three baselines of the radio
linked interferometer network, we were able to confirm this prediction
by making maps of the emission from OH 127.8-0.0 in several velocity
intervals (Booth et al., 1981).

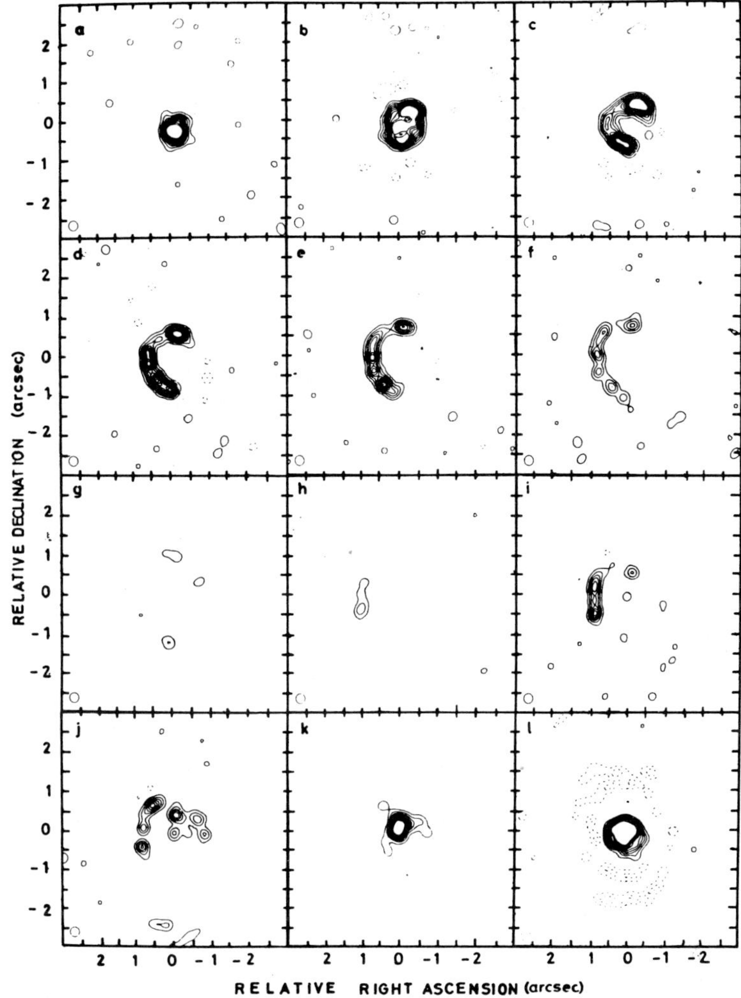

Fig. 2: MERLIN maps of the spatial distribution of the OH emission
of OH 127.8-0.0 in the velocity intervals marked a to l in
Fig. 1.

2. THE SPATIAL STRUCTURE OF THE OH EMISSION

During the past 2 years we have used the Jodrell Bank MERLIN instrument
to map 16 OH/IR stars e.g. Norris, Diamond and Booth (1982) and Diamond
et al. (in prep.). Complementary VLA measurements have been made by
Spencer, Bowers and Johnston, (in prep).

Spectral line aperture synthesis is limited by the size of the correlator in the system since for each baseline we need a large number of
delays in order to achieve a high frequency resolution. The MERLIN
correlator currently has 1024 delays and so we limit the number of tele-

scopes in the array to 4 (6 baselines) when we are able to achieve 160 delays per baseline (limited because of the modular construction of the correlator) or 80 resolution channels across the band giving a typical resolution of $\sim$ 1 km s^{-1}. The VLA has no such restriction but it is limited in spatial resolution. Norris, Booth and Diamond (1982) have outlined the MERLIN mapping technique applied to line work.

An example of a set of MERLIN maps of an OH/IR star is shown in Fig. 2. The source is OH 127.8-0.0 and its spectrum is shown in Fig. 1. The individual maps correspond to the velocity intervals marked on that spectrum. Typical features of OH/IR maps demonstrated in Fig. 2 are the barely resolved structures corresponding to the strongest spectral peaks (a & 1) and the gradual evolution of a partial shell as we proceed through the spectrum, each succesive ring of emission corresponding to a lower line of sight velocity relative to the star. Full shells of emission are rarely observed and this is not entirely unexpected since turbulence and streaming motions in the circumstellar envelope will break up the constant velocity paths along the line of sight, destroy- ing coherence. From observations of a number of sources, Diamond et al. (in prep.) have shown that an upper limit to the turbulent velocity in the shell is $\sim$ 1.7 km s^{-1}.

This may not be the only explanation for the incompleteness of the shells, however. Recent observations of the main-line emission from OH 127.8-0.0 (Diamond, Booth and Norris, in prep.) have shown that 1667 MHz emission is strongest in that segment of the shell which is missing in the 1612 MHz maps in Fig. 2. This would suggest variations in pump conditions caused, for example, by varying IR optical depths in the shell.

A final reason for asymmetries in the spatial structure in the shells may simply be the absence of OH in some regions. Goldreich and Scoville (1976) and Huggins and Glassgold (1982) have suggested UV photodis- sociation of H_2O as a means of producing the OH in the outer regions of the circumstellar envelopes. Thus an asymmetric UV radiation field could cause the observed effect. Diamond, Norris and Booth (1983b) have attributed a major asymmetry in the NML Cygnus shell to enhanced OH production in the direction of the Cygnus OB association which is now believed to be at the same distance as NML Cyg. (see Morris and Jura, 1983).

Although expansion is the dominant dynamical process in the circumstel- lar shells, several maps show evidence of other dynamical effects e.g. rotation in OH 104.9 + 2.4 (Norris, Diamond and Booth, 1982) and Dia- mond (this volume) has evidence of a bipolar expansion in IRC 10420 (Diamond, Norris and Booth 1983a).

3. MASS LOSS

The MERLIN and VLA measurements referred to in section 2 have produced maps of 16 OH/IR stars. This data is summarized in Table 1. In general

TABLE 1: Properties of the OH/IR stars

Source	Distance (kpc)	Expansion veloc. (km s^{-1})	Shell radius (arcsec)	Mass loss rate ($\times 10^{-5} M_\odot$ yr^{-1})
IRC10011	0.5(0.5)	19.3	4.4*	2.5
OH127.8-0.0	5.8(3.3)	10.9	1.53	23.0
OH138.0+7.3	2.4	9.0	> 0.5	> 0.35
OH141.7+3.5	3.7	12.2	> 0.38	> 0.65
IRC50137	0.8	17.7	3.0*	2.7
VYCMa	1.5	38.9	2.1*	10.3
OH17.7-2.0	5.4	14.0	1.25	17.1
OH26.5+0.0	0.85(2.2)	14.8	3.25	3.0
OH32.8-0.3	3.2(4.3)	17.7	2.2*	23.6
OH39.7+1.5	0.64(1.4)	17.0	2.0	0.75
R Aql	0.3	8.2	4.0*	0.31
IRC10420	3.4	37.0	1.25	18.0
OH53.6-0.2	0.8	13.5	> 0.5	> 0.58
R R Aql	0.9(0.4)	6.8	1.0*	0.15
NML Cyg	2.0 (see text)	34/20 (double shell)	1.5/2.5	13.4
OH104.9+2.4	2.0(2.5)	14.3	1.44	3.2

Kinematic distances are given for the unidentified sources, except where
a phase lag measurement of the shell linear diameter is available (Herman & Habing, 1981; Herman (private communication). In that case the
distance is estimated directly using the angular diameter measurement;
the kinematic distance is then given in parenthesis.
Angular radii were measured with MERLIN (Norris et al., 1982, Diamond
et al. 1982a,b and in prep., Chapman et al., 1983) and VLA (Spencer et
al., in prep). VLA data is indicated with *.
Mass loss rates estimated as in section 3.

the circumstellar shells are shown to have angular diameters of a few
arcsecs and linear radii $\sim 10^{16}$ cm. Since the expansion velocity v, is
known from the OH spectrum, the mass loss rate, $\dot{M}$, can be calculated
from $\dot{M} = 4\pi r^2 nv$ where r is the shell radius and n is the gas density. n may be estimated theoretically, e.g. Goldreich and Scoville
(1976), and turns out to be $\sim 10^4$ cm^{-3} at these radii, giving $\dot{M} \sim$
10^{-5} $M_\odot$ yr^{-1} in a typical case.

Such high mass loss rates cannot be supported for very long and so stars in this evolutionary state must be short lived ($\sim 10^3$ yrs). Bowers, Johnston and Spencer (1981) have noted a correlation between the quantity rv^2 and mass loss rate, based on the requirement of a minimum OH optical depth for the maser process to occur. This correlation can thus be used to derive mass loss rates from the measured shell size and the expansion velocity.

4. DISTANCE MEASUREMENT

An important application of the OH mapping results is in the accurate determination of the distances to the OH/IR stars. Jewell et al. (1980) and particularly Herman and Habing (1981) have shown that by careful monitoring of the OH spectra the phase lag between the blue and red shifted emission peaks may be measured and a "light travel" shell diameter may be determined to $\sim$ 10 % accuracy. Dividing this by the shell's angular diameter measured with the interferometer gives the distance to the source. Herman (private communication) estimates that an overall accuracy of $\sim$ 10 % may be achieved. Since this is a direct estimate of distance involving no assumptions it will clearly become a valuable technique in the measurement of the galactic distance scale.

REFERENCES

Booth, R.S., Kus, A.J., Norris, R.P., and Porter, N.D.: 1981, Nature 290, pp. 382-4.
Bowers, P.F., Johnston, K.J., and Spencer, J.H.: 1981, Nature, 291, pp. 382-5.
Chapman, J., Cohen, R.J., Norris, R.P., Diamond, P.J., and Booth, R.S.: 1983, Mon.Not.R.Astr.Soc. (in press).
Diamond, P.J., Norris, R.P., and Booth, R.S.: 1983a, Astron. Astrophys. (in press).
Diamond, P.J., Norris, R.P., and Booth, R.S.: 1983b, Mon.Not.R.Astr.Soc. (in press).
Goldreich, P. and Scoville, N.: 1976, Astrophys.J., 205, pp. 144-54.
Harvey, P.M., Bechis, K.P., Wilson, W.J., and Ball, J.A.: 1974 Astrophys. J.Suppl. 27, pp. 331-57.
Herman, J. and Habing, H.J.: 1981, Physical Processes in Red Giants, pp. 383-90 eds. Iben, I. and Renzini, A.
Huggins, P.J. and Glassgold, A.E.: 1982, Astr.J., 87, pp. 1828-35.
Jewell, P.R., Weber, J.C., and Snyder, L.E.: 1981, Astrophys.J.Lett. 242, pp. L29-31.
Johansson, L.E.B., Andersson, C., Goss, W.M., and Winnberg, A.: 1977, Astron.Astrophys.Suppl. 28, pp. 199-210.
Morris, M. and Jura, M.: 1983, Astrophys.J. 267, 179-83.
Norris, R.P., Booth, R.S., and Diamond, P.J.: 1982, Mon.Not.R.Astr.Soc. 201, 209-22.
Norris, R.P., Diamond, P.J., and Booth, R.S.: 1982, Nature 299, pp. 131-4
Reid, M.J., Muhleman, D.O., Moran, J.M., Johnston, K.J., and Schwartz, P.R.: 1977, Astrophys.J., 214, pp. 60-77.
Schultz, G.V., Sherwood, W.A., and Winnberg, A: 1978. Astron.Astrophys. 63, pp. L5-7.

OBSERVATIONS OF EXCITED STATE OH MASERS AT 6.0 AND 13.4 GHz

A. Baudry[1], S. Guilloteau[2], M. Walmsley[3], T.L. Wilson[3], A. Winnberg[4]
[1] Observatoire de l'Université de Bordeaux, 33270 Floirac, France
[2] Université Scientifique et Médicale de Grenoble, CERMO BP 68,
    38402 Saint-Martin-d'Hères Cedex, France
[3] Max-Plank Institut für Radioastronomie, Auf dem Hügel 69,
    D-5300 Bonn 1, R.F.A.
[4] Onsala Space Observatory, 43900 Onsala, Sweden

This is a brief report on a sensitive search for excited OH absorption
and emission sources made with the Effelsberg 100 m telescope towards
several compact HII regions. Our data are discussed elsewhere (Guilloteau
et al., 1983) and we mainly present here our $\Pi_{3/2}$, J = 5/2 and 7/2 OH narrow emission results (masers). In the J = 5/2 state, 14 sources were
observed with a typical 1 σ sensitivity of 0.2 Jy. Among 13 positive
detections, 4 and may be 5 are new maser sources : G43.1, W48, K3-50,
ON3 and G34.3 (?). With the exception of W3(OH) the 6035 MHz line is
always much stronger than the 6030 MHz line. Comparison with spectra
published by others suggests time variability in M17, W51 and Orion A.
In W3(OH) a weak broad emission was also detected at 6049 MHz. In our
j = 7/2 observations the peak to peak noise was the range 0.1 to 1 Jy.
No masers were detected even in the strong 6035 MHz sources associated
with M17 and W33 cont. Thus the F = 4-4 emission line in W3(OH) seems
to be really exceptional (Baudry et al., 1981).

$\Pi_{3/2}$ excited state masers are expected from the collisional pumping
model of Flower and Guilloteau (1982) provided that the kinetic temperature is larger than the dust temperature. Because the collisional mechanism tends to overpopulate the upper levels of the $\Pi_{3/2}$ Λ-doublets, collisional excitation is likely to be a dominant process in $\Pi_{3/2}$ excited OH
emission sources. Present collision rates do not incorporate the dependence on hyperfine quantum numbers and thus one expects nearly equal
inversion of the 6035 an 6030 MHz lines. However model calculations show
that, in agreement with observations, W3(OH) excepted, inversion is more
efficiently quenched at 6030 MHz than at 6035 MHz. One would also expect
from a collisional model to detect J - 7/2 masers associated with strong
J = 5/2 masers. Because J = 7/2 masers appear to be rare whereas J = 7/2
absorption features are often detected collisional models may also depend
on ΔJ = 2 collisional cross-sections. Despite these uncertainties we derive, using the results of Dewangan and Flower (1982), that collisions
are dominant if (with $T_K \simeq 100$ K) $n_{H_2} > 5 \times 10^6$ and 10^8 cm^{-3} for the $\Pi_{3/2}$,
J = 5/2 and 7/2 states respectively.

319

R. Fanti et al. (eds.), VLBI and Compact Radio Sources, 319–320.
© *1984 by the IAU.*

From various arguments including estimates of OH rotation temperatures we think that the OH *absorbing* region associated with a compact
HII region lies close to the ionization front (Guilloteau *et al.*, 1983).
Maser lines tend to appear on one side of the absorption features. This
suggests that the absorption and maser sources are in close relationship
but does not tell, without interferometric measurements, whether the
masers are dense clumps within the OH absorbing region.

References

Baudry, A., Walmsley, C.M., Winnberg, A., Wilson, T.L. : 1981, Astron.
     Astrophys. 102, 287.
Dewangan, D.P., Flower, D.R. : 1982, Monthly Notices Roy. Astron. Soc.
     199, 457.
Flower, D.R., Guilloteau, S. : 1982, Astron. Astrophys. 114, 238.
Guilloteau, S., Baudry, A., Walmsley, C.M., Wilson, T.L.,Winnberg, A. :
     1983, Astron. Astrophys., in press.

MERLIN OBSERVATIONS OF THE OH MASER EMISSION FROM IRC+10420

P.J.Diamond
Onsala Space Observatory
S-439 00 Onsala
Sweden

MERLIN observations of the 1612 and 1665 Mhz OH masers from the supergiant
IRC+10420 reveal several major departures from the usual morphology of
OH/IR stars.

IRC+10420 is one of the most unusual OH maser sources yet
found. It is an F8 supergiant with strong OH maser emission at 1612,
1665 and 1667 Mhz. A north-south optical nebulosity with an angular
size of 2 arcsec has been detected (Thompson and Boroson,1977) and its
peculiar IR properties are reminiscent of Eta Carina. These characteristics
make IRC+10420 unique among Type II OH/IR stars.

Diamond et al.(1983) have made MERLIN maps of the 1612 Mhz OH
masers and found that the structure is very axisymmetric with the most
blueshifted masers (42.5 - 50.0 km/s) forming a north-south double
separated by 0.5 arcsec, while the masers from 50 - 60 km/s form an
almost edge-on disc at a position angle of 108°. The disc has an angular
radius of 1.25 arcsec. Diamond et al.(1983) suggest from these and other
data that IRC+10420 is a bipolar nebula. Bowers (1983) has observed the
1612 and 1667 Mhz OH masers with the VLA. His maps of the blueshifted
1612 Mhz masers agree with those of Diamond et al.(1983), the redshifted
1612 Mhz masers, which were below the sensitivity of MERLIN, are spread
over a larger area. The VLA observations of the 1667 Mhz OH masers reveal
an unusual anomaly in that these masers appear to be situated at between
3 and 5 arcsec from the central star. This result is inconsistent with
models of maser pumps (e.g.Elitzur et al.,1976)

In Fig.1 we show a) the 1665 Mhz OH spectrum from IRC+10420
and, b) the MERLIN maps of these masers. The 1665 Mhz spectrum covers
the same velocity range as the 50 - 60 km/s 1612 Mhz masers and the maps
show the same general configuration, namely a disc like stucture oriented
along approximately the same position angle. The flux density of both
the 1612 and 1665 Mhz disc components have been increasing at a rate of
1.25 Jy/yr since 1976 (Benson et al.,1979) suggesting that at this

R. Fanti et al. (eds.), VLBI and Compact Radio Sources, 321–322.
© 1984 by the IAU.

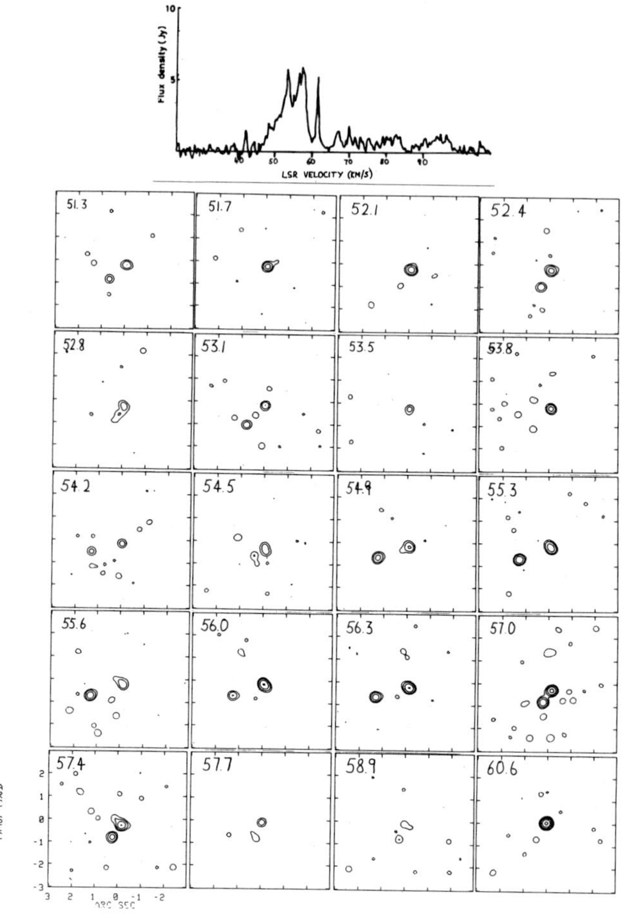

position in the envelope the 1612 and 1665 MHz disc component masers are
situated in the same region and/or have a common pump source. Two
anomalies are present in the 1665 MHz maps: i) there is no increase in
overall emission size as the velocity approaches the stellar velocity
and, ii) there is an unresolved point component at 60.6 km s^{-1} instead
of at the extreme velocities as is usual in OH/IR stars.

These results are not inconsistent with the suggestion by Diamond
et al (1983) that IRC+10420 is a bipolar nebula but the simple geometrical picture requires modification.

REFERENCES

Benson, J.M., Mutel, R.L., Fix, J.D., Claussen, M.J., 1979, Ap.J. 229,
    L87-90
Bowers, P.F., 1983, preprint
Diamond, P.J., Norris, R.P., and Booth, Astron.Astrophys. (in press)
Elitzur, M., Goldreich, P., Scoville, N., 1976, Ap.J., 205, pp 384-396
Thompson, R.I. and Boroson, T.A., 1977, Ap.J., 216, L75-77

VLBI SYNTHESIS OBSERVATIONS OF CIRCUMSTELLAR OH MASERS [+]

R.P.Norris,[1] R.S.Booth,[2] P.J.Diamond,[2] D.A.Graham,[3] L-A.Nyman[2]
1. NRAL, Jodrell Bank, Macclesfield, Cheshire, England
2. Onsala Space Observatory, S-43900 Onsala, Sweden
3. Max Planck Institut fur Radioastronomie, Bonn, W.Germany

ABSTRACT

A technique for mapping spectral line sources using closure phase
VLBI techniques is described, and some initial results on circumstellar
shell sources are presented. The maps show an unresolved point source
on the nearside of the shell, but no corresponding source on the far-
side of the shell. This is interpreted in terms of amplification of
the stellar thermal emission. Preliminary results are also presented
on another source, OH17.7, which show correlated flux outside the
velocity range of the known OH masers, and it is suggested that this
represents high velocity ($\sim$600 km s^{-1}) violent motions around the central
star. If confirmed, this may represent a new class of source, represen-
ting violent phenomena associated with red giant stars.

INTRODUCTION

The OH maser emission from OH/IR stars is characterised by a double
peaked spectrum which has been shown by MERLIN and VLA observations to
emanate from an expanding shell surrounding a late type star (usually
presumed to be an M giant or supergiant). A review of these observations
is given by R.S. Booth elsewhere in these proceedings and so will not be
duplicated here. Instead, we present the results of VLBI observations,
which are designed to investigate the compact structure of these sources
and we consider a mechanism which may produce the observed structure.

TECHNIQUES AND OBSERVATIONS

The technique, which is described in detail by Norris (1982)
includes corrections for correlator-dependent phase offsets including a
retarded-baseline effect and digitisation errors. The amplitudes are
then calibrated using simultaneous autocorrelation spectra. One strong
spectral feature, which need not be unresolved, is mapped using the
closure phase technique (Cornwell & Wilkinson 1981). Other spectral

---

+ Discussion on page 461

R. Fanti et al. (eds.), VLBI and Compact Radio Sources, 323–326.
© *1984 by the IAU.*

channels are then phase referenced to this feature, and their phases corrected for the structure of the reference. The data in these channels are then CLEANed to produce maps of the brightness distribution integrated over suitable velocity intervals. The observations presented here were made in February 1981 (OH127.8 and OH104.9) and March 1983 (OH17.7) using 4 telescopes of the European VLBI network (EVN): Jodrell Bank, Effelsberg, Dwingeloo and Onsala.

RESULTS AND DISCUSSION

Fig.1 shows a closure phase map of the sharp blue shifted feature at -66 km s^{-1} in the source OH127.8. The emission from this feature appears substantially unresolved on even the longest EVN baselines. In contrast, all other components are found to be resolved by the EVN with a characteristic size $\sim$0.1 arcsec, as shown in Fig.2. We suggest this represents the size of a turbulent cell.

A similar result has been obtained for the more complicated source OH104.9. In this case, there are three compact features at the blue-shifted end, but all features elsewhere in the spectrum are resolved. Preliminary processing on other sources indicates similar results.

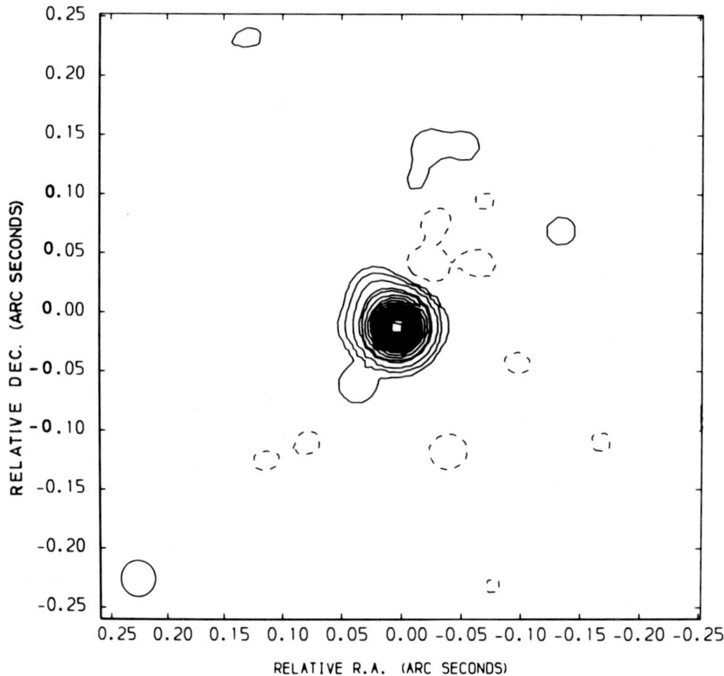

Fig.1 Closure-phase map of the -66 km/s feature of OH127.8, using Jodrell Bank, Effelsberg, Dwingeloo and Onsala. Lowest contour is 1%.

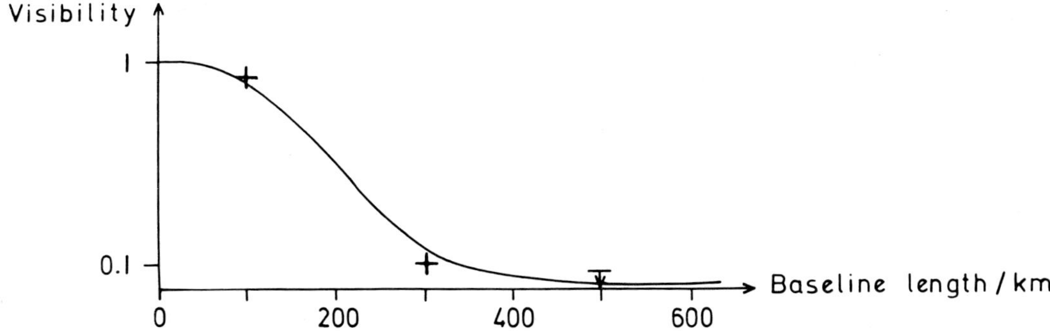

<u>Fig.2</u> Typical visibility curve of a feature at the red-shifted end of the spectrum of OH127.8. The curve drawn represents a gaussian of width ∿0.1 arcsec.

MERLIN maps of OH127.8 and OH104.9 (Norris et al. 1982) show that the blue shifted peak of emission emanates from a compact component which is spatially coincident with the star. Fig.1 shows that this component is unresolved even on European VLBI baselines. The corresponding red-shifted emission appears resolved on the same baselines. We suggest that the compact blue-shifted component may represent the stellar thermal emission which is being amplified by the OH masers in the circumstellar shell. In this case, the OH compact component will remain exactly spatially coincident with the star, regardless of any inhomogeneities or disturbances in the circumstellar shell. That such amplification may occur, even if the maser is saturated, can be shown by considering the direction, as well as the intensity, of the saturating emission (Norris et al. 1983).

This hypothesis has two important consequences:

i)   It provides a situation in which an easily observed radio emitter is known to be coincident with an optical emitter to an accuracy of a few milliarcsec. Several OH/IR stars have bright optical counterparts, and they should be invaluable in relating the optical and radio reference frames, and will be observed in conjunction with the HIPPARCOS project. In addition perhaps a similar result may be obtained for the H_2O and SiO masers around late type stars.

ii) The sizes (and spectral types) of the central stars of sources like OH127.8 and OH104.9 are unknown, although they are usually assumed to be M giants or supergiants. Since our hypothesis implies maser sizes equal to the stellar diameter, global VLBI measurements may be used to determine, or place limits upon, the stellar diameters of some of the nearest sources.

PROBABLE DISCOVERY OF BROADBAND EMISSION ASSOCIATED WITH AN OH/IR STAR

Recent EVN observations of the OH/IR star OH17.7 produced an

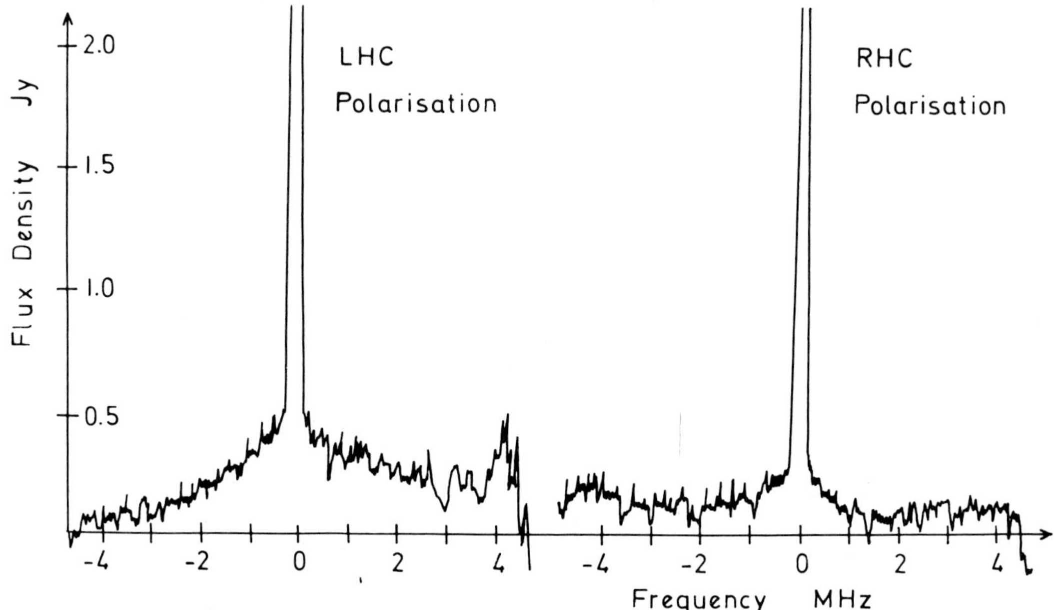

Fig.3 Autocorrelation spectra of OH17.7, measured in June 1983 with the
Jodrell Bank MK IA telescope. Because of the wide bandwidth (10 MHz) the
characteristic double peaked spectrum appears as a single spike in the
centre of the band. No attempt has been made to remove a baseline from
the spectrum. An indication of triangular wings may be seen in both spectra.

unexpected result in that correlated flux was detected not only within
the OH lines but also across the entire bandwidth of 250 kHz. The data
have not yet been completely analysed, but a flux density of the order
of 0.5 Jy is implied. Subsequent spectra (Fig.3) imply that this emission
is due to broad weak wings extending 600 km s^{-1} either side of the strong
maser lines. Observations with the Effelsberg telescope allow an upper
limit of ∿10 mJy to be placed on any 6 cm continuum flux from this source
although the source was recently detected by IRAS as a strong far-IR
emitter (IRAS working group, 1983).

We emphasize that the radio detections have yet to be confirmed by
further observations. However, we speculate that, if they are confirmed,
they represent a new type of violent phenomenon associated with late
type stars.

REFERENCES

Cornwell,T.J. & Wilkinson,P.N., 1981.Mon.Not.R.astr.Soc.,196,1067-1086.
IRAS Working Group 1983. Nature 303, 480.
Norris,R.P.,1982. in 'VLBI Techniques' pp341-345.ed.F.Biraud,CNES Toulouse
Norris,R.P.,Diamond,P.J. & Booth, R.S. 1982. Nature 299, 131-134.
Norris, R.P., Booth,R.S., Graham, D.A., Nyman, L-A., 1983. In preparation

THE STRUCTURE OF OH MASER CLOUDS ASSOCIATED WITH LATE-TYPE STARS

P. F. Bowers[1,2], K. J. Johnston[1], and J. H. Spencer[1]
[1]E. O. Hulburt Center for Space Research, Naval Research
Laboratory, Washington, D. C. 20375
[2]Sachs/Freeman Associates, Bowie, Maryland 20715

The absolute positions and the arcsec structure of OH maser clouds
surrounding 20 Mira variables and late-type supergiant stars have been
measured using the Very Large Array in a spectral line mode at 1612 MHz.
The stars observed are listed in Table 1 which indicates that the angular radii θ of the maser clouds range up to 4". The linear radii R
range from < 100 AU for the Mira variable U Ori to 10^4 AU for the supergiant IRC+10420 and are correlated with the stellar mass loss rates.

With the exception of VY CMa, for which there is evidence of a
complex disk-like or ellipsoidal geometry, the stars in Table 1 appear
to be surrounded by predominantly spherical, expanding envelopes. For
nine stars the low and high velocity peak emission features are unresolved knots positionally coincident to within 0".1. For well resolved
sources, gas at intermediate velocities is distributed in portions of
ring-like structures whose sizes increase as the velocity approaches the
stellar velocity; an outline of a complete circular shell is seen in
some cases (OH 127.8-0.0, OH 26.5+0.6, and OH 32.8-0.3).

Nevertheless, there can be significant deviations from the idealized
spherical case. 1) The emission distributions are clumpy at most velocities, indicating density/velocity perturbations. 2) Positional offsets between the low and high velocity peak emission features are observed for some stars whose underlying geometry is clearly spherical
(e.g. OH 26.5+0.6). 3) The peak emission can be extended (OH 32.8-0.3).
4) There is often significant emission outside the peak velocities, not
predicted for a spherical shell expanding at a constant velocity.

These effects are probably caused by a combination of density clump-
ing ($\sim 10^{15}$-10^{16} cm) and streaming motions ($\sim$ 1-2 km/s) superposed on
the radial expansion. Streaming motions are larger for stars located in
close proximity to OB associations (VY CMa and NML Cyg), indicating that
the interstellar ultraviolet radiation field is important in determining
the observed OH distribution.

Most oxygen-rich, Mira-type stars appear to lose mass in a spherical
outflow at a rate < 10^{-4} M$_\odot$/y, supporting radiation-driven or shock-
driven mass loss models and suggesting these stars are responsible for
the optical haloes surrounding some planetary nebulae. Non-spherical

327

R. Fanti et al. (eds.), VLBI and Compact Radio Sources, 327–328.

geometries and larger mass loss rates for related evolved objects
(carbon stars, bipolar nebulae, non-spherical planetaries) may be the
result of evolution in a close binary system. A more complete discussion is given by Bowers, Johnston, and Spencer (1983).

Table 1

PROGRAM STARS

Star	Position (±0″.1)		Distance	θ	R	$\dot{M}$
	α(1950)	δ(1950)	(kpc)	(″)	(10^3AU)	($10^{-6}M_\odot$/y)
IRC+10011	$01^h03^m48^s.09$	+12°19′51″.4	0.5*	4.4	2.2	22
OH 127.8−0.0	01 30 27.63	+62 ·11 31.2	3.3	1.9	6.3	80†
OH 138.0+7.3	03 20 41.48	+65 21 32.8	2.4	≤0.8	≤1.9	≤ 7.6†
OH 141.7+3.5	03 29 23.65	+60 10 04.4	3.7	≤0.9	≤3.3	≤ 22†
IRC+50137	05 07 19.68	+52 48 53.9	0.8	3.0	2.4	11
U Ori	05 52 50.92	+20 10 06.0§	0.2	<0.6	<0.1	1.2
IRC+40156	06 29 45.03	+40 45 08.2	1.4	1.8	2.5	13†
VY CMa**	07 20 54.74	−25 40 12.4	1.5	2.1	3.2	230
IRC−20197	09 42 56.55	−21 47 54.4	0.7	≤1.4	≤1.0	4.1
WX Ser	15 25 31.98	+19 44 13.0	1.0	<1.1	<1.1	5
OH 17.7−2.0	18 27 39.77	−14 31 03.9	3.4*	0.9	3.1	20†
OH 26.5+0.6	18 34 52.47	−05 26 37.1	0.5*	3.5	1.6	5
OH 25.1−0.4	18 35 33.36	−07 12 35.2	9.1	<1.4	<12.7	<310†
OH 32.8−0.3	18 49 48.16	−00 17 53.5	3.2*	2.2	7.0	32
OH 39.7+1.5	18 56 03.88	+06 38 49.8	1.4*	2.0	2.8	28
OH 39.9+0.0	19 01 42.90	+06 08 44.2	7.7	0.9	7.2	29
R Aql	19 03 57.67	+08 09 07.7§	0.3	4.0	1.2	0.8
IRC+10420**	19 24 26.74	+11 15 10.9	3.4	3.0	10.2	200†
RR Aql	19 55 00.30	−02 01 17.1	0.9*	1.0	0.9	1.8†
NML Cyg**	20 44 33.84	+39 55 57.1	1.8	3.0	5.4	160

*Distance determined from angular radius and phase lag measurements.
†Mass loss rate derived from radius of maser region.
§Position precessed from epoch 1980.9 with no correction for known
proper motion.
**Supergiant.

REFERENCE
Bowers, P. F., Johnston, K. J., and Spencer, J. H. 1983, <u>Ap. J.</u>, in
press.

THE SPATIAL STRUCTURE OF SILICON MONOXIDE MASERS[+]

Adair P. Lane
National Radio Astronomy Observatory
Edgemont Road
Charlottesville, Virginia 22901

VLBI observations of stellar masers provide important constraints on models for maser excitation and for the dynamics and physical conditions in circumstellar envelopes. The spatial distribution and apparent sizes of SiO masers toward several late type variable stars and toward the star-forming region in Orion have been measured using the 75 km baseline between the 37 m telescope of Haystack Observatory and the 14 m telescope of the Five College Radio Astronomy Observatory. The fringe spacing is 0.02 arcsec and the transitions observed were the $v = 1$, $J = 1 - 0$ and $v = 2$, $J = 1 - 0$ lines at 43 GHz. The data were obtained in September 1979 and April 1982.

Assuming a Gaussian brightness distribution for individual features, the apparent sizes of the smallest features in the $J = 1 - 0$ lines toward Mira variables are $2 - 6 \times 10^{13}$ cm (FWHM); for the supergiant VX Sgr the sizes are $2 - 4 \times 10^{14}$ cm. The maser sizes and flux densities imply the masers are saturated. The peak brightness temperatures are in the range $5 \times 10^9 - 2 \times 10^{10}$ K.

Maps of the spatial distribution of maser features at different velocities have been obtained by analysis of relative phase as a function of time. Toward the Mira variable R Cas the $v = 1$, $J = 1 - 0$ masers occur out to distances of at least 4 R_* (1.5×10^{14} cm). The

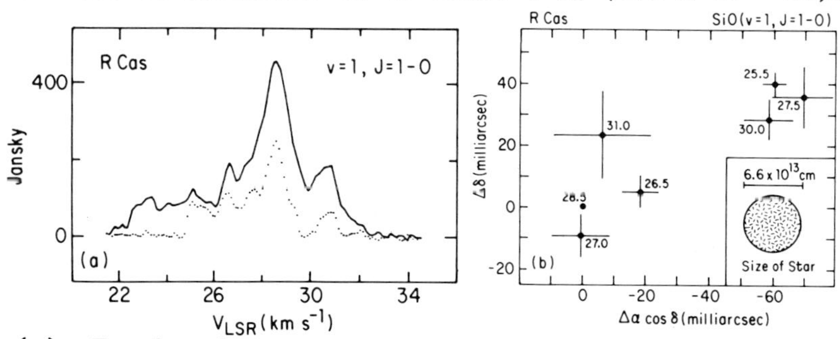

Fig. 1 – (a) Total and cross power spectra toward R Cas. ($V_* = 24.7$ km/s). (b) Map of SiO maser features toward R Cas.

+ Discussion on page 463

329

R. Fanti et al. (eds.), VLBI and Compact Radio Sources, 329–330.

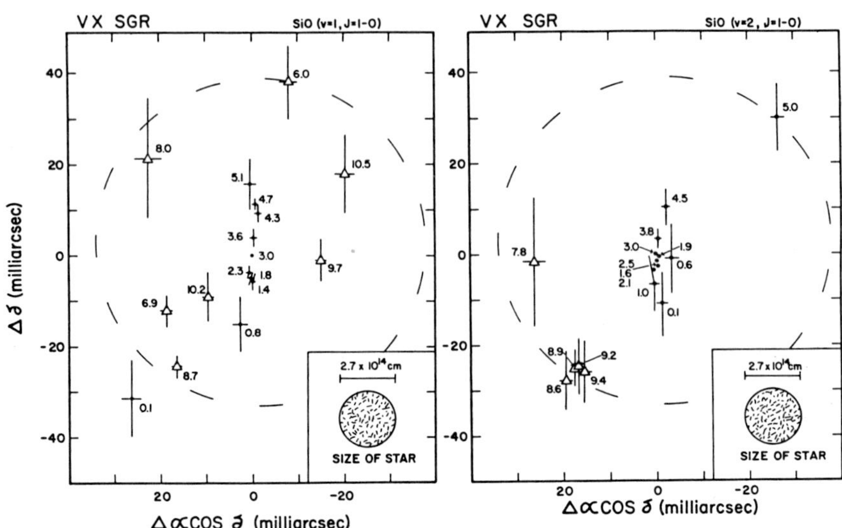

Fig. 2 - Maps of features in two SiO transitions from VX Sgr.
($V_* = 6$ km/s; Δ = redshifted features; $\bullet$ = blueshifted features).

spatial distributions of the v = 1, J = 1 − 0 and v = 2, J = 1 − 0 masers toward VX Sgr are roughly similar to each other and are consistent with location of the masers in an expanding circumstellar shell of radius 6 R_* (~8 x 10^{14} cm). Toward the Mira variable R Leo the masers are clustered closer to the stellar surface (within 2 R_* or 10^{14} cm) and do not show any clear kinematic pattern. A spatial offset of ~10^{15} cm was measured between the two major groups of features toward Orion/IRc 2.

The data are inconsistent with a model locating the masers in convective cells in the stellar photosphere; instead it is more likely the masers are radiatively or collisionally pumped clumps of gas in the innermost turbulent regions of the circumstellar envelope.

The assistance of J. Moran, M. Reid, R. Predmore, and D. Clemens in obtaining the data is gratefully acknowledged.

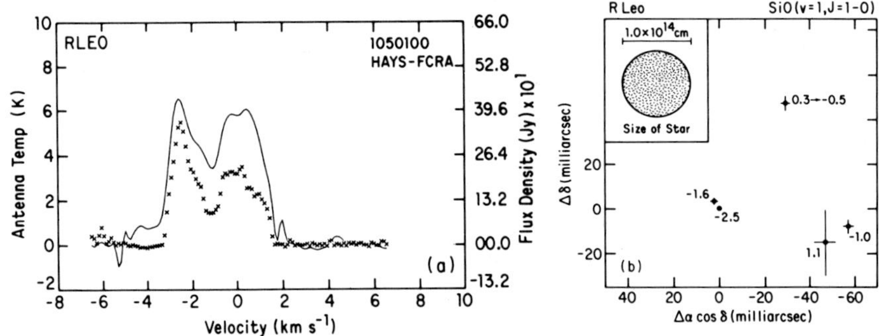

Fig. 3 - (a) Total and cross power spectra toward R Leo.
($V_* = -1.0$ km/s). (b) Map of SiO maser features toward R Leo.

STRUCTURE OF H_2O MASER OUTBURST IN ORION KL NEBULA

L.I. Matveyenko
Space Research Institute
Moscow, USSR.

Abstract
    The H_2O outburst in Orion KL had been observed by the VLBI technique
in 1979-1983. The line profile is asymmetrical. The flux density and
line width are variable and correspond to a partially saturated maser.
The outburst region had a compact and a less compact one whose sizes are
equal to 0.3 and 2.5 a.u. respectively. The fine structure of the profile corresponds to knots whose size is smaller 0.2 a.u. The brightness
temperature of the region, $T_b \simeq 10^{17}K$, implies a high emission directivity.

    In August-September 1979 a strong H_2O maser outburst happened in
Orion KL nebula. Its velocity was ~ 8.1 km/s, the maximum flux density
about 2×10^6 Jy. On September 25, 1979 we observed the outburst on
Simeiz-Pushino interferometer. The fringe size was equal to 2.5-4 mas
and frequency resolution 2.2 kHz [1]. On November 16, 1979 Orion KL
was observed on the Simeiz-Effelsberg and Green Bank-Haystack interferometers. The fringe size was 1.3-10 mas and the frequency resolution
25 KHz. The observations repeated each year include Simeiz-Evpatoria-
Pushino interferometer, [1]. The outburst emission was measured in
the linear and left circular polarizations.

    The line profile of the outburst is asymmetrical and has a high
frequency tail. The fine structure of the profile is complex and variable. The flux density and the line width are changing with time and
correspond to a partially saturated maser.

    The fringe visibility of the outburst region changes with the inter-
ferometer hour angle, polarization, line velocity and time. The fringe
visibility corresponds to complex structure of the outburst. The central
and the high velocity part of the profile were radiated by a region whose
size is equal to ~ 1 mas (~ 0.5 a.u.). The flux density of this region
$F \simeq 0.65\ F_{tot}$, the line width $\Delta f \simeq 20$ kHz and polarization $P \simeq 60\%$.
The brightness temperature of the region $T_b > 10^{16}K$.

    The main part of polarization emission was determined by a region
331

R. Fanti et al. (eds.), VLBI and Compact Radio Sources, 331–332.
© *1984 by the IAU.*

whose size is equal to $\sim$ 0.5 mas (0.25 a.u.), and corresponds to the central part of the line. The polarization is equal to P $\stackrel{\sim}{>}$80% and bandwith to Δ f < 20 KHz.

The low velocity emission of the outburst was determined by a region whose size is equal to 6 mas, the flux density F $\simeq$ 0.35 F_{tot}, Δ f $\simeq$ 50kHz and P $\simeq$ 50%.

The fine structure of the line profile perhaps connects with knots which are located inside the compact region. The line width of the features Δ f $\simeq$ 5 kHz and corresponds to a low temperature of the knot medium T_k $\simeq$ 60 K.

In 1982 the line width increased Δ f = 40-50 KHz and the size of the compact region increased too. Perhaps the line width Δ f is determined by the differential velocity of the emission region. The extra brightness temperature of the outburst region assumes high beaming of emission.

Reference

1. Matveyeñko L.I., Loran D.M., Genzel R., 1982, Pis'ma Astron.Zh., Vol
     8, No 12, p.711

MAGNETIC FIELD STRUCTURE OF STAR FORMING REGIONS:
VLBI SPECTRAL LINE RESULTS

J.A. Garcia-Barreto[1,2], B. F. Burke
R.L.E., Massachusetts Institute of Technology, U.S.A.
M.J. Reid, J.M. Moran and A.D. Haschick[3]
Harvard-Smithsonian Center for Astrophysics, U.S.A.

I. INTRODUCTION

Magnetic fields play a major role in the general dynamics of astronomical phenomena and particularly in the process of star formation. The magnetic field strength in galactic molecular clouds is of the order of few tens of μG [6]. On a smaller scale, OH masers exhibit fields of the order of mG [7, 9] and these can probably be taken as representative of the magnetic field in the dense regions surrounding protostars. The OH molecule has been shown to emit highly circular and linearly polarized radiation [11, 2]. That it was indeed the action of the magnetic field that would give rise to the highly polarized spectrum of OH has been shown by the VLBI observations of Zeeman pairs of the 1720 and 6035 MHz by Lo et. al. [7] and Moran et. al. [9]. VLBI observations of W3(OH) revealed that the OH emission was coming from numerous discrete locations [8] and that all spots fell within the continuum contours of the compact HII region [1, 3]. The most detailed VLBI aperture synthesis experiment of the 1665 MHz emission from W3(OH) was carried out by Reid et. al.[10] who found several Zeeman pairs and a characteristic maser clump size of 30 mas. In this work, we report the results of a 5 station VLBI aperture synthesis experiment of the 1665 MHz OH emission from W3(OH) with full polarization information. We produced VLBI synthesis maps of all Stokes parameters of 16 spectral features that showed elliptical polarization. The magnitude and direction of the magnetic field have been obtained by the detection of 7 Zeeman pairs. The three dimensional orientation of the magnetic field can be obtained, following the theoretical arguments of Goldreich et. al. [4, 5], from the observation of π and σ components.

II. RESULTS

Reid et. al. [10] showed, more recently, that in the case of W3(OH) the OH maser radiation was coming from different subregions within the OH emission complex. We detected maser emission from 17 subregions. In five of these, labeled 1, 2, 4, 5 and 8 we detected 16 elliptically polarized components. They showed a percent of linear polarization in

R. Fanti et al. (eds.), VLBI and Compact Radio Sources, 333–334.
© *1984 by the IAU.*

the range from 6 to 44% and polarization position angles in the range
from 97 to 210° (E of N) depending on the particular feature. The percent
of total polarization was found to be in the range from 23 to 99%. We
were able to detect 7 Zeeman pairs which appeared in subregions 2, 3, 5,
6, 7, 8 and 9. The field strength of about 5.5 mG was remarkably uniform
and in all cases it was pointing away from Earth. In subregion 5, there
is a RCP σ, an elliptically polarized feature(assumed to be a π component)
but the companion LCP σ feature was not reliable identified since it was
at the same frequency as a strong LCP σ line in subregion 7. It is worth
noting that the velocity of the LCP σ feature in subregion 5 completely
agrees with the expected velocity given the values found for the magnetic
field from other Zeeman pairs. All 3 components are observed to belong
to the same maser clump and they may be the first complete Zeeman pattern
being detected. The projection of the magnetic field on the plane of the
sky may be obtained from the polarization position angles of the features
that showed elliptical polarization, while the angle that the magnetic
field makes with respect to the line of sight can be deduced from the
observation of π and σ components. This experiment has been the first
attempt to do so and for this reason we have conducted another VLBI
experiment designed to observe other maser sources which will provide us
with more information on the 3 dimensional orientation of the magnetic
field in other regions of recent star formation.

JAGB would like to thank CONACYT (Mexico) and the IAU for their
partial support that allowed him to attend this symposium.

III. REFERENCES
1. Baldwin,J.E.,Harris,L.S. and Ryle,M. 1973 Nature 241, p 38.
2. Davies,R.D.,Jagger,G. de and Verschuur,G.L. 1966 Nature 209, p 274.
3. Dreher,J.W. and Welch,W.J. 1981 Ap. J. 245, p 857.
4. Goldreich,P.,Keeley,D.A. and Kwan,J.Y. 1973 a Ap. J. 179, p 111.
5. Goldreich,P.,Keeley,D.A. and Kwan,J.Y. 1973 b Ap. J. 182, p 55.
6. Heiles,C. and Troland,T.H. 1982 Ap. J. Letters 260, L23.
7. Lo,K.Y.,Walker,R.C.,Burke,B.F.,Moran,J.M.,Johnston,K.J. and Ewing,M.S.
       1975 Ap. J. 202, p 650.
8. Moran,J.M.,Burke,B.F. and Barrett,A.H. 1968 Ap. J. Letters 152, L97.
9. Moran,J.M.,Reid,M.J.,Lada,C.J.,Yen,J.L.,Johnston,K.J. and Spencer,J.H.
       1978 Ap. J. Letters 224, L67.
10. Reid,M.J.,Haschick,A.D.,Burke,B.F.,Moran,J.M.,Johnston,K.J. and
       Swenson,G.W. Jr. 1980 Ap. J. 239, p 89.
11. Weinreb,S.,Meeks,M.L.,Carter,J.C.,Barrett,A.H. and Rogers,A.E.E.
       1965 Nature 208, p 440.

[1] Present Address: Instituto de Astronomia, Universidad Nacional
                    Autonoma de Mexico.
[2] Partially supported by CONACYT, Mexico.
[3] Present Address: Haystack Observatory, Massachusetts, U.S.A.

PROPER MOTIONS AND DISTANCES OF WATER MASER COMPLEXES [+]

M. H. Schneps, M. J. Reid, J. M. Moran
Harvard-Smithsonian Center for Astrophysics

R. Genzel
University of California, Berkeley

D. Downes
Institut de Radio Astronomie Millimetrique

B. Rönnäng
Onsala Space Observatory

We report preliminary results of a long term spectral line VLBI experiment to observe internal proper motions of water maser sources in the vicinity of newly formed stars. This technique yields a picture of the three-dimensional kinematics of the region and a measure of the distance to the source. First results from the galactic center source SGR B2 are presented.

1. INTRODUCTION

Water masers have been observed to occur in diverse astrophysical settings ranging from the gaseous atmospheres of evolved stars to the turbulent environs of newly formed stars (cf. review of Reid and Moran 1981). Although the excitation mechanism is poorly understood, the maser's tremendous intensity and point-like emission characteristics may be exploited to study the dynamics of the maser region. In regions of active star formation, the cloudlets of maser gas are swept along by the supersonic flows from a young star, and these motions may be traced by making successive VLBI images of the region over a period of one or two years.

A very important result of such observations is that a direct distance to the object can be obtained. The current precision of VLBI allows distances to be obtained for maser complexes lying at any distance within the galaxy. These maser complexes typically contain 50

R. Fanti et al. (eds.), VLBI and Compact Radio Sources, 335–338.
© *1984 by the IAU.*

to 100 individual emission cloudlets. The distance is obtained from a statistical comparison of the angular motions with the radial velocities (obtained from doppler shifts of the maser line) of the maser spots. Since conditions in these complexes are highly turbulent, ordered flows are generally overwhelmed by random motions, and the distance determined is relatively insensitive to systematic motions within the complex. The uncertainty in the distance is inversely proportional to the square root of the number of proper motion vectors observed. Efforts are therefore made to detect as many maser cloudlets in any given source as possible.

In previous work for Orion KL (Genzel et al 1981a) a distance to the source of 480 +/- 80 pc was derived, respectively, in excellent agreement with optical determinations. A systematic outflow emanating from IRc2 was observed. For the sources W51 Main (Genzel et al 1981b) and W51 North (Schneps et al 1981), which are considerably more turbulent, distances of 7 +/- 1.5 kpc and 8 +/- 2.5 kpc, were obtained.

Encouraged by the early results, we undertook a new experiment covering sources of interest for galactic dynamics: Cepheus A, Orion KL, Sgr B2, W49 North, W3(OH), and G12.2+0.1. This second generation experiment employed improved software and observing procedures with the intent of maximizing the number of maser features detected. The observations, made at five epochs (1980 DEC, 1981 APR and DEC, 1982 MAR and JUN), were closely spaced in time in order to follow the motions of the masers during their one to two year lifetimes. Care was taken to maintain the same observing procedures among the recording sessions in order to reduce errors due to systematic instrumental effects. The observations involved six stations: Onsala, MPIFR, Haystack, NRAO, VLA, and OVRO. All epochs have been correlated, a task which required 50 percent of the NRAO Mk II correlator for two years, and calibration and mapping is underway.

2. PRELIMINARY RESULTS FOR SGR B2

A subset of the data for the maser complex in the galactic center, SGR B2, has been analyzed for the 1981 APR and 1982 MAR epochs. The northern and middle clusters (cf. Elmegreen et al 1980), separated by 44 arc seconds in a nearly north-south direction, were observed simultaneously in the telescope beams. Fringe rate maps (fig 1a: northern cluster) show the masers to be grouped in tight clusters, sparsely distributed over a three arc second region for each of the two clusters. Approximately 80 emission features were identified in these fields. Figure 1b shows a detail from fig. 1a with LSR velocities of the features indicated in km/sec.

Using the fringe rate maps as a guide, subfields containing emission features were imaged with synthesis techniques for each frequency channel. This showed that many of these maser emission spots to be small groups of masers distributed over several milliarcseconds, roughly doubling the number of maser features identified. Fig. 1c

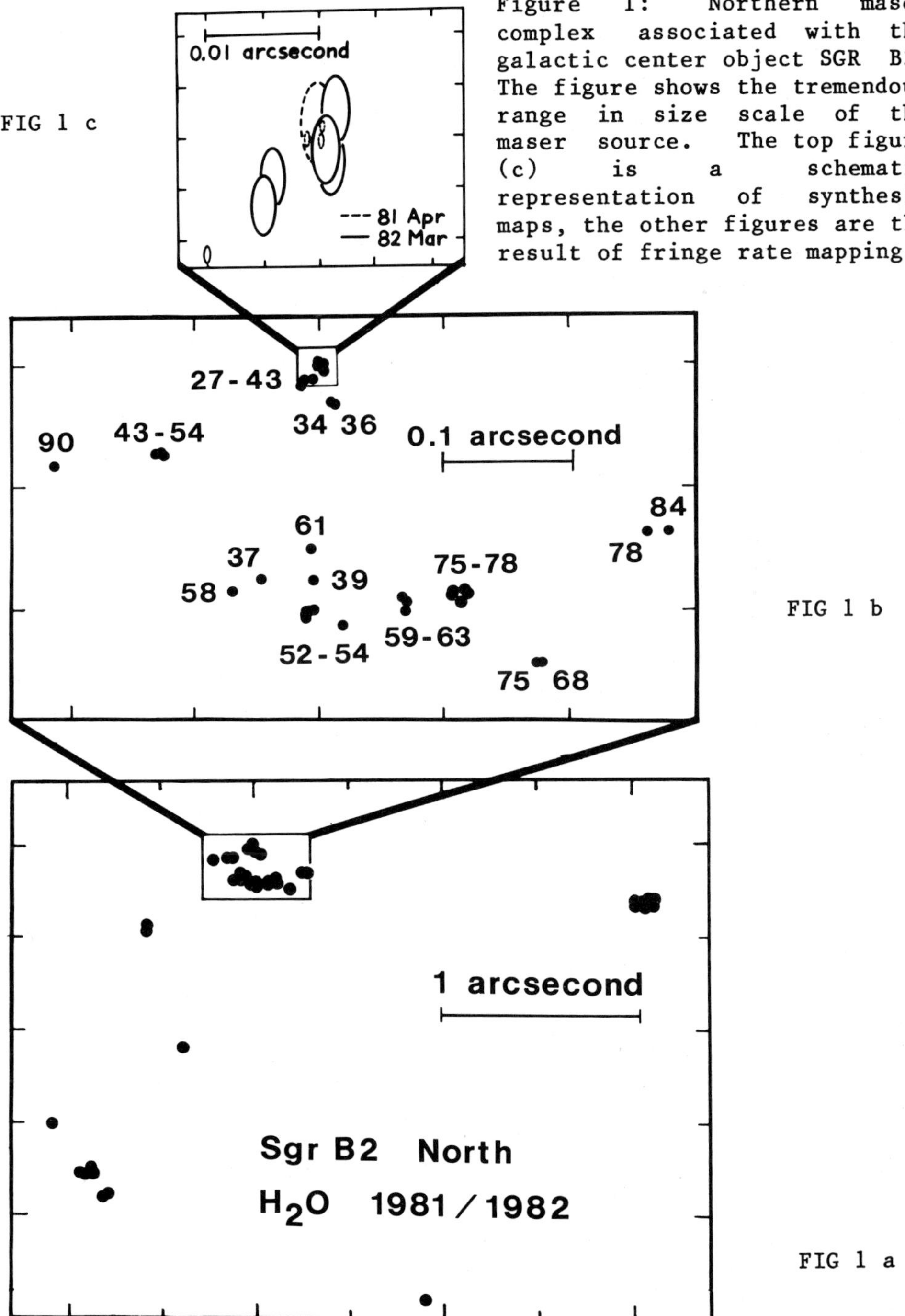

Figure 1: Northern maser complex associated with the galactic center object SGR B2. The figure shows the tremendous range in size scale of the maser source. The top figure (c) is a schematic representation of synthesis maps, the other figures are the result of fringe rate mapping.

schematically summarizes 20 synthesis maps, cover a 5 km/s velocity range, for a subfield of fig. 1b. The size of the features in the figure indicate the relative intensities.

3. EXPECTED PRECISION OF GALACTIC DISTANCES

Unlike previous experiments, the measurement errors in maser positions should be a negligible contribution to the distance uncertainty. The resolution of the synthesized beam is approximately 0.4 millarcsecond. The measurement precision of the position of the emission centroid is limited by the dynamic range (about 100:1) for strong features, and by the signal-to-noise ratio for weak features. We thus expect that for the majority of maser features, the position of the emission centroid can be determined to about 10 microarcseconds. We anticipate that systematic errors due to baseline uncertainties, etc., will be less than this. The latter source of error will be further reduced (to second order) when observations of one epoch are compared against another in a differential measurement of the proper motion. The magnitude of the position shifts typically exceed 100 microarcseconds, which is an order of magnitude greater than the position uncertainties.

The statistical errors in the distance depends primarily on the number of maser features we can identify between any two epochs. If motions of all the features found in SGR B2 can be traced over several epochs, the distance to the galactic center could be determined to perhaps 10 percent. However, if only the motions of maser clusters are traced, the uncertainty will be closer to 20 percent.

REFERENCES

Elmegreen, B. G., Genzel, R., Moran, J. M., Reid, M. J., Walker, R. C.: 1980. Ap.J. 241, 1007

Genzel, R., Reid, M. J., Moran, J. M., and Downes, D.: 1981a, Ap J., 224, pp. 884.

Genzel, R., Downes, D., Schneps, M. H., Reid, M. J., Moran, J. M., Kogan, L. R., Kostenko, V. I., Matveyenko, L. I., and Rönnäng, B.: 1981b, Ap. J.,247, pp. 1039.

Reid, M. J., and Moran, J. M.: 1981, Ann. Rev. Astron. Astrophys, 19, pp. 231.

Schneps, M. H., Lane, A. P., Downes, D., Moran, J. M., Genzel, R., and Reid, M. J.: 1981, Ap. J., 249, pp. 124.

RADIO ASTROMETRY [+]

K. J. Johnston
E. O. Hulburt Center for Space Research, Naval Research
Laboratory, Washington, D. C. 20375

ABSTRACT
    Radio astrometry is now the premier astrometric technique for mea-
suring the positions of celestial objects. The precision with which
absolute positions can be determined is approaching a few milliarcseconds. The progress towards establishing an almost inertial reference
frame based upon the positions of extragalactic radio sources is reviewed as of June 1983. The outlook for relating this reference frame
to optical reference frames is also reviewed.

INTRODUCTION
    Astrometry is the measurement of the precise positions and motions
of celestial objects. It is usually divided into the measurement of the
relative positions of objects over small ($<$ 1°) and large ($\sim$ 360°)
angles. The latter is usually referred to as the absolute position of
the coordinates of the object in right ascension (relative to the first
point of Aries) and declination. This paper reviews the status of
astrometric positions as regards to large angles. However before
reviewing large angle astrometry, let me address how astrometry is
associated with many aspects of this symposium.
    Stellar kinematics and dynamics may be considered a subfield of
astrometry. Stellar kinematics deals with the space motions of stars
leading to studies of precession, solar motion, galactic rotation,
statistical and secular parallaxes. Stellar dynamics deals with tidal
effects in clusters and associations, and the effects of the overall
gravitational potential on objects. Papers in this symposium volume
deal with the motions of plasma ejected in a strong gravitational field
from a central compact object. The measurement of relative positions
over very small angles (milliarcseconds) is the topic of many papers in
this symposium discussing "superluminal" motion. They also deal with
the evolution of "beams" from these objects which in the case of SS433
display precessional motion. It is unfortunate that as yet there is no
strong spectral line in the radio spectrum which can yield the radial
velocities of this ejected plasma. Other papers in this symposium deal
with maser emission. These may be considered studies of the "stellar"
dynamics of clusters where the internal motions lead to the determination of the distances to the maser cluster.

[+] Discussion on page 464   339

R. Fanti et al. (eds.), VLBI and Compact Radio Sources, 339–346.

AN INERTIAL REFERENCE FRAME

Extragalactic radio sources, since they are the furtherest known celestial objects, display very little motion on the celestial sphere. As such, a catalog of precise radio source positions of extragalactic sources may be used to define an almost inertial reference frame against which motions of objects on the earth, motions of the earth, objects in the near earth environment (satellites), and objects on the celestial sphere (planets, stars, and galaxies) may be defined. This leads to applications in the fields of astrophysics, astronomy, and geophysics.

Celestial reference frames are based upon many stars. The FK4 system contains 1535 stars while the proposed FK5 system will contain over 5000 stars. There is an ample number of compact radio sources for a fundamental reference frame (Johnston and Ulvestad 1982). A working group of IAU Commission 24 for the Identification of Candidate Radio/Optical Sources has selected 236 sources of flux density > 1 Jy (Argue et al. 1983). The distribution of these sources is shown in figure 1. There is a paucity of sources south of $-40°$ declination, but work is progressing on selecting sources. The initial reference frame will probably consist of $\sim$ 100 sources of flux density > 1 Jy. If a denser grid of sources is needed, over a thousand sources of flux density 0.2 Jy are available.

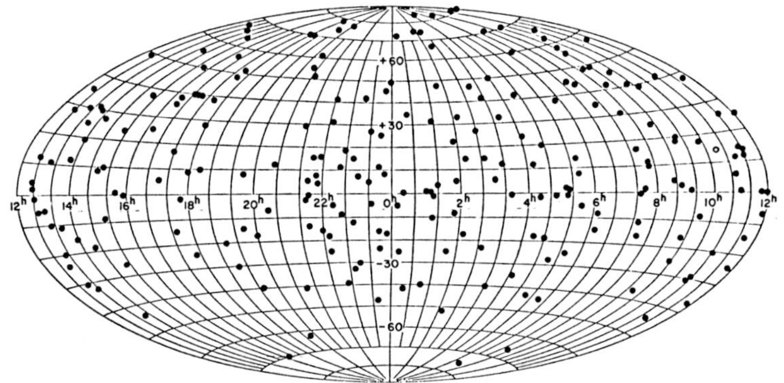

Fig. 1 The distribution of radio/optical sources proposed by the working group of IAU Commission #24 to establish an almost inertial reference frame (Argue et al. 1983).

PROGRESS TOWARD MEASURING PRECISE POSITIONS

The measurement of the positions of compact extragalactic radio sources has progressed very well over the past ten years. The promise of high accuracy for Very Long Baseline Interferometric measurements made in the late 60's are now being fulfilled. In 1972, positional accuracies of 0."1 were considered good. At this time precision is approaching 0."001. Here we will deal only with precision because radio interferometric measurements of celestial source position are made with respect to the instantaneous pole of rotation of the earth at the time of the measurements. The celestial positions reported depend upon the models for earth motions such as polar motion, spin axis motion such as

precession and nutation, revolutionary motions about the earth moon
barycenter, as well as solar system and solar galactic motions. Until
these motions are known on the 0."001 level, it is difficult to compare
the accuracy of catalogs.

TABLE 1

CATALOG OF RADIO SOURCE POSITIONS

Authors	Instrument	Number of Sources	Precision	Observing Epoch	Reference Epoch
Connected Element Interferometry					
Elsmore & Ryle (1976)	Cambridge	55	0."03	1973.1; 1974.2	1950
Elsmore (1982)	Cambridge	25	0.03	1979.8	1950
Hilldrup et al.(1982)	VLA	29	0.02	1980.0	1950, 2000
Kaplan et al. (1982)	Green Bank	16	0.01	1979.9	1950, 2000
Perley (1982)	VLA	393	0.05	1981.0	1950
Wade and Johnston (1977)	Green Bank	34	0.03	1975.4	1950, 2000
Ulvestad et al. (1981)	VLA	250	0.10	1979.10	1950
Very Long Baseline Interferometry					
Clark et al. (1976)	US-Europe	18	0.04	1973.9	1950
Purcell et al. (1980)	Madrid-Goldstone-Tidbinbilla	117	< 0.01	1978.0	1950
Shaffer et al. (1982)	US-Europe	48	0.005	1981.5	2000
Ma et al. (1983)	US-Europe	21 Mk II	0.002	1975.2	2000
		71 Mk III	0.001	1981.2	2000
Sovers et al. (1982)	Madrid-Goldstone-Tidbinbilla	117	0.002	1977; 1978	2000

The improvement in positional precision and in the number of sources
measured since 1970 is shown in Table 1. Included are catalogs with a
large number of sources with accuracies better than 0."1 or catalogs of
higher accuracies with over ten sources. The precision of the catalogs
varies from 0."1 to 0."001. The radio frequencies of the measurements are
between 2-9 GHz. Earlier measurements were made at only one frequency.
The later catalogs of Hilldrup et al. (1983), Kaplan et al. (1982),
Purcell et al. (1980), Shaffer et al. (1982), and Ma et al. (1983) were
made at two frequencies between 1.4-2.6 and 4.9-8.4 GHz in order to
eliminate the delay path length in the ionosphere as a cause of systematic error. It is very difficult to compare the catalogs because the
earlier work was expressed in terms of the standard B1950 epoch. The
constants involved in precession, nutation, other earth rotation parameters, as well as time scales are not expressed as exactly in this
system, which is based upon the fundamental catalog FK4, as in the new
FK5 system which has the new standard epoch J2000. For example, the
precession constant in the earlier system differs by 0."01/year from that
of the FK5 system. The IAU adopted resolutions during the General
Assemblies of 1976 and 1979 establishing the FK5 fundamental reference
frame. However exact transformation to the FK5 system cannot be performed until the catalog containing the stars defining the reference

frame is available.   One can see from the epoch of observations of the
different catalogs that this effect is very important.   Before attempting to directly compare catalogs, one should look for rotations between
the coordinate reference frames.   This will occur even when positions
are expressed in the standard J2000 epoch of the FK5 reference frame.
These rotations will lead to improved knowledge of the effects of earth
rotation parameters and of motions of the earth, sun, and galaxy.   The
precision of the "best" catalogs in Table 1 is about a milliarcsecond.
These VLBI catalogs were made from observations obtained over a period
of several years.   In contrast, the connected element catalogs are made
from only three or less observing epochs spaced at most by a year.   The
catalog claiming the highest precision (Ma et al. 1983) was reduced in a
manner compatible with the J2000 reference epoch.   The positions of the
common sources from the Ma et al. (1983) catalog compares favorably with
those of other catalogs within the errors of the other catalogs.   For
example, the root sum of the squares of the differences for sixteen
common sources in the Ma et al. (1983) and the Sovers et al. (1982)
catalogs is at the 3 milliarcsecond level.

Evidence for the stability of the reference frame defined by extragalactic radio sources can be demonstrated by the repeatability of the
positions in the various catalogs.   The position of the very variable
source BL Lac appears to be stable at the 2 milliarcsecond level for the
period 1978-1982.   The flux density from this source has recently undergone variations in intensity by a factor of 5, indicating several outbursts in the 1980-1981 time frame.   The mas structure has been shown by
Phillips and Mutel (1983) to be that of a typical superluminal source in
which the radio core dominates the emission with emission moving away
from the central core with an apparent velocity in excess of the velocity of light.

REFERENCE POINT OF RIGHT ASCENSION

The celestial coordinate system based upon our rotating earth
uniquely defines the source declination but leaves the zero point of
right ascension to be defined.   Radio interferometric measurements of
source positions, as long as they are carried out by earth based antennas also need to define a zero point of right ascension.   In order to
align the radio coordinates with the zero point of right ascension as
defined by optical methods, the position of a radio source, or several,
are made to coincide with their optical counterparts.   This has been
done in some cases by using the position of quasars such as 3C273B or of
radio stars such as Algol.   More will be said later about the detailed
relationship and suitability of sources for relating the optical and
radio reference frames.

NEEDS FOR FUTURE VLBI MEASUREMENTS

As stated by Johnston (1983), future VLBI measurements aimed at the
high precision astrometric work should be made as follows:
1.   These observations should be expressed in a uniform manner on
the standard J2000 reference epoch of the FK5 reference frame defined by
the IAU system of astronomical constants as stated in the resolutions

passed at the 1976 and 1979 IAU General Assemblies. A convenient definition of this system for radio astronomers is found in Kaplan (1981).
The data should be expressed so that the complete observable can be
reconstructed. This data must contain the epoch of observations, the
delay, and delay rates, etc. This is extremely important because only
with this knowledge can the data be analyzed for precise determination
of earth rotation parameters, precession, nutation, etc.
     2. The observations should be made using: a) dual radio frequencies
to eliminate ionospheric effects, b) wide radio bandwidths (400 MHz) to
insure highest accuracy, and c) incorporate water vapor radiometers and
weather stations to calibrate atmospheric effects.
     3. Source structure should be measured at timely intervals in order
to remove this effect from the catalog positions.
     4. A core list of approximately fifty sources should be establish-
ed. These sources should be primarily used as the calibrators for
measuring precise baselines and comparing the reference frames defined
by individual source catalogs.
     5. The right ascension zero point should be established by a single
extragalactic source.

RELATIONSHIP OF OPTICAL/RADIO REFERENCE FRAMES

     As stated earlier, the optical FK4 or FK5 reference frames may be
related to the radio reference frame by measuring the positions of
objects that display both optical and radio emission. Some objects that
have been proposed are a) stars that display associated radio emission,
b) solar system objects such as asteroids, planetary satellites, and
artificial orbiters, and c) optical counterparts of compact extragalactic radio sources.
     The problem with the radio emission associated with stars is that
the emission mechanisms are not thermal and are located in a circumstellar region that may have a diameter of several A.U. Optical positions are centered on the stellar photosphere. Flare stars such as
the RS CVn variables appear to be one of the best candidates for coincident optical/radio emission. Recent VLA measurements by Johnston,
Wade, and Florkowski (1983) indicate that for the star HR1099 the optical and radio emission is probably coincident to 0".01. Further measurements will indicate whether this can be extended to 0".001.
     Maser emission from OH, H_2O, and SiO are also associated with late
type Mira variable stars. The position of the OH and H_2O masers can be
measured using the VLA. Recent VLA measurements by Johnston, Spencer,
and Bowers (1983) of R Aql and RR Aql show the H_2O emission is distributed over 0".1-0".2. The H_2O emission from the symbiontic star RX Boo is
shown in figure 2 and is distributed over 0".3. Since these masers are
pumped by energy from the star, the most intense masers hopefully will
be located in close proximity to the star ($\sim$ few A.U.). The star's
optical emission may be located in the maser cloud by modeling the
geometry and velocity gradients of the H_2O maser emission hopefully to
an accuracy of 0".02. The SiO masers may be superior to H_2O masers as
they are located closer to the star (Soulie and Baudry 1983).

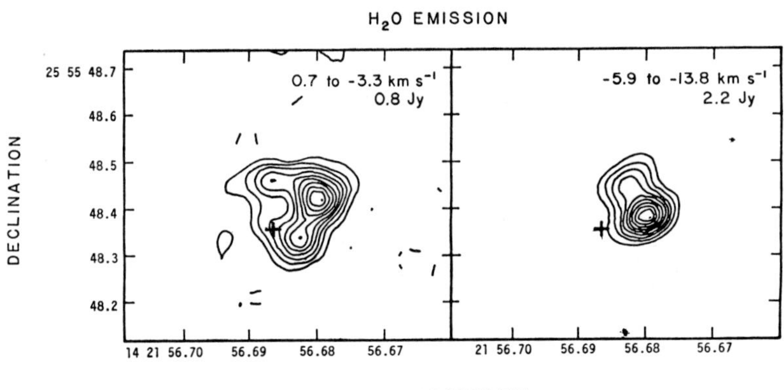

Fig. 2   The H$_2$O emission associated with the late type star RX Boo.
This is the emission averaged over the velocity ranges 0.7 to -3.3 km
s^{-1} and -5.9 to -13.8 km s^{-1}. The position of the strongest spectral
feature at 6 km s^{-1} is marked by a +.

Mass loss from early type stars allows these objects to be radio
sources. This emission is from ionized hydrogen close to the star.
However in most cases this emission displays spatial structure on arc
second scales. Again here one may obtain radio positional accuracies of
a few hundredths of an arc second.

The dynamical reference frame of the planets may be related to
extragalactic radio sources through radio measurements of solar system
objects. The asteriods and satellites of planets have finite sizes of
order 0."1 to 1."0 and are also not very intense at radio frequencies.
Measurements of the centroid of emission are possible at the 0."01 level.

Direct optical measurements of the radio sources making up the
reference frame have also been made. The accuracy of the radio
positions presented in Argue <u>et al</u>. (1983) are already adequate. Precise optical positions of 28 extragalactic radio sources north of 0°
which are referred to the FK4 system through the AGK 3RN indicate that
there are no significant differences larger than up to 0."2 (de Vegt and
Gehlick 1982) between the radio and FK4 reference frame. The accuracies
of the optical positions are at the 0."05 level. The positions of more
optical counterparts are needed.

Therefore at this time measurements are progressing which will
relate the optical to radio reference frame for individual sources at
the 0."05 level (the accuracies of optical positions). Improvement in
the optical reference frame can be made through comparison with the
radio reference frame. Future improvements in optical astrometric
methods such as HIPPARCOS or optical interferometry may achieve individual optical positional accuracy at the milliarcsecond level. The
radio positions relative to background extragalactic radio sources of
approximately 100 stellar objects at $\sim$ the 0."01-0."001 level should be
available by 1988.

CONCLUSION

    Radio astrometry now appears to have the capability of determining celestial positions over large angles to precisions of a few milliarcseconds. Over small angles (< 1°), relative positions are at the sub-milliarcsecond level. A large number of radio sources ($\sim$ 100) have now been measured with high precision and thus an almost inertial reference frame based upon extragalactic source positions is close to reality. This reference frame may be related to the optical reference frame at the 0".05 level which is now the accuracy of optical positions. Future observations of sources which display coincident optical/radio emission may improve this relationship as the precision of optical position determination improves.

REFERENCES

Argue, A.N., Elsmore, B., Fanselow, J., Harrington, R., Hemenway, P., Johnston, K.J., Kühr, H., Kumkova, I., Niell, A.E., de Vegt, Ch., Walter, H. and Witzel A.: 1983, submitted to Astr. Astrophys.

Clark, T.A., Hutton, L.K., Marandino, G.E., Counselman III, C.C., Robertson, D.S., Shapiro, I.I., Wittels, J.J., Hinteregger, H.F., Knight, C.A., Rogers, A.E.E., Whitney, A.R., Niell, A.E., Rönnäng, B.O., and Rydbeck, O.E.H.: 1976, A.J. 81, p. 599.

de Vegt, Ch. and Gehlich, U.K.: 1982, Astron. Astrophys. 223, p. 213.

Elsmore, B.: 1982, Abhandlungen aus der Hamburger Sternwarte Band X het 3, 129.

Elsmore, B. and Ryle M.: 1976, Mon. Notices Royal Astron. Soc. 174, p. 111.

Hilldrup, K., Fomalont, E.B., Wade, C.M., Johnston, K.J.: 1983, submitted to A.J.

Johnston, K.J.: 1983, "Multidisciplinary Uses of the Very Long Baseline Array", Report of the National Research Council, Washington, D. C.

Johnston, K.J., Seidelmann, P.K., and Wade, C.M.: 1982, A.J. 87, p. 1593.

Johnston, K.J., Spencer, J.H., and Bowers, P.R., submitted to Ap. J.

Johnston, K.J. and Ulvestad, J.S.: 1982, NOAA Technical Report NOS95, NGS24 Proceedings of Symposium #5 Geodetic Applications of Radio Interferometry, IAG, Tokyo, Japna, 7-8 May 1982, pp. 7-18.

Johnston, K.J., Wade, C.M., and Florkowski, D.: 1983, IAU Colloq #76.

Kaplan, G.H.: 1981, USNO Circular #163.

Kaplan, G.H., Josties, F.J., Angerhofer, P.E., Johnston, K.J., and Spencer, J.H.: 1982, A.J. 87, p. 570.

Ma et al.: 1983, in preparation.

Perley, R.: 1982, A.J. 87, p. 859.

Phillips, R. and Mutel, R.: 1983, private communication.

Purcell, G.H., Fanselow, J.L., Thomas, J.B., Cohen, E.J., Rögstad, D.H., Sovers, O.J., Skjerve, L.J., and Spitzmesser, D.J.: 1980, "Radio Interferometry Techniques for Geodesy", NASA Conference Publication 2115.

Shaffer, D.: 1982, presented at the URSI meeting January 1982.

Shapiro, I.I., Wittels, J.J., Counselman III C.C., Robertson, D.S., Whitney, A.R., Hinteregger, H.F., Knight, C.A., Rogers, A.E.E.,

Clark, T.A., Hutton, L.K., and Niell, A.E.: 1979, A.J. 84, p. 1459.
Soulie, G. and Baudry A.: 1983, Astron. Astrophys. Suppl. Ser. 52, p. 299.
Sovers, O. et al.: 1982, private communication.
Ulvestad, J., Johnston, K., Perley, R., and Fomalont, E.: 1981, A.J. 86, p. 1010.
Wade, C.M. and Johnston, K.J.: 1977, Astron.J. 82, 791.

PULSAR ASTROMETRY [+]

Joseph H. Taylor, Carl R. Gwinn,
Joel M. Weisberg and Lloyd A. Rawley
Joseph Henry Laboratories and Physics Department
Princeton University
Princeton, N. J. 08544 U.S.A.

High precision measurements of the celestial coordinates of pulsars are desirable for a number of reasons. If carried out at several epochs, the measurements can yield angular proper motions; together with distance estimates based on dispersion measure, the proper motion of a pulsar reveals two of three components of its space velocity, and consequently provides important kinematic information on pulsar ages (see, for example, Manchester, Taylor and Van 1974; Lyne, Anderson and Salter 1982; and references therein). Direct measurements of annual parallaxes are also possible in principle, and are marginally feasible with present techniques for a few of the closest pulsars. Model independent distances obtained from parallax measurements, together with observed pulsar dispersion measures, yield the electron density along the line of sight to the pulsar. Knowledge of the interstellar electron density in the solar neighborhood provides a calibration of the dispersion-based distance scale that is complementary to the calibration derived from neutral hydrogen absorption measurements of more distant pulsars (Weisberg et al. 1980), and permits appropriate statistical analyses to be made of the local space density of pulsars and their birthrate (e.g. Taylor and Manchester 1977). Finally, pulsar astrometry can be expected to yield important information on the relative orientations of fundamental reference frames. In particular, pulse timing observations yield positions in a reference frame based on motions of the planets, while interferometric position measurements are based on an Earth-equatorial system. At present the relative orientation of these two coordinate systems is known to only $\sim 0\overset{''}{.}2$ accuracy, though the potential precision of both types of measurements is much higher.

Pulse arrival times can be measured with typical accuracies of $0.1 \lesssim \Delta t \lesssim 1$ ms. Therefore, if a series of timing observations extending over a year or more are available, a pulsar's position can be estimated with an accuracy of approximately c $\Delta t/(1$ AU$)$, or typically a few tenths of an arc second or better. When data spanning several years are available, proper motions can also be measured in this way (Gullahorn and Rankin 1978a,b; Helfand et al. 1980; Downs and Reichley 1983). Some pulsars have intrinsic timing irregularities which

R. Fanti et al. (eds.), VLBI and Compact Radio Sources, 347–353.
© *1984 by the IAU.*

accumulate to several milliseconds or more after a few years; these
irregularities are unpredictable, and produce systematic errors that
limit the astrometric accuracies attainable. In the best cases,
however, the pulsar clock remains "perfect" (within measurement errors)
even after several years; for these pulsars, position accuracies of
< 0.1 and proper motions of a few milli-arc seconds per year are
achievable (Downs and Reichley 1983).

Pulsar astrometry by connected-element interferometric techniques
has been done most notably by Backer and Sramek (1981, 1982) and by
Lyne, Anderson and Salter (1982). Backer and Sramek measured the
positions of five pulsars to accuracies of 0.2 (absolute) or 0.03
(relative to nearby comparison sources), and the proper motions of the
same pulsars to accuracies of 3 to 40 mas/yr. In addition, they placed
an upper limit of 0.004 on the annual parallax of one object,
PSR 1929+10. The more extensive project of Lyne et al. used a phase
referencing technique, with calibration sources within the same beam
area. They measured the proper motions of 26 pulsars, with accuracies
of 1 to 20 mas/yr, and the annual parallaxes of nine sources, with
uncertainties of 2 to 9 mas. Only one of the parallaxes was
significantly nonzero, however; for PSR 1929+10, they obtain a
parallax of 0.022 ± 0.008, considerably larger than the upper limit of
Backer and Sramek. Thus, the unsatisfactory situation regarding pulsar
parallax measurements is summarized in Table 1:

Table 1. Parallax Estimates for PSR 1929+10

Annual Parallax of PSR 1929+10	Reference
0.022 ± 0.008	Lyne et al., 1982
< 0.004	Backer and Sramek, 1982
0.011	Expected value, from dispersion measure

It is clear that VLBI techniques should be capable of better
results than those already discussed. The greatest practical difficulty
in doing pulsar astrometry by VLBI involves the compromise that must be
made in choice of observing frequency: the pulsars are relatively weak
sources with steep spectra, suggesting a low observing frequency;
however, at frequencies below ~ 1 GHz, ionospheric irregularities pose
severe problems. Thus, when we embarked about two years ago on a
program designed to measure the parallaxes of several nearby pulsars,
we chose to observe at 1.6 GHz and to use the large collecting area of
the Arecibo telescope.

The other stations involved are Green Bank and Owens Valley. We
use standard Mark II recording techniques, except that pulsar pulse
phase information is encoded on the tapes recorded at Arecibo. This
information is used to inhibit correlation during the off-pulse periods,
thereby increasing the signal-to-noise ratio by the inverse square root
of the pulsar duty cycle -- a factor of three to five. Each pulsar is

observed alternately with a comparison source, with cycle times of 4-7
minutes, for as long as the sources are in Arecibo's tracking range, or
about 90 minutes. As shown in Table 2, we are observing six pulsar-
quasar pairs, in addition to three quasar-quasar pairs, for calibration
purposes.

Table 2. Source List

Source	Reference Source	Angular Separation (degrees)
PSR 0823+26	0822+272	0.5
PSR 0950+08	0938+119	4.8
PSR 1133+16	1119+184	4.0
PSR 1237+25	1222+217	4.9
PSR 1919+21	1923+210	1.3
PSR 1929+10	1947+079	5.4
1421+122	1427+109	2.1
1548+115	1551+130	1.8
1756+237	1751+288	5.2

After cross-correlating the recorded data, we slow the fringe
rates as much as possible by using the best available values of station
coordinates, source coordinates, polar motion, and UT1-UTC. We also
remove the Arecibo clock drift, calculated from the change in delay on
one source observed at the same sidereal time on two consecutive days.
Finally, we remove an ionospheric path length, calculated from
satellite Faraday rotation data, and a simple troposheric model (Reid
1983).

The Faraday rotation data are obtained from the World Data Center
in Boulder, Colorado. Data are available from earth satellite
receiving stations in Boulder, Sagamore Hill (Massachusetts), and
Ramey (Puerto Rico), presented as hourly values of integrated electron
content at the zenith over a geographic point near the receiver (for
details see Klobuschar 1975). The Ramey location is very close to
Arecibo, so these data were used directly; for Green Bank and Owens
Valley we used the Sagamore Hill and Boulder data, respectively,
retarded by 1/2 hour to account for the effective longitude differences.
We interpolated the hourly measurements, and applied geometric factors
appropriate for a spherical ionosphere at a height of 400 km. We
believe that this procedure usually gives the ionospheric phase delays
to within a few turns in the daytime, and < 1 turn at night. Thus, the
differential ionospheric phase delays between members of a source pair,
caused by large-scale ionospheric structure, should be known to $\lesssim$ 0.1
turns. None of our observations were taken during times of severe
ionospheric disturbance, and on most days we see little evidence of
localized, short term ionospheric fluctuations.

After the ionospheric corrections have been applied, the residual
fringe rates are typically less than 3 mHz. Fluctuations of the

residual phase about a straight line vary from 0.1 to 1 turn over 1.5
hours, and are dominated by the behavior of the Rubidium clock at
Arecibo and by the unmodeled part of ionospheric activity. We fit the
residual fringe rates to a first or second order polynomial. The
polynomial is then integrated and used to rotate the phases by the
appropriate integral number of turns, for phase connection.

We then fit a simple model to the fringe phases. The model
includes the position offset of the source, relative to the calibrator;
a polynomial, typically of degree five, which models the Arecibo clock
and residual ionosphere; a polynomial of order two or three which absorbs
variations due to the Owens Valley ionosphere and calibrator position
offsets; and clock offsets for two of the three stations. The position
offset is sensitive to the difference of the fringe phases; all other
parameters are sensitive to their sum. Thus, the position offset is
only weakly covariant with the other parameters, and the details of the
rest of the fitted model are largely unimportant. In most of our data,
it is easy to detect deliberately (or accidentally) misconnected fringes
by inspecting the post-fit residuals.

The short tracking time available at Arecibo limits our u-v plane
coverage so much that several positions may have approximately equal
values of chi-squared after the fit. Such points are separated by
integral numbers of turns of phase on the Arecibo-Green Bank and
Arecibo-Owens Valley baselines. Therefore, we have assumed that the
pulsars' proper motions are described approximately by the results of
Lyne, Anderson, and Salter (1982) and that the quasars move by much
less than a fringe between sessions. Using these assumptions, we
translate the measured position offsets to the appropriate fringe.

In Fig. 1 we show our results for the quasar pair 1421+122 and
1427+109 on four different dates. A formal solution for the relative
proper motion of 1421+122 over this seven month span yields
$\mu_\alpha = 3 \pm 5$ mas/yr, $\mu_\delta = - 3 \pm 6$ mas/yr. Thus, our results are fully
consistent with zero relative motion between the sources, as expected.

For the pulsar-quasar pair PSR 0950+08 and 0938+119, we have
obtained usable data on seven dates spanning ten months. By resolving
fringe ambiguities so as to be most nearly consistent with the proper
motion given by Lyne, Anderson and Salter (1982), and then fitting our
data for proper motion and parallax, we obtain the results shown in
Figure 2 and in the top line of Table 3. We also tried shifting all of
the measured positions by up to 2 fringes in each direction of
ambiguity, and found only one other solution even marginally consistent
with the Jodrell Bank data and a reasonable value of parallax. This
alternative solution is listed below our preferred solution in Table 3,
followed by the Jodrell Bank results for proper motion.

The error bars plotted in Figs. 1 and 2 are considerably larger
than the formal standard errors from the fits for position offset, and
represent our best estimates of the systematic errors in the model

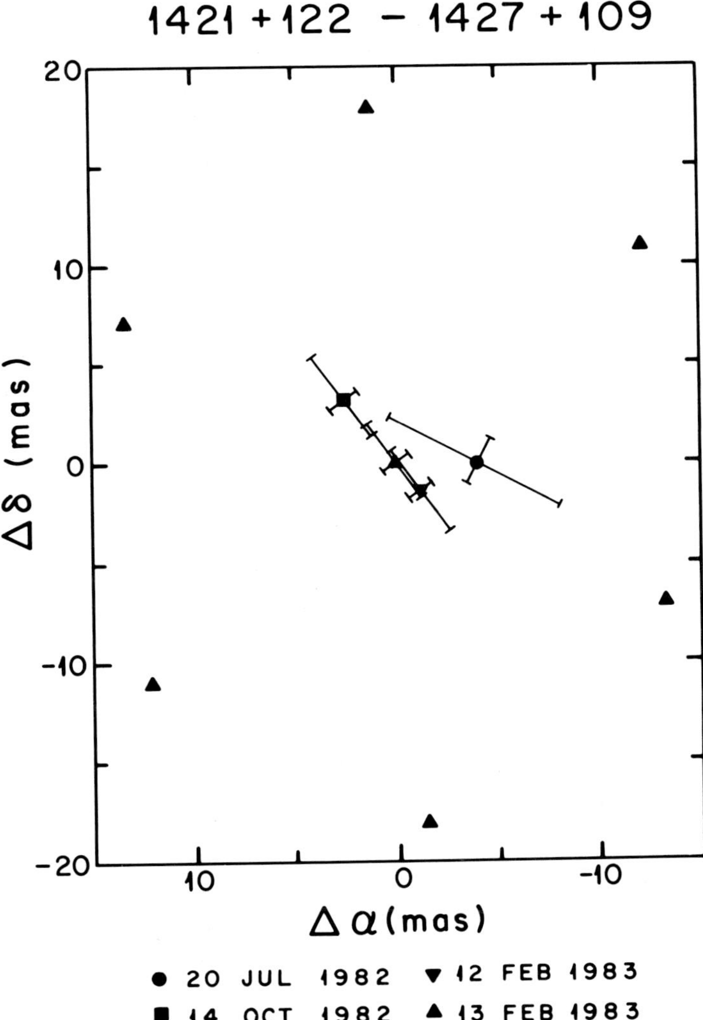

Fig. 1.   Measured position of the radio source 1421+122, relative to
that of 1427+109, on four different dates. The black
triangles show the closest six lobe-shifted positions
corresponding to the 13 February point. The much larger
error bars for the 20 July measurement are the result of
having no ionospheric data from Boulder on that day.

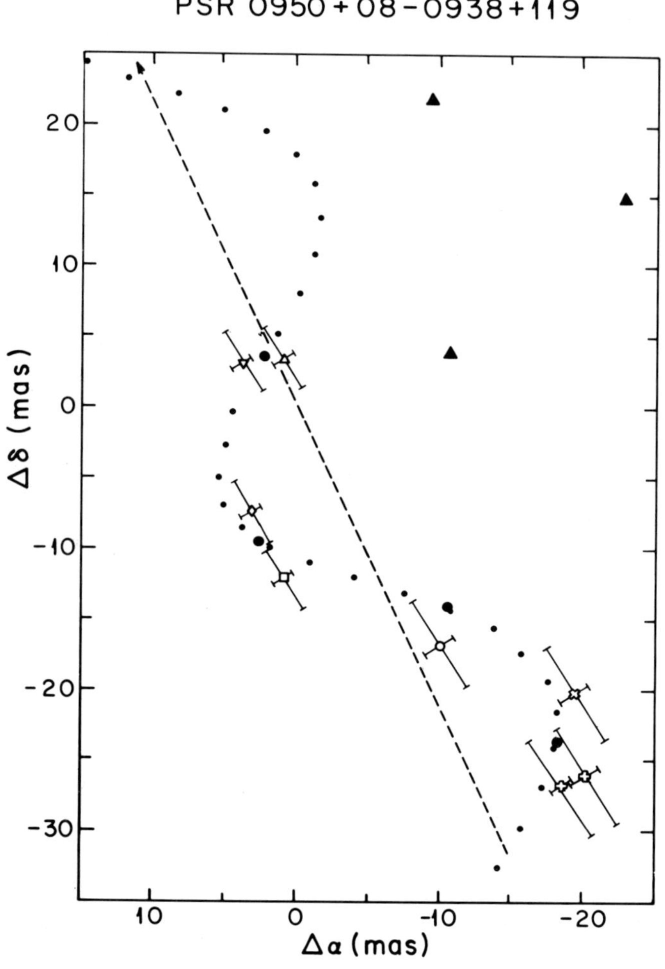

Fig. 2. The positions of PSR 0950+08, relative to 0938+109, on seven dates. Small block dots illustrate the path of the pulsar, according to our best-fitting solution for proper motion and parallax; large black circles give the expected position at the times of our four sessions. The dashed line gives the proper motion alone, and is in good agreement with that measured by Lyne, Anderson, and Salter (1982).

Table 3. Astrometric Data for PSR 0950+08

	$\mu_\alpha \cos \delta$ (0$''$001/yr)	μ_δ (0$''$001/yr)	π (0$''$001)
Preferred solution	17 $\pm$ 4	35 $\pm$ 5	7.5 $\pm$ 0.8
Alternative solution	27 $\pm$ 4	26 $\pm$ 5	14.6 $\pm$ 0.8
Lyne et al. (1982)	15 $\pm$ 8	31 $\pm$ 5	---

ionosphere, UT1-UTC, and polar motion. Further observations should
help to ascertain whether these estimates are always realistic.

We provisionally conclude, at the time of this progress report on
an ongoing project, that the distance to PSR 0950+08 is 130 $\pm$ 15 pc.
Since the dispersion measure is 2.969 cm^{-3} pc, the average density of
free electrons along the line of sight must be $<n_e>$ = 0.022 $\pm$ 0.003.
This value is in good agreement with the electron densities determined
along the lines of sight to more distant pulsars through HI absorption
measurements (Weisberg et al. 1980).

REFERENCES

Backer, D..C. and Sramek, R. A.: 1981, Pulsars, IAU Symposium No. 95,
     ed. W. Sieber and R. Wielebinski, Dordrecht: Reidel, p. 205.
Backer, D. C. and Sramek, R. A.: 1982, Astrophys. J. 260, 512.
Downs, G. S. and Reichley, P. E.: 1983, Astrophys. J. (Supplement
     Series), in press.
Gullahorn, G. E. and Rankin, J. M.: 1978a, Astron. J. 83, 1219.
Gullahorn, G. E. and Rankin, J. M.: 1978b, Astrophys. J. 225, 963.
Helfand, D. J., Taylor, J. H., Backus, P. R. and Cordes, J. M.: 1980,
     Astrophys. J. 237, 206.
Klobuschar, J. A.: 1975, Air Force Cambridge Research Laboratory,
     report number AFCRL-TR-75-0502.
Lyne, A. G., Anderson, B. and Salter, M. J.: 1982, Mon. Not. Roy.
     Astron. Soc. 201, 503.
Manchester, R. M. and Taylor J. H.: 1981, Astron. J. 86, 1953.
Manchester, R. M., Taylor, J. H. and Van, Y.-Y.: 1974, Astrophys. J.
     (Letters) 189, L119.
Reid, M. J.: 1983, Personal Communication.
Salter, M. J., Lyne, A. G. and Anderson, B.: 1979, Nature 280, 477.
Taylor, J. H. and Manchester, R. N.: 1977, Astrophys. J. 215, 885.
Weisberg, J. M., Rankin, J. and Boriakoff, V.: 1980, Astron. Astrophys.
     88, 84.

A NEW WAY OF TYING TOGETHER THE HIPPARCOS FRAME TO THE VLBI EXTRAGALACTIC
FRAME : ASTROMETRIC OBSERVATIONS OF STELLAR MASER SOURCES +

A. Baudry, J.M. Mazurier, Y. Requième
Observatoire de l'Université de Bordeaux et LA 352 du C.N.R.S.
33270 Floirac, France

The Hipparcos (H) reference system will contain 100 000 stars whose positions and proper motions will be known with an accuracy of $0\overset{''}{.}002$ and $0\overset{''}{.}002$/yr respectively. Because a residual rotation of the H system may exist and because a VLBI extragalactic frame is being constructed one must (i) determine an eventual rotation of the H rigid frame, (ii) link the H system to the VLBI extragalactic frame. VLBI observations at JPL (Fanselow *et al.*, 1981) have provided a catalogue of about 100 quasars with internal accuracy of $\sim 0\overset{''}{.}001$.

One major difficulty in tying together the H reference system to the VLBI frame follows from the fact that extragalactic radio sources have faint optical counterparts : fainter than 17 mag in general, with one well known exception, 3C273B. Because objects fainter than ~ 12 mag are not accessible to Hipparcos bright link stars must be observed with respect to faint extragalactic sources. This can be achieved by Space Telescope. However, simulations (Froeschlé and Kovalevsky, 1982) have shown that a direct link to the VLBI frame by means of radio stars also observable by Hipparcos is a more promising method. An accuracy of $\sim 0\overset{''}{.}002$ is expected provided that stars and radio sources coincide.

Radio *continuum* stars have been proposed to link optical and radio systems (e.g. Walter, 1977). Non thermal emitters associated with RSCVn type stars are a priori good candidates because the radio emission size is thought to be comparable with that of the optically unresolved binary system. However, these link stars may not be always appropriate : they show unpredictable radio flares ; most of them are weak radio sources (S_{5GHz} is often < 5-10 mJy) ; their spatial structure is unknown at the milliarc second level.

Maser *line* sources associated with late type stars may also be used to tie together the optical and radio frames of reference. Several Mira variables and late type supergiants exhibit strong (fluxes > 10-50 Jy) OH, H_2O and SiO maser emission. Unification of the H and VLBI systems is achieved as follows : (1) optical measurements of maser stars with ground-based astrometric facilities and during the Hipparcos mission at a favou-

+ Discussion on page 465   355

R. Fanti et al. (eds.), VLBI and Compact Radio Sources, 355–356.
© *1984 by the IAU.*

rable period of the light cycle ; (2) radio interferometer position measurements with respect to extragalactic objects of those maser bright spots that are closely related to the photospheric activity of stars. In two previous papers (Baudry *et al.*, 1979, and Soulié and Baudry, 1983) we have suggested that observations of H_2O and SiO masers are especially promising.

In a preliminary work the 33 cm astrograph of the Bordeaux Observatory was used to derive with respect to AGK3 reference stars positions of 11 late type stars (Soulié and Baudry, 1983). The Bordeaux automatic meridian circle is now used for repeated observations of radio continuum and maser stars. In this paper we present our very first results concerning maser stars. Positions are referred to FK4 stars. (About 30 fundamental stars are observed each night.) Taking into account all measurements

Name	α(1950.0)	Epoch	Number of obs.	δ(1950.0)	Epoch	Number of obs.
VY CMa	$7^h20^m54^s.667$	1983.14	4	$-25°40'13''.96$	1983.13	3
R Cnc	8 13 48.478	1983.16	10	11 52 51.95	1983.16	10
R LMi	9 42 34.756	1983.17	5	34 44 33.62	1983.17	5
R Leo	9 44 52.222	1983.16	6	11 39 40.43	1983.16	6
S Vir	13 30 23.129	1983.30	7	- 6 56 18.35	1983.27	6
RX Boo	14 21 56.729	1983.40	10	25 55 47.24	1983.40	10
S CrB	15 19 21.549	1983.42	10	31 32 45.61	1983.42	9
RU Her	16 08 08.485	1983.44	5	25 12 00.23	1983.45	4
U Her	16 23 34.695	1983.44	10	19 00 17.67	1983.44	9

the internal mean errors are : $\varepsilon_\alpha \cos\delta = 0''.117$ and $\varepsilon_\delta = 0''.151$. The position accuracy is of the order of $\varepsilon/\sqrt{N}$ where N is the number of observations for each star. Comparison of our astrograph and meridian circle measurements gives position differences in the range : $|\Delta\alpha| \simeq 0^s.001$ to $0^s.020$ and $|\Delta\delta| \simeq 0''.03$ to $0''.30$. For VY CMa the differences are larger and mostly due to the presence of a faint companion and of a small nebulosity.

References

Baudry, A., Delannoy, J., Lequeux, J. : 1979, J. Optics 10, 359.
Fanselow, J.C., Sovers, O.J., Thomas, J.B., Bletzacker, F.R., Kearns,T.J., Purcell, G.H., Rogstad, Jr., D.H., Skjerve, L.J., Young, L.E. : 1981, Reference Coordinate Systems for Earth Dynamics, p. 351 (eds. Gaposchkin, E.M., Kolaczek, B., Reidel, Dordrecht).
Froeschlé, M., Kovalevsky, J. : 1982, Astron. Astrophys. 116, 89.
Soulié, G., Baudry, A. : 1983, Astron. Astrophys. Suppl. Ser. 52, 299.
Walter, H.G. : 1977, Astron. Astrophys. Suppl. Ser. 30, 381.

INTERRELATION OF PRESENT OPTICAL AND RADIO REFERENCE FRAMES [+]

Chr. de Vegt

Hamburger Sternwarte

Abstract: Comparison of optical and radio positions in the northern
hemisphere yields local systematic differences up to o."2, mainly due to
combined systematic errors of current optical reference frame and contributing main catalogues. Interrelations of radio/optical frame and
future developments are discussed.

The requirements to obtain an unique interrelation of the independent
radio - and optical reference frames by a suitable net of carefully
selected extragalactic and galactic objects which display optical
counterparts provides some difficulties which are partly due to the
different underlying principles by which the traditional optical fundamental reference system is constructed.

As a consequence, the satisfactory alignment of radio/optical maps of extended objects or the correct identification of a radio point source
emitter can be achieved only if the systematic differences of both
reference frames in that particular sky region are known ; at least
with the same accuary as the single position determinations themselves.

RADIO REFERENCE FRAME

An extragalactic radio reference frame based on about 25o extragalactic
compact sources with optical counterparts is under investigation by a
working group of IAU Com. <u>24</u> (1). To a certain extent such a frame is
already provided by several radio source position catalogues obtained
from VLBI and CEI techniques. Accuracies of about 1o mas have been

R. Fanti et al. (eds.), VLBI and Compact Radio Sources, 357–360.
© *1984 by the IAU.*

obtained and further improvements to better than 5 mas are expected
soon. Due to its extragalactic nature this new reference frame should
be free of any systematic motion for a foreseen timescale, provided,
all astronomical constants (precession, nutation ...) involved are
determined definitely and all instrumental effects are correctly calibrated, these problems however need further investigation and cannot
be considered as having been solved at present.

OPTICAL REFERENCE FRAME

The current optical reference frame is based on the FK4 system, realized by the positions and proper motions of 1535 bright stars m_V < 6.5
which provide an uniform coverage of the sky with about 1* / 28 sq.deg.
The FK4 will be soon replaced by the FK5 which will have an increased
stellar density and is extended to fainter magnitudes $\sim$ 8. Furthermore, the FK5 will be based on the new precession constant, a different
equinox and an improved system of positions and proper motions. (2)
For an all sky comparison of the radio/optical reference frame, the
stellar density of the FK4 and FK5 is too low.
Comparisons of both systems are based at present on the IRS catalogue
(3) , which is represented on the northern hemisphere by an improved
version of the AGK3R - catalogue: the AGK3RN (4). The southern hemisphere
will be soon covered by the SRS catalogue, probably available at the
end of 1983. The complete IRS catalogue will contain about 50 000 stars
on system FK4 or probably FK5 and provides then a stellar density of
1* / sq.deg. in the magnitude range m_V $\sim$ 7 - 9.

EXAMPLE OF COMPARISON OF REFERENCE FRAMES

The following comparison is based on about 30 extragalactic and galactic
sources ; optical positions were determined in the system of the AGK3RN
catalogue at average epoch 1980 (see (5)) . Optical positions have been
obtained mainly from prime focus plates of large reflectors using a
system of fainter reference stars based on the AGK3RN system (6). In

contrast to the radio system, any optical catalogue system differs at epoch $t \neq t_o$ from its initial system (t_o) mainly due to the influence of systematic errors of proper motions. Therefore the differences $\Delta \delta$ "optical - radio", shown in the diagram (only δ will be discussed in this context) represent the combined effect of the FK4 system at $t \sim 1980$ and the AGK3RN at t ; an unique separation of both contributions is not possible at present. Concerning the FK4 part, a sky averaged total error of about $0\overset{\prime\prime}{.}1$ may be assumed (7). From the comparison one therefore can estimate the differences of the radio/optical frames based on the combined AGK3RN/FK4 systems. The diagram shows that local systematic differences may reach about $0\overset{\prime\prime}{.}2$. Results from additional 4o sources will be soon available to improve the sky coverage.

FUTURE OPTICAL REFERENCE FRAME

Mainly due to the influence of proper motions, the optical reference frame has to be improved continuously. Substantial progress will be expected during this decade both, from groundbased and space activities. <u>Groundbased</u>: the IRS will be reobserved by USNO pole-to-pole, new or improved photoelectric transit circles have been set in operation (Bordeaux, Tokyo, Washington 7") extensive work on new photographic catalogues is underway (Hamburg Obs., RGO, ...) As a result new positions of IRS stars with m.e. $\sim 0\overset{\prime\prime}{.}08$ are expected, photographic positions and proper motions with m.e. of $\approx 0\overset{\prime\prime}{.}05$ and $\sim 0\overset{\prime\prime}{.}003$/y will be added for fainter stars, providing then a dense net for faint radio source position work. <u>Spaceborn</u>: The HIPPARCOS astrometry-satellite-mission (2) will provide positions and proper motions of 1oo ooo stars with m.e. of $\sim 0\overset{\prime\prime}{.}002$ and $\sim 0\overset{\prime\prime}{.}002$/y at mean epoch 199o, probably all IRS stars will be included. This high precision stellar net will be adjusted finally to the extragalactic radio reference frame. Thus a close cooperation between radio and optical astrometry will be needed to establish a new global reference frame for the benefit of astronomy and neighbouring disciplines.

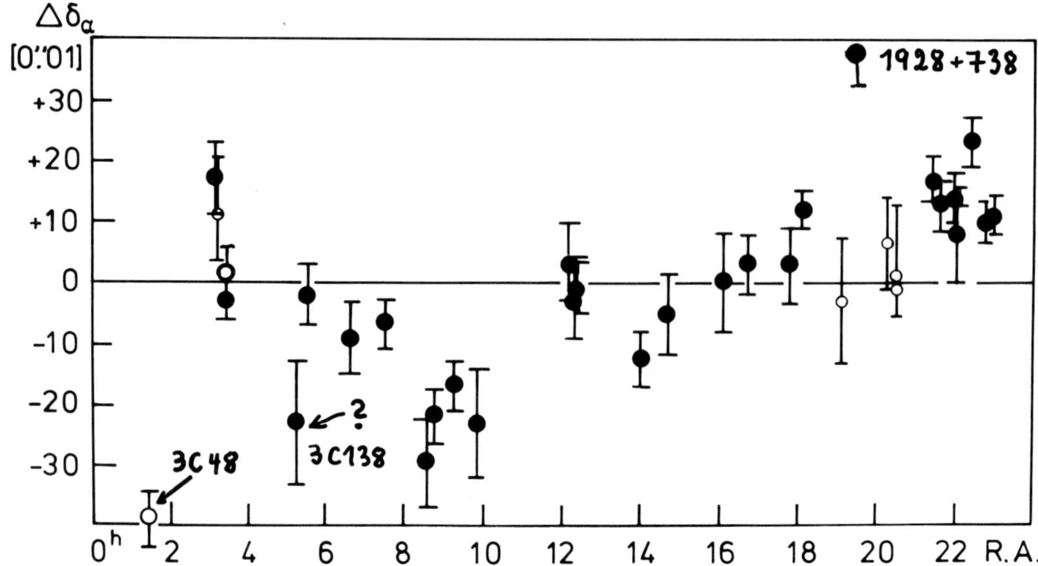

Differences $\Delta\delta$ "optical-radio" for AGK3RN/FK4 based optical system and corresponding radio data. Open circles show radio stars, filled extra-galactic sources. Indicated displacement of particular sources is due to source structure.

REFERENCES

(1) Argue A.N., de Vegt Chr., 1982, Abh.Hamb.Stw. 1o/3

(2) Fricke W., 1982, ESA-SP-177 , 43

(3) Smith C., 198o, Mitt.Astron.Ges. 48, 95

(4) Corbin T., de Vegt Chr., 1981, Astron.Astrophys. 1o4, 88

(5) de Vegt Chr., Gehlich U.K., 1982, Astron.Astrophys. 113, 213

(6) de Vegt Chr., 1979, Proc. IAU Coll. 48, 1o1 Vienna

(7) Lederle T., 1978, CDS Inf.Bull. 14, 62

Acknowledgement

This research has received financial support from BMFT, Project o1ooo13/8.

NEARLY MICROARCSECOND PRECISION DIFFERENTIAL ASTROMETRY [+]

J. Marcaide[1], I. Shapiro[2], J. Ball[2], N. Bartel[2], T. Clark[3],
B. Corey[4], M. Gorenstein[2], R. Preston[5], M. Ratner[2], A. Rogers[4],
and A. Whitney[4]

[1]Max-Planck-Institut für Radioastronomie, Bonn, F.R.G.
[2]Harvard-Smithsonian Center for Astrophysics, Cambridge, Mass.
[3]Goddard Space Flight Center, Greenbelt, Maryland
[4]Haystack Observatory, Westford, Mass.
[5]Jet Propulsion Laboratory, Pasadena, Calif.

In studies of extragalactic radio sources with multiple compact
components the determination of which components, if any, are stationary
and which moving is of importance. In order to learn about the radio
properties of the individual components it is also relevant to be able to
register maps made at several wavelengths. Both tasks are usually not
possible with VLBI because of the irrecoverable corruption of the fringe
phase introduced by the propagation medium and the instrumentation. However, when two or more compact radio sources are separated by only a small
angle from each other difference techniques can be used to help tackle
both questions.

In this contribution we present relative position determinations of
the radio sources 1038+528 A, B (see Marcaide _et al_. in these Proceedings)
at several wavelengths for a series of observations from November 1979
through March 1981. The separation of the sources was measured between
reference points chosen in each quasar map defined by the positions of
the strongest CLEAN components. The observable used was the difference
phase-delay corrected for the source structure and the fringe ambiguity.
The details of the technique have been discussed by Marcaide (1982) and
Marcaide and Shapiro (1983).

The results of our experiments are shown in Fig. 1. The most precise
relative position determination was obtained from data from March 1981
at $\lambda 3.6$ cm: The standard deviation was under 4 microarcseconds, which
corresponds to a phase-delay rms postfit residual under 2 picoseconds, or
equivalently to about half a millimeter of light travel path. A comparison of this separation with that obtained from data from November 1979 at
the same wavelength shows no indication of relative motion of the reference points in the maps with an upper limit of 18 microarcsecond/year.
Such a limit is consistent with the cosmological interpretation of the

+ Discussion on page 467   361

R. Fanti et al. (eds.), VLBI and Compact Radio Sources, 361–362.
© _1984 by the IAU._

redshifts. However, such a small separation rate computed from only two epochs has to be viewed with caution (see Marcaide and Shapiro (1983) for a related discussion).

Using this broader data base we confirm the result presented by Marcaide et al. elsewhere in these Proceedings that the separation between the reference points in the maps of the A and B quasars is less at longer wavelengths of observation than at short wavelengths. We think this result is due to a wavelength-dependent location of the peak of brightness of the A quasar.

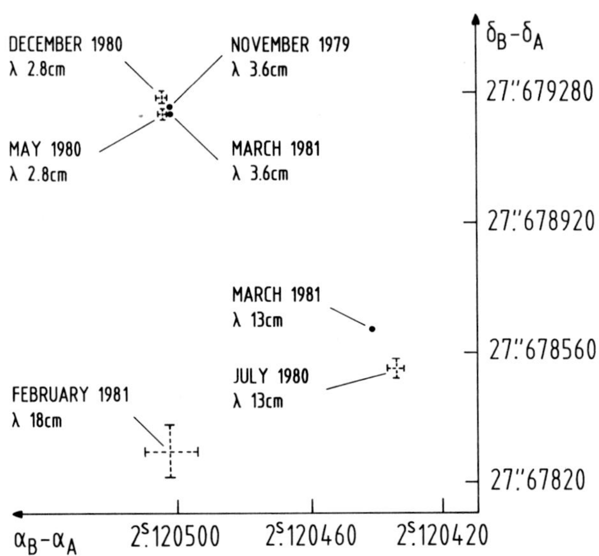

Figure 1:
Positions of the reference points chosen in the map of 1038+528 B relative to the reference points chosen in A (at the origin). The dashed error indicates that some assumption on the quasar structures was needed due to lack of information from the experiment itself. The displacement of the solution for λ18 cm observations with respect to the solutions for λ13 cm is likely due to use of an extrapolation of the structure at λ13 cm to that at λ18 cm based on an incorrect spectral index estimate for one source feature.

REFERENCES

Marcaide, J.M.: 1982, Ph.D. thesis, Massachusetts Institute of Technology
Marcaide, J.M. and Shapiro, I.I.: 1983, Astron. J. (to appear in August
     1983 issue)

VLBI GEODESY: TECHNIQUES AND RECENT RESULTS [+]

David B. Shaffer
Interferometrics Inc.

INTRODUCTION

From its inception, it was realized that VLBI had the potential for
very precise determinations of baselines between antennas. Since the first
early, crude results (Cohen and Shaffer 1971), the data collection systems
and analysis techniques have been continuously refined, and VLBI is now
producing baseline measurements to accuracies of a few centimeters over
trans- and inter-continental distances. Since the Mark III system and the
basic techniques of delay and delay rate baseline determination have been
previously described (Rogers et al. 1983; Shapiro 1976), and there have
been several recent meetings that dealt exclusively with VLBI geodesy
techniques (NASA 1980; NOAA 1982), I will concentrate in this review on
recent developments in equipment and procedures, followed by presentation
of new results. This review is based primarily on the efforts and results
of the so-called "East Coast VLBI Group" comprised of staff and support
contractors at NASA Goddard Space Flight Center (NASA/GSFC), the NEROC
Haystack Observatory, MIT, and the Harvard-Smithsonian Center for Astrophysics.
Additional input comes mostly from staff of the National Geodetic Survey
(NGS), Onsala Space Observatory, and the Jet Propulsion Laboratory.

THE NASA CRUSTAL DYNAMICS PROJECT

During the 1970's, development of VLBI and laser techniques made it
clear that distance measurements accurate to a few centimeters were feasible
over very long distances. NASA undertook developments in both these areas,
leading to the launch of the laser geodesy satellite (LAGEOS), improvement
of laser ranging systems, and the Mark III VLBI system. This new equipment
provided the expected precision, and NASA subsequently formed the Crustal
Dynamics Project (CDP).

The goal of the CDP is to acquire data for, and to make available the
results of laser and VLBI observations to, the geodesy and geophysics
community. Measurements of continental drift, distortion and stability of
individual crustal plates, and baseline changes that might precede or follow
earthquakes in geologically active regions such as southern California are
among the aims of the Project. To achieve scientifically useful results,

+ Discussion on page 468   365

R. Fanti et al. (eds.), VLBI and Compact Radio Sources, 365–374.
© 1984 by the IAU.

certain measurement accuracies must be attained during the 1980-1988 time frame of the CDP. Table 1 shows the accuracies that the CDP has assured will be available, and goals for further improvements.

Table 1
CRUSTAL DYNAMICS PROJECT MEASUREMENT REQUIREMENTS

	Commitment	Goal
	(all one sigma)	
GEODETIC PRODUCTS		
Intersite Velocity Accuracy	1 cm/year	0.4 cm/year
Baseline Length Precision	6 cm	2 cm
Earth's Pole Position Precision	10 cm	4 cm
INSTRUMENT PERFORMANCE		
Laser-Range/VLBI-Delay rms (3-minute normal point)	4 cm	1 cm

THE MARK III VLBI SYSTEM

The Mark III system can be regarded as a set of hardware and software for VLBI geodesy. The hardware is not just the IF processor and tape recorder, but includes hydrogen maser frequency and time standards, dual-frequency, broad-band receivers, a tone injection system with cable length calibration, meteorological sensors (temperature, pressure, and humidity), and water vapor radiometers. The contribution of several of these subsystems to the overall system accuracy is discussed below.

Broad-band Receivers

The precision of a group delay measurement is directly proportional to the spanned bandwidth over which it is determined. At X-band, the CDP receivers and Mark III terminal record signals in the range 8210 to 8570 MHz, using a bandwidth synthesis technique (Rogers 1970). With this bandwidth, single measurement delay accuracies are better than about 100 picoseconds, or 3 cm equivalent range, even on the weakest detectable sources. Strong sources have delay precisions of just a few picoseconds, or only a millimeter or so of range. At S-band, the recorded band covers 2215 to 2305 MHz.

The delays and delay rates measured at X- and S-band are combined to eliminate the $1/f^2$ dispersive effects of the ionosphere. For instance, the true group delay is:

$$\tau_{true} = \frac{f_x^2}{f_x^2 - f_s^2} \, \tau_x - \frac{f_s^2}{f_x^2 - f_s^2} \, \tau_s \, ,$$

where f_s and f_x are the effective S- and X-band frequencies, and τ_s and τ_x are the group delays measured in the two bands. The S-band delay and its associated error come in only at about the 10% level because of the f-squared terms. Hence, the smaller spanned bandwidth at S-band does not degrade the final measurement.

The ionospheric correction is now routinely applied to all data that are collected and analyzed by the East Coast group. An experiment in September/October 1980 clearly showed the improvement provided by dual-frequency observing. These data, taken during a period of unusually high ionospheric total electron content, were analyzed with and without the correction. The uncorrected results gave baselines that were too long by 15 to 20 centimeters when compared to results obtained at times of lower electron content and the corrected results. (Error sources such as the ionosphere and the troposphere that cause extra path delay biases always have the effect of making the baselines seem too long. The antenna that is further from the source will be observing at a lower elevation angle, and thus a greater atmospheric/ionospheric path, with a concommitant larger delay bias. The parameter estimation program tries to match the excess delay by lengthening the baseline.) The uncorrected results also show much greater scatter in the residuals within a single day's analysis and in the day-to-day means. (Two seven-day sessions separated by two weeks comprise these observations, which were part of the MERIT preliminary campaign.)

The active ionosphere during the MERIT observations also pointed up a deficiency in the bandwidth synthesis frequency array that was in use at S-band: a 3-frequency minimum redundancy array that spanned 75 MHz with a 25 MHz unit spacing. The resultant delay ambiguities were 40 nanoseconds, far less than the day/night range of ionospheric delays (almost 100 nanoseconds). Determination of the correct delay ambiguity was rather difficult. This difficulty was the incentive for several improvements to observing and analysis techniques. The previous 7-frequency Mark III mode ("Mode B", with 4 frequencies at X-band and 3 frequencies at S-band) was replaced by a 14-frequency mode ("Mode C", with 8 frequencies at X-band and 6 frequencies at S-band). These frequencies are used in arrays derived by D. S. Robertson of the NGS that have zero redundancy but some missing spacings, moderate sidelobes (50 to 60%), and small unit spacings (5 MHz at S-band and 10 MHz at X-band). The corresponding ambiguity spacings are 200 nanoseconds at S-band and 100 nanoseconds at X-band. The widely separated ambiguity lobes greatly simplify ambiguity resolution. The baseline analysis program was modified to check closure delays on arbitrary triads of stations, so that incorrect ambiguity resolution can be spotted quickly.

The multiple frequency channels are calibrated by a 1 MHz rail that is injected at the receiver, ahead of the low noise amplifiers. These "phase cal" tones are extracted by the correlator and the resultant phases are used to correct for dispersion in the RF to digital path and to coherently combine the channels. The phase calibration rail also provides the fundamental time reference for the observations. (It can be thought of as injecting one microsecond ticks into the receiver.) The system is driven by the station maser, and cable length variations between the maser and the front end are monitored by a reflectometer length measuring system (the "cable cal" system).

The current CDP receivers are uncooled parametric amplifiers and have rather high system temperatures: about 100 K at S-band and 200 K at X-band. Efforts are now underway to upgrade the receivers by replacing the paramps with cooled GaAs-FET amplifiers. The first of these systems has been installed at the new Mojave base station at Goldstone. The new system temperatures are about 70 K at both S- and X-band. (The S-band temperature includes some 30 to 40 K of spillover that results from rather poor illumination provided by the dual-frequency feed.)

Meteorological Sensors and Water Vapor Radiometers

Propagation delays which arise in the troposphere currently set the limits on the accuracy of VLBI baseline determinations. These delays are caused by the dry and wet components of the atmosphere.

The dry component of the atmosphere causes a zenith path delay of about 7 nanoseconds, or about 2 meters equivalent length. The exact value depends on the altitude of the station and the local temperature and pressure, which determine the mass of the air overhead. Temperature and pressure are recorded for every VLBI scan, and an atmospheric model is used to calculate a correction for all the observations.

The wet component of the atmosphere is less amenable to modelling based on ground-level humidity readings, since the water vapor is not well mixed. However, the 22.2 GHz emission from water shows promise of allowing a direct measurement of the path delay caused by water vapor. The brightness temperature in the water line is roughly proportional to the excess path length. Existing water vapor radiometers (WVR's) are two-channel devices, with one channel at about 21 GHz and the other near 31 GHz. The theory and operation of two WVR's are described by Elgered and by Resch and Miller (both in NOAA 1982).

Several WVR's of the Resch design are in use by the CDP, but because of reliability problems, no convincing demonstration of their ability to improve VLBI results has yet been made. In tests at the Very Large Array (VLA) in New Mexico, Resch _et al_. (1983) have shown that the use of WVR's did improve the phase stability, sometimes dramatically. In the VLA tests, the WVR's used the VLA cassegrain optics, and sampled the same air column as was being used for the VLA measurements. As used at most CDP VLBI sites, however, the WVR's are often at a distance of 100 meters or more from the antennas which are tracking the radio sources. This offset and the 7-degree beam of the WVR's means that the radiometers are not necessarily sampling the same air column as the VLBI observations. On very cloudy and/or rainy days, the differences could be significant. I believe that future WVR improvements should include mounting them on the VLBI antennas, with modest size dishes (1 to 2 meters) of their own, so that the water vapor delay is measured more accurately.

Hydrogen Masers

Masers are currently the best readily available time and frequency standards. Recent experience has shown that the masers are rather sensitive to temperature changes. This fact was not properly appreciated in the past, and the masers were often operated in telescope control rooms with poor temperature regulation. The high sensitivity of the Mark III revealed daily, cyclical variations of a nanosecond or more in the residuals from some experiments in which the masers had inadequate temperature control. Most of the CDP stations are now outfitted with very well insulated and temperature-controlled "isolation boxes" for the masers. These boxes regulate the maser temperature to a few hundredths of a degree Celsius.

The best VLBI baseline solutions have an rms scatter of group delay residuals of 50 to 70 picoseconds about the best-fit solution for 24 hours of data. This scatter is comparable to the Allan variance expected for two masers over a 24-hour period. Thus, after the wet component of the troposphere, maser behavior is a significant contribution to the quality of VLBI baseline determination. A possible way to overcome the limitations of the masers, and perhaps even allow the use of less stable clocks, is the use of a "clock source" at a declination greater than about 70 degrees that is visible throughout the day for all (Northern!) stations. The clock source would be observed every other scan, and all other observations would be referenced to it. By analyzing only differences with respect to the clock source, variations of the time standards with periods longer than the clock source observation interval would not affect the data. Use of this technique has been hampered by the lack of suitably strong sources and the long slew times of most antennas. The new generation of receivers will overcome the source strength problem, and the implementation of faster slewing antennas (Mojave and Wettzell slew at 1 degree per second or more) will allow this technique to be more thoroughly tested.

Radio Source Structure

The ideal geodetic source is strong and unresolved on all baselines. As shown by nearly everyone at this Symposium, many compact sources are far from simple in structure. The brightness temperature limit of about 10^{12} K (Kellermann and Pauliny-Toth 1969), set by inverse Compton scattering, means that the strongest sources will not be the most compact. Structure in a source causes phase gradients in the (u,v)-plane. These gradients cause phase slopes over the width of a bandwidth synthesis frequency range since the individual frequency channels sweep out slightly different paths in the (u,v)-plane. These phase slopes are indistinguishable from the phase slope ($d\phi/d\omega$) that determines group delay. Hence, source structure can be a cause of error in the VLBI measurements.

Errors due to structure can be greatly reduced by correcting for the phase effects as predicted by a map of the source. The strongest phase gradients occur near the minima of visibility functions, so large errors can be avoided by not observing at low visibilities (Thomas 1980). The easiest way to avoid structure problems is to observe point sources. The East Coast group has followed this approach, and nearly eliminated sources with significant structure from their set of reference sources. Table 2 shows the set of sources used for trans- and inter-continental VLBI by the East Coast group.

Table 2
VLBI REFERENCE SOURCES

0106+013		1226+023	(3C273)*
0234+285		1404+286	(0Q208)
0300+471		1641+399	(3C345)**
0528+134		1741-038	
0552+398		2216-038	
0851+202	(OJ287)	2251+158	(3C454.3)
0923+392	(4C39.25)*		

* Significant structure
** Slight structure

Unfortunately, 3C273 has been chosen to define the right ascension zero point. This source has complex, time variable structure (Pearson *et al*. 1981) and in retrospect is a poor choice for this purpose. Better choices would be 0106+013 or 0851+202, which have simple structure, lie near the ecliptic, and are visible from the southern hemisphere.

Table 3 shows the expected contribution of the various error sources to the error budget for measurements of intermediate and intercontinental baselines. These estimates were arrived at after two days of discussion in October 1982 by most of the USA practitioners of geodetic VLBI. Some of these estimates are pessimistic, since they were enlarged to include worst cases. For instance, VLBI instrumentation has been used to measure baselines of 600 and 1200 meters, to accuracies of millimeters (Lundqvist in NOAA 1982; Rogers *et al*. 1978). Source structure was found to cause errors of less than 1 mm on a Haystack-NRAO-HRAS-OVRO set of baselines (Shaffer 1982).

Table 3 indicates that VLBI is meeting the CDP commitment of 6 cm for baseline length accuracy, and is well on the way to the 2 cm goal shown in Table 1. Baseline orientation, especially in the vertical location of the stations, is more difficult to determine than baseline length, since length is invariant to orientation. Height errors in a station location arise from atmospheric/ionospheric propagation errors. Excess path lengths mimic the signature of a lowered elevation. This effect is more severe for short

baselines where the propagation delays are highly correlated for the several
stations and do not give rise to strong station-dependent signatures.

Table 3
VLBI ERROR SOURCE SUMMARY

| Error Source | 400 km Baseline | | | 4000 km Baseline Length |
| | Vertical | Horizontal | | |
		Transverse	Length	
Instrumentation	1.4 cm	0.9 cm	0.9 cm	1.0 cm
Ionosphere	0.2	0.1	0.1	0.2
Dry troposphere	1.5	0.6	0.8	0.5 - 1.0
Wet troposphere*	4.0/<9.0	0.5/1.0	0.6/1.2	1.5/<3.0
UT1/Polar Motion+	0.6/1.3	0.6/1.3	$\cong$ 0	$\cong$ 0
Source Structure	0.3	0.3	0.3	0.3
Aggregate rss Estimate (cm)	4.6-9.3	1.4-2.0	1.4-1.7	2.0-3.3

* First number assumes WVR, second assumes the use of modeling or
surface meteorology data.

+ Assumes 10-20 cm _a priori_ data, such as from POLARIS, or
satellite or lunar laser ranging.

RECENT RESULTS

The densest data set of baseline length determinations comes from the
National Geodetic Survey POLARIS project (Robertson and Carter, in NOAA
1982). This program makes one measurement per week between the Westford
and GRAS (ex-HRAS) antennas. The goal of POLARIS is high precision
determination of polar motion and UT1 variations, with baseline lengths as
a by-product. Figure 1 shows the results of two and a half years of
measurements between Haystack and GRAS, including POLARIS results. [The
Westford-GRAS baseline has been referenced to Haystack by adding the
Haystack-Westford baseline determined by Rogers _et al_. (1978).] Typical
errors on a length measurement are a few centimeters (see Figure 2). The
rms scatter of all the baseline length measurements is about 2 cm on this
3136 km baseline. The various experiments were analyzed by the NASA/GSFC
group with a set of source coordinates in the J2000 system (Kaplan 1981).
(This set is designated SSC(GSFC)82 R 01 in the annual report of the BIH

for 1982.) Slightly different analysis procedures applied to these data
by the NGS investigators result in a comparably small scatter. However,
there is a small offset of about 2 cm between the mean baseline lengths
determined by the two groups, probably caused by different source catalogs.
This offset highlights the fact that VLBI is somewhat better suited to
measuring changes in baseline length rather than absolute baseline lengths.
Length variations are nearly unaffected by any biases as long as the analysis
procedures are consistent from experiment to experiment.

The Haystack-GRAS measurements show no significant changes in baseline
length over the two and a half year period shown in Figure 1. This result
is not surprising, since both stations are on the North American plate;
and there are no areas of recent geological activity between the two
stations, although the baseline does cross the New Madrid, Missouri area,
site of large earthquakes in the early 1800's. The formal velocity error
is comparable to the CDP goal given in Table 1 and will improve as the data
span grows over the next few years.

Figure 2 shows baseline length measurements of the Haystack-OVRO
baseline for a six year period. The errors on each solution are shown,
and are typically a few centimeters. The average error is somewhat reduced
from 1979 onward, when the Mark III went into service. The fact that the
errors did not decrease dramatically, despite the high precision of the
Mark III, shows the predominant effect of external error sources on the
VLBI reductions, mostly from the troposphere.

The Haystack-OVRO baseline also shows no change in length, with a
formal "speed limit" that is better than the CDP goals and shows the quality
of data that can be obtained during a moderately long observing campaign.
Although both stations are again on the North American plate, it is not so
clear what level of stability should be expected on this baseline. OVRO
is located in a geologically active area. It is only a few kilometers from
a large, active fault that runs the length of the Owens Valley at the base
of the Sierra Nevada mountains. The baseline crosses the Basin and Range
province, an area of expected extension. Sites in Nevada, Utah, and Colorado
will be visited by mobile VLBI systems during the life of the CDP in order
to determine whether the Massachusetts-California baseline is truly invariant,
or if the apparent stability is a coincidence caused by offsetting motions.

The Haystack to Onsala, Sweden baseline has been measured some 40
times from mid-1980 through early 1983 by the NASA/GSFC and NGS groups.
The formal baseline length results show a possible lengthening of the
baseline at a rate of 1.8 ± 0.6 cm/year. This change is in the direction
and about the right size predicted by continental drift theory, but the
experimenters are reluctant to claim a true detection of plate motion.
The random and systematic errors are somewhat larger on this 5600 km baseline
than on the intra-USA baselines, and a larger set of data is required to
establish the USA to Europe velocity.

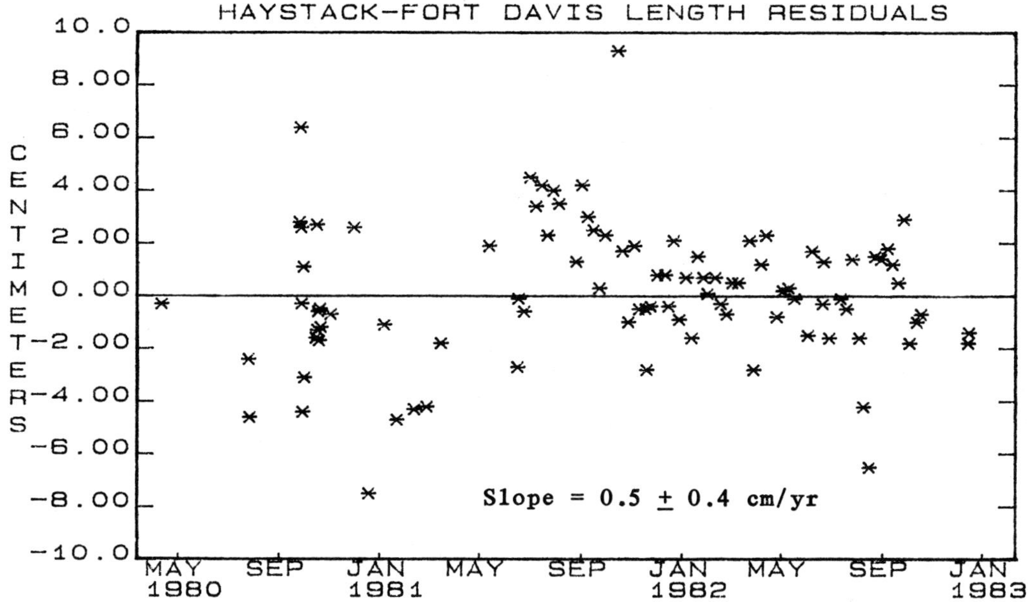

Figure 1.   Haystack to GRAS length residuals
about 3 135 841.02 meters.

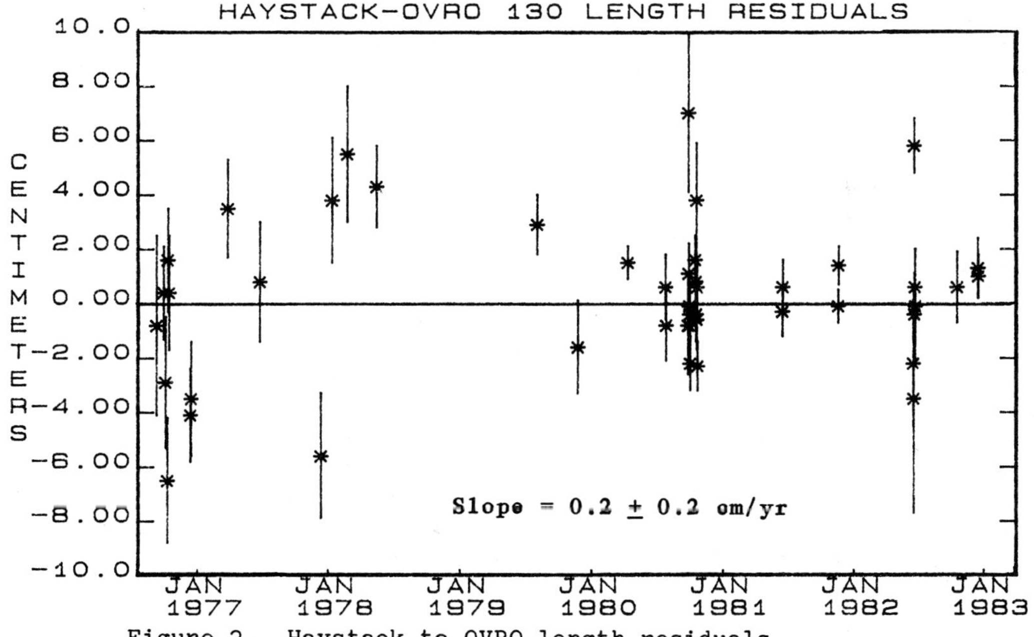

Figure 2.   Haystack to OVRO length residuals
about 3 928 881.63 meters.

I thank my colleagues for discussions and advice in the preparation of this paper, especially Dr. C. Ma of NASA/GSFC.

REFERENCES

Cohen, M. H. and Shaffer, D. B.: 1971, *A. J.* **76**, pp. 91-100.
Kaplan, G. H. (editor): 1981, U. S. Naval Observatory Circular No. 163.
Kellermann, K. I. and Pauliny-Toth, I. I. K.: 1969, *Ap. J.* **155**,
    pp. L71-L78.
NASA: 1980, Conference Publication 2115, "Radio Interferometry Techniques
    for Geodesy".
NOAA: 1982, Technical Report NOS 95 NGS 24, "Geodetic Applications of
    Radio Interferometry".
Pearson, T. J. *et al.*: 1981, *Nature* **290**, pp. 365-368.
Resch, G. M., Hogg, D. E., and Napier, P. J.: 1983, *Radio Science*,
    (submitted).
Rogers, A. E. E.: 1970, *Radio Science* **5**, pp. 1239-1247.
Rogers, A. E. E. *et al.*: 1978, *J.G.R.* **83**, pp. 325-334.
Rogers, A. E. E. *et al.*: 1983, *Science* **219**, pp. 51-54.
Shaffer, D. B.: 1982, paper at the Crustal Dynamics Project investigators
    meeting.
Shapiro, I. I.: 1976, *Methods of Experimental Physics*, M. L. Meeks,
    editor (Academic Press, New York), Vol. 12, part C, pp. 261-276.
Thomas, J. B.: 1980, JPL Publication 80-84.

SET UP AND TESTING OF A SOFTWARE PACKAGE FOR THE GEODETIC ANALYSIS
OF VLBI DATA +

A. CAPORALI
TELESPAZIO S.P.A., ROMA
G. SYLOS LABINI
ISTITUTO DI FISICA, UNIVERSITA' DI BARI

1. INTRODUCTION

The need for centimetric accuracies set by the application of the VLBI
technique to geodesy implies a considerable computational effort, becau
se of the intrinsic complexity of the model and because there is an increasingly large number of calibrations and corrections which can be ac
counted for only via software. The program VLBI 3 (Robertson, 1975) has
been developed for the geodetic and astrometric analysis of VLBI data.
It includes an accurate theoretical model of the observables and is sup
ported by a number of routines for parameter fitting and input/output o
perations with data and results. The original VAX version of VLBI 3, due
to N. Bàrtel and M.I.Ratner, runs in batch mode and requires routines
which are in general unavailable in standard VAX systems. We have prepa
red (Caporali and Sylos Labini, 1982) a modified VAX version of VLBI 3.
This version runs on our standard VAX/VMS computers and contains a number of changes in the FORTRAN source which allow to the user a real ti
me interaction with the program. In addition, having a Tektronix graphic station at our disposal, we decided to replace the existing plot
package - which used the line printer - with a "ad hoc" graphic program
which permits interactive display of the results of each run of VLBI 3.
Our work was mostly concerned with the input/output sections of VLBI 3.
The theorical model of the VLBI observables has, for the moment, been
left unchanged. We have, however, noticed that the theorical model could
be updated and made more precise, e.g. in the computation of the nutation terms and of the aberration. This updating and a more extensive geo
detic and astrometric analysis of VLBI data are planned to be done next.

2. DESCRIPTION OF THE PROGRAM AND OF THE RESULTS

The flow diagram of VLBI 3 is given in Fig. 1. The first three modules
(SETUP, INPUT and NRMSET) initialize the program by reading the input da
ta and control flags. FERMTR computes the theoreticals and partial deri-
vatives. NORMAL formes and solves the normal equations. ADJUST computes
and prints the adjustments to the a priori values of the "solve for" pa
rameters. If requested, VLBI 3 produces a plot file which serves as input to the graphic program to display the results. We analyzed data ta
ken on May 17-19, 1978 with the Haystack (Mass.), Onsala (Sweden), Owens
Valley (Calif.), NRAO (W. Wirginia) radiotelescopes, and ten radio sour
ces with declinations ranging from -5° 31' (3C278) to 50° 49' (NRAO 150).

R. Fanti et al. (eds.), VLBI and Compact Radio Sources, 375–376.
© *1984 by the IAU.*

Because the data were taken at only one frequency (7850 MHz), the base-line estimates are affected by the systematic errors due to the charged component of the propagation medium. Fig. 2, obtained with our graphic package, shows that systematic effects are present in the residuals, but also indicates that the self consistency of the adopted solution is at the subdecimeter level.

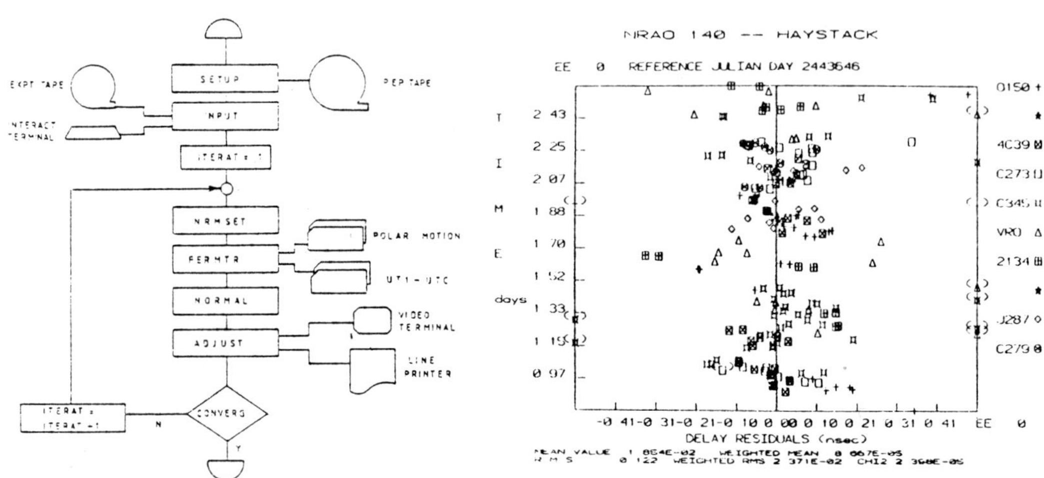

FIG.1 - BLOCK DIAGRAM OF VLBI 3   FIG.2 - SPREAD OF POSTFIT RESIDUALS

Table 1 summarizes our baseline estimates:

	OWENS VALLEY	ONSALA	NRAO
HAYSTACK	3 928 881.67	5 599 714.66	845 129.94
	± 0.02	± 0.03	± 0.01
OWENS VALLEY		7 914 131.17	3 324 244.18
		± 0.04	± 0.02
ONSALA			6 319 317.76
			± 0.04

TABLE 1 - BASELINE RESULTS IN METERS   WITH 1σ FORMAL ERROR

They are consistent with the results obtained by Herring and coworkers (Herring, 1981) from the analysis of the same data set.

A. CAPORALI and G. SYLOS LABINI (1982): Telespazio Internal Report.

T.A. HERRING and 16 co-workers (1981) : J.G.R. 86 1647 - 1651.

D.S. ROBERTSON (1975): Ph. D. Thesis NASA GSFC X-922-77-228.

THE VLB ARRAY +

K. I. Kellermann
National Radio Astronomy Observatory

ABSTRACT

The VLB Array (VLBA) is a synthesis radio telescope which has been designed to extend the resolution of the VLA in order to allow sub-milliarcsec studies of compact galactic and extragalactic radio sources over a wide range of wavelengths and spectral resolution.

I. BACKGROUND

Shortly after the first observations made with independent oscillator-tape recording interferometers (Very Long Baseline Interferometer or VLBI) (Broten et al. 1967; Bare et al. 1967), it was realized that VLBI was a powerful tool which could be used to study a wide variety of problems in galactic and extragalactic astronomy, including fundamental astrometry, terrestrial geodesy and geophysics, as well as in fundamental physics (e.g. Gold, 1967; Shapiro, 1967; Cohen et al. 1968; Burke 1969).

It soon became apparent that a multi-element radio telescope array with dimensions comparable to the size of the Earth would be needed to study the complex structure found in compact radio sources (e.g. Swenson and Kellermann 1975), and in 1974 the NRAO began to investigate the feasibility of constructing a dedicated VLB Array to complement and extend the Very Large Array (VLA) then being built on the Plains of San Augustin in New Mexico. By 1977 it was clear that the technology developed for the VLA and being used for VLBI experiments around the world was sufficient to construct a radio array of truly global dimensions, and the specifications and conceptual design for such an array was circulated as an NRAO report, An Intercontinental Radio Telescope (ed. K. Kellermann). In 1978 a Canadian group reported on a similar array planned for Canada (ed. T. Legg). The concept of a dedicated VLB Array was further developed in the U.S.A. at Caltech and at NRAO, and two further reports were issued, in 1980 A Transcontinental Radio Telescope (Caltech, ed. M. Cohen), and in 1981 Design Study for the Very Long Baseline Array (NRAO, ed. K.

377

R. Fanti et al. (eds.), VLBI and Compact Radio Sources, 377–382.
© 1984 by the IAU.

Kellermann), which further specified the system design and set performance specifications. Likewise, several updates to the Canadian concept were made during the period 1978 to 1983.

In 1982 following a three-year study, the Astronomy Survey Committee ("Field Committee") of the U.S. National Academy of Science recommended the construction of a Very Long Baseline Array Radio telescope as the highest priority for major new ground-based astronomical facilities, and in May 1982, the NRAO submitted to the National Science Foundation a proposal for the design, construction, and operation of the VLBA.

This proposal, which was based on the earlier design studies, was the result of a ten-year effort in which more than 60 scientists and engineers throughout the country had contributed. Detailed engineering design and prototyping of individual subsystems will continue through 1984, and construction of the VLBA is expected to start in 1985. Partial operation is scheduled to begin in 1987 when the first antennas are completed, and full operation is expected by the end of 1988.

II. PERFORMANCE SPECIFICATION AND SYSTEM DESIGN

The number and location of the VLB Array elements has been chosen to give the best possible resolution and image quality consistent with budgetary constraints. To simplify the operation and management of the Array, it has been decided to place all elements on U.S. territory. The Array will contain 10 antenna elements spaced throughout the United States from Hawaii to Puerto Rico. Two of these will be in the state of New Mexico, close to the VLA; the others being in Massachusetts, Iowa, Washington, California, Arizona, and Texas. The two New Mexico antennas will be located in such a way that the baselines between these antennas and the VLA will to some extent fill in the spacings intermediate between the VLA and VLB Array. It is hoped that in the future three more antennas may be added in this area to give essentially continual coverage between the compact VLA "D configuration" and the full 8000 km VLBA dimensions.

The local oscillators at all stations will be synchronized by hydrogen maser frequency standards, and the IF signal will be recorded on magnetic tape for later playback in a central processing facility. One of the New Mexico antennas will be about 70 km south of the center of the VLA and will be connected to the VLA Control Center via a microwave data link. At all times this antenna will be remotely operated with the local oscillator reference and tape recorder located at the VLA Control Center. For some applications it will be used as a real time coherent extension of the VLA, effectively doubling the resolution of the VLA at the cost of degraded dynamic range. On the other hand, for VLBA problems requiring a relatively large field of view, one VLA antenna (or in some cases all three outer antennas of the A configuration) can be used together with the 10 VLBA elements;

and for special cases all 27 VLA antennas can be used together with the VLB Array to obtain very high sensitivity. All of the elements will be controlled and monitored in real time by a single array operator via leased telephone lines, and each antenna will normally be unattended except for changing magnetic tapes.

The basic system specifications for the Array are outlined in Table I. The wide geographic coverage of the array and maximum operating wavelength gives the best possible resolution which can be obtained from the surface of the Earth consistent with current antenna and receiver technology as well as consideration of a reliable and cost effective operation. The number of antenna elements allows about 80 percent of the amplitude and phase information to be obtained from self-calibration procedures and is sufficient to give good image quality (dynamic range) over a wide range of declination.

TABLE I

DESIGN SPECIFICATIONS

Number of Elements	10
Size of Elements	25m
Overall Size	8000 km
Wavelength Coverage (10 bands)	0.7 cm to 90 cm
Resolution	0.2 to 24 milliarc sec
Sensitivity	0.1 milli Jy
Polarization	Linear and circular
Spectral Resolution	0.2 Hz to 50 kHz

Antenna Elements. 25-meter diameter antennas were chosen, as antennas of this size with the desired accuracy can be readily fabricated with conventional techniques, and gives a good compromise between collecting area and shortest operating wavelength. Each element will be a conventional wheel and track structure designed for reliable low maintenance operation and will have a shaped paraboloid primary surface to give a high efficiency. It is hoped to build a reflector surface with an rms surface accuracy $\lesssim$ 0.45 mm to allow operation at frequencies $\gtrsim$ 40 GHz. At frequencies above 1 GHz, operation will be from the Cassegrain focus, and an asymmetric secondary reflector will be rotated to illuminate the eight feed horns located at the Cassegrain focus. Prime focus feeds will probably be used at the two lowest frequencies. Dual frequency operation can be provided with dichroic reflectors and is initially planned for the S/X wavelength bands commonly used for the NASA and NGS geodetic programs.

<u>Radiometer Systems</u>. Receivers for the two lowest frequencies will use relatively simple feed systems and ambient temperature GASFET amplifiers. In the six intermediate wavelength bands, GASFET amplifiers cooled to 20K will be used to give the best possible sensitivity consistent with reliable operation and economic construction, while MASER amplifiers are being considered for the two shortest wavelengths to give state-of-the-art sensitivity. Rapid change of the observing wavelength will be possible from the Operations Center, allowing flexibility in observing programs as well as minimizing the impact of receiver failures or poor weather conditions.

The sensitivity and resolution in each of the 10 planned wavelength bands is shown in Table II. The values given for noise fluctuations represent the noise in each picture element provided that there is a reference feature typically ten times stronger visible on all baselines which is sufficiently strong to phase the array in a typical coherence time of 10 minutes. Such reference features are most conveniently used if they are within the primary beam of the antenna pattern; but phase referencing to nearby sources can also be used. Water vapor radiometers will be installed at each element to measure the water vapor content in the atmosphere in order to minimize the phase variations due to fluctuations in atmospheric water vapor along the line of sight. Use of the full VLA will also allow self calibration of the VLBA on sources weaker by a factor of five than indicated above.

<u>TABLE II</u>

VLB ARRAY
SENSITIVITY AND RESOLUTION

FREQ. (GHz)	RCVR	SYSTEM TEMP-K	RMS NOISE 8 Hours-mJy	RESOLUTION MAS
0.32	FET	65	0.16	24
0.61	FET	55	0.1	13
1.4/1.7	FET	29	0.035	5
2.3	FET	31	0.035	3.5
5	FET	37	0.04	1.6
8.4	FET	40	0.05	0.9
10.7	FET	45	0.05	0.7
15	FET	65	0.06	0.5
22	MASER	45	0.06	0.35
43	MASER	75	0.16	0.2

<u>I.F. and Recording System</u>. The I.F. system is being designed to accommodate up to 32 frequency channels each with a bandwidth select-able between 125 kHz and 16 MHz. The recording system will operate in a 2-bit or 4-bit mode at a normal rate of 100 Mbps (50 MHz bandwidth), and for limited periods at rates up to 200 Mbps (100 MHz bandwidth) or more for high sensitivity continuum observations, or at lower rates as appropriate for narrow band spectral line observations. The system is expected to be transparent to the specific recording medium, to allow for improvement in this rapidly developing area with a minimum of system retrofits, and to keep, where appropriate, compatibility with older VLBI recording systems which are in present use.

Initially, the record/playback system will be based on inexpen-sive consumer type Video Cassette Recorders. Each recorder will be modified to write at a 12.5 (or 25) Mbs data rate, and a bank of 8 (or 4) recorders will be used to obtain the full 50 MHz (100 Mps) band-width. A second recorder rack will normally be available as a spare at each station, but for limited periods can be used to double the recorded bandwidth, at the expense of increased tape consumption. Unattended operation of each station for at least 24 hours between tape changes is planned.

<u>Processor System</u>. The correlator system will be able to handle the input from up to 19 antenna elements in the normal continuum mode, and 14 antennas with full polarization processing. In the spectro-scopic mode, up to 512 frequency channels will be available with a frequency resolution down to 62 Hz.

III. <u>FUTURE EXPANSION</u>

Like all arrays, the VLBA can be expanded to improve the sensi-tivity, resolution and dynamic range. The addition of a single element in South America would greatly improve the resolution in the north-south direction, while extensions to the Pacific will also be of interest to supplement and extend the new Australia Telescope, as well as the new mm dish in Japan and the dedicated VLBI dishes planned in China. The future placement of a large antenna in Earth orbit will even further extend the power of the VLBA (e.g. Burke, this volume).

On more immediate time scales, the use of other large radio tele-scopes such as the VLA, Arecibo, and Bonn 100-m will be used together with the VLBA, as will the new VLBI antennas being constructed in Italy and planned in Canada. This will increase both the angular resolution and image quality, as well as greatly enhance the sensi-tivity. Of particular interest will be the intermediate scale high sensitivity baselines available from European antennas.

IV. <u>OPERATION</u>

Normally each antenna element of the VLB Array will run entirely under control from an Operations Center. A few technician/operators will be available at each site, however, for inspection, routine maintenance, and the simpler unscheduled repairs of malfunctioning equipment. The local staff will also be responsible for updating operating systems at the local control computer, for changing and shipping the data tapes to the Operations Center, for security and precautionary oversight, for emergency intervention and for routine start-up and shutdown procedures.

The Operations Center will provide for major maintenance and repair requiring personnel with special skills, special equipment, or major replacement parts. However, since it is planned to replace complete modules in the case of failure, many such replacements can be easily performed by the local site personnel. Defective modules will be returned to the Operations Center for repair. This procedure, while requiring a somewhat larger than normal inventory of spare parts, will reduce travel and personnel costs, and keep Array down-time to a minimum.

The Array will be operated using a preplanned program under the control of a central computer, which will simultaneously monitor the performance of the antennas and receivers as well as the meteorological conditions at each site. An Array Control Operator will be present at all times at the Operations Center to intervene when necessary and to carry out various housekeeping tasks. From time to time, brief samples of the received signal at each antenna will be sent to the Operations Center via the telephone lines and correlated in nearly real time to check that all components of the Array are functioning properly and to monitor meteorological effects on the data.

Many individuals from the U.S. university community as well as at NRAO are actively involved in the design of the VLB Array, and in particular the radio astronomy laboratories at Caltech and MIT are working with NRAO in all phases of the design and construction. When completed, the VLB Array will be operated by the National Radio Astronomy Observatory as a national facility available to all qualified scientists. As with other NRAO facilities, observing time will be based on scientific merit without regard to institutional or national affiliation. The National Radio Astronomy Observatory is operated by Associated Universities, Inc., under contract with the National Science Foundation.

REFERENCES

Bare, C. et al. 1967, _Science_, _157_, 189.
Broten, N. W. et al. 1967, _Science_, _156_, 1593.
Burke, B. 1969, _Physics Today_, _22_, 54.
Cohen, M. H. et al. 1968, _Science_, _162_, 88.
Gold, T. 1967, _Science_, _157_, 302.
Moran, J. et al. 1967, _Science_, _157_, 676.
Shapiro, I. 1967, _Science_, _157_, 806.
Swenson, G. et al. 1967, _Science_, _188_, 1263.

THE CANADIAN LONG BASELINE ARRAY[+]

T.H. Legg,
National Research Council of Canada, Ottawa

On 1983 May 31, the National Research Council of Canada
gave formal approval to a new national facility, the
'Canadian Long Baseline Array'. The instrument will be made
up of eight radio telescopes spaced across southern Canada,
with a ninth antenna, primarily for geophysics, in the
north. Though the project is still some way from being
funded, the NRC approval was an important and encouraging
step in a process that started five years ago.

The Canadian Long Baseline Array (CLBA) began in 1978 with
the submission of a proposal to a committee of the Canadian
Astronomical Society (CAS). Soon after, the CAS adopted an
expanded version of the proposal, with the recommendation
that the project be given the highest priority for a new
astronomical instrument for Canada. In the years following,
the CAS has continued its strong support for the project.

Since 1980, the array has been the work of the 'CLBA
Planning Committee' of the CAS. This committee has drawn
on the efforts of more than 45 radio and optical
astronomers, geophysicists, and engineers in producing a
design for the array. The work is summarized in a report
published in 1982*, which includes design, reliability and
cost studies done under contract by Canadian industry. The
description of the CLBA that follows is largely abstracted
from this report.

* The report: 'The Canadian Long Baseline Array', dated
1982 September 27, is to the Natural Sciences and
Engineering Research Council and the National Research
Council, from the Canadian Astronomical Society.

+ Discussion on page 468   383

R. Fanti et al. (eds.), VLBI and Compact Radio Sources, 383–389.
© *1984 by the IAU.*

1. The Overall Telescope

Two features that are especially emphasized in the design of
the CLBA are (i) sensitivity, which is reflected in the
choice of the largest antenna affordable, and (ii) image
forming over as wide a range of angular scales as possible,
a feature discussed further under 'Configuration'.
Estimated sensitivities, with receivers attainable in 1982,
are shown in Table I, together with the frequencies covered.
The two sets of figures are for a single pair of antennas
(N=2), and for coherent operation of the linear array of
telescopes (N=8), which may be possible at the longer
wavelengths or with improved atmospheric measurements.

The beamwidths listed are for the $\sim$ 5000 km maximum
Canadian baseline. These figures would be reduced to 2/3
of their value with the addition of an antenna in France.
Considerable interest has been expressed amongst French
scientists, particularly those in the Groupe de Recherches
Géodésie Spatiale, in having their planned 32m antenna
used, as first priority, in cooperation with the CLBA. The
French site is essentially ideal from the point of view of
the U-V plane coverage.

TABLE I

Frequency (GHz)	N=2 (mJy)		N=8 (mJy)		Beamwidth marcsec
	1 sec	15 min	15 min	12 hours	
0.611	90	3.0	0.57	0.082	12.2
1.5	41	1.4	0.26	0.037	5.0
2.25	45	1.5	0.28	0.041	3.3
5.0	51	1.7	0.32	0.046	1.5
8.3	70	2.3	0.44	0.064	0.9
10.7	76	2.5	0.48	0.068	0.7
22.0	473	16.	3.0	0.43	0.34

2. The Configuration

Eight of the CLBA antennas will form a nearly linear array,
running from Newfoundland to British Columbia, but slightly
tilted from an exact east-west line. The linear configuration was chosen instead of a two-dimensional array only
after much weighing of relative advantages. Scientifically
the choice seemed evenly balanced between a wide field-of-
view (i.e. sensitivity to a wide range of angular scales) on
the one hand, and coverage of low declinations on the other.
The more economical maintenance of a line of southern
stations turned out to be the deciding factor.

In arriving at suitable configurations, the image forming
properties of different arrays were compared quantitatively
by simulating the observation of a number of test sources.
Two configurations selected in this way are shown in Fig. 1,
with corresponding U-V curves in Fig. 2 and an example of
the imaging tests in Fig. 3. Other test sources used were
an extragalactic source with jets, and multiple sources,
for simultaneously testing the large and small scale imaging
properties of the arrays.

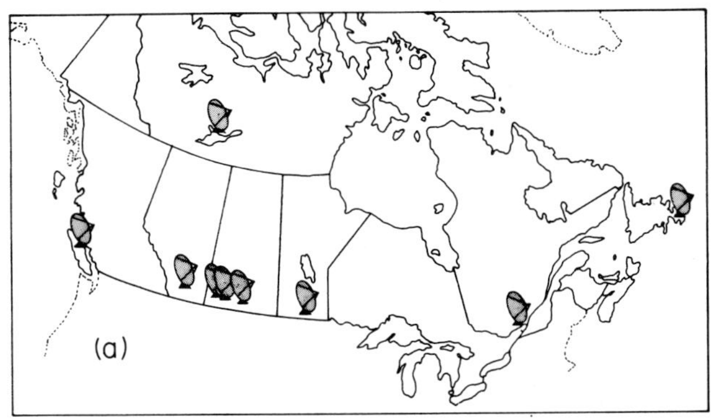

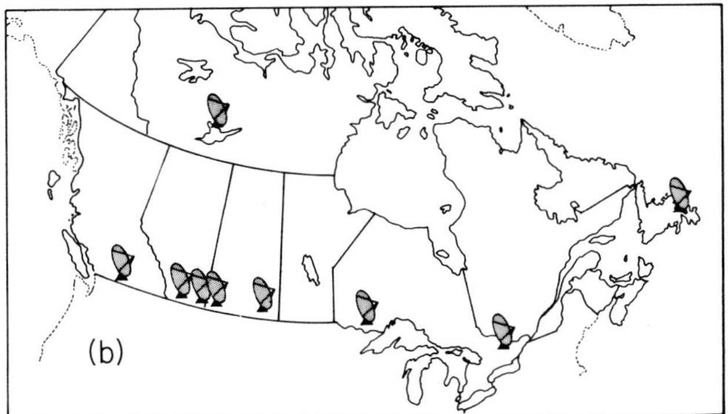

Fig. 1 – Two array
configurations desig-
nated (a) J3M and (b)
P10, that fulfill the
CLBA design aims.
Array P10 uses
existing radio
telescope sites (but
not the telescopes) at
Penticton, B.C. and in
Algonquin Park,
Ontario.

A primary aim of the imaging tests was an array with the
widest field-of-view consistent with good imaging. Configurations that best satisfied this aim were found to be
those with roughly an exponential distribution of baseline
lengths. The arrays of Fig. 1 are examples. These arrays
have minimum baselines of about 75 km, and are able to map
sources as large as 1.0 arcsec, at 2.2 GHz (0.1 arcsec at
22 GHz) with the dynamic range (peak brightness to peak
brightness error) remaining above 300.

Other advantages of exponential configurations are: geometrically similar sampling of the visibilities of similar
sources of different angular scale, and a near uniform
'filling-in' of the U-V plane for a moderate (say $\sim$ 10%)

bandwidth. The tilt of the arrays of Fig. 1 from an east-
west line also slightly improves the filling-in of the U-V
plane. After the first 12 hours of observations, the U-V
tracks interleave instead of overlapping.

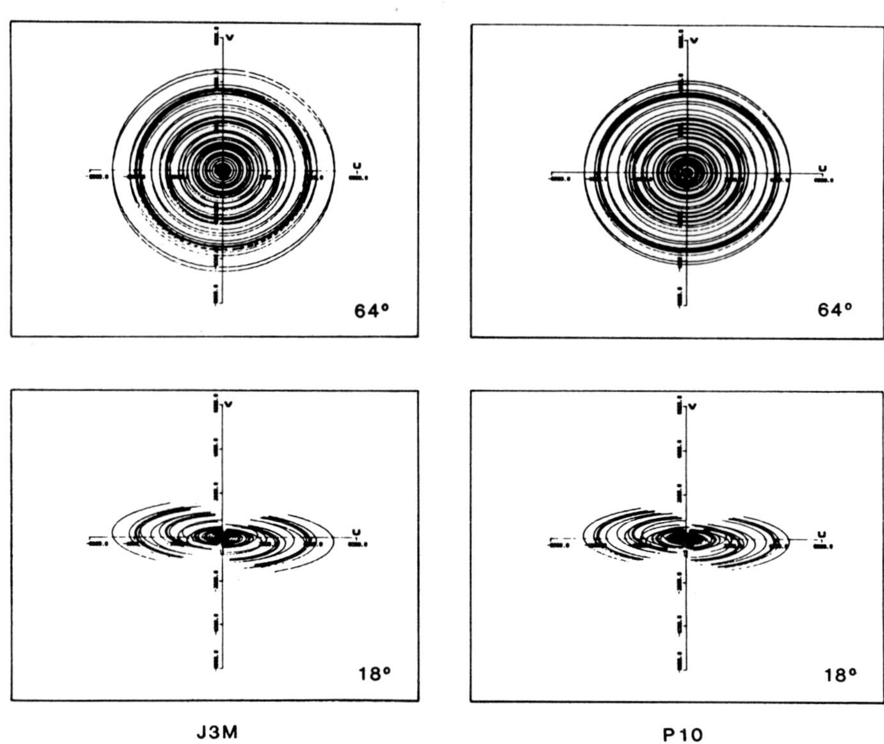

Fig. 2 - The U-V tracks at two declinations for the arrays
of Fig. 1, excluding tracks involving the northern antenna.
The dimensional tick under the 'U' indicates 6000 km.

The eight southern antennas of the linear array will be
used primarily for astronomy, while the ninth antenna, at
Yellowknife, Northwest Territories, will provide north-
south baselines which are a priority of Canadian geophysicists. A second geophysical priority will be
satisfied by having baselines on, and close to, the geologically stable Canadian shield.

3. Antennas

The 32m dia antennas of the CLBA will be upgraded versions
of a standard wheel-and-track communications antenna. The
antenna will have 'shaped' main and secondary reflectors to
give near uniform illumination (i.e. little apodization)
and, consequently, high efficiency ($\gtrsim$ 70%). At the time
of the CLBA report, the highest frequency considered was

22 GHz. Since then, the possibility of improved panels, to
give a useful collection area at 43 GHz, has been considered. Any such improvements would have to be made, however,
to fit within the overall cost estimates of the 1982 CLBA
report.

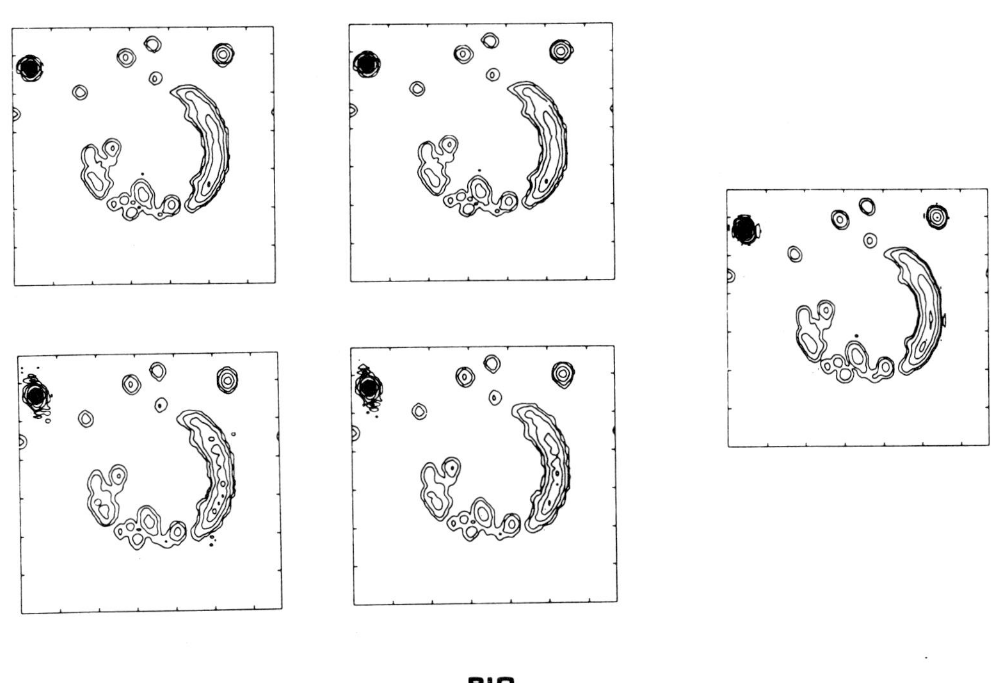

Fig. 3 - Reproduced images from simulated observations with
arrays J3M and P10, for declinations 64° (top) and 18°
(bottom). The original, a DRAO map of CTB1, is on the
right. The lowest contour level is 0.002 of the peak
brightness.

4. <u>Antenna Feeds</u>

Dual polarized feeds for each of the frequencies listed in
Table I will be arranged at the vertex of the 32m antenna,
as shown in Fig. 4. The feed will be selected by appropri-
ately tilting the antenna subreflector. A dual frequency
feed will be used for the 2.25 and 8.3 GHz bands. The 611
MHz feed, not shown in Fig. 5, will be a large uncorrugated
horn of square aperture, placed to one side of the vertex.
The length of this horn is kept manageable by having a
flare angle that varies with distance from the throat.

5. Receivers

Dual GaAs FET amplifiers will be used at all of the
frequencies listed in Table I, and cooled (except at 611
MHz) to 20K. The development of such amplifiers is underway at the University of Alberta and the National Research
Council, Ottawa. The local oscillators will be phase-
locked to a station hydrogen maser. Wider intermediate

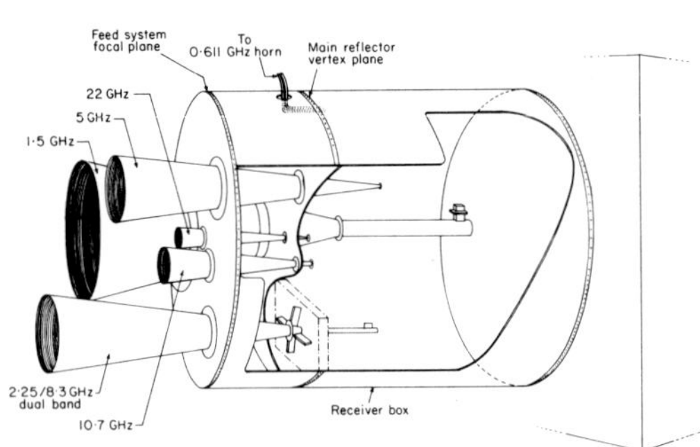

Fig. 4 - The feed
arrangement at the antenna
vertex. The 611 MHz feed,
not shown, is located
further off axis.

frequency bandwidths than the 200 MHz specified in the CLBA
report are now being discussed.

6. Recorders

Video recorders with automatic cassette changers have been
specified since the 1978 proposal. The 1982 design calls
for two channels, of 4 recorders each, recording a total of
96 M bits sec^{-1}. These would run unattended for 24 hours.
This bandwidth will possibly be increased as a result of
rapid advances in the recorder field. Reliability of the
cassette machines, which was already good, has been
improved by the development of automatic equalizers by
J.L. Yen at the University of Toronto.

7. Correlators

The correlators will receive data streams from up to 10
antennas. Normally, 32 lags will be produced for each of
four polarization products. Fractional-bit delay correction and three-level 'fringe rotation' schemes will be
incorporated. Up to 2048 spectral line channels will be
provided by recirculating the data to the correlators.

8. Image Processing

Image processing for the CLBA is designed to give the non-
expert user a source image, a spectrum, or geophysical data
without necessarily any direct intervention on his part.
The processing is divided into levels in which, typically:

(i) edited and calibrated correlation coefficients are produced; (ii) images are produced, and (iii) images are enhanced in various ways. The design centers upon a VAX 11/780 computer and three PDP 11/24 computers controlling array processors.

9. Operation

The antennas will be controlled over telephone lines from a central computer and will be unattended except for a few hours a day. Teams from the headquarters will undertake nearly all of the maintenance.

10. Staff and Costs

The staff of the CLBA will total 75, as indicated in Table II; and construction costs will total an estimated 70.1×10^6, distributed as shown in Table III. Yearly operating costs will be 7.7×10^6 of which salaries will make up 3×10^6 and the purchase of new receivers, or other equipment to upgrade the instrument, 1.5×10^6.

TABLE II
CLBA STAFF

Operations	31
Engineering support	6
Computer programming	8
Administration and services	9
Development	13
Scientists	8
TOTAL	75

TABLE III
ESTIMATED COSTS
(Millions of 1982 Canadian dollars)

Antennas and feeds	31.7
Receivers and electronics	10.2
Correlation and processing	6.4
Headquarters	5.2
Sites	4.4
Design, management, other	5.8
Contingency	6.4
TOTAL	70.1

PROPOSED VERY LONG BASELINE INTERFEROMETRY AT 103 MHz IN INDIA[+]

R.V. Bhonsle, S.K. Alurkar, S.S. Degaonkar,
A.D. Bobra and R. Sharma
Physical Research Laboratory
Ahmedabad-380009, India.

Abstract: Three radio telescopes operating at 103 MHz are being installed
at Ahmedabad, Rajkot and Surat separated by about 200 km from each other
for observing interplanetary scintillations (IPS) of compact radio sources for study of solar wind plasma dynamics as well as radio source size
measurements for cosmological studies. Of these, two radio telescopes
at Ahmedabad and Rajkot have been commissioned and started synchronous
daily observations of IPS of a few compact radio sources with relative
time accuracy of about $\pm$ 1 millisec. The third telescope at Surat is
expected to go in operation by the end of 1983. As soon as all the three
telescopes go in for simultaneous operation, it is proposed to (1) aug-
ment the telescope sensitivity so as to detect sources with flux density
$\sim$ 1 Jansky (2) incorporate better time and frequency standards at each
station which can be synchronised to better than μs relative time accuracy (3) develop suitable receivers and data acquisition system for generating interference fringes using a general purpose computer and (4) take
advantage of the availability of three telescopes to incorporate 'closure
phase and amplitude' techniques which eliminate undesirable atmospheric
and ionospheric phase distortions.

The paper describes (1) scientific objectives and motivation for
carrying out VLBI measurements of compact radio sources at meter wavelengths and (2) conceptual system design.

1. INTRODUCTION

During the last couple of decades, high angular resolution observations of radio galaxies and quasars have been extensively carried out using Very Long Baseline Interferometry technique (Miley 1980 and references
therein). At centimetric wavelengths the angular resolutions achieved
so far have been of the order of 10^{-4} arcsec. In recent years two-dimensional mapping of some radio sources has also been successfully carried
out and their structural changes that occur from year to year have also
been detected. Study of angular structure of radio sources gives information about the mechanism of production of these radio sources, their
+ Discussion on page 469   391

R. Fanti et al. (eds.), VLBI and Compact Radio Sources, 391–395.
© *1984 by the IAU.*

their physical environment and cosmology. Among the problems which have
been studied so far include diffuse emission, "hot spots", cores and jets
that constitute these radio sources. However, most of these angular structure studies have been carried out with a network of radio telescopes
operating at decimetre and centimetre wavelengths using VLBI technique.
Similar studies at meter wavelengths need to be carried out for the understanding of the outer envelopes of compact radio sources wherein initially
accelerated electrons loose their energy by radiation and begin to radiate
predominantly at longer wavelengths. Hence there is a strong case for
making systematic observations at metre wavelengths using VLBI. Unfortunately, longer wavelength radiations (metrc and decametre $-\lambda$) are adversely
affected in intensity and direction of arrival due to scattering and refraction during their propagation through earth's ionisphere, interplanetary and interstellar medium, which limits the angular resolution of a
terrestrial radio telescope and introduces positional errors. It has
recently been demonstrated that if one has 3 or more radio telescopes
available for observations, then one can exploit 'closure Phase' technique
(Wilkinson et al. 1977) which eliminates the effect of turbulence in the
propagation medium, thereby allowing determination of direction of arrival
of radiation unambiguously.

This paper describes an outline of the proposed plan to initiate VLBI
technique at a frequency of 103 MHz in India using radio telescopes being
set up for interplanetary scintillation studies.

2. DESCRIPTION OF AN IPS RADIO TELESCOPE AT 103 MHz

The Physical Research Laboratory, Ahmedabad (India) has been engaged
in setting-up an observatory consisting of three radio telescopes operating at 103 MHz, for studies in solar wind plasma and radio source structures using IPS technique. Under this project, the three telescopes
which are separated from each other by about 200 km would be set-up in
Western India, in the State of Gujarat as shown in Figure 1. Of these,
two radio telescopes at Ahmedabad and Rajkot have been commissioned and
started daily simultaneous observations of IPS of a few compact radio
sources with a relative accuracy of $\pm$ 1 ms. The third telescope is expected to go in operation at Surat by the end of 1983. The description
of one of the IPS telescopes has been published (Alurkar et al. 1982).

Figure 2 shows the block diagram of the IPS radio telescope. Briefly,
the telescope consists of a full-wave dipole array of 64 rows, each one
consisting of 16 dipoles. The antenna array has a rectangular aperture
of 5000 square metres; 200 metres in the North-South and 25 metres in
the East-West direction. The aperture is divided into two halves and
the telescope is operated as a correlation interferometer thereby enabling recording of sine and cosine fringes. The sensitivity of the telescope is such as to detect sources with 5 Janskys at signal to noise
ratio of 5. The telescope has a beam width of 2° in declination at the
zenith and 7.5° in right ascension. The telescopes will be operated as
a transit instruments each day for recording interplanetary scintillat-

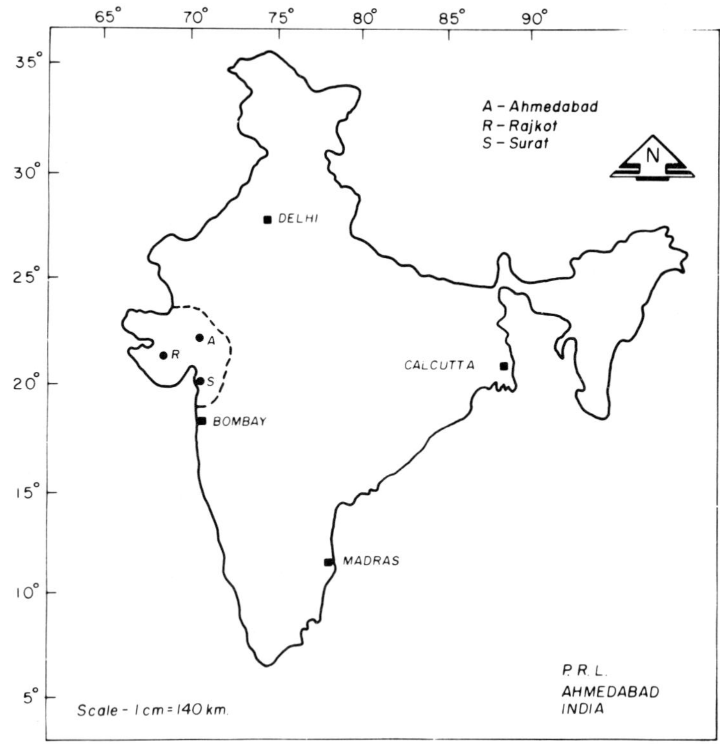

Figure 1. Location of three radio telescopes for IPS experiment in India.

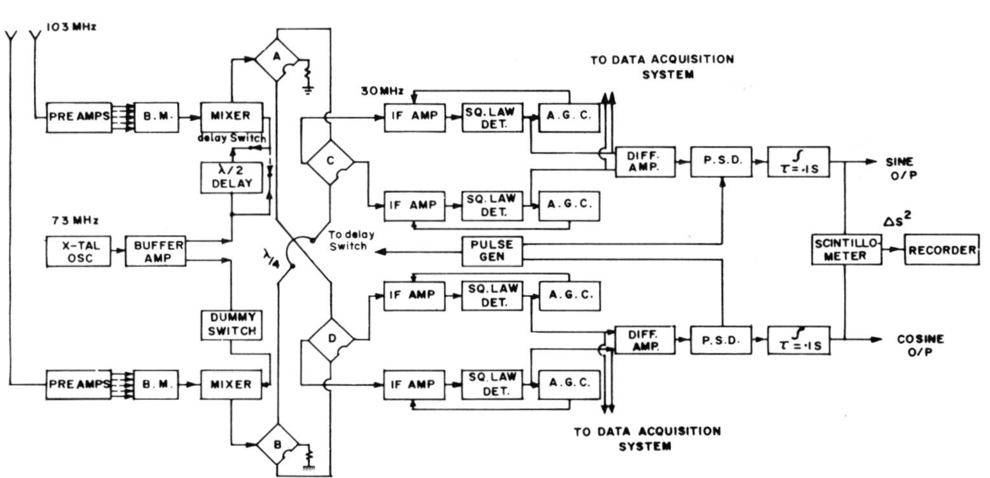

Figure 2. Block diagram of an IPS radio telescope at 103 MHz.

ions of selected radio sources simultaneously from 3 stations. Presently, the relative time accuracy is maintained within few milliseconds.
The correlation interferometer data (sine and cosine outputs) are recorded on charts for monitoring purposes and digitized and recorded on
magnetic tapes, compatible with a general purpose computer. The data
from three stations will be auto- and cross-correlated for computation
of the solar wind velocity. Using single station IPS data, radio source
structure studies have also been initiated.

3. PROPOSED MODIFICATION OF IPS RADIO TELESCOPE FOR VLBI

Since three identical radio telescopes will be available by the end
of 1983, we propose to modify this system to operate in the VLBI mode
to yield an angular resolution of 3 arcsec at 103 MHz as follows:

1. Since the presently available time accuracy of the order of few
milliseconds will be insufficient, it will be necessary to incorporate
Rubidium Vapour time and frequency standard, which will yield relative
time accuracy between two telescopes of about a microsecond.

2. The local oscillator voltages will be derived from these Rubidium
frequency standards.

3. The IF output will be digitized using 1-bit digitization scheme.
These data will be recorded on video tape recorders with frequency response up to 2 MHz. These data will be cross-correlated to produce interferometer fringes using a general purpose computer and will be used
to attempt closure phase and amplitude techniques.

The block diagram of the proposed VLBI terminal is shown in Figure
3. Most of the specifications of this terminal are similar to the NRAO
mark II Tape Recorder system (Clark 1973, Cohen 1973). The antenna is
connected to a low noise preamplifier (LNA) in order to overcome the
losses due to long cable lengths from antenna to the receiver. The local
oscillator voltage is derived from the highly stable Rubidium Vapour
frequency standard. The amplified signal from antenna is then mixed
with LO to get an IF output which is in turn further amplified. The IF
output is then amplified and clipped, by fast saturating amplifiers, sampled and recorded on a video tape transport along with timing information. The time at VLBI station will be synchronised using a potable atomic clock to an accuracy of about 1 μsec. The data so recorded will be
brought back to a central computer for processing to search for interference fringes.

After demonstrating the feasibility of 'closure phase and amplitude'
techniques with the modified set up, additional radio telescopes with
tracking capability will be installed to obtain various baselines to
give adequate uv coverage for mapping of radio sources at metre wavelenghts.

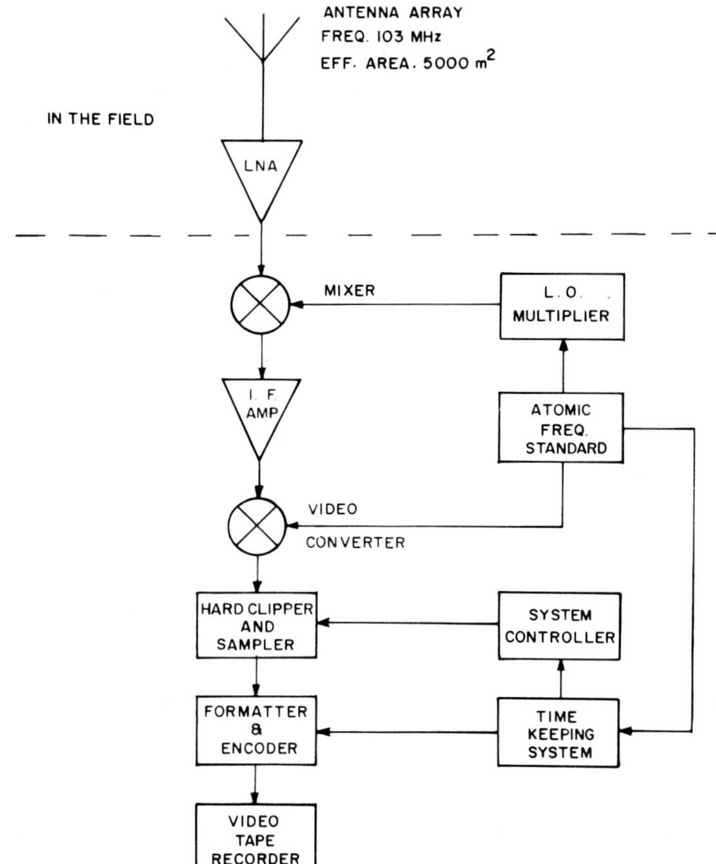

Figure 3.
Proposed VLBI
terminal at 103 MHz

Acknowledgements: The authors thank Professors D. Lal and S.P. Pandya
for encouragement. The IPS project is financially supported by Department of Science and Technology and Department of Space, Government of
India.

REFERENCES

Miley, G.K.: 1980, Ann.Rev. Astron. Astrophys. 18, pp. 165-218.
Wilkinson, P.N., Readhead, A.C.S., Purcell, G.H. and Anderson, B.: 1977,
     Nature, 269, pp. 764-768.
Alurkar, S.K., Bhonsle, R.V., Sharma, R., Bobra, A.D., Sohanlal, Nirman,
     N.S., Venat, P. and Sethia, G.: 1982, J. Inst. Elec. & Telecom.
     Engrs. 28, pp. 577-582.
Cohen, M.H.: 1973, Proc. IEEE., 61, pp. 1192-1197.
Clark, B.G.: 1973, Proc. IEEE, 61, 1242-1248.

ORBITING VLBI: A SURVEY[+]

Bernard F. Burke
Massachusetts Institute of Technology
Cambridge, Massachusetts

ASTRACT

The Very-Long-Baseline Interferometry technique is not limited by
the size of the Earth. Near-Earth-orbiting space vehicles can carry
radio telescopes that can serve as VLBI stations using existing technology. By proper use of ground VLB arrays, a single orbiting VLBI
satellite can yield radio maps with 2-dimensional coverage and high
dynamic range at all declinations. A single orbiter can be used out
to orbits that yield an effective aperture greater than two Earth
diameters. Interstellar scintillations are a limiting factor only in
the micro-arc-second range.

I. SUMMARY

Radio Interferometry has provided ever-increasing angular resolution, progressing from the first hard-wired, two-element systems
through radio-linked, multiantenna arrays, and culminating in Very-
Long-Baseline Interferometry, for which the Earth's size is the current
limit. At the same time, a parallel requirement for complete Fourier
coverage has generally developed, since rough angular sizes are useful
as a first guide, but eventually maps of high dynamic range are needed
to settle the crucial questions. The VLA was developed in response to
this need, and a strong demonstration of the usefulness of this capability is illustrated by Perley's recent Cygnus A maps, which show in
striking detail the nature of the source and reveal, for the first time,
the relativistic beam that powers the double-lobed radio source. The
necessary high dynamic range is only possible when there is complete
coverage of the Fourier transform.

When orbiting radio telescopes were first contemplated as a means
of extending VLBI baselines beyond the confines of the Earth, it quickly
became apparent that there were three significant advantages to the
technique: firstly, the desired improvement in angular resolution was
achieved, secondly, it would be possible, using the orbiting VLB
station in conjunction with ground-based arrays, to obtain "snapshots"

+ Discussion on page 470

R. Fanti et al. (eds.), VLBI and Compact Radio Sources, 397–403.

of rapidly changing sources such as SS433 and Cygnus X-3 that were
otherwise ambiguously studied by Earth-rotation synthesis; thirdly, a
single orbiting VLB station in an inclined orbit, used in conjunction
with several ground stations, naturally gave complete coverage of the
u-v plane for all declinations. Finally, it has become evident that
ground-based VLB arrays and orbiting systems are essentially complementary.

The concept of Orbiting VLBI, OVLBI for short, originated in
several places within the first few years of VLBI development. The
subject was discussed on two occasions in 1971 at Byurakhan, in
conversations that involved N.S. Kardashev, J.S. Shlovskii, L.I.
Matveyenko, K.I. Kellermann, and B.F. Burke. Transformation from
concepts to proposals occurred in the 1970's. Work was started in 1973
by a group under N.S. Kardashev, using the Salyut space station as a
mount for a 10 meter antenna, to be used at 75 cm wavelengths. In
1976, a group headed by B.F. Burke proposed the use of a 4-m telescope
on the Spacelab 2 mission: this was then re-proposed in 1978 with the
participation of a broadly-based team consisting of T.A. Clark, M.H.
Cohen, K. Johnston, K.I. Kellermann, J.M. Moran, R. Preston, A.E.E.
Rogers, and I.I. Shapiro, with B.F. Burke as principal investigator.
The project was jointly submitted by MIT, JPL, and Goddard Space
Flight Center. The antenna, a 3.7-m carbon-epoxy paraboloid, would
have been used at wavelengths of 3.8 and 18 cm.

In 1979, the first international proposal was made by B.F. Burke
and N.S. Kardashev, who proposed to make use of NASA's VOIR spacecraft
during the cruise phase of its mission to study the plasma fluctuations
of the interstellar medium by making VLBI observations of pulsars.
The spacecraft had been planned to use a 5-m paraboloid for wide-band
telemetry, and was to have a high-capacity memory on board: Kardashev
and Burke proposed to use the paraboloid first in receiving mode to
store pulsar signals on board, and then to re-transmit the data to
Earth. Subsequent re-scoping of the mission eliminated the VLB possibility, but VLBI astronomers should remain alert to such possibilities
in the future.

More recently, several studies have been published, taking a detailed look at the potential problems and technical feasibility of
OVLBI. A brief paper on OVLBI was presented in 1976 by Preston,
Hagar, and Finley at a meeting of the American Astromonical Society.
An early study by V.I. Bujakas _et al._ was presented in 1973 to the 23rd
IAF Congress, giving the outline of a proposed "infinitely built-up
space radio telescope", together with an analysis of the potential
for OVLBI. In 1980, Kardashev, Pogrebenko, and Tsarevsky published
an analysis of OVLBI with million-kilometer baselines by Andrezanov and
Kardashev(1981). A detailed analysis of a Space Shuttle mission
carrying a 50-m paraboloid was carried out by a Technical Working Group
composed of scientists and engineers from Marshall Space Center, Jet
Propulsion Laboratory, Naval Research Laboratory, Draper Laboratory,
and the Massachusetts Institute of Technology, the report being

published in 1982. A related report, investigating the general
capabilities of the Space Shuttle for OVLBI and the technical applicability of existing VLB systems, was published in 1982 by Roberts,
Doxsey, and Burke.

The major planning exercises known to be under way at the present
time are:
1) The Shuttle-attached OVLBI studies of NASA, now concentrating
on antennas in the 15-meter size range.
2) The QUASAT concept, originated at JPL and now the subject of a
joint study by NASA and ESA, currently conceived of as a free-flying
15-meter paraboloid in an eccentric orbit, using ground-based VLB
arrays to fill in the holes in the u-v plane. Frequently coverage
would range from 1.6 to 23 GHz.
3) A low-Earth-orbit mission planned by the Soviet Academy of
Sciences, comprising a 30-m paraboloid for use at λ = 18 cm.
4) A million-kilometer eccentric-orbit mission of the Soviet
Academy of Sciences, utilizing a 10-m antenna at 18 cm wavelength.

II. TECHNICAL AND SCIENTIFIC CONSIDERATIONS

The work of the NASA Technical Working Group provides a useful
reference point for the guidance of future studies. A particular
mission was studied in detail: a 50-meter antenna, to be carried in
low-Earth-orbit by the Space Shuttle. The concept is illustrated in
Figure 1. The questions addressed were (1) Could such an antenna be

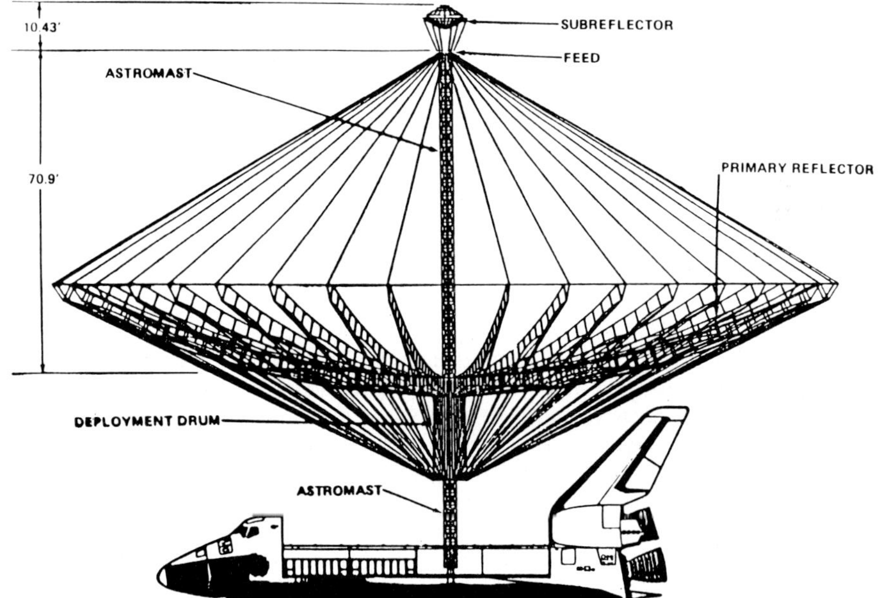

Figure 1. The 50-meter OVLBI concept, shown deployed from the Space
Shuttle. (MSC)

built to work at 3 cm wavelength; (2) Could the antenna system be
pointed with sufficient accuracy to serve OVLBI purposes; (3) Were
the mechanical properties of the system stable; (4) Were the required
electronic systems technically ready; (5) Could the data flow be
handled; (6) Could the Mark III VLB system be used to reduce the data;
and (7) Was the scientific program viable? The answers to all
questions were favorable.

The required electrical performace of the antenna turned out to
be well matched to the current state-of-the-art. Presently, the best
achievable ratio of antenna diameter to rms tolerance, D/ε is 1.5×10^4.
This implies that an antenna of 50-m diameter would reach its Ruze
maximum in gain at 10.2 GHz. Ordinarily, one might suppose that the
Ruze maximum, where the antenna efficiency is $(1/4\pi)$ of its ideal
value, imposes too high a price. For an interferometer pair, however,
the effective area is the geometric mean of the two antenna areas, so
the penalty paid by having one antenna at the Ruze maximum is only
-2 db, a price well worth the ability to make X-band observations.

The pointing requirements appeared initially to be severe; the
solution was to recognize that the antenna pointing error had to be
known, and then simply entered into the reduction procedure. The
dynamical properties of the antenna were a challenging problem, since
antenna pointing was complicated by the coupling between the large
structure and the Space Shuttle to which it was attached. Simulation
studies by Draper Laboratory showed that the potential instabilities
could be controlled by proper specification of standard Space Shuttle
control parameters. The work served as an example of dynamic interaction of the Shuttle flight control system with a deployed payload,
Sackett and Kirchway, 1982 <u>AIAA Guidance and Control Conference</u>,
Aug. 9-11, San Diego, CA.

There were few problems concerning electronics, data handling, and
data reduction. Some software modifications were needed for the Mark
III VLB data system, but no fundamental difficulties were found. The
conclusion was that there was a complete state of technical readiness
for OVLBI.

The 50-m radio telescope exercise was not likely to become an
active mission, but served as a useful planning exercise. There was
clear realization in the Technical Working Group that there were two
concepts that were far more realistic: 15-m low-Earth orbiting payload
on a space shuttle, a space platform, or a free-flying satellite. The
QUASAT concept, put together jointly by a group of US and European
scientists, is an example of a free-flying satellite that makes full
use of the ground-based VLBI arrays. It is currently under intensive
study, and in this conference is described in detail by Schilizzi <u>et al</u>.

III GENERAL MISSION CONSIDERATIONS

Given the technical feasibility of operating a VLBI terminal in

space, one might still be wary of the problems that arise from the
rapid motion of the spacecraft. The concern is easily answered,
however: even normal ground-based VLBI stations are moving at 1500 km/
sec because of the Earth's rotation, a problem that is routinely
solved. A spacecraft in low-Earth orbit moves at 16 times this velocity, giving fringe rates (i.e. Doppler shifts) that are only slightly
more than an order of magnitude higher than those normally encountered
in practice. The additional complication can be compared to the
changes that occured over a decade ago when VLB frequencies moved
upward from L-band to K-band.

The low-Earth orbit mission (such as the shuttle-based concept)
and mid-orbit mission (such as QUASAT) cover the u-v plane with such
density that truly high quality maps should result from OVLBI observations. The improvement over ground-based arrays is demonstrated by
the comparison in Figure 2 between the capability of the U.S., VLB

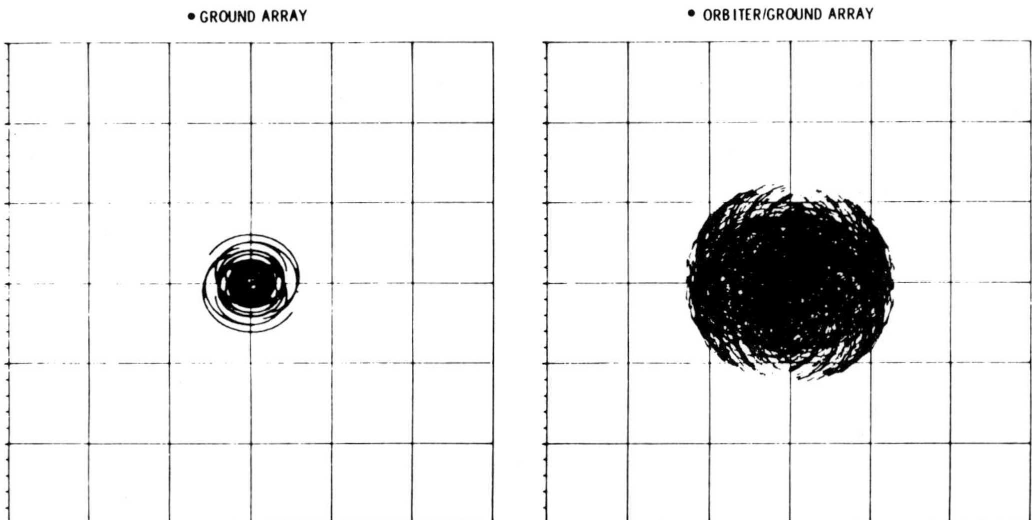

Figure 2. Comparison of u-v coverage by the US VLB Array, compared to
the coverage when combined with a low-Earth-orbit OVLBI terminal.

and a single low-Earth orbit OVLBI station used in conjunction with the
ground-based array. A few basic principles determine the system capabilities; the best angular resolution in the low-Earth orbit case is
given when the source under observation is normal to the orbital plane,
and complete coverage requires that the OVLBI satellite orbit passes
over the ground station at some point. D. Roberts showed that an ideal
orbital inclination exists: if one defines a figure of merit as the
number of beamwidths summed over the entire sky, the optimum orbital
inclination is 57°, a convenient orbit for launching such vehicles.
The optimum is broad, so the exact inclination is not critical, but
neither polar nor equatorial orbits have desirable characteristics.

The QUASAT studies now underway indicate that optimal orbits can
also be specified for the mid-orbit case. When one goes to still higher
orbits, however, holes develop in the u-v coverage. Both the studies
of Kardashev, Pogrebenko, and Tsarevsky and of Roberts, Doxsey, and
Burke concluded that high-quality mapping for orbits in the $10^5 - 10^6$

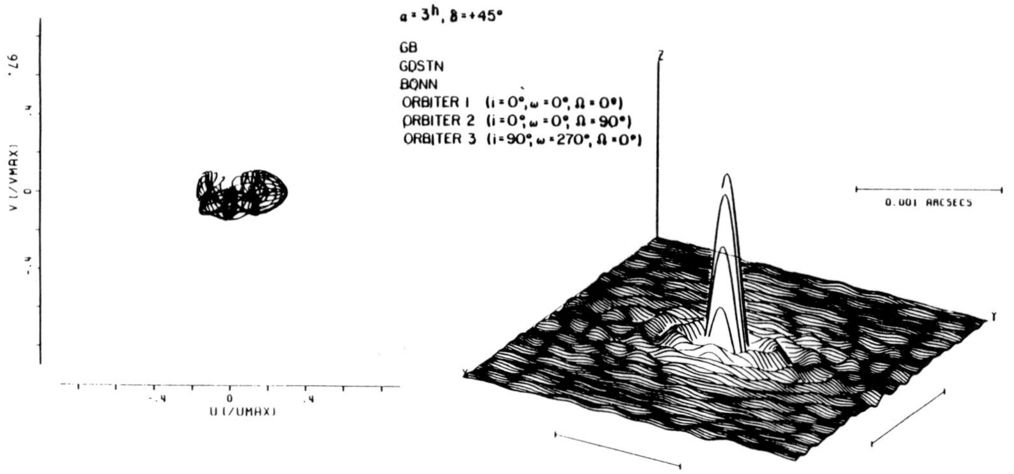

Figure 3. Fourier coverage of an OVLBI system with three 200,000-km
orbiters, combined with the US VLBI array.

km range required more than one satellite. An example of the coverage
yielded by a system of three satellites in eccentric orbits is given in
Figure 3. The u-v coverage is less dense than in the low-and mid-Earth
orbit case.

IV. THE FUTURE

Even the missions under study at the present time are a number of
years away, but planning for space missions necessarily involves long
lead times. These missions are of the scale commonly known as "Explorer
Class" and involve budgets of the order of 100 million dollars, a class
referred to as "moderate missions" in the recent Astronomy Survey
Committee Report by G. Field et al. If one assumes that these initia-
tives are successful, however, it is not too soon to contemplate the
next step.

A variety of arguments lead to the expectation that interstellar
scintillation and scattering is not a major limitation for orbit
reaching to a million kilometers. At galactic latitudes above 20°, the
mean scattering angle is of the order of 10^{-6} arc-seconds at 1 cm wavelength, or half the resolution of the million-km baseline referred to
above. One therefore concludes that a million kilometer OVLBI system
would not be seeing limited at wavelengths of the order of 2-4cm and
shorter. A more conservative system, composed of 3 satellites in
200,000 km orbits, would function at L-band; the ultimate choice of

scale must await further discussion. The project would certainly be an international one, and could be a viable mission before the turn of the millenium.

A look into the still more distant future shows exciting prospects within the framework of our present knowledge of physics. Interstellar scintillation is indeed a problem, if interferometry over interplanetary distances is contemplated. Kardashev and Burke, in the VOIR proposal, drew on recent image-restoration work that shows promise for such long baselines. Some of the possible goals of interplanetary VLBI are dramatic: the entire universe can be brought within the near field of a 1 cm interferometer system that spans 10 astronomical units; the emitting region of pulsars could be resolved; and radio holography might be performed on objects like the Crab Nebula, giving 3-dimensional representations. The methods would require very large apertures, such as the "infinitely built-up space radio telescope" of Bujakis et al., in order to achieve the necessary signal-to-noise ratio. The required technology is beyond the current state-of-the-art in some areas, and the mission costs are formidable.

On a more practical note, the main focus at present must certainly be on the modest missions. It is not too soon, however, to contemplate the next step: the 10^5 to 10^6 km missions that could take place in the later 1990's. A mission model could be developed as a world-wide collaborative effort, and its certain success would lay the groundwork for the work of another generation in another millenium.

HISTORICAL NOTE

Prior to the Byurakhan conversations, the subject of OVLBI had been discussed in the USSR and the USA. In 1970, a Space Science Board study on priorities in space science took place at Woods Hole, with the astronomy panel recommending that an exploratory "Very Long Baseline Interferometer" be built. There were similar proposals made by Soviet radio astronomers at about the same time.

REFERENCES

1971. "Priorities for Space Research 1971-1980, p. 86. Space Science Board, NAS-NRC, Washington.

1973. B.I. Bujakis, A.S. Gvamichaya, L.A. Gorchkov, et al., Proceedings, XXVIII Congress, International Astronautical Federation, Prague.

1975. "Opportunities and Choices in Space Science, 1974", p. 59. Space Science Board, NAS-NRC, Washington.

1976. R.A. Preston, H. Hagar, and S.G. Finley, Bull. Am. Astron. Soc. **8**, 497.

1980. N. Kardashev, S. Pogrebenko, and G. Tsarevshky, Astr. Zh. **57**, 634.

1981. V. Andreyanov and N. Kardashev, Kosm. Issl. Ak. Nauk **19**, 763.

1982. S.H. Morgan and D.H. Roberts, eds. "Shuttle VLBI Experiment" NASA TM-82491.

1982. Quasat: A VLBI Observatory in Space, B. Anderson et al., Mission Proposal submitted to ESA, November 12; 1982.

ORBITING VERY LONG BASELINE INTERFEROMETER DEMONSTRATION USING THE
TRACKING AND DATA RELAY SATELLITE SYSTEM +

G. S. Levy, C. S. Christensen, J. F. Jordan, R. A. Preston
Jet Propulsion Laboratory, California Institute of Technology,
Pasadena, 91109

B. F. Burke
Massachusetts Institute of Technology, Cambridge, 02139

ABSTRACT   A proposal has been made to use the Tracking and Data Relay
Satellite System (TDRSS) as an orbiting element for a very long baseline
interferometry (VLBI) demonstration.   The TDRSS is a satellite system
designed to coherently track and relay data between other satellites and
the central ground station.   This system could also be used to coherently
observe a celestial radio source.   A ground-based frequency standard
would be used to coherently drive the spacecraft receiver local oscillator and transmitter.   The data will be telemetered to a ground station,
where it will be recorded on a Mark III terminal.

The purpose of this proposal is to demonstrate the technology required for a free-flying orbiting very long baseline interferometry
(OVLBI) observatory.   A successful demonstration of the techniques required for an OVLBI may improve the chances of obtaining an early approval of a dedicated free-flyer mission.

Conventional ground-based VLBI requires independent frequency standards with instabilities that will not contribute more than a small fraction of a cycle of phase error during an integration interval.   The
interferometric data are recorded in real time on a wide-band VLBI terminal such as the Mark III.   For a low-cost free-flying OVLBI system,
such as the proposed "Quasat" OVLBI mission, (Schilizzi, et al., 1983,
and Preston, et al., 1982), the frequency reference is supplied from a
ground-based standard, and the interferometric data is telemetered to
the ground station for recording.

The electronic configuration of Quasat can be simulated fairly
closely by the TDRSS.   Figure 1 shows the proposed system configuration
of the TDRSS.   The initial tests will be performed using a ground beacon
to obtain overall system stability data.   The ground station uses a cesi-
um standard as a frequency reference, which is multiplied up to approximately 15 GHz and transmitted to the spacecraft as a pilot tone.   The
spacecraft accepts the uplink carrier and uses it to drive a frequency
synthesizer, which supplies coherent references to all the onboard

+ Discussion on page 470   405

R. Fanti et al. (eds.), VLBI and Compact Radio Sources, 405–406.
© *1984 by the IAU.*

oscillators. A 5-meter steerable, unfurlable parabolic antenna is used
to receive the celestial or beacon signals at S-band. These signals are
coherently amplified, frequency translated, and transmitted to the ground
station on a telemetry link. The ground station receives the signal and
sends a coherent intermediate frequency version to the VLBI trailer,
where it will be recorded on a Mark III VLBI terminal. A sample of the
station frequency standard will also be sent to the trailer, where it
will be compared to a hydrogen maser and a difference signal will be
recorded for use in generating corrections in the correlator.

If the first series of tests with a ground beacon shows that the
characteristics of the overall system are not adequate to support VLBI
experiments, the investigation will be terminated. If the stability of
the complete system is adequate (as our initial analysis indicates),
then celestial VLBI will be demonstrated in conjunction with ground-based
radio astronomy observatories.

REFERENCES:

Schilizzi, R., et al. (1983) IAU Symp. 110, Bologna.
Preston, R. A., et al. (1982) Proc. International Conf. of VLBI Tech-
    niques, Toulouse, CNES, Cepadues Editions.

ACKNOWLEDGEMENT:

The research described in this paper was carried out by the Jet
Propulsion Laboratory, California Institute of Technology, under contract with the National Aeronautics and Space Administration.

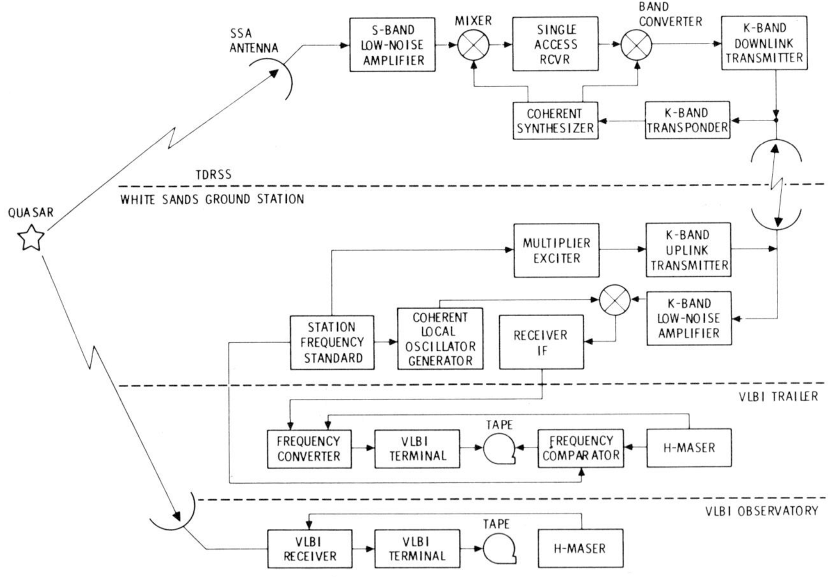

Figure 1. Demonstration block diagram.

THE QUASAT PROJECT

R.T. Schilizzi[1], B.F. Burke[2], R.S. Booth[3], R.A. Preston[4],
P.N. Wilkinson[5], J.F. Jordan[4], E. Preuss[6], D. Roberts[7]

[1] Netherlands Foundation for Radio Astronomy, Dwingeloo, NL
[2] Massachusetts Institute of Technology, Cambridge, Mass. USA
[3] Onsala Space Observatory, Onsala, Sweden
[4] Jet Propulsion Laboratory, Pasadena, Calif. USA
[5] Nuffield Radio Astronomy Laboratories, Jodrell Bank, UK
[6] Max-Planck-Institut für Radioastronomie, Bonn, FRG
[7] Brandeis University, Waltham, Mass. USA

SPACE VLBI AND QUASAT

Preston et al (1976) and Burke (1982, these proceedings) have long
extolled the virtues of launching a radio telescope into space to
increase VLBI baseline lengths and thus angular resolution, and to
provide a much enhanced image formation capability. The scientific
motivation for this has been covered in a number of memoranda referenced
by Burke in these proceedings, and by Anderson et al (1982). Efforts to
mobilise western astronomical support for space VLBI met with success in
late 1982 at a meeting of US and European radio astronomers in Toulouse,
France, at which a decision was taken to propose a joint mission to ESA
and NASA. Shortly thereafter, a formal proposal was made to ESA
(Anderson et al 1982) for a free flying satellite in an elliptical orbit
out to 15000 km from the Earth, designed to observe in concert with the
major ground-based VLBI networks and arrays. The mission, dubbed QUASAT,
was received favourably in both ESA and NASA, with the result that
formal Assessment Studies are scheduled to begin in both agencies in
October 1983.

THE ADVANTAGES OF ORBITING ANTENNAS FOR VLBI

A space VLBI system (Figure 1), and QUASAT in particular, offers
several crucial advantages over ground-based VLBI (see also Morgan et al
1982, Preston et al 1983):

1) angular resolution - Baselines to QUASAT are three times longer than
the maximum baselines on Earth, thus providing at least a factor of
three increase in angular resolution (or an order of magnitude decrease
in the area of the equivalent "seeing disk"). "Super-resolution" should
increase the resolution by a further factor (see point 2b below).

R. Fanti et al. (eds.), VLBI and Compact Radio Sources, 407–414.
© 1984 by the IAU.

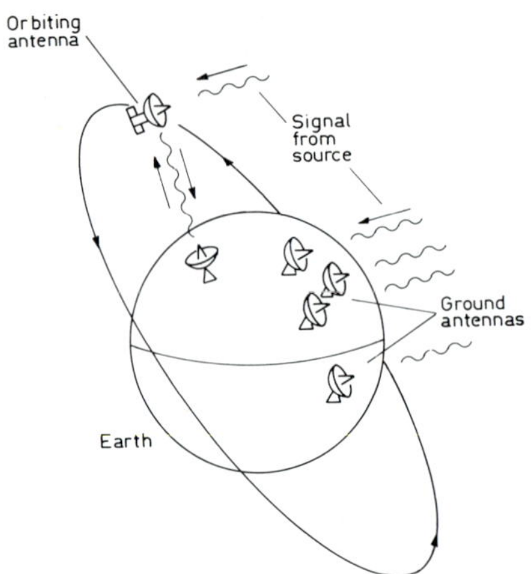

<u>Figure 1</u>   Schematic diagram of QUASAT VLBI system

2) filled u-v plane - The relative motion of QUASAT and the Earth provides for exceptionally dense coverage of the aperture plane of the interferometer system. As long as the QUASAT orbital period is not commensurate with the Earth's rotational period, each pass of the satellite contributes additional spatial frequencies in the u-v plane. This is important in reducing the usual ambiguities and uncertainties in image restoration. Particular points to note in this regard are:
a) The field of view governed by the size of "holes" in the u-v plane is substantially enlarged compared with that of a ground-based array alone. The other factors affecting the field of view are discussed later in this paper.
b) Super-resolution becomes possible. In the standard image restoration algorithms, the final map is smoothed with a gaussian function whose FWHM is of the order of $1/r_{max}$ radians where r_{max} is the maximum baseline length in wavelengths. This is a very cautious approach, corresponding as it does to weighting down the contribution of the data from the longest baselines by about 10 compared with that from the shortest baselines. With only a few baselines such caution is justified. Obtaining higher resolution effectively means making estimates of the values of the visibility function in areas of the u-v plane beyond r_{max}. With the well-filled u-v plane of QUASAT, an enhancement in resolution of at least 30% for complex sources can be obtained (A.C.S. Readhead, private communication); a greater enhancement can be expected for strong compact sources.
c) The lower sidelobe level of the synthesized beam in the QUASAT system reduces the effect of residual calibration errors in the data. The major effect of the unknown data errors is to change the size and position of the sidelobe responses in the map away from their expected positions in unpredictable ways; they can thus not be "subtracted out" with full precision. However because of their reduced amplitude compared with the ground-based array the maps will be "de-sensitised" to calibration errors.

3) Sky coverage - The ground networks planned for operation in Europe, USA and the southern hemisphere by the end of the decade have maximum values of east-west separation of telescopes which are about twice that of the maximum north-south separation. In the equatorial band $|\delta| < 30°$, the u-v tracks become progressively more linear, so that less of the u-v plane is filled, and in addition the synthesized beam becomes more elongated north-south, reducing the resolution in that direction by up to a factor of two. Since half of the sky is located in this band, this is a substantial problem. The QUASAT VLBI system creates north-south spacings comparable to the east-west whatever the declination of the source, and thus provides excellent u-v coverage over the whole sky.

4) Rapid Mapping - Since a single pass of the QUASAT antenna working with a ground array results in fair u-v coverage a crude map can be constructed from a single pass alone. This property is likely to prove useful for monitoring the structural evolution of flaring systems whose structure can vary significantly in time periods of less than a day (e.g. SS433).

THE QUASAT MISSION

Summary

        The proposed mission is a free-flying spacecraft carrying a 15m radio frequency antenna in an elliptical orbit with perigee of about 4000 km and apogee of about 15000 km. The space-borne antenna will observe radio sources at frequencies of 22,5 or 1.7 GHz in conjunction with networks of antennas on the ground, and relay the intermediate frequency data via the equivalent of a 40 Mbit/s (or perhaps a 100 Mbits/sec) link directly to telemetry stations (NASA DSN or ESA networks) on the ground. A clock reference for the antenna in space, stable to 2 or 3×10^{-14}, will be based on hydrogen maser oscillators on the ground and relayed directly to the satellite from the telemetry stations in turn. All communication with the space VLBI antenna will be through the telemetry station network. The two way link provide range-rate data for orbit determination.

        After transmission to the ground, the IF data will be recorded on magnetic tape on the ground in exactly the same way as for the ground based elements of the array. The tapes will then be brought together at the central processing facility (CPF) of the European, US or southern hemisphere VLB array and correlated by local CPF staff. After correlation and calibration, the data will be sent to the appropriate principal investigators. A block diagram of the system is shown in Figure 2.

        The mission operational lifetime is expected to be at least two years.

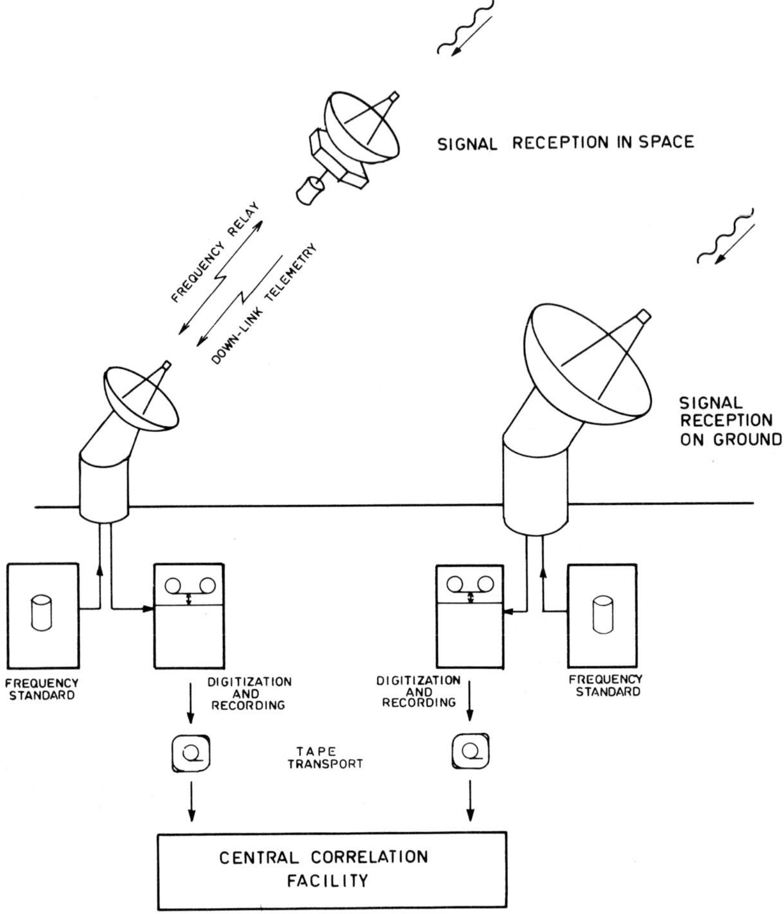

Figure 2   Block diagram of QUASAT VLBI system

Requirements on the Mission

The following are some of the technical requirements on the mission derived from the QUASAT proposal.

Receivers. Observation frequencies are to be K-band 22 GHz, C-band 5 GHz, and L-band 1.7 GHz; simultaneous observation at two frequencies is desired. The receivers should be passively cooled so that the system temperature is ≤ 200 K at K-band, ≤ 75 K at C-band and ≤ 40 K at L-band.

Polarisation. The QUASAT receivers must detect both hands of circular polarisation simultaneously at 1.7, 5 and 22 GHz.

I.F. System. The I.F. bandwidth should be at least 20 MHz, but the system should be capable of carrying 50 MHz through the I.F. chain for extra sensitivity in the event that the required spectrum space can be arranged. A dual channel IF system is desired to allow for simultaneous

observation at 2 frequencies or with both hands of circular
polarisation.

<u>Desired Orbit Parameters.</u> The orbit semi-major axis must be large
enough to provide space-ground baselines long enough to significantly
improve on the resolution obtainable from earth, but not so large that
substantial holes in the U-V plane coverage result. A reasonable goal is
to provide baselines of at least 20,000 km. Therefore the height of
apoapsis should be at least 15,000 km. The periapsis of the orbit must
be low enough to provide intermediate length space-ground baselines down
to intercontinental distances. Thus the height of periapsis should be
within 5000 km. The inclination of the orbit should be ~ 45° so that
observations in directions ± 45° from the perpendicular to the orbit
plane will cover the bulk of the entire sky as the orbit plane
precesses. Preliminary analysis indicates a range of 45° to 65° is
acceptable for the inclination.

<u>Orbit knowledge.</u> Determination of the orbit should be made with errors
not greater than 100 m in spacecraft position, 10^{-2} m/sec in spacecraft
velocity and 2×10^{-7} m/sec² in spacecraft acceleration so that the
residual fringe rates and delays on QUASAT baselines can be brought
within standard windows in the central processors. Corrections in real
time for the several hundred kHz fringe rates on QUASAT baselines will
then be possible by offsetting an LO synthesizer at the telemetry
station.

     It is proposed that the even more accurate orbit knowledge required
for phase referencing observations (position errors ~ observing
wavelength) be determined, post-hoc, from the radio astronomy data
itself.

<u>Mapping time.</u> The typical mapping period will be 24 to 48 hours, but
depending on the scientific goal, some observations will last only a few
hours.

<u>Observing Direction.</u> Typical observing directions will be approximately
perpendicular to the orbit plane.

<u>EXPECTED SCIENTIFIC PERFORMANCE OF QUASAT</u>

<u>U-V Coverage with the proposed orbit</u>
     Figure 3 shows the u-v coverage resulting from 24 hours of
observation with QUASAT in conjunction with (a) the European array plus
affiliate stations, and (b) the US dedicated array, for a source at δ =
60°. The coverage within a radius of ~ 8000 km of the U-V plane origin
results almost exclusively from the ground array alone while the
coverage outside that radius results almost exclusively from baselines
to the satellite.

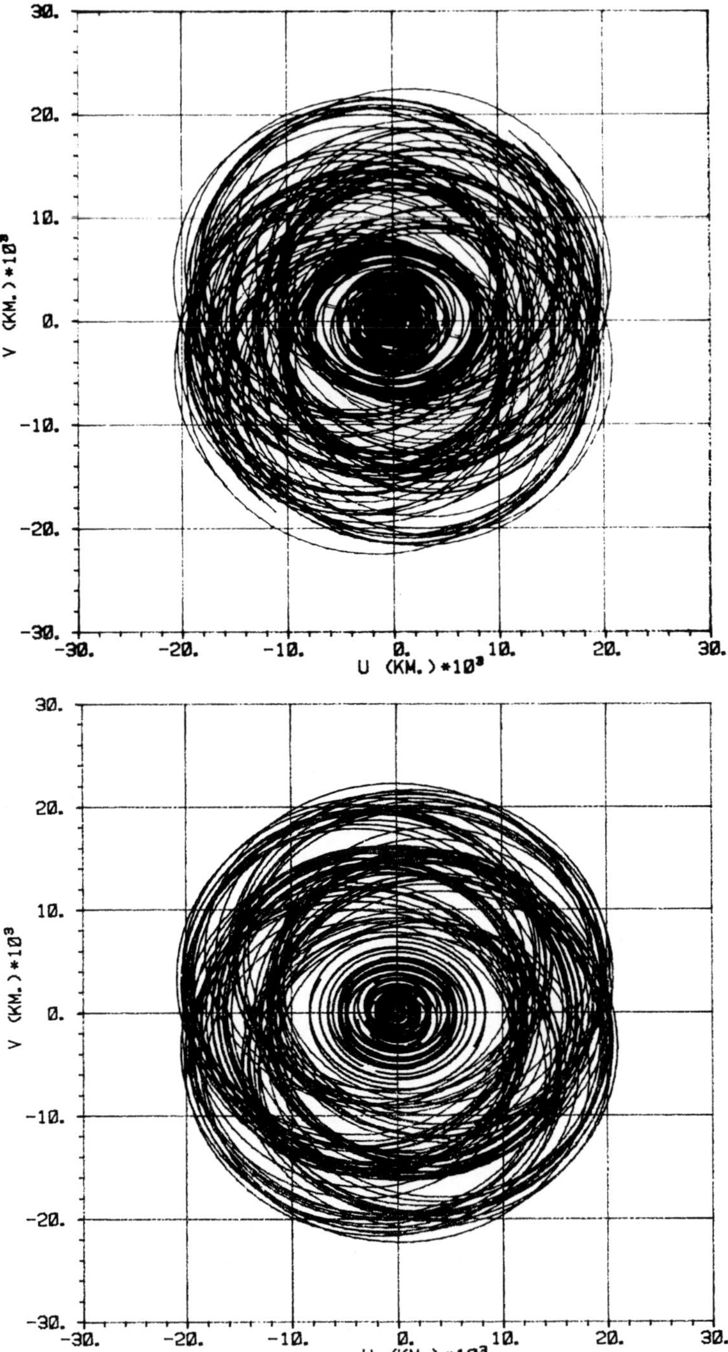

Figure 3   U-V coverage for a source at $\delta = 60°$ with QUASAT and (a) the US dedicated array and (b) the European VLBI Network and affiliate stations.

Linear resolution

Observations at 1.35 cm will be made with a beam of between 60 and 80 microarcsec. This corresponds to

    1 A.U. at the Galactic Centre
    2×10^{16} cm (10 light days) at M87
    10^{17} cm (50 light days) at NGC1275
    1.5×10^{18} (0.5 parsec) at a redshift of 1,
    assuming H_o = 75 km s^{-1} Mpc^{-1} and q_o = 0.05

Sensitivity

With the system parameters for QUASAT given in the previous section, and assuming standard parameters for ground radio telescopes, a 100 to 1 dynamic range map should be achieved for 250 mJy sources at 1.35 cm and 50 mJy sources at 6 and 18 cm after a 24 hour observation. Source count data indicate that ~ 1000 compact sources should be mapable at 1.35 cm and some 10000 at 6 and 18 cm. In a 2 year mission, some 250 to 300 full 48^h observations will be possible, so careful selection of objects to be observed will be necessary.

Field of view

Four factors affect the field of view of the QUASAT system: (a) holes in the u-v plane coverage, (b) integration time per data point in relation to the ($\leq$ 7 km s^{-1}) motion of the satellite through the aperture plane, (c) the bandwidth of the individual IF channels, (d), the maximum map size that can be handled by the image processing system. Factors (b) and (c) individually lead to a worst-case field of view of 20 milli-arcsec (with 10% decorrelation at the field edge) for an integration time of 10 seconds, and a bandwidth of 64 MHz respectively. The latter does not pose a problem, but the former does, since typical integration times to detect fringes on weak sources will be 100 to 300 seconds. However it may be that use of the global fringe fitting algorithm (Schwab and Cotton 1982) for sources which have a compact component in the field of view determined by the 100 to 300 second integration, will allow a wider field of view to be obtained, perhaps as large as that corresponding to integration times of 1 second. This is under study at present. One should also note that a 1024x1024 map at 1.35 cm covers a field of view of 20 milli-arcsec; however computer limitations to the size of maps are likely to diminish in the next 10 years.

MISSION OPTIONS AND ALTERNATIVES

Although the general profile of the QUASAT mission is now established there are a number of aspects which will be reviewed for their potential impact on the mission. A partial list of these follow: consequences of a 5 to 10 year mission, 50 MHz link bandwidth, analogue or digital link, cryogenically cooled receivers, on-board stable oscillators, alternative modes for down linking of telemetry (e.g. TDRSS), phase referencing strategies, astrometry and geodesy applications, other receiving frequencies (e.g. 43 and 0.408 GHz).

PRESENT AND FUTURE ACTIVITIES

Work has already begun on defining various aspects of the project
in more detail. In the USA, A.C.S. Readhead, R.A. Preston and B.F. Burke
are leading a group at Caltech and JPL studying the image formation
properties of the QUASAT mission, whilst in Sweden R.S. Booth and B.O.
Rönnäng (Onsala Space Observatory) are coordinating a preliminary study
of the characteristics of the radio astronomy feeds on QUASAT. QUASAT
Study Teams have been formed in Europe and the USA which will act as
consultants to the ESA and NASA Assessment Studies. A QUASAT Workshop is
planned for Summer 1984.

These activities all point to a growing interest being taken in
QUASAT in the astronomical community and in the space agencies, an
interest that needs to evolve into strong support at national and
international level to ensure selection of the mission in three or four
years time.

REFERENCES

Anderson, B. et al (1982) Mission Proposal to the European Space Agency.
Burke, B.F. (1982) Astronautics and Aeronautics, October issue.
Morgan, S.H. et al (1982) NASA Doc. TM-82491.
Preston, R.A. et al (1976) Bull. Amer. Astron. Soc. $\underline{8}$ (No. 4).
Preston, R.A. et al (1983) Proc. Conf. on VLBI Techniques, Toulouse p.
     417.
Schwab, F.R., and Cotton, W.D. (1983) Astron. J. $\underline{88}$, 688.

DISCUSSION OF THE PAPERS PRESENTED AT THE SYMPOSIUM

DISCUSSION ON THE PAPER BY BROWNE (p. 1)

SAUNDERS : If BL Lacs are to be considered as radiogalaxies seen
end-on, shoudn't we see emission lines in BL Lacs ?

BROWNE : I am in fact suggesting that BL Lacs are counterparts of the
weak Fanaroff-Riley class-I radiogalaxies which, of course, do not have
strong emission lines.

SCHEUER : The evidence for the slow non relativistic motion of double
radio structure is mostly based on old data and refers chiefly to the
diffuse structure. I think it is still true that, for all we know so
far, the very compact hot-spots could be moving at a few tenths of c, so
that the fact that jets point at the most compact hot spots may not be
evidence for the intrinsic one-sidness of jets. We need more evidence
on this point.

BROWNE : Yes, I agree with you. If hot spot emission is beamed, it
might also alleviate some of the problems such as the confinement of
very compact hot spots in some lobes.

BURKE : You cannot invoke statistics in the case of a single example
like 4C32.69. Single observationally selected objects are often
remarkable and improbable. Even with a carefully framed a priori
hypothesis, a single measurement gives an estimate of the mean, and no
estimate of the dispersion.

MILLER : In the unified scheme for BL Lac-type objects, presumably the
optical emission must be relativistically beamed. In the quasars'
unified scheme, the optical emission should not be so beamed that the
emission-lines are drowned out (the line-to- continuum ratios are
similar in compact and extended quasars). Are the two schemes
compatible?

415

R. Fanti et al. (eds.), VLBI and Compact Radio Sources, 415–470.
© 1984 by the IAU.

BROWNE :   Yes, provided the beamed component of quasar optical   emission
never exceeds by very much the unbeamed continuum.   Some optical beaming
is necessary to explain why core-dominated quasars are more likely to be
OVV's.

VAN BREUGEL :   If BL Lac-objects   are   relativistic   jets   seen   end-on,
wouldn't you expect that VLBI-jets would point towards those lobes?

BROWNE :   Yes.

VAN BREUGEL :   Schilizzi and I have found a 25   milli-arcsecond   jet   in
Mk501   which   is   nearly   perpendicular   to its large scale structure as
found using the WSRT (e.g., van der Laan et al., this volume).

BROWNE :   You are talking about one object.   On average jets   and   lobes
line up,   but if there are small bends these will be amplified in end-on
sources.

BRIDLE :   If a jet with bulk relativistic motion also   wiggles,   it   may
then appear as only a few bright knots along its path, due to variations
in the Doppler boosting.   It   may   then   not   be   termed   a   "jet"   by
observers.   This   makes it difficult to infer the statistics of whether
"jets" brighten or dim where they bend.   The   ones   we   call   jets   are
selected to be the ones which show the most continuous emission.

DISCUSSION ON THE PAPER BY CAWTHORNE (p.   7)

WRIGHT :   What is the median redshift of the QSOs   in   your   sample   for
which redshifts are available?

CAWTHORNE :   The   red-shift   distribution   is   strongly   peaked   in   the
interval 1.0 - 1.5, with some 20 % above 1.5.

JAUNCEY :   Accurate radio and   optical   position   measurements   will   be
particularly   important   in identifying the radio component that lies in
the nucleus   of   the   optical   object.   These   should   be   made,   where
possible.

CAWTHORNE :   Unfortunately MERLIN   does   not   give   absolute   positions.
However   good   radio   and   optical positions would help us to understand
some of the more mysterious sources such as 3C454.

DISCUSSION ON THE PAPER BY VAN DER LAAN (p. 9)

READHEAD : Have you done any simulations of the self calibration scheme
at Westerbork.   I ask because in a number of your examples there are
north-south extensions which may not be real.

VAN DER LAAN : Yes, De Bruyn and Noordam have done many simulations and
can assess the realiability of such features. I refer you to De Bruyn's
paper.

READHEAD : The dominant p.a. of 3C371 is east-west in VLBI scales.
The hint of structure in p.a.-30 deg, which can be seen in our first
survey map, is probably due to poor (u,v) coverage since we do not see
it in maps at both 6 cm and 2 cm.

VAN DER LAAN : The large scale faint feature in p.a.-37 deg is one we
are very confident about.   It is, moreover, consistent with a larger
scale unpublished 50 cm Westerbork map showing a halo of about 3'. At 6
cm this is resolved out except for its brightest ridge.

PERLEY : My VLA observations of 3C371 do not show any trace of a
feature extending from the core near p.a.- 40 deg. My map has 16"
resolution at 20 cm wavelength.

VAN DER LAAN : The feature is present only at well below the 1% of peak
flux level. Your dynamic range is marginal for this purpose, although I
see a hint in your map near p.a. +140 deg.

DISCUSSION ON THE PAPER BY PEARSON (p. 15)

JAUNCEY : The properties of the "Compact Doubles", i.e. two widely
separated components with roughly equal brightness, and their high
association with galaxies, make them good candidates for gravitational
lensed objects.   Our observations of 1934-63 (see Preston's talk)
support this.

SUBRAMANIAN : We have tried to model the "compact-doubles" as examples
of gravitational lenses. The small separations in these doubles are not
easy to model. Two kinds of model roughly fit :   1) a model with 3
images, with 2 of the images very close together and a faint third image
farther away (about 1"); 2) a model with a black hole of 10 Mo
producing two images. Our work is as yet very preliminary.

BARTHEL : Concerning the compact doubles: what limit on the size of extended structure can you give? did the VLA find all the flux density?

PEARSON : The VLA data have been published by Perley (1982, Astron.J., 87, 859). The resolution was about 0".4 and the upper limit on extended structures was 0.2% of the peak brightness.

VAN BREUGEL : Does the galaxy identifed with 2021+614 have forbidden narrow line emission?

PEARSON : Bartel has obtained a spectrum of 2021+614. He tells me that it is a narrow-line radio galaxy with a redshift of 0.22.

WRIGHT : I would like to comment that all the known radio-selected QSO's with red-shift larger than 3 show evidence of a "humped" radio spectrum.

MUTEL : Hodges, Phillips and I have recently completed a VLBI Survey of 10 sources selected on the basis of a 'peaked' spectrum (between 0.5 and 2.0 GHz). We found that only 3 of them possibly have simple double structure at 18 cm. This confirms that not all peaked spectrum sources are compact doubles (but probably all compact doubles have peaked spectra).

SHAFFER : Is it clear which component in the 5 GHz map of 3C390.3 is really the core? In some sources, the brightest component is not the core.

PEARSON : The weak north-western component in the 5 GHz map is partially resolved; it has also been detected in Linfield's 10.6 GHz observation. Whether the bright, unresolved component is "really" the core, one cannot say, however.

DISCUSSION ON THE PAPER BY WILKINSON (p. 25)

READHEAD : I just want to add a comment on the class of steep spectrum compact sources you discussed. In two cases (3C147 and 3C309.1) Simon has found evidence based on the low X-ray emission, for relativistic motion towards us. This may therefore be a common property in this type of source.

WILKINSON : Whyborn in the next talk will show new evidence on the

3C147 core. The situation regarding these "predictions" of superluminal motion is clearly more complicated than we realised.

VAN BREUGEL : Steep spectrum cores in galaxies tend to have simpler radio structures. I would expect that these cores are in gas poor environments i.e. ellipticals. Could you specify what type of galaxies you mean?

WILKINSON : They are ellipticals.

REES : You showed evidence that the gas in quasars was more disturbed (and at higher pressure) than on the radio galaxies. Although this could indicate that quasars involve mergers, could it not simply indicate that the quasar activity has itself heated and disturbed the gas?

WILKINSON : The only direct evidence that these objects may be associated with mergers is in the specific case of 3C48 which resembles the 0351+026 system. This system is clearly a violent merger of 2 rich gas galaxies. The arguments for the other quasars being in mergers is therefore purely circumstantial. The evidence is consistent with the hypothesis but does not prove it.

HUTCHINGS : Optical imaging data exist on 3C48 and should be compared with your 1" resolution radio maps.

WILKINSON : Yes! We would like to do this but haven't yet.

NEFF : Are the distorted quasars in clusters, i.e. are there surrounding galaxies with which to merge?

WILKINSON : They are all very distant objects, with red-shifts larger than 0.5, apart for 3C48, so very little is known about their environments as yet.

DISCUSSION ON THE PAPER BY WHYDORN (p. 29)

REES : To display superluminal motion, a source must not only involve relativistic outflow, but also fluctuate (or be "blobby") on the scales that can be observed. The absence of measured superluminal motions is not therefore in itself incompatible with relativistic outflow. There is therefore no inconsistency with inverse Compton arguments for

relativistc motion (e.g. those of Simon et al., 1983). The only strong
argument against relativistic outflow in the M87 jet would come if
significant limits could be set to the proper motion of the knots.

PREUSS : What Rees said is in my view the reason why one should avoid
"predicting" superluminal motion from evidence for relativistic outflow.

COHEN : The archetypical superluminal sources 3C273 and 3C345 both show
the brightest component to be in the middle, not at one end, at low
frequencies (2.3 GHz) where the "core" is self-absorbed. Thus 3C147 is
not necessarily different from the superluminals, except for the NW
bulge. How certain are you that the NW bulge is real?

PREUSS : We carefully looked at the data and we are quite certain that
it is real. In fact it is just here, along the direction of this NW
extension, where the only structural changes may have taken place
between April 1981 and December 1982.

SIMON : 1) The prediction of $v/c = 9$ by Simon et al (1983) assumed
maximum possible angle to line of sight of 8 degrees. In that model,
$v/c \lesssim 0.5$ implies $\theta \lesssim 2$ degrees. 2) There were significant changes at 18
cm over 6 years, but it is not possible to unambiguously interpret these
changes as superluminal motion (Simon, Readhead, and Wilkinson, this
meeting). 3) Model of Simon et al (1983) was of course very simplified;
the calculation of bulk relativistic motion is still good, although
simple superluminal motion is apparently not occuring.

DISCUSSION ON THE PAPER BY BACKER (p. 31)

SCHILIZZI : Could you remind us what the spectra for the sources are?

BACKER : Both 3C84 and 3C345 have slightly negative logarithmic indexes
at mm wavelengths ; both have been decreasing slowly over the past year
or two. 3C273 has been varying rapidly since 1981; its spectrum at 89
GHz was inverted during the early stages of the 1983 outburst.

DISCUSSION ON THE PAPER BY ROBERTS (p. 35)

JAUNCEY : I would like to ask the speakers to use the properly
designated IAU names for their sources.

MUTEL : Could you not eliminate the time-variable phase difference (RL) problem at the VLA by using the 'VX' mode (i.e. apply a previously obtained gain solution) instead of phasing in real time?

ROBERTS : The right-left phase drift typically takes place on the scale of hours. Thus the VA-mode is suitable for each coherent MKIII scan. The scan-to-scan variations are removed from the VLBI by use of the RR/LL ratio on the VLA-Green Bank baseline. At 6 cm this can be done on a timescale of minutes if needed, as the signal-to-noise in the parallel fringes is necessarily very high (1000s) if we are able to detect the cross fringes reliably.

MOFFET : Is it clear that the cross-polarization coefficients (the D's) are constant in time? For example, they depend on antenna pointing errors, which are certainly variable under windy conditions.

ROBERTS : As yet we have no information on the time-constancy of the instrumental constants, although we are aware of the possibility of the kind of problem you suggest. With more dual-polarization equipment it would be possible to check for this possibility.

DISCUSSION ON THE PAPER BY GRAHAM (p. 43)

SHAFFER : Both 3C 236 and 3C 298 look like they are consistent with Larry Rudnick's suggestion of alternating injection in that the brightness peaks on one side of the core would "fill in the gap" on the other side.

MENON : Which component do you suggest produces the low frequency turnover?

GRAHAM : The small, bright western hot-spot will obviously turn over, but it may well do this much higher than 100 MHz.

READHEAD : Surely 3C298 is a steep spectrum compact source, and the turnover at 100 MHz is from structure 0.5" in size which you don't see on your VLBI map. If observed with low dynamic range 3C147 would look very like this.

ANANTHAKRISHNAN : The source has a large percentage of its flux in a diffuse component of 400 mas size since IPS at 327 MHz shows that only about 20% of its flux is in a component smaller than 400 mas. One should be able to see this in low resolution low frequency VLB.

GRAHAM : The VLB map shows about 70% of the total flux at 18 cm. The total extent of the source at lower frequencies is about the same as at 18 cm, so there is no very large diffuse component.

DISCUSSION ON THE PAPER BY MENON (p. 45)

VAN BREUGEL : The emission measure for free-free and optical line emission are similar. One might expect therefore that low frequency turnovers for bright emission-line objects occur at higher frequencies. Wouldn't it be possible to explain the red-shift dependence of the low frequency turnovers as being due to free-free absorption ?

MENON : The measured angular sizes, where available, agree very well with the sizes computed on the basis of synchrotron self-absorption model. It is possible that for the smallest sources the shape of the spectrum below the turnover frequency is influenced by free-free absorption.

DISCUSSION ON THE PAPER BY JONES (p. 47)

SHAFFER : Apropos of the fact that your dynamic range is considerably less than expected from SNR considerations, one should remember that amplitude closure may not hold at high frequencies, where non-linear phase variations within an integration may affect the several baselines differently.

JONES : I agree. All of the assumptions that various errors remain constant during an integration interval are less valid at higher frequencies, and consequently the current self-calibration techniques will not necessarily assure closure as well as at lower frequencies. Baseline-dependent errors may be limiting the dynamic range of VLBI maps at any frequency, and are very difficult to correct without good point sources.

DISCUSSION ON THE PAPER BY BRIGGS (p. 49)

ROBERTS : Could you comment on the relation of the changes in the VLBI visibility and the source outbursts? How does this relate to Wolfe's model for the source?

BRIGGS : The model that has the source continuum driving the cloud spin temperatures, and therefore optical depths, was lightly constrained by the observations taken in the period 1976-1981 (Wolfe, Davis, Briggs, 1982). Since then the source continuum has varied without the line depths varying in the manner predicted by the model. The model has been abandoned.

SIMON : Based on 1661 MHz observations of 0235+614 (used in calibrating a VLBI experiment), the size is roughly 1 mas, with no extended emission to the 0.5 % level.

BRIGGS : The source may be a little different at lower frequencies and may change with time. The model may work well for a 1 mas source. The gradients in opacity will simply need to be steeper.

DISCUSSION ON THE PAPER BY WEISTROP (p. 51)

JAUNCEY : Do you have any Merlin map at 1" resolution of this object to match with your optical photo ?

WEISTROP : No, we don't, but de Bruyn will show tomorrow a map made at Westerbork.

JAUNCEY : Radio astronomers making these type of high resolution surveys dicussed at this Symposium, should remember that 4 m optical telescopes make perfectly acceptable maps at 1" resolution and should be included more often.

NEFF : Is there any evidence for spiral structure in the optical nebulosity ?

WEISTROP : We have not yet looked for such structure in our data, but plan to.

HUTCHINGS : We have no optical data on this object. If the galaxy is a triaxial ellipsoid, or an inclined disk, the projection on the sky of an orthogonal radio jet need not be at right angle to the optical axis.

WEISTROP : If the disk has symmetry with respect to the minor axis, an orthogonal radio jet will appear orthogonal projected on the sky.

BIRKINSHAW : The "optical" structure relevant to the VLBI jet is not
that of the nebulosity of 1219+28, but rather the very central optical
structure (hidden by the BL Lac light). A comparison of the morphology
of the fuzz and the VLBI structure is not very meaningful.

WEISTROP : If the central optical structure is aligned with the minor
axis of the fuzz , the comparison is meaningful.

DE BRUYN : I will show a low-resolution WSRT redundancy map tomorrow,
indicating that 1219+28 is one of the best "point sources" (on arcsec
scale) known sofar, but that there is a very faint extended emission
region to the south-west in the direction of the companions.

DISCUSSION ON THE PAPER BY COTTON (p. 53)

MENON : Optical quietness should not be decided on the basis of the
single epoch plates of the PSS since variability is a characteristic of
these sources.

COTTON : For a single source this is true but there are a number of
sources in my sample and none appears on the PSS prints.

WEISTROP : Are there any plans for deep optical observations of these
sources using CCD detectors and large optical telescopes? Such
observations can go several magnitudes fainter than the Palomar Sky
Survey.

COTTON : The possibility of this type of observations is being
explored.

ZANINETTI : I would like to know why you have considered inverse
Compton losses and not synchrotron losses?

COTTON : Both inverse Compton and synchrotron losses were included in
the detailed models; in all of the examples I have shown that inverse
Compton losses are more significant than synchrotron losses.

COHEN : Bulk relativistic motion towards the observer apparently makes
the problem of explaining the spectrum more difficult. Why not look
into bulk relativistic motion away from the observer?

COTTON : Relativistic motion away from the observer will greatly reduce the observable flux density. Since several of these sources are relatively bright (about 5 Jy at 400 MHz) they would be extremely bright sources if viewed from the forward direction.

SAUNDERS : Your theme seems to be that there may be a new class of object with compact radio properties which are optically "quiet", the criterion for optical silence being no identification on the PSS. In fact, several sources are known (from surveys such as Peacock and Wall, 1982, M.N.R.A.S., 198, 843) with the same radio properties which have no PSS identification but which, on deeper investigation, simply turn out to be faint galaxies. Their radio and optical properties are unexceptional. Is there any reason to think that your sources are different from those?

COTTON : The main point I was trying to make is that these sources radiate via synchrotron radiation, though probably have significantly weaker magnetic fields than the flat spectrum sources. On the basis of the current data for 2147+145 a simple synchrotron model predicts more optical inverse Compton than is observed and very short relativistic electron lifetimes. None of the observed properties of these sources are peculiar. Only an attempt to determine the physical conditions indicates a difference between these sources and the better studied flat spectrum sources.

DISCUSSION ON THE PAPER BY FANTI (p. 57)

SIMON : A curiosity in 3C138 is that there is a 200 mJy component 1-2 minutes of arc away, along the jet, to the south-west. It may be related; a VLA map is still in production.

WILKINSON : Just to clarify the point about whether galaxies or quasars have different structures, there is a range of radio structures visible in both optical types but the "pathological" cases where the source is really distorted are only found among the quasars.

MUTEL : It appears that the sources with a dominant double structure may be in the same class as the so-called 'compact doubles' (Phillips and Mutel, 1982, A.A., 106, 21) but with larger linear sizes (1-3 Kpc) and lower peak frequencies. There appears to be much correlation between the wavelength of maximum flux density and linear size. Also, all of the doubles reported in your paper and most of the previously known VLBI doubles are galaxies (or EF).

FANTI : Resolution effects can be very important. 3C138 would look like a close double with Merlin.

VERON : Sometimes ago Roland et al.(1982, Astron.Astrophys.,116,60) have observed with Merlin four compact very steep spectrum sources ($\alpha > 1.3$). All of them turned out to be doubles with separations in the range of 2 - 10 arcsec with unresolved components (smaller than 0.8"). Three of these sources are as yet unidentified, the fourth is identified with a faint galaxy with red-shift about 0.5.

DISCUSSION ON THE PAPER BY VAN BREUGEL (p. 59)

SHAFFER : In most objects, the low-frequency turnovers are consistent with synchrotron self-absorption since compact components are seen with the VLBI. NGC 1068 is a rather unique in having free-free absorption as the dominant cause of the low-frequency turnover.

VAN BREUGEL : I agree that high surface brightness and spectral curvature can be explained by synchrotron self-absorption. I disagree that the free-free absorption in NGC 1068 is just a special case. There is mounting evidence that a large fraction of quasars are embedded in spiral-type galaxies and also many radio and optical properties of Seyferts and steep spectrum cores in quasars are similar. Thus steep spectrum radio cores in quasars are probably also embedded in dense environments. For typical Seyfert parameters one than expects also free-free absorption at low frequencies.

DE BRUYN : I agree that the flattening of the spectrum of NGC 1068 is due to free-free absorption. Yet in the steep spectrum cores, like 3C 147, the turnover is at the frequency where you would expect synchrotron self absorption to become important on the basis of the brightness temperature (as derived from size).

VAN BREUGEL : I believe that both synchrotron self-absorption and free-free absorption contribute to the spectral turnovers in compact radio sources (see also answer to Shaffer's question).

DISCUSSION ON THE PAPER BY PHILLIPS (p. 63)

PORCAS : How strong is the central component of CTD 93 at 5 GHz ?

PHILLIPS : About 0.05 Jy.

BACKER : Can we clear up the rate of discovery of compact doubles in a recent survey : is it 1/10 as you report or 3/10 as reported by Mutel this morning ?

PHILLIPS : A difference in semantics. We found one clear double and 2 "candidates" that may be triples, out of ten.

DISCUSSION ON THE PAPER BY PRESTON (p. 67)

BACKER : I realize that the galactic center is not a quasar, but can you comment on the single baseline result for this object?

PRESTON : The galactic centre appears to be extended in a general east-west direction, though with only one baseline the size and shape is very model dependent.

MUTEL : Can you comment on the variability of 0438-43?

NICOLSON : 0438-43 had a strong outburst in mid 1960's. This died away smoothly over about 10 years. The 13 cm flux decreased from 8 Jy in 1970 to 4 Jy in 1980. The 13 cm flux density of 1934-63 has not varied significantly since 1967.

SHAFFER : Although 0438-43 may be the first very high redshift QSO to be mapped, other high z quasars, like OH471 and OQ172, definitely have structure detectable with VLBI.

PRESTON : VLBI maps of five very high redshift QSO's were made by Walker (unpublished). Four of the sources show little structure (including OH471). The other source (OQ172) does show some weak extended structure.

READHEAD : You have to be very careful in classifying sources as having double morphology on the basis of observation at a single frequency. Core sources often have "double" structure at one frequency. You really need to know the spectra.

DISCUSSION ON THE PAPER BY ZAMORANI (p. 85)

SCHEUER : Have you tried correlating the X-ray luminosity of steep-spectrum radio quasars with their core radio flux only? and, if so, what is the result ?

ZAMORANI : Yes, this has been done by Tananbaum et al. (1983a), for the complete sample of 33 3CR quasars, for most of which good measurements of the core radio flux are available. While there is no evidence within the sample for a dependence of X-ray luminosity on the total radio luminosity, they find a significant correlation between X-ray luminosity and radio luminosity of the central radio component for the 3CR quasars which have a "triple" radio structure.

MILLER : In answer to Scheuer's question, we have found a strong correlation between the X-ray luminosity and the radio cores of powerful double radio galaxies. The cores of double extended quasars lie on this correlation.

BEGELMAN : I am aneasy about one aspect of the logic which leads one to reguard the X-rays as a second order consequence of the radio flux in the radio-loud sources. While the presence of strong radio emission is sufficient to guarantee a strong X-ray flux, the converse is not true. Therefore it might be sensible to explore the hypothesis that the presence of X-rays is necessary for the production of the radio emission. One possibility is that Compton heating by X-rays is necessary to reduce the free-free absorption in a thermal wind, in order for the radio to get through. Have you considered this alternative approach ?

ZAMORANI : No, I have not examined this possibility.

PANAGIA : Concerning Begelman's question, since to produce enough free-free absorption one needs an high emission measure, one should expect a clear anti-correlation between radio flux and hydrogen recombination line intensity. To my knowledge such a relationship is not found from actual observations.

REES : If you are right in conjecturing that the X-ray component correlation with the radio emission comes from a region larger than 10E19 cm, then the radio loud quasars would be less likely than the radio-quiet quasars to show large amplitude X-ray variability on short timescales. Is there any evidence of that ?

ZAMORANI : A systematic study of the X ray variability properties of a large number of quasars observed with the Einstein Observatory has been recently completed (Zamorani et al. 1983). On a time scale of about one day, they found four variable quasars; two of them are radio-quiet and two are radio-loud (flat spectrum). Among the 31 quasars analyzed on this time scale , 24 are radio-loud (14 with flat radio spectrum and 10 with steep radio spectrum) and 7 are radio quiet. Hence, the percentage of variable objects in the two classes of quasars would seem to suggest, if taken at its face value, a higher frequency of short timescale variability in radio quiet quasars.

COHEN : The very large dispersion in surface brightness seen by VLBI is mainly due to the fact that most components are optically thin. In the few cases where there are reasonable diameter measurements near the synchrotron maximum, the brightness temperatures are between 10E11 and 10E12 K.

WRIGHT : In any luminosity-luminosity plot it is important to consider selection effects. I would like to see the effect of randomizing the red-shifts between objects in your plots.

ZAMORANI : I have not applied the "randomization" test that you are suggesting. However, I am confident that the correlations discussed in this paper are real for two different reasons. First, correlations are present in all our samples even if we use the observed flux densities, instead of the luminosities; second, you would be right in worrying about the reality of a luminosity-luminosity correlation if the objects were selected independently from flux limited surveys at two different frequencies. In fact, in this case there would be no information at all for those objects which lie beyond the detection limit in one of the bands, but not in the other. But this is not our case : the objects were pre-selected on the basis of their radio and/or optical properties and then observed with the Einstein Observatory at X-ray frequencies. The use, that we make, of both the X-ray detections and the X-ray upper limits assures that any correlation found in a comparison of radio and/or optical data with X-ray luminosities is real.

HUTCHINGS : There is a correction that needs to be made to L_{opt} to remove the luminosity of the QSO galaxy. This ranges from zero to more than one magnitude in the objects we have resolved. In our sample of 33 objects there is a correlation, with slope 1, between the X-ray luminosity and the red band luminosity in the QSO rest frame.

DISCUSSION ON THE PAPER BY UNWIN (p. 105)

JAUNCEY : How sure are you that, within a given source, the components
have significantly different velocities ?

UNWIN : In 3C 273 we measure the same speed to within the errors for
two components (C3 and C4). The new expansion in 3C 279 is at 1/3 the
rate found in 1971, but the motions are not seen in the source
simultaneously.

WALKER : Components with velocities that are different by a factor of
two are seen in 3C 120. These components were seen at different times.
Components seen at the same time may have similar velocities.   These
results are presented in another paper.

REES : There is a class of models, invoking induced scattering in a
foreground medium, which can generate apparent superluminal motions even
in the absence of bulk motion, provided that the brightness temperature
is such that kT is less than $m_e c^2$. Your peak contour levels are well
above this at 5 GHz, but may be not at 10 GHz.   Could the case for
superluminal effects be proven on the basis only of observations where
the contour levels are below $m_e c^2$? If the answer is " yes ", then the
class of models based on induced scattering can be discarded.

UNWIN : In both 3C 273 and 3C 345, all the superluminal motion is
measured relative to the 'core' whose brightness is well above $m_e c^2$ at
both 5 and 10 GHz; however the moving components in 3C 273 have 10^8 GHz
brightness less than $m_e c^2$. Such models cannot therefore be eliminated
at present.

DISCUSSION ON THE PAPER BY MOORE (p. 109)

KELLERMANN : Could your observations be explained if the relativistic
ejection occurred from an engine which itself was in motion around some
other body? This would allow you to have rectilinear motion with
apparently different origin, but, since the same engine is involved,
there is no problem in explaining the equality of the fluxes.

MOORE : Both components must be strongly boosted.   Therefore the
ejection angle for the core and the western component would still have
to be nearly the same, even though ejection occurred at different points
in the two engine's motion. Also the core is always present and does

not move relative to NRAO 512 (Bartel et al., this Conference), so it
seems likely that the core engine is stationary.

ALLER : Could another source component have appeared between your first
two epochs at 1 cm? The cm flux curves indicate the possibility of a
new component appearing in late 1981.

MOORE : The trajectory has been sampled fairly frequently at 1.3 cm and
2.8 cm and it is very continuous. It is unlikely that what we observe
is due to another component. There could be a new component which is
still unresolved in the core.

ALLAN : As you have position differences between 1.3 and 2.8 cm
observations, you must be very careful of optical depth effects in
interpreting the observations. Taking optical depth effects into
account, can you make a model with simple linear motion?

MOORE : Optical depth effects (in either component) may well account
for the slight difference in radial separation observed at 1.3 cm and
2.8 cm. However, I doubt these effects alone could explain the
non-radial motion or acceleration.

MARCAIDE : I'm surprised to note that the ratio of strengths between
the core and the outgoing component does not change monotonically in
time. Could you comment on that ?

MOORE : First, I would say that the second epoch map may have a 180
degree reversal as it was made with a different processor and software
than the first and third epoch maps. While this would not affect our
results of non-radial motion or acceleration, the change in the flux
ratio would than be monotonic. Second, the absolute calibration of 1.3
cm maps is not yet completed, so it is unclear how the flux of each
component has changed. Untill these questions are addressed, a definite
answer is premature. It may be necessary to invoke intrinsic
variability of the components.

BEGELMAN : Could you explain the variation of component strength, as
well as the trajectory, in terms of a precession model ? The apparently
moving material might not in fact be the same gas, but rather material
that happens to be brightest at any given time. The picture I have in
mind is that of a lawn sprinkler which swings into and then across our
field of view. Coupled with relativistic effects, this could involve
large change in both position angle and brightness of the apparently
moving component.

MOORE : A precession model in which the western component is discrete and ejected linearly from the core can be ruled out by the non radial motion. However, the model you suggest is feasible (as long as there is relativistic motion). It depends on the dynamic range of the maps and the time scale for decay of previously emitting regions. Our dynamic range is good (about 40 to 1) and the data are well fitted by only two components, so the decay time scale would have to be short.

DISCUSSION ON THE PAPER BY BARTEL (p.   113)

JOHNSTON : Could the 200 micro-arcsec scatter in your observations be due to self absorption effects in the core source? That is, the position of the core source in 3C 345 is not stable to 200 micro-arcsec with time.

BARTEL : Yes, that is a possibility.

ALLER : The epoch of the 1980 position measurement is coincident with the large outburst at cm wavelengths. The source had a substantial opaque region at 3.6 cm at that time, so that the "peak" in the intensity may appear down the jet where the optical depth is about 1. This is the same point that Johnston brought up.

MARCAIDE : Can you tell what your chi square per degree of freedom is and, if not unity, can you comment on the reason ?

BARTEL : If we consider the observed fringe phases to be independent samples with the statistical standard errors calculated from the signal-to-noise ratio only, chi square per degree of freedom is definitely not unity. The reason is the effect of systematic errors which dominate fringe phase uncertainties. We used the root-mean-square of residuals to estimate the errors of our observed fringe phases, so that the final chi square per degree of freedom of the phase differences is unity. The statistical standard errors we quote come from this least-square analysis. They do not take account of possible correlation in the errors of the observables.

VAN DER LAAN : Is NRAO 512 completely quiescent? If it is flux variable, indicative of changes in opacity structure, its centroid may move about at your remarkable levels of accuracy.

BARTEL : NRAO 512 has an inverted and time variable spectrum between 2.3 and 8.3 GHz, so that changes of the position of the brightness center, caused by changes in the opacity, are conceivable. However, the source structure at 10 GHz appears to be nearly pointlike (FWHM less than 0.7 mas) at contour levels greater than 5 % of the peak. Hence it is questionable whether the peak moves around by as much as 200 micro arcsec.

GORENSTEIN : Regarding the 200 microarcsec error in the difference of positions, do you anticipate that you will be able to account for this error with systematic effects? Do you have any indications now that the 200 micro arcsec "jitter" between epochs might represent true center of brightness changes as suggested by opacity effects?

BARTEL : Neglected effects, such as errors in the earth orientation or unmodeled changes in the atmospheric delay, may account for position errors substantially larger than the quoted uncertainty of , e.g., 20 micro-arcsec for the July 1980 epoch. However, the position at this epoch is offset by about 200 micro-arcsec from the mean, perhaps a bit too much to be caused by neglected effects only. Changes in position of the center of brightness may well have taken place.

WILKINSON : Can you confirm that 2-frequency observations were crucial to get the accuracy which you have reported ? In other words what would have been the accuracy if you had only observed at 3.6 cm?

BARTEL : If we had only observed at 3.6 cm, the relative deduced position in right ascension of 3C 345 would have changed by about 75 micro-arcsec in July 1980, about 140 micro-arcsec in march 1981 and about 200 micro-arcsec in june 1981. Hence our ionospheric-free positions, obtained from dual frequency observations in the 1980's, are more accurate than those obtained from single-frequency observations alone, like those at 3.8 cm in the 1970's.

DISCUSSION ON THE PAPER BY MUTEL (p. 117)

DE BRUYN : Do you believe in the reality of the features to the north of the peak ?

MUTEL : Yes. It appears that the component " A ", in fact, has been expanding preferentially along the 10 degree p.a. channel, with no evidence for expansion orthogonal to the channel. The one-dimensional cuts, however, show emission to the north of component " A " at all four epochs, so there appears to be long-lived emission in the channel unrelated to components " A " or " B ".

DISCUSSION ON THE PAPER BY ALLER (p.   119)

MUXLOW : A comment on the identification of the core within the VLBI structure. There is a suggestion of a weak component of few mJy at 1.67 GHz, about 0.4" to the south, derived from Merlin calibration data.

DISCUSSION ON THE PAPER BY WALKER (p.   121)

ALLER : Historically there have been a range of position angles given for the VLBI structures. What are the best current estimates of the range in position angles of the different events?

WALKER : One of the earliest observations gave a rather different position angle (about 60 degrees as I recall) than all subsequent observations and may have been wrong. The current observations give position angles between about - 100 and - 110 degrees. Any fluctuations could be a result of measurement uncertainties for this low declination source. The structure from about 10 to 200 mas reported by Benson in the next paper clearly has a different position angle (- 92 degrees).

SCHEUER : Is there evidence for different apparent velocities at different times but at given position in a source ?

WALKER : The various moving components in 3C 120 have been followed through similar distances from the core although the faster, more recent components, were only seen at one or two points in the common distance range. The velocities of the faster components appear to be constant over their full range, but the constraints are not strong within the region over which the slower features were observed.

UNWIN : The early data on 3C 279 gave a much higher rate than more recent data on a different component at the same distance. The data on this source is not extensive at this time.

JAUNCEY : The 3C 120 expansion plot that you showed seems to me, as a skeptic, to be totally consistent with a constant expansion velocity fitting all events.

WALKER : The rates of the 1972.5 - 1974.5 component are about 1.51 $\pm$ 0.13 mas/year. For the more recent components we found the following rates. Feature A : 2.57 $\pm$ 0.09 mas/year; feature B : 2.34 $\pm$ 0.2 mas/year; feature C : 1.65 $\pm$ 0.4 mas/year. The rates of features B and C will be refined by future observations. Some of these rates would have to be in error by a few sigma for them to be the same.

ALLAN : Do you see any evidence in your maps for components running into each other?

WALKER : There is no clear evidence yet for features running into each other.

DISCUSSION ON THE PAPER BY BENSON (p. 125)

SCHILIZZI : On the model proposed by de Bruyn and I for the relation of the small and large scale structure in the superluminal sources, the wiggles in the 200 mas jet in 3C 120 should be regarded as a nutation on the nutation on the long-term precession!

BRIDLE : It is encouraging to see a radio "jet" that is very long and very thin, with positive evidence for something moving along it at its base. It is meaningful to call a thing like this a " jet ". The term has been applied much to loosely in other papers here and in the VLBI literature to denote "barely resolved structure".

DISCUSSION ON THE PAPER BY BAATH (p. 127)

SAUNDERS : I note you say Cambridge was used as a VLBI station. I thought we were supposed to be too elitist to do VLBI ! Perhaps it was before my time.

BAATH : Yes, we did use Cambridge in 1978 and did found fringes even though the coherence time was short.

ECKART : Do you see any motion in Mk421 and 1749+70? Do the secondary components of these sources become stronger when the "core" does so ?

BAATH : It looks like a radio component was squeezed out through the jet in Mk421. The superluminal limit at the red-shift of Mk 421 (z =

0.03) is 1 mas/year, so the motion is only weakly superluminal if any.
There is no evidence for motion in 1749+701. The peak outside the core
actually moved inward but this is probably due to the birth of a new
component closer to the core. Both sources showed similar behaviour in
that the core brightened between the first and the second epoch and than
stayed bright during the third. During this time secondary components
appeared and moved through the jet of Mk 421 and died out in 1749+701.
What we observe in 1749+701 fits very well with the flux density
monitoring made by Geldzahler et al. at NRL.

DISCUSSION ON THE PAPER BY MUXLOW (p. 141)

HUTCHINGS : Our work (Gower et al. 1982) of modelling sources like
these suggests that there is a class of one-sided jets which cannot be
fit with a simple precession model, but all of which can be fit with one
in which the core angle increases with time.

BURKE : Instead of using arbitrary parameters, why not use physics like
the allowed solid-body motion? Have you tried nutation as well as
precession?

MUXLOW : I have not as yet tried nutation in addition to precession.
This may account for the structure. There is likely to be a great deal
of uncertainty, however, in the type of solid-body which may be
involved.

BRIDLE : I have a general comment on precessing-jet models in which
the cone angle is allowed to vary. These open up a huge parameter space
in which a very wide range of structures may be matched with very little
uniqueness. To make these "fits" convincing one should have either more
cycles of the structure on one side of the core, or evidence of the
shape of the structure on the other side - to show the expected S
symmetry. I doubt very much that the present "fits" force us to believe
in opening cone angles, as Hutchings suggested.

MUXLOW : I agree. Some form of additional complexity is required over
and above simple precession in order to account for the jet in 3C418.
What form this takes is open to question. I can, however, say that in
the absence of other complications I do not think that it takes the form
of a variable jet speed. I have found that allowing the jet to slow
within 1 arcsec of the core degrades the quality of fit. Beyond 1
arcsec from the core some improvement results. With just 2 cycles of

precession visible and no counterjet we are clearly not forced to believe in opening cone angles. This has however been postulated in other radio sources and seems the most reasonable suggestion at this time.

VAN BREUGEL : Is 3C418 variable? If so, this might provide constraints on the precession cone-angle.

MUXLOW : The core has a flat spectrum and is probably variable. Unfortunately, very little is known about the nature of this variability since the source lies at low galactic latitude and has thus escaped the attention of many workers.

BARTHEL : You mentioned a pressure gradient in order to get a good fit for your precessing jet with constant cone angle. From high-resolution VLA observations Barthel, Miley and Schilizzi (in prep.) argue that such pressures may be present in high-redshift quasars ($z > 1.5$). I note with interest that 3C418 is at high redshift.

DISCUSSION ON THE PAPER BY REID (p. 145)

GORENSTEIN : Do you have a limit to flux density of counter-jet?

REID : In both epoch maps any counter jet must be weaker than 0.02 of the peak intensity and less than about 0.05 of the intensity approximately 15 mas down the jet.

NEFF : Is there evidence for wiggles in the VLBI jet of M87, and if so are they in agreement with the optical wiggles reported by Nieto?

REID : The second epoch map has less pronounced wiggles than the first epoch. Higher spatial resolution may be required to conclusively establish the wiggles. Even if the wiggles in the VLBI jet are established, they would be quite different from those seen optically with less than 1/100th of the angular resolution.

DISCUSSION ON THE PAPER BY NIETO (p. 147)

VAN BREUGEL : Is there a difference in the radio-optical spectral indices of the knots and inter-knot regions?

NIETO : No real accurate quantitative work has been done on the electronographic material, nor on the photographic ones (Nieto and Lelievre, 1982). Therefore it is difficult to say. But at first no major difference appears between the last VLA map (Biretta et al. 1983) and the restored photographic images at a 0.2" scale (Lorse and Nieto, 1983), which suggests a rather constant radio-optical spectral index for the knots throughout. The index of the inter-knot regions require an even more careful quantitative study since they are quite faint. In all cases the comparison is not easy because the radio and the optical data have different resolutions and different signal to noise ratios.

DISCUSSION ON THE PAPER BY PERLEY (p. 153)

REES : If the bends in the outer parts of the jet and counter-jet of NGC 6251 were due to "environmental" effects rather than ballistic motions, then the absence of a 4 to 1 asymmetry in scale would surely not be evidence against relativistic jet speeds.

PERLEY : My comment refers to jets in which identical knots or blobs can be identified as having been ejected at the same time and which move ballistically, or through identical media. If bends are due to environmental effects, the analysis will not apply.

DISCUSSION ON THE PAPER BY DE BRUYN (p. 165)

BRIDLE : Is it fair to compare source sizes from maps having very high dynamical range with source sizes from maps having much lower dynamical range? Perhaps, if the comparison sources were also mapped at very high dynamic range you could find faint outer extensions to them as well (this may be particularly the case if the extended structures are edge-darkened, i.e. Fanaroff/Riley class I).

DE BRUYN : The comparison of largest angular sizes of superluminal sources and the other 3CR sources can be used to infer information about their relative intrinsic sizes and "precession" angles only within the context of Doppler boosted kinematic models (see Schilizzi and de Bruyn, 1983). If the cores in the superluminal sources are intrinsically weak cores, the dynamic range in our maps does not really differ from that in the maps of other presumably unboosted 3CR sources. If the cores are unboosted, the superluminal sources are very different type of sources and the comparison is not meaningful.

SAUNDERS : Your deprojected superluminal sizes are so large that such
giant sources, if they exist, may have been missed in surveys because :
a) low luminosity giants have very low brightnesses and may anyway be
resolved out; b) high luminosity giants would have their tails removed
by synchrotron and inverse Compton losses, so that one would have to
guess that two widely separeted hot-spots were a single source. The
requirement for precession is than not that the deprojected sizes of
your sources would otherwise be too large, but that so far you have not
seen any superluminals with deprojected sizes of about 100 Kpc, the size
that countless doubles have.

DE BRUYN : The deprojected angular sizes of most of the sources
(excluding 3C 120) are not so large that such sources would have been
missed or failed to be identified. Also the largest angular sizes that
we measure in our and other maps of the superluminal sources really do
refer to fairly sharp edges in the sources, suggesting that loss
mechanisms have had little effect on the sizes of these sources.

MILLER : If the "precession" modification to the unified scheme is
correct, then we expect a large fraction of the steep-spectrum extended
doubles to show similar precession (typically over 30 degrees). Yet in
those sources the VLBI jets are well aligned with the outer lobes and
hot-spots. Are there samples large enough to provide a definite
statement of this inconsistency ?

DE BRUYN : I think the superluminal source sample is large enough to
make our result a significant one. One has to bear in mind, however,
that the superluminal sources have intrinsically relatively faint lobe
emission. So when comparing with steep spectrum doubles one should
select doubles of the same absolute lobe power. I am not sure whether
such a comparison can already be made.

DISCUSSION ON THE PAPER BY PADRIELLI (p. 169)

READHEAD : The observations of symmetric sources are very interesting
because they suggest that the basic jet is intrinsically two-sided. Can
you be confident that the strongest component is the core and not a knot
embedded in a core-sided jet in which the core is invisible ?

PADRIELLI : The definition of core is not easy without high frequency

VLBI observations, because it could be self-absorbed at 18 cm. However in the case of 0859-14 we are quite confident that the two elongated features in north and south direction correspond to a symmetric structure about the core. This is also confirmed by the VLA arcsec structure.

DISCUSSION ON THE PAPER BY DENNISON (p. 177)

JAUNCEY : Is there any sign of opacity effects below 300 MHz ?

DENNISON : That is a very important question, to which we do not have an answer at the moment. I suspect that at frequencies somewhat below 300 MHz, dominance by more extended components will dilute the effect. However, accurate flux measurements and monitoring at these frequencies are urgently needed to understand what is going on.

ANANTHAKRISHNAN : At Ooty we have observed a sample of 50 LFV sources at 327 MHz for compact components by interplanetary scintillation. We find: i) all the low frequency variable sources in our sample contain components smaller than 300 mas; ii) 80 % of compact components have sizes smaller than 30 mas; iii) for 22 of the sources in ii) for which VLBI measurements at 18 cm or high frequencies exist, the VLBI flux is less than the scintillating flux by about a factor two. Therefore it seems to me that only the compact scintillating flux could be involved in the variability phenomenon and their spectra are likely to be quite flat.

DENNISON : That is consistent with our observations and, as I pointed out in reference to Jauncey's question, at quite low frequencies (below 300 MHz) the dominant components are likely to be more extended and therefore less variable. Indeed, there is possible evidence that the variations are weaker at 318 MHz when compared with the 430 and 606 MHz.

DISCUSSION ON THE PAPER BY LEGG (p. 183)

JOHNSTON : BL Lac had a large outburst in the early 1970's. Could the decay differences in the outbursts between the 1960's and 1970's be due to multiple bursts during the large outburst in the early seventies versus single bursts for the 1960's ?

LEGG : No. I think it is clear that there is a genuine difference in the decay rates of individual events after february 1970. Decay rates after this date are consistently a factor of two smaller than decay rates before.

DISCUSSION ON THE PAPER BY MILLER (p. 189)

BURKE : There is a strong evidence on other grounds against the gravitational lens model. Roberts, Turner, Gott and I examined 25 high luminosity quasars with the VLA and the data imply that lensing is not a frequently occurring phenomenon

MILLER : I agree. However many of the arguments against gravitational lensing are of a statistical nature, whereas these comparisons place constraints on the physical parameters which would be required for these sources.

KONIGL : Could you comment on the correlation displayed specifically by the OVVs and BL Lac objects ? Do they differ from the general compact sources that you considered ?

MILLER : I have not included BL Lac type objects, as many do not have measured red-shifts, and presumably they would not lie on the correlation of fig. 1, as they are X-ray bright but have only weak or absent emission-lines. Consequently there is no limit on the amount of beamed X-ray emission. However, I do not believe that they necessarily form a completely separated class, as the correlations between X-ray, optical, and radio continua are the same as for the compact quasars. I have not been able to investigate the OVV's, as there are few objects with measured H-beta, radio, and X-ray luminosities.

SETTI : In case of BL Lac's, how did you take into account both of the optical and X-ray variability ?

MILLER : For the known variable sources I have either taken the radio flux density measured at the closest time to the X-ray observations , or else the mean of the observed extremes. It is not too important in this case, since the correlation extends over a wide range in luminosity.

PRESTON : Luminosity – luminosity plots of data from flux-limited samples can produce apparent correlations due to selection effects. Would you comment on such selection effects in your analysis?

MILLER : The selection effect arises in data containing a much wider range of luminosities than the flux densities. The effect is to spread the data along a line of slope unity. There are two methods of ensuring that this selection effect is not significant. The first is to check that there is also a correlation between flux densities. The second method is to observe all the sources in the sample with no a priori bias, and to include, as worst case, any non detections. Both approaches show that these correlations reflect intrinsic physical associations between the variables.

SAUNDERS : The correlation between X-ray luminosity and (clearly unbeamed) H-beta is tight, but is there in fact any other evidence that the X-ray emission is unbeamed ?

MILLER : It would always be hard to prove that there is not a component of beamed X-ray emission. However, fig. 1 shows that the total X-ray emission could not be enhanced due to any beamed component by more than a factor about 3. This would be consistent with previous findings that radio-loud quasars are, on average, about a factor 3 brighter in X-rays than radio-quiet QSO's.

DISCUSSION ON THE PAPER BY SAUNDERS (p. 193)

REES : Could there be so much dust in Cygnus A that even the near infra-red, e.g. Paschen-alfa, does not get out in the plane ?

SAUNDERS : The required extintion is about 50 magnitudes in the visual. The extintion has to occour inside the NL region because we see narrow Balmer lines (thus the well known dust lane in Cygnus is irrelevant). Putting the required amount of dust in the broad lines region and assuming a galactic gas to dust ratio, would completely wipe out the nuclear X-ray emission that is observed from NL radio galaxies. A different argument is that sources with small Pnucl/Ptot and faint broad lines, e.g. 3C381, have Eb-v less than 1 magnitude; if there is continuity of properties between NL and BL radio galaxies, it is hard to see how Cygnus, with Pnucl/Ptot only a little less than that of 3C381, could have so much more extintion.

Panagia : Your test on the broad Paschen-alfa line is based on a single object observation. Are you planning to extend your observational sample to make your conclusion more general ?

SAUNDERS : Yes. But for this purpose, Cygnus A is a perfectly representative NL radio galaxy and there is no reason to think other NL radio galaxies would give a different result: Cygnus A will always give the clearest result because of its high radio and narrow lines fluxes.It seems clear that we cannot account for the known nuclear properties of classical double radio galaxies with a relativistic beaming model.

DISCUSSION ON THE PAPER BY ALLAN (p. 195)

HUTCHINGS : Do the statistics of observed bends tell you anything about the frequency of relativistic and non relatistic cases ?

ALLAN : The frequency of the two cases can indeed be derived from the observations, and I am doing that at present.

ROMNEY : What are the effects of change in velocity or brightness at the bend ?

ALLAN : If you believe that you know the change in intrinsic velocity or brightness at the bend, than this will simply alter the value of the dynamic range parameter D.

DISCUSSION ON THE PAPER BY SCHEUER (p. 197)

ROBERTS : Would you come to comment on the gravitational lens model for super-luminal motion, since you did'nt get a chance to say anything about it ?

SCHEUER : Magnification by gravitational lenses will occour from time to time, particularly for sources at large red-shifts, but I doubt whether there will be enough to affect any statistical test significantly.

BEGELMAN : Another problem with the tapered cone model is that if there are any inhomogenieties or time dependence at all, then unless the emissivity drops off very rapidly with angle from the jet axis (with an angular scale length less than $1/\gamma$), you are likely to see the compact jet pointing in the opposite direction from that of the large scale jet

WEILER :   As you mentioned, your "fringing beam" or "computer controled Christmas tree" has similarities to Sanders or Bachall and Milgrom models, except that yours is one sided. Their model predicted, among other things, high degree of circular polarization of opposite signs with net cancellation when the relevant parts of the source are unresolved in normal polarization observations. Would not your single sided jet then produce much higher degrees of circular polarization than are observed because it lacks the cancellation?

SCHEUER :   The Bachall and Milgrom model contained a very specific radio emission mechanism, of electrons going out along magnetic field lines at small pitch angles, and that produces the large polarization. I do not postulate any particular emission mechanism, but would prefer to think of diverging streams of plasma emitting ordinary incoherent synchrotron radiation.

PREUSS :   How crucial really will be the (desired) proper statistics of super-luminal motion for the survival of the relativistic beam model? Suppose a considerable number of nuclei in extended radio sources should turn on to be "superluminal", would this mean the end of the relativistic beam model ?

SCHEUER :   Nearly all of the superluminal sources are the nuclei of extended radio sources and of course that does not mean the end of the relativistic beaming model, even in its simplest form. To reject that model by statistics one must measure a sample which is free from orientation bias, and that requires a lot of care.

ALLAN :   You rule out the tapered cone model on the basis of the fat emission predicted. Have you calculated the predicted emission on your "computer controlled Christmas tree" model ?

SCHEUER :   In the tapered cone model we observe emission from material that has always travelled at an angle of about $1/\gamma$ to the line of sight. In the "computer controlled Christmas tree" the emitting material has travelled most of the way at angles much larger than $1/\gamma$ to the line of sight; γ can be arbitrarily large and the observed jet can be arbitrarily narrow.

DISCUSSION ON THE PAPER BY REES (p. 207)

BEGELMAN :   You have some serious observational constraints if you wish to put a screen with substantial Thomson optical depth at about 10E20 cm. In order to polarize the lines without washing them out, the temperature of this gas must be considerably less than 10E7 K. However you would then have problems with soft X-ray absorption, so you have to be very careful with your geometry.

REES : I agree that there are constraints both on the electron temperature and on the geometry (free-free absorption in the radio sets another constraint).

DE BRUYN : Does induced Thomson scattering destroy intrinsic synchrotron polarization of VLBI jets ? We do find radio polarization in the compact cores.

REES : Generally not. You would need a Thomson depth exceeding unity to destroy high polarization. This probably cannot occur on the VLBI scale, but could be important on the smaller scale of the optical continuum (especially if there is an opaque " false photosphere " of e+ e- pairs, as Guilbert, Fabian and I have recently discussed).

VAN DER LAAN : Barthel yesterday reported results of the first round in a "crucial esperiment" : large double radio galaxies with VLBI jets which were found not to exhibit superluminal motion. What is required is theoretical attempts to make relativistic beams produce sheaths of slow bulk motion and high synchrotron emissivity. Such "stationary jets" would look like bona-fide jets when opaque, but might be limb brightened when transparent. Would you comment please ?

REES : A relativistic jet has less inertia, for a given energy flux, than a slow one. It is therefore more vulnerable to entrainment and to some kind of instability. It is my impression that it would be all to easy to dissipate energy in a sheath as the jet plough into the emission-line region. The surprising thing is that it is not stopped completely.

ALLAN : Do you think that VLBI polarization measurements at different frequencies can in principle rule out the electron-positron model ?

REES : A pure e+ e- plasma obviously gives no net Faraday rotation. Indeed it was to account for the lack of Faraday depolarization (which would occur if mildly relativistic or sub-relativistic particles were present in number exceeding those of the high-gamma radiating particles) that Jones and O'Dell several years ago invoked an e+ e- plasma. On the other hand, it would be hard to find convincing an unambiguous evidence for Faraday rotation (i.e. against an e+ e- plasma), because any apparent wavelength-dependence of the polarization vector could be due to unresolved substructures and spectral index variations within the source.

ULRICH : There are ojects, like 3C390.3, where the lines and the continuum are similarly polarized, but there are other objects such as NGC4151, where the continuum is polarized, but the broad emission lines do not have the same angle and percentage of polarization.

REES : I agree that electron scattering is only one possible cause of linear polarization (and perhaps not the most likely one in most cases). However it is certainly possible to have optical depth of about 0.1 due to electron scattering by hot gas in the BL emitting region (though there would be excessively broad wings to the lines if the gas were too hot). The gas would then produce some optical polarization. Also, as I mentioned, the effect of induced scattering would then complicate the interpretation of variable components with $KT \gg 10E11$ K.

DENNISON : Regarding the effects of Compton drag on an electron-positron plasma, what happens in the case in which the radiation field is highly anisotropic and, in particular, directed radially outwards?

REES : A test particle is accelerated by the Compton effect, provided that the radiation flux comes predominantly from behind when transformed to the particle rest frame. The reason the acceleration cannot achieve high γ s is that the aberration and Doppler effects tend to reduce the flux of radiation coming from behind, and to enhance the drag effect of photons coming from transverse or forward direction.

DISCUSSION ON THE PAPER BY BlANDFORD (p. 215)

SCHEUER : In your introductory remark you mentioned the claim that radio emission is just the smoke and therefore radio astronomers cannot tell us basic truths about quasars. If one goes one step deeper and works out the power flowing through VLBI jets, using the classical minimum energy formula for synchrotron radiation, one finds powers similar to the X-ray and optical luminosities.

BlANDFORD : Of course, I agree. In fact, in the most powerful radio galaxies the jet power dominates the other luminosities. Nevertheless in the majority of active galactic nuclei radio jets probably carry no more than a small fraction of the total power and are therefore reguarded as secondary phenomena by optical and X-ray astronomers. My main point was just the truism that in our attempts to comprehend the innermost workings of a quasar, we should all use the evidence from the complete electromagnetic spectrum and not fall prey to " spectral chauvinism ".

JAUNCEY : It is important to see how the observed low frequency variability and the large γs implied, fit into these models. Do you have any suggestion ?

BLANDFORD : Lorentz factors γs of about 10 are necessary to avoid catastrophic inverse Compton losses if, as is typically the case, the variability brightness temperature is about $10E15$ K. This is comparable with those suggested by observations of apparent superluminal expansions. There is a further point. When devising a model to explain low frequency variability, you must be carefull to make it radiatively efficient. Some models require unreasonably large powers.

REES : A comment on low frequency variability. The field strength near a black hole may be $10E3$ gauss, implying a cyclotron frequency in the GHz band. We know that it is easier to get coherence for cyclotron than from synchrotron radiation, so conceivably detectable flux of radio emission could emerge from radii smaller than $10E15$ cm, permitting very rapid variability. The main problem is , of course, how this radiation could survive synchrotron absorption and induced scattering further out from the nucleus. An evacuated channel would certainly be required .

BLANDFORD : I certainly agree. The Sun and Jupiter assure us that there are many alternative mechanisms to synchrotron radiation for producing radio outbursts.

DISCUSSION ON THE PAPER BY BEGELMAN (p. 227)

ALLAN : Do you think that we get anywhere near the high densities and temperatures relevant to your nuclear limit in real objects ?

BEGELMAN : The existence of nuclear-limited tori depends on the material in the center of the torus having a very small viscosity, corresponding to α values in the range of $10E-10$ to $10E-6$ for a $10E8$ M$\odot$ hole and the standard α parametrization. It is an open question whether α can be this small although the Sun appears to be characterised by a very small α . A second requirement for a long lived stady-state is that mass be supplied to the torus at a sufficiently high rate.   If material is poured into the torus without being drained away, then the pressure and density in the center should approach the nuclear limit.

HUTCHINGS : Can you make any predictions on the regions of the $\dot{M}/M$ plane where you expect your mechanism to be observably different ? Do you have any BL region predictions ?

BEGELMAN : My hypothesis is that the BL clouds will form only when $\dot{M}$ is large compared with the Eddington value, and the viscosity is sufficiently small. If BL clouds or filaments are produced by a torus which is up against both the photospheric and nuclear burning limits, then the model "predicts" that they should always be observed at a density of about $10E10$ cm^{-3}, with an ionization parameter $\Xi = 10 - 30$ and full width of velocity dispersion of $8000 \times \dot{M}_8$ Km/sec. At this stage I would not like to predict detailed line profiles, but it appears that the model may be able to explain the systematic line shifts (relative to the NL, and between low- and high-ionization lines) claimed by Gaskell (1982, Ap.J.,263,79). The model predicts a 'grey body' spectrum with a colour temperature of about $(1 - 3)$ $10E5$ K, consistent with Malkan and Sargent's spectral decomposition (1982, Ap.J., 254, 22). However, to explain the observed line ratios, it is probably necessary to involve an additional hard (non-thermal) component of the continuum. Such continuum may come from magnetically driven flares on the surface of the torus.

DISCUSSION ON THE PAPER BY BIRKINSHAW (p. 229)

BEGELMAN : Provided that jets are indeed highly supersonic, I am not convinced that one should be too concerned about the jet braking up when one of your instabilities becomes non linear. Supersonic flow must dissipate energy if it is forced to follow too curved a trajectory. This dissipation may prevent the instabilities from growing beyond the marginally non linear phase, without being so severe that it slows down the jet in a short distance.

BIRKINSHAW : Naturally, this calculation cannot consider large momentum losses by the beam. Rather the results describe the length-scale for production of a significant sheath. The effect of Kelvin — Helmoltz instabilities on the sheath is another story.

FERRARI : Shoudn't you be able to fix a lower limit to unstable wavelengths introducing gradients across your cylindrical beam ? This is in fact the result of previous linear analysis in growth rates.

BIRKINSHAW : With plausible lower limits the growth length of the instability at low n is about 100 R in the equal density case.

DISCUSSION ON THE PAPER BY FERRARI (p. 233)

ALLAN : What is the maximum velocity in your wind solution ?

FERRARI : The asymptotic velocity in the wind model depends essentially on two factors : the temperature at the basis of the flow and the momentum deposition along the flow. The highest are obtained when momentum deposition lasts very long and basis temperature is high. If one uses $L \gg L_{edd}$, collimation factors $\varepsilon > 0.8$ and $T \gtrsim 10E9$ K, asymptotic velocities can approach the velocity of light. Obviously these assumptions must then be justified in terms of the disk model, but for large absorption of radiation, as in Begelman's model, they are perfectably reasonable.

DISCUSSION ON THE PAPER BY SALVATI (p. 239)

COHEN : How do you explain the contraction seen in 4C39.25?

SALVATI : Contractions are a typical signature of phase effects. Should this finding be confirmed in a number of cases, phase models would become much more attractive.

ALLER : Can you account for the Doppler boosting apparently required to account for the high intensity derived for low frequency variability, or for the broad frequency spectra (radio to X-ray)?

SALVATI : Relativistic bulk motion and Doppler boosting can be added to this geometry. However, the feature which you allude to depend partly on light-travel-time effects; i.e., a rapidly propagating wave in a slowly moving emitting material does at least part of the job.

DISCUSSION ON THE PAPER BY GORENSTEIN (p. 243)

ROBERTS : We have observed 0957+561 with the VLA at 6 cm, giving angular resolution of 0.3" and dynamic range of 1 to 1000. We find that the small source G is resolved (0.3"x0.15") so that it cannot be wholly the third quasar image. In addition, it is difficult to model the structure of G when one includes a point source of flux and position required by the interpretation of your third VLBI component G' as the third image. We suspect, as you suggested as one possibility in your paper in "Science", that G' is too close to the optical nucleus of the lens galaxy G1 for it to be the third image (it is observed to be too bright, relative to the lens models of Young et al., 1981 and Greenfield, PH.D. Thesis 1981). The intrinsic radio properties of the G radio source are within a factor of two of those of M 87, for scales between arcminute and milliarsec.

DISCUSSION ON THE PAPER BY MARCAIDE (p. 247)

ROBERTS : Could you give the separation of the S and X band brightness peaks in the A quasar in milliarcsec?

MARCAIDE : 0.7 ± 0.1 mas.

BURKE : In a formal way, you can derive an upper limit for the mass of the foreground quasar. What do you get from your present limit on a 3rd image ?

MARCAIDE : The present observational limit on the third image places a poor limit of 10E14 to 10E15 M⊙. The apparent lack of gravitational distortion on B places a lower limit of several 10E13 M⊙ as given by Dyer and Roeder(1980, Ap.J. 238, L67). I would bring down, perhaps to 10E13 M⊙, the above limit using their method, although I am skeptical of the basic assumption of the method.

PORCAS : Could you tell us what the story is on possible flux density variations in the two quasars ?

MARCAIDE : We do not have evidence of any total flux density variation in any of those two quasars. The structure of A may have changed from november 1979 to march 1981, but without clear associated flux density change to approximately 10 % level.

DISCUSSION ON THE PAPER BY SUBRAMANIAN (p. 249)

PORCAS : Moore at Jodrell Bank has also done modelling of the 0957+561 system, including use of the same VLBI data. His time delay is much longer than your estimate of 1 year. How unique are these models ?

SUBRAMANIAN : The time delay that we calculate does come out to be about 1 year for a variety of models, which differ basically in where one locates the center of the cluster, provided we fit all the observed constraints. Since I have not seen Moore's calculations it would be difficult for me to say exactly why they differ.

GORENSTEIN : Refsdal has shown that the bendind law of the lens determines the time delay and it is unnecessary to decompose the delay into geometrical and potential terms. If one does decompose the delay for each image into these two terms, then the sign of the terms will usually be opposite and therefore the contribution will cancell.

SUBRAMANIAN : I agree that it might be unnecessary to split the time delay into geometrical and potential terms. However our reason for doing this was to understand exactly why we were getting a time delay of about 1 year while Young et al. got a time delay of 5 - 6 years. The reason turned out to be that there was a sign mistake in Young's et al. potential time delay (they got 3 + 2 years rather than 3 - 2 years).

ROBERTS : Fluctuations in the B quasar which are due to mini-lensing by stars in the galaxy G1 can be separated from intrinsic variations in the object quasar by their wavelength independence. Our monitoring of 0957+561 does not show the same kind of variation in A and B as are seen in the optical. In addition, Young showed that the timescale for mini-lensing changes in B is very long (about 100 years) at least for the kind of stars to be considered to be important.

SUBRAMANIAN : I agree that flux variations which are due to mini-lensing can be distingushed from those due to intrinsic variation. However the situation is complicated by the fact that mini-lensing need not affect the radio emission and the optical emission in the same way, because of the differences in size of the radio and optical emitting regions. Thus you can have optical variation due to minilensing without corresponding variation in the radio. Finally the time scale for mini lensing changes depends on the mass of the lens. For a star of mass M, the time scale for variation is roughly $50 \times (M/M_\odot)^{0.5}$ years. So if low mass stars are present, say in the halo of the galaxy, one can have variation in flux over time scales of a few years (Gott 1981, Ap.J.

243, 140).

DISCUSSION ON THE PAPER BY CRANE (p. 259)

BARTEL : You associated the large circular component in the central region of M81 with the NL emission region. What are your arguments against this component being a SNR?

CRANE : Component 4 coincides with the known optical NL emission region and at least 10% of the observed flux density at 6 cm is thermal. Also the suggestion of a shell-like structure is probably an artifact of fitting a continuous range of structure with two gaussian components.

DISCUSSION ON THE PAPER BY ANANTHAKRISHNAN (p. 261)

SIMON : In 1975, 3C84 was observed at 92 cm, 3-station VLBI. Model fitting to this data indicates a compact component less than 20 mas which has a more extended component (about 50 mas) to the north.

ANANTHAKRISHNAN : I think that is probably what we are seeing.

DISCUSSION ON THE PAPER BY NEFF (p. 263)

BRIDLE : It was not obvious to me which feature in your map was the core and which was the "jet". There is a real problem of whether to call all trains of knots "jets". They might be the brightest features of blobby jets (with fainter connecting emission), or they might not. I feel we should be cautious about applying the jet terminology too quickly, as it carries such strong prejudices about the underlying physics.

NEFF : I am in complete agreement with this comment. We hope to justify our presumptuousness in higher frequency, higher resolution and higher dynamic range observations. I wish to point out that for Seyfert galaxies the two contending explanations for radio emission are star-bursts and 'mini-jets'. We feel that this observation supports the latter hypothesis, and therefore we refer to it as a jet.

WEISTROP : Are large HI envelopes observed in other Seyfert galaxies or galaxies with active nuclei?

NEFF : Hawarden et al. (A.A. 74, 230) report large HI densities envelopes around early type galaxies, two of which (NGC 1512 and NGC 5921) are not known to be active galaxies. Bergeron et al. (preprint) have detected a very large envelope around 2251-178, a low z quasar, which may be the result of an HI envelope interacting with a strong continuum source.

JAUNCEY : Mk348, with its rotation axis perpendicular to the plane of the sky, would perhaps be expected to show super-relativistic motions on the simple beaming model.

NEFF : We hope to search for bulk motions with higher resolution observations. Even if the motion of material in the jet is "slow", we can measure such motion in a reasonable time because of the galaxy's proximity.

MARCAIDE : I find risky to deduce the core nature only from its compactness without spectral information.

NEFF : I agree. We hope to determine the location of the core in this galaxy by combining our (future) 6 cm observations with those reported here at 21 cm.

DISCUSSION ON THE PAPER BY LO (p. 265)

TAYLOR : With your accretion model, don't you have some difficulty in explaining the constancy over nine years of the flux density of the compact source?

LO : The infall time scale is about 10E4 years which refers to infalling from one parsec to near the center (0.1pc). What happens close in is not clear. To fall into a black-hole, the matter would have to lose its angular momentum and the time scale would depend on detailed physical conditions close in. The relative constancy of the compact source may just reflect the fact that there is not much material near the black hole to be accreted. Presumambly, the matter seen to be falling now will be accreted eventually, but over a long time scale.

MUTEL : Compared with most other galactic radio cores, the Galactic

Center source is remarkably bare (no jets or multiple components). How wide a field of view have you searched for other non-thermal components which may be associated with this source?

LO : The lower resolution maps of Sag A by Ekers et al., over at least 10'x10', do not show much bright features except Sag A East and Sag A West.

KELLERMANN : Is an elliptical brightness distribution consistent with interpreting the size at 3.8 cm as the result of scattering?

LO : It is still possible to do so; one would have to appeal to elongated structures in the scattering medium. The experts tell me that there may be evidence for this.

MARCAIDE : From the data in the Goldstone--Greenbank interferometer can you place a new limit to the detection of the very compact component once detected by Kellermann et al.? In view of your limit can you comment on the nature or existence of that component?

LO : The upper limit on the $0''.001$ component is less than 0.01 Jy.

WRIGHT : I'd like to comment that our galaxy is, of course, an interacting system. And most models of the Magellanic Stream imply, as a by-product, quite large amounts of gas falling directly into the Galactic Centre. This may be relevant to Galactic Centre models.

SCHILIZZI : Can you be sure that the compact radio source is at the dynamical centre of the galaxy? Could you invent a model in which the compact source is not at the dynamical centre?

LO : I am reasonably certain that the compact radio source is very close to the dynamic center of the Galaxy, if not defining it. The dynamic center is, of course, not well defined observationally. IRS16, the best candidate for the central star cluster, is not completely understood, and its absolute position is uncertain by about $1''$. As mentioned in my talk, the modest radio luminosity of the compact source does not preclude it as a possible stellar object. Reynolds and McKee (1980, Astrophys. J, 239, 893) suggested that it is a pulsar moving through the center. On the other hand, circumstantial evidence is overwehlming that the compact radio source is a unique object in the Galaxy.

DISCUSSION ON THE PAPER BY BARTEL (p.   275)

ROBERTS : Could you clarify for me exactly what it is that you measure,
i.e. what geometrical properties of the pulsar emission region are
determined? I am thinking of resolution across the emission region
versus resolution along the line of sight.

BARTEL : If the pulse structure is due to a temporal intensity
modulation of the emission emanating from a single region at the pulsar,
then we would not expect any significant differences between the
interferometer phase-delays of the three components. However, if the
pulse structure is due to emission from regions which are spatially
distributed in a plane perpendicular to the line of sight – as is
conceivable if scattering effects are involved or if several rotating
beams of radiation are pointing towards the observer from slightly
different locations – then differences between the interferometer phase
delays of the components may become observable. We determined upper
limits on any such phase-delay differences and hence on any position
offsets.

DISCUSSION ON THE PAPER BY MUTEL (p.   277)

HUTCHINGS :   One of your objects, LSI + 61 303, has a 26-day radio cycle
and is a unique X-ray source. Any information on the radio structure
would be of great interest.

PORCAS : How rapidly does HR 1099 vary? In particular, if it changes
during the course of the VLBI observations, how do you get over the
problem of interpreting the visibility function?

MUTEL : The shortest time scales we have seen are several minutes. For
the VLBI data reported here, the sources had nearly constant flux
levels. In any case, we measure the correlated flux one each minute for
sources with sufficient signal-to-noise ratio

NEFF : Using the VLA, Brown, Broderick and Neff have detected (with
eclipse data) a position offset between the two senses of circular
polarization. Do you see any evidence for such a separation in your
dual polarization work?

MUTEL : We have not yet observed in the dual polarization VLBI mode.

DENNISON : Is there any possibility that the observed gaussian brightness distribution is caused by interstellar scattering, and have interstellar scintillation been searched for in this object?

MUTEL : Interstellar scattering and scintillation effects are very unlikely because the stars are, in general, less than a few hundred parsecs away.

DISCUSSION ON THE PAPER BY GELDZAHLER (p. 283)

KONIGL : Could you comment on the interpretation of the SCO X-1 lobes in terms of magnetic pinches, as proposed by Achterberg, Blandford, and Goldreich?

GELDZAHLER : This is an interesting idea. The authors make two predictions: 1) that the shape of the lobes should be conical with the apex pointed away from the central component, and 2) deals with Faraday roation in the lobes. Our present maps really aren't sufficient to confirm or deny (1), but our improving 2 cm VLA observations may be able to do so, at least for the NE lobe. We have data in hand to check the Faraday rotation and are currently doing so.

RUSK : Has there been a search for circumstellar material around Sco X-1 which the "beam" might be impinging upon? Old outbursts may have increased the density of the interstellar medium about Sco X-1 so that it may not be the "typical" $n = 0.1 - 1 \text{ cm}^{-3}$.

GELDZAHLER : Yes, there has been such a search. For example, the 408 MHz Jodrell Bank - MPI survey of Haslam et al. reveals no evidence for any low level shell structures near the outer lobes. Also, deep optical plates taken by Dave Malin at the AAT show no such structures.

BRIDLE : Lower-resolution VLA observations of Sco X-1, particularly at 20 cm, could be valuable as they might detect diffuse emission connecting the "lobes" to the central source. This could confirm the physical association of the three emitting regions.

GELDZAHLER : Hjellming and Wade have already made such observations with the C-array of the VLA and no such connecting emission was found.

DISCUSSION ON THE PAPER BY NICOLSON (p. 285)

ALLER : Did the triple flares exhibit a characteristic modulation frequency?

NICOLSON : The triple flares appear to be a superposition of consecutive outbursts. This is supported by 2 cm measurements at Parkes which show distinct outbursts.

BORIAKOFF : Do you have any polarization information on the flares?

NICOLSON : Measurements at Parkes for one flare at 6 cm failed to detect polarization.

HUTCHINGS : A comparison of the radio properties of Circinus X-1 and LSI+61303 (0236+61) may be interesting.

NICOLSON : A comparison would require data at optical, infrared and X-ray wavelengths as well as radio. Another similar objects may be AO538-66 in the LMC. We have some limited data for this object. The upper limit for radio flares is 50-100 mJy at 6 cm.

WEILER : Do your VLBI observations tell you anything about the size of the radio outbursting region as opposed to the size of the binary ystem?

NICOLSON : At a distance of 10 Kpc the linear size is greater than 25 AU whereas the binary dimension is about 1 AU. However the longest baseline data is probably limited by interstellar scattering.

DISCUSSION ON THE PAPER BY WRIGHT (p. 287)

MUTEL : How many stars are known like HI-36 which have steeper than expected optically thick radio spectra?

WRIGHT : About 10.

KONIGL : 1)What is the spectral index in the steep portion of your spectrum? 2) Is it possible to invoke acceleration at large radii due to the formation of grains in the flow?

WRIGHT : 1)The spectral index varies in different objects from about +0.8 up to about +1.5 in Vy 2-2. 2) Yes it's possible: see Marsh

(1976) Astrophys. J., 201, 190.

DISCUSSION ON THE PAPER BY SCHILIZZI (p.   289)

PORCAS : Is it true that, in addition to an arbitary registration on
the sky, there is also a 180 deg orientation ambiguity in your December
1982 sequence of maps?

SCHILIZZI : There were only 3 hours of usable closure phase data on one
epoch (JD2445310). This constrained the model to the orientation shown.
We assumed the same orientation for the following epochs.

HUTCHINGS : In addition to the direction jitter seen here, there is
jitter in the moving line radial velocity. If, as seems likely, they
are connected, it should be possible to determine the place of origin of
the optical lines and a 3 dimensional picture of the jitter.

SCHILIZZI : We have not yet checked the detailed correspondence of
jitter in the radio position angles and that in the optical radial
velocities.

JOHNSTON : A comment and then a question. Hjellming and I have been
observing SS433 at 6 and 2 cm for over three years now. At 6 cm we can
see radio radiation out to a distance of three turns in the corkscrew.
Over this range, we do not see any deviation from the predicted
positions of the observed values. At 2 cm we also see at least 2 turns
of the corkscrew and also see no deviation. My question is how often do
you see emission that is in a different position from the prediction?

SCHILIZZI : Only on 9 December 1981 do we see emission in a very
different position angle (ΔPA about 20 deg.) to that predicted by the
kinematic model. At other epochs we see deviations of a few degrees
only.

SPENCER : Similarly there is only one outburst seen on Merlin maps
(occuring in April/May 1982) which is in an anomalous position angle.

DISCUSSION ON THE PAPER BY SPENCER (p.   297)

BEGELMANN : To avoid kinetic energy fluxes in excess of 10E42 — 10E43

erg/sec it is necessary to assume that the line-emitting gas has a very
small filling factor.   It may be the interstitial gas that is
responsible for the radio emission and, if this is the case, then it
would not be too surprising if flares at the central engine could
sometimes eject this gas at a higher velocity than the line emitting
gas. Therefore, the anomalous ejection may not show up in optical
observations.   Could you put a lower limit on the amount of energy
contained in relativistic particles plus magnetic fields in one of these
outbursts?

SPENCER : The power required to replace the radio jets in a period of
164 days is about 10E39 ergs/sec, assuming equipartition between high
energy electrons and magnetic field in the jets.   This is comparable
with some of the lower estimates of the kinetic energy flux in the
optical clouds, suggesting that the kinetic energy fluxes are indeed
higher.

DISCUSSION ON THE PAPER BY KONIGL (p.   299)

MORIMOTO : Molecular line observation of IR sources in the bipolar flow
molecular clouds showed existence of rotating disks of dense molecular
gas of several thousand AU in diameter.

VAN BREUGEL : Regarding the similarities between galactic bipolar-flows
(i.e. Herbig-Haro objects) and extra-galactic jets I must mention that,
while Herbig-Haro objects are probably shock excited, the only
unambiguous result from optical line emission associated with radio jets
(Coma A=3C 277.3) shows that there is photo-ionization (presumably by
non-thermal emission from shocks in the jet).

KONIGL : I agree with you that unambiguous data on shock excitation in
extragalactic jet sources is still lacking, although the work of Brodie
et al. (1983), referenced in my talk, suggest that the spectroscopic
properties of the radio and line-emitting knots in CenA are similar to
those of Herbig-Haro objects. The analogy between these objects and
optical emission clumps in extra galactic jets could, however, hold even
if the latter are not directly associated with shocked material.   In
this connection I would like to mention recent photo-inization models
(e.g., Ferland and Netzer, 1983, Astrophys.J.264, 105) which showed that
the emission in a partially ionized medium with a low ionization

parameter could be very similar to the emission from a shock. The photo-ionization source could be the low shock attached to an accelerating clump in the jet. Such shocks are invoked also in accounting for the continuum emission from Herbig-Haro objects (e.g., Schwartz, 1981, Astrophys.J., 243, 197).

LÖ : What is the source of energy for the jet? And is there a problem with accounting for the energy and momentum of the outflowing molecular gas?

KONIGL : The source of the energy is most likely the gravitational binding or rotational energy of the central star, or even energy liberated in accretion onto the star. As I mentioned, there is now direct evidence for outflows from the immediate vicinity of some embedded sources, but the exact mechanism which transfers the energy and momentum to the wind is not yet clear. One of the difficulties is that very large momentum discharge rates are required to power the outer lobes - much larger than available from radiatively driven optically thin winds accelerated by the observed photon fluxes from these sources, and in fact also larger than those inferred from the direct measurements in the immediate vicinity of the sources. I would like to emphasize, however, that the jet formation and collimation mechanisms that I mentioned do not depend on the precise nature of the "central engine" which powers the wind, but only on the presence of a wind and of a flattened ambient mass distribution.

DISCUSSION ON THE PAPER BY CORDES (p. 303)

SHAFFER : The sources NRAO150 ($|b|$=2 degrees) is probably affected by interstellar scattering at 18 and 13 cm, in a non-symmetric fashion.

CORDES : The way to test whether ellipticities in visibility functions are due to interstellar scattering is to check that the axial ratio is independent of wavelength. Ellipticities would be caused by non spherically symmetric irregularities in electron density if there is some preferred alignment of irregularities somewhere along the line of sight. I recall that interplanetary scintillations indicate that asymmetric blobs exist in the solar wind plasma with alignment in the direction of the local wind velocity.

SIMON : What is the minimum scattering size expected at 329 MHz? For 3C147 an upper limit to the size of the core of 8 mas is observed.

CORDES : The pulsar observations suggest that θ_{min} is about 7 mas at 327 MHz for b=90 deg but the error on this prediction is at least a factor of two. I would not be surprised if there are holes in the large scale height component of the scattering medium but to date there are no measurements that show such holes.

KONIGL : Could you comment on the possible physical nature of the two components that you used in your model-fitting?

CORDES : I suspect that there are several kinds of regions where electron density "turbulence" is produced. The Crab and Vela pulsars both show enhanced scattering, presumambly related to their associated supernova remnants. The low-scale height component has a filling factor compatible with an extreme population I class of objects, such as HII regions or stellar winds associated with O and B stars. Gesansky (1980, An. Rev. Astron. Astrophys., 18, 289) has argued that the maintenance of the turbulence may be important for the energy balance of the interstellar medium. The large-scale height component might be maintained by supernovae or infalling material.

ANANTHAKRISHNAN : Pramesh and I had presented evidence, at the 1982 IAU General Assembly in Patras, that between latitudes 0° and 4° and longitudes 0° and 40° there is a total absence of IPS indicating substantial enhancement in scattering towards the galactic centre. In the same paper we had also stated that such enhancement is not seen in the anticentre direction. Therefore it is clear that Simon's observation of 3C147 should not be seriously affected by interstellar scattering. Thus scattering is not only latitude dependent but also longitude dependent; i.e., one must look at it in terms of galactocentric radius.

DISCUSSION ON THE PAPER BY R. NORRIS (p. 323)

REID : It is very unlikely that circumstellar OH masers are saturated. The brightness temperature needed to saturate the 1612 MHz transition is T_{sat} = 10E11 (0.01/Ω) K where Ω is the beam solid angle of the emission. This temperature is several orders of magnitude greater than typical brightness temperatures. Also, the extremely bright features you observe in the blue shifted OH peak cannot be saturated if they amplify the stellar continuum, since the beam pattern would be the solid angle of the star at the distance of the OH maser which is very small (e.g., Ω less than 10E-8).

NORRIS : In reply to your first point, I would say that T_{sat} is rather uncertain because it depends on a poorly known quantity, the thermalisation rate. Empirically, the smoothness and linearity of the OH intensity variations in response to the varying IR flux suggest that the masers are saturated. Regarding your second point, I don't agree that this is the correct solid angle to use. The central star may be very small, but the appropriate beaming solid angle is determined by the geometry of the shell and observer. I think you have to be rather careful, as the results of my model are rather non-intuitives, and really require careful calculations, which are in progress.

JOHNSTON : Bowers, Spencer and I (this volume) have mapped about twenty late type stars in 1612 OH using the VLA. The spatial structure of this emission is consistent with central outflows in almost all cases although there are probably small deviations from this simple model in all cases. For the supergiant stars and some others such as IRC 10420 more complex models are needed to fit the data.

NORRIS : Yes, we find the same result. Most sources are generally consistent with a simple expanding shell model, although small perturbations from this model are nearly always required. Sources such as IRC10420 and the supergiants represent a small class of interestig exceptions.

MUTEL : The source OH17.7-2.0 has been mapped previously by VLBI. Would you comment on why the continuum source has not been previously seen?

NORRIS : Two possible reasons suggest themselves. One, which is more exciting, is that we are witnessing a short lived transient event. We shall, of course, be closely monitoring the source to investigate this possibility. The other possible reason, which is more mundane, is that only the channels containing the strong maser lines were mapped. Furthermore, it should be emphasized that since the broadband source is only a fraction of a Jy in intensity, the previous observations may not have had sufficient sensitivity.

BOOTH : The fact that OH17.7 was not detected in a previous VLBI experiment suggests that we are indeed observing a transient phenomenon. If the high internal motions suggested by the new spectrum are confirmed and if OH17.7 corresponds to an M-supergiant it is conceivable that we are beginning to witness a pre-supernova build up in the star.

DISCUSSION ON THE PAPER BY LANE (p. 329)

NORRIS : Do you have any information on the structure of the SiO masers that are resolved out by your measurements? Could you perhaps use the Hat Creek interferometer for this as your measurements on VX Sgr show the resolution effects to be the same in the two SiO transitions?

LANE : Roughly half of the observed total-power flux in both lines at 43 GHz is resolved out with the 75 Km baseline. Presumably this flux originates either from extended halo components surrounding the hot spots detected by the interferometer or from an ensemble of many weak maser components, small in size, but spread over an area larger than the fringe spacing. Clearly, shorter baselines, preferably with aperture synthesis capabilities, are needed to locate this emission. An effort to map stellar sources in the V=1, J=2-1 SiO line at 86 GHz with the Hat Creek interferometer would be very useful.

BOOTH : Do your observations suggest that the SiO masers may be at the same radial distance as the H_2O masers and what are the implications for pump process?

LANE : VLBI measurements toward several Mira's and semi-regular variables suggest the H_2O masers occur at somewhat greater radial distances than the SiO masers, as is expected from the lower excitation requirements of H_2O. Toward VX Sgr, for example, the radius of the H_2O maser shell is about 2 x 10E15 cm, compared to 8 x 10E14 cm for the SiO J = 1-0 masers. The implications for SiO pump processes are not easy to specify since pump models must also take account of the fact that maser lines from higher J may be formed in different regions (as comparison of J=1-0 and J=2-1 profiles suggests).

DISCUSSION ON THE PAPER BY SCHNEPS (p. 335)

GENZEL : Can you say anything about the distance to the galactic center?

SCHNEPS : We just processed a portion of the SGR B2 data for two epochs almost a year apart. This source is too variable to permit a quick distance determination based on so little data: misidentification of features is a problem with only two epochs processed. We can already see from the data that the source must be at least 5 Kpc distant. A better number will have to wait for further processing.

DIAMOND : How can you be sure that the observed proper motions are indeed motions and not just a misidentification of various features due to their intrinsic variability?

SCHNEPS : This was demonstrated well by the Orion KL data (Genzel et al. 1981a). Observations over five epochs showed the maser spots to move linearly with time, which demonstrate that the kinematic motions are being observed, rather than a "Christmas Tree" effect. Misidentification of features is a serious problem if only 2 epochs are observed. However the new experiment observes five epochs over 2 years, enabling us to follow variable features over several epochs.

DE BRUYN : How certain are you that the masers are distributed spherically, which is a necessary condition to derive distances from a comparison of radial velocity and proper motion distributions?

SCHNEPS : Departures from isotropy can be measured to a large extent since the distribution of masers on the plane of the sky, as well as the distribution of motions in three dimensions, are observed. Only the spatial coordinate along the line of sight is indeterminate. Allowance can generally be made for observed asymmetries by modelling the errors in distribution, as was done for W51 by Genzel et al. (1981).

MARCAIDE : The very accurate distance to Orion was determined using an isotropic flow model. How does Erickson et al. finding of the bipolar flow in that source (1982, Ap.J., 261, L103) affect the determination?

SCHNEPS : Were the masers aligned in a highly collimated jet an ambiguity could arise, but we know that this is not the case. The distribution of masers on the sky is not simply jet-like. In such a poorly collimated flow no errors are introduced by assuming that the masers are irregularly distributed on an expanding shell, as opposed to an expanding cone.

DISCUSSION ON THE PAPER BY JOHNSTON (p. 339)

CAPORALI : You mentioned that in giving astrometric coordinates of radio sources one should specify the frequency of observation, because the source structure may change with frequency. Is there any indication that not only structure, but also the position of the center of emission changes with the observing frequency?

JOHNSTON : Yes, the apparent position of many quasars may change by up to 2 mas or more in measuring the peak emission over frequencies from 5 to 22 GHz because the majority of the cores of the sources are self absorbed.

COHEN : The "core" in 3C273 is strongly self absorbed at 2.3 GHz and the centroid has a one or two mas difference in right ascention between 2 and 5 GHz. It would be better to us a BL Lac for a right ascension reference source.

JOHNSTON : I agree with your suggestion. My basic point was that only one source located near the equator should be used as the right ascension zero point.

DISCUSSION ON THE PAPER BY TAYLOR (p. 347)

SHAFFER : What are the ambiguity spacings on the PSR0950+08 proper motion/parallax determination and what effects might they have on the results?

TAYLOR : Between 15 and 20 mas. Incorrect identification of fringe numberings is unlikely in our data for PSR 0950+08.

DISCUSSION ON THE PAPER BY BAUDRY (p. 355)

JAUNCEY : Because of the problems with galactic radio sources, proper motions, structures, position co-incidence etc., it is important to compare directly the Hipparcos and EGRF. Just what is the magnitude limit of Hipparcos, since there are several southern BL Lacs within one magnitude of 3C273?

BAUDRY : The nominal Hipparcos precision, about 2 mas, is expected for objects brighter than about 9-10 magnitude. The magnitude limit is somewhere between 12 and 13 magnitude. Thus 3C273 should be accessible to Hipparcos.

JOHNSTON : The relative positions of maser sources to the optical may be difficult. VLA observations at H_2O masers show emission spread over 0.2" to 0.3". The location of stellar optical emission to this maser emission to a relative accuracy less than 0.01" will require that the structure of the H_2O emission be modeled in some detail.

NORRIS : Regarding the positional coincidence of the masers and stars, the model I discussed in my talk may well apply to SiO and H_2O masers. If so, then when observing with the longest baselines all emission may be resolved except that which represents the amplified stellar thermal emission, which is of course coincident with the star. We find, with VLBI observations of OH masers, that this one unresolved spot, which on the shorter baselines is not significantly different from the other components, is all that is seen on the longest baselines.

BAUDRY : VLBI observations of stars in H_2O and SiO must be made in order to find the unresolved spots similar to those observed in OH. They must also be observed more or less regularly in order to find out how they behave with respect to the stellar position.

MUTEL : It is not obvious that there are any OH masers for which the optical stellar position and any individual maser feature coincide within 10 mas. Only in those stars for which a simple spherical shell geometry is applicable can one expect the line edges to be coincident with the star's line of sight, but these models are clearly not viable for supergiant stellar masers (as mapped by VLBI) and even the Mira variables show significant deviations from spherical symmetry (as mapped at the VLA). Some OH/IR stars (e.g. OH127.8 0.0) appear to have spherical symmetry, but they, of course, have not optical counterparts.

BAUDRY : Geometry in SiO stars (Miras) has to be investigated. For this we intend to use the IRAM interferometer, at 86 GHz.

DISCUSSION ON THE PAPER BY DE VEGT (p. 357)

CAMPBELL : How far did you make sure that the precession/rotation models used in establishing the radio and optical catalogues are the same? The sinusoidal signature in the residuals of your comparison reminds one of a systematic difference between these models.

DE VEGT : The influence of the model is about 10 mas.

JAUNCEY : Which optical catalogue should we now be using in the South?

DE VEGT : The final catalogue for southern hemisphere is SRS (available at the end of 1983, probably). An intermediate frame is provided by WL50 (Washington-Leoncito)

DISCUSSION ON THE PAPER BY MARCAIDE (p. 361)

SHAFFER : How did you place a Gpc distance limit on the quasar? Low tranverse velocities are possible for local quasar models.

MARCAIDE : I said under reasonable assumptions, i.e. that at least for one quasar its transverse velocity be a fraction of its radial velocity if the red-shift were due to local doppler-shift. Possibly a better way to use the astrometric limit would be to assume transverse speeds of about 300 Km/sec and then the distance to the A quasar would be in the Mpc range.

BARTEL : You determined the separation between the two quasars with nearly microarcsecond precision. Suppose you detect a significant difference in the separation of the two quasars of, say, 50 microarcsec in the next epoch observation. Would you be inclined to believe to have detected proper motion?

MARCAIDE : I would try to understand a bit better possible systematics. Using the maps deduced from the second epoch with the phase-delay observable from the first epoch, and vice versa, should help in the research. I encourage you to read our paper which will appear in the A.J. August '83 Issue, where we wrote several related comments.

DISCUSSION ON THE PAPER BY NIELL (p. 363)

CAPORALI : You mentioned that the nutation model used in the astrometric reduction of VLBI data is probably not enough accurate for the precision of the VLBI measurements. Have enough VLBI data been accumulated to permit that the nutation constant is included in the least squares solution?

NIELL : Our dual frequency data covers only the period 1978-1982. When we solve for a nutation-like departure from the 1984 system, any change is not significant.

DISCUSSION ON THE PAPER BY SHAFFER (p. 365)

BARTEL : You selected sources with simple brightness distributions. Do
you nevertheless make attempts to correct for structure phase?

SHAFFER : We do not normally correct for structure, but reductions
which do correct for structure show that there is only a small effect
much less than 1 cm, presumably because the effect of structure shows up
on only a few data points out of the 100 or more data points per
baseline in a given experiment.

KELLERMANN : Your results show little, if any, improvement when MKIII
was used instead of MKII. Indeed, even without increased sensitivity
one would expect some improvement over the past 10 years due to more
refined techniques.

NIELL : The sensitivity of MKIII is required to make the mobile VLBI
geodetic system possible, since the antennas are small.

KELLERMANN : Yes, but my question was directed to Shaffer and his
results using large fixed antennas.

GORENSTEIN : As comment to Kellermann and Shaffer, I believe the
sensitivity of the Mark III system as compared to Mark II is needed to
study the systematic effects that limited geodetic accuracy.

SHAFFER : Yes, systematic errors show up quite obviously in the
residuals of each baseline solution. Each source tends to show its own
systematic trends. Unfortunately, we have not had time to properly
analyze these residuals.

KELLERMANN : You are both avoiding an answer to my question: How has
the improved sensitivity of MKIII improved the result?

DISCUSSION ON THE PAPER BY LEGG (p. 383)

SAUNDERS : This is a question directed to you and Kellermann.   Why
don't the U.S. and Canada cooperate fully on a VLB array? It's a waste
of money to have separate arrays - shouldn't you as scientists be
persuading your governments of this? After all, even the English and
French manage to cooperate with each other in the European Space Agency!

LEGG : In principle, yes, I would agree that a well-planned joint array would be superior scientifically. There seem to be difficulties, however, in getting agreement on how to build such an instrument, and in how to operate it. One difficulty is possibly the success of the U.S. effort to have their array funded. There may be an understandable reluctance to delay, or even de-rail, their project by considering collaboration.

DE BRUYN : You mentioned the possibility of improving the UV-coverage by observing over an 8% relative bandwith. Have you investigated whether large spectral index changes in the sources (known to be present) would not deteriorate rather than improve your dynamic range?

LEGG : One could make sure that the dynamic range was not deteriorated by measuring spectral effects. This could be done by covering exactly the same U-V track (in λ/d) at different wavelengths. I.e., imagine a particular track at wavelength λo produced by two telescopes. Another two telescopes, producing the next longest track at λo, could also produce the original track at wavelength λo$+\Delta\lambda$. The spectral differences could therefore be measured, independently of visibility structure.

WRIGHT : Do you think that the STARLAB project will be in funding conflict with the CLBA?

LEGG : We are told that there should be no direct interference because these funds would come from different sources. At some level, however, there would presumably be only one pot of money, and possibly some conflict, though we hope that there might be enough difference in timing to avoid this.

DISCUSSION ON THE PAPER BY BHONSLE (p. 391)

SHAFFER : I think this is a good frequency and array for monitoring sources. Do you have additional plans for mapping or other uses?

BHONSLE : Yes, I do have additional plans to utilize our three IPS telescopes, which are separated by | 200 Km. First, I would like to incorporate Mark II VLBI terminal at each of the three stations and calibrate angular size measurements of radio sources made with the IPS technique. Next, I shall incorporate antenna tracking of radio sources and attempt mapping using "closure phase and amplitude" techniques.

DISCUSSION ON THE PAPER BY BURKE (p. 397)

WEISTROP : 1) Are there plans for a VLBI experiment on the successor to
VOIR (VOIR has not been funded)? 2) When do you think the first OVLBI
experiment will be done?

BURKE : 1) Not likely; 2) 1990 or shortly thereafter.

DISCUSSION ON THE PAPER BY LEVY (p. 405)

SHAFFER : What baseline length (40,000 Km?) was used for the S-Band VLBI
sensitivity calculations? TDRSS is at geosynchronous.

LEVY : The standard list of sources was used. By adjusting the angle of
observation the baseline can be varied down to much less than an earths
diameter.

GALACTIC OBJECTS

EXTRAGALACTIC OBJECTS